Knowledge Base for Post-Fire Safe-Shutdown Analysis

Draft Report for Comment

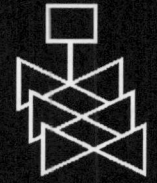

U.S. Nuclear Regulatory Commission
Office of Nuclear Reactor Regulation
Washington, DC 20555-0001

AVAILABILITY OF REFERENCE MATERIALS
IN NRC PUBLICATIONS

Knowledge Base for Post-Fire Safe-Shutdown Analysis

Draft Report for Comment

Manuscript Completed: November 2003
Date Published: January 2004

Prepared by:
M.H. Salley

S.D. Weerakkody, NRC Project Manager

Prepared for:
Division of Systems Safety and Analysis
Office of Nuclear Reactor Regulation
U.S. Nuclear Regulatory Commission
Washington, DC 20555-0001

COMMENTS ON DRAFT REPORT

Any interested party may submit comments on this report for consideration by the NRC staff. Comments may be accompanied by additional relevant information or supporting data. Please specify the report number (**Draft NUREG-1778**), in your comments, and send them — by the end of the 60-day comment period specified in the *Federal Register* notice announcing availability of this draft — to the following address:

Chief, Rules Review and Directives Branch
Mail Stop: T6-D59
U.S. Nuclear Regulatory Commission
Washington, DC 20555-0001

You may also submit comments electronically using the NRC's Web site:

http://www.nrc.gov/public-involve/doc-comment/form.html

For any questions about the material in this report, please contact:

Mark H. Salley
Mail Stop: O-11A11
U.S. Nuclear Regulatory Commission
Washington, DC 20555-0001
Phone: (301) 415-2840
Email: MXS3@nrc.gov

ABSTRACT

Every operating nuclear power plant is required to have a program that demonstrates the capability to safely shut down and maintain the reactor in the event of a fire. The U.S. Nuclear Regulatory Commission (NRC) initially issued its requirements in the Fire Protection Rule set forth in Title 10, Section 50.48, of the *Code of Federal Regulations* (10 CFR 50.48) and Appendix R to 10 CFR Part 50. The NRC has since issued numerous related generic communications over the past 20 years. The purpose of this document is to facilitate understanding of this technically challenging process and the regulatory framework upon which it is based by compiling all essential information in a single source. This document also lays the groundwork for future risk-informed activities in the post-fire safe-shutdown area.

This page intentionally left blank.

CONTENTS

CONTENTS (continued)

CONTENTS (continued)

Appendices

Figures

Tables

THE AUTHOR

Mark Henry Salley is a Fire Protection Engineer in the Office of Nuclear Reactor Regulation (NRR) of the U.S. Nuclear Regulatory Commission (NRC). Mr. Salley holds Master and Bachelor of Science degrees in fire protection engineering, both from the University of Maryland at College Park. He is a registered professional engineer in fire protection engineering and a member of the National Fire Protection Association (NFPA), American Nuclear Society (ANS), and Society of Fire Protection Engineers (SFPE).

Prior to joining the NRC, Mr. Salley was the Corporate Fire Protection Engineer for the Tennessee Valley Authority Nuclear (TVAN) program. There, he was responsible for the overall TVAN Fire Protection and Fire Safe-Shutdown Program. Mr. Salley worked on the restart of Sequoyah Nuclear Plant, Units 1 and 2; Browns Ferry Nuclear Plant, Units 2 and 3; and the completion of construction, licensing, and startup of Watts Bar Nuclear Plant, Unit 1.

Mr. Salley has an extensive background in fire protection engineering, including firefighting, design engineering, fire testing, and analytical analysis. Mr. Salley has authored a number of papers in the area of fire protection engineering.

This page intentionally left blank.

ACKNOWLEDGMENTS

This NUREG-series report began as a training tool used in the quarterly workshops that the U.S. Nuclear Regulatory Commission (NRC) sponsors for new regional fire protection inspectors. The technical basis for the workshops was a draft letter report prepared principally by Mr. Kenneth Sullivan of Brookhaven National Laboratory (BNL) under contract to the NRC. In light of the NRC's move toward a more risk-informed, performance-based approach, Mr. Steve Nowlen of Sandia National Laboratories (SNL) contributed insights to the integration of deterministic criteria and risk-informed approaches.   The NRC's regional fire protection inspectors and staff from the NRC's Office of Nuclear Reactor Regulation (NRR) provided numerous comments in the training sessions, which were factored into BNL's final draft Revision 1 of the letter report. Naeem Iqbal, a Fire Protection Engineer in the NRR Plant Systems Branch (SPLB) must also be acknowledged for helping to compile this information into a single coherent resource, and for providing peer review.

This page intentionally left blank.

EXECUTIVE SUMMARY

As a result of a major fire that occurred at the Browns Ferry Nuclear Power Plant in 1975, the U.S. Nuclear Regulatory Commission (NRC) significantly revised its regulatory framework to enhance fire protection programs (FPPs) at operating nuclear power plants (NPPs). The revised criteria used in this framework had three main objectives to (1) prevent significant fires, (2) ensure the capability to shut down the reactor and maintain it in a safe-shutdown condition, and (3) minimize radioactive releases to the environment in the event of a significant fire.

Recent studies by Sandia National Laboratories (SNL) have shown that the revised criteria are beneficial to safety. Plant design changes required by the new regulatory framework have been effective in preventing a recurrence of a fire event of the severity experienced at Browns Ferry. In addition, according to a 1989 study performed by SNL, plant modifications made in response to the new requirements have reduced the core damage frequencies (CDFs) at some plants by a factor of 10.

The NRC's regulatory framework provides several options for ensuring that structures, systems, and components (SSCs) important to safe shutdown are adequately protected from the effects of fire. Because of the potentially unacceptable consequences that an unmitigated fire may have on plant safety, each operating plant must perform a documented evaluation to demonstrate that, in the event a fire were to initiate and continue to burn (in spite of prevention and mitigation features), the performance of essential shutdown functions will be preserved and radioactive releases to the environment will be minimized. The document that describes this evaluation process and its results is commonly referred to as a "safe-shutdown analysis" (SSA).

Fire protection for NPPs is a complex subject. The purpose of this document is to facilitate understanding of the regulatory framework of the Fire Protection Program by compiling the related knowledge into a single document. This document assumes that the reader has had little or no involvement in the development and/or implementation of fire protection criteria, post-fire safe-shutdown analysis, or any of its related engineering disciplines. The criteria and assumptions described in this document are based on the NRC's regulatory framework for fire protection, as it was in place at the time of this writing. This document only clarifies existing criteria. This document does not contain any new or different staff positions and does not impose any new requirements. The knowledge base documented in this NUREG-series report must be used within the context of the licensing basis of each individual plant and with due consideration for the NRC's Backfit Rule, as specified in Title 10, Section 50.109, of the *Code of Federal Regulations* (10 CFR 50.109).

This page intentionally left blank.

ABBREVIATIONS

Φ	Phase
AC	Alternating Current
ACRS	Advisory Committee on Reactor Safeguards (NRC)
ADS	Automatic Depressurization System
AFW	Auxiliary Feedwater
ANS	American Nuclear Society
ANSI	American Nuclear Standards Institute
AOV	Air-Operated Valve
APCSB	Auxiliary and Power Conversion Systems Branch
AWG	American Wire Gauge
B&W	Babcock and Wilcox
BL	Bulletin
BFN	Browns Ferry Nuclear Power Plant
BNL	Brookhaven National Laboratory
BTP	Branch Technical Position
BWR	Boiling-Water Reactor
BWROG	Boiling-Water Reactor Owners Group
CCDF	Conditional Core Damage Frequency
CCDP	Conditional Core Damage Probability
CDF	Core Damage Frequency
CE	Combustion Engineering
CFR	Code of Federal Regulations
CLB	Current Licensing Basis
cm	Centimeter
CMEB	Chemical and Mechanical Engineering Branch
CO_2	Carbon Dioxide
CPT	Control Power Transformer
CRGR	Committee for Review Generic Requirements
CS	Core Spray
CSR	Cable Spreading Room
CSPE	Chlorosulfonated Polyethylene
CST	Condensate Storage Tank
CT	Current Transformer
DC	Direct Current
DHR	Decay Heat Removal
DID	Defense-in-Depth
ECCS	Emergency Core Cooling System
EDG	Emergency Diesel Generator
EDO	Executive Director for Operation (NRC)
EDS	Electrical Distribution System
FHA	Fire Hazard Analysis
FPP	Fire Protection Program
ft	Foot (or Feet)
ft^2	Square Foot (or Square Feet)
EOP	Emergency Operating Procedure
EPRI	Electric Power Research Institute

FSAR	Final Safety Analysis Report
GDC	General Design Criterion
GE	General Electric
GL	Generic Letter
HIF	High-Impedance Fault
HP	Horse Power
HPCI	High-Pressure Coolant Injection
HVAC	Heating, Ventilation, and Air-Conditioning
IEEE	Institute of Electrical and Electronic Engineers
IN	Information Notice
IPEEE	Individual Plant Examination of External Events
IR	Insulation Resistance
kV	kilo-Volts (1,000 Volts)
LCO	Limiting Condition of Operation
LERF	Large-Early Release Frequency
LOCA	Loss-of-Coolant Accident
LOOP	Loss of Offsite Power
LPCI	Low-Pressure Coolant Injection
LPI	Low Pressure Injection
m	Meter(s)
m^2	Square Meter(s)
MCC	Motor Control Center
MCM	One Thousand Circular Mils
MCR	Main Control Room
MHIF	Multiple High Impedance Faults
MOV	Motor-Operated Valve
MSIV	Main Steam Isolation Valve
NEC	National Electrical Code®
NEI	Nuclear Energy Institute
NEMA	National Electrical Manufacturers Association
NFPA	National Fire Protection Association
NPP	Nuclear Power Plant
NRC	U.S. Nuclear Regulatory Commission
NRR	Office of Nuclear Reactor Regulation (NRC)
NSSS	Nuclear Steam Supply System
NUREG	Nuclear Regulatory
OPL	Omega Point Laboratories
P&ID	Piping and Instrument Diagram or Process and Instrument Diagram
Pa	Pascal
PASV	Pressurizer Auxiliary Spray Valve
PE	Polyethylene
PFSSD	Post-Fire Safe-Shutdown
PORV	Power-Operated Relief Valve
PRA	Probabilistic Risk Assessment
PU	Polyurethane
PVC	Polyvinyl Chloride
PWR	Pressurized-Water Reactor
RCIC	Reactor Core Isolation Cooling
RCS	Reactor Coolant System

RES	Office of Nuclear Regulatory Research (NRC)
RG	Regulatory Guide
RHR	Residual Heat Removal
RMS	Root-Mean Square
RSP	Remote Shutdown Panel
RWST	Refueling Water Storage Tank
SDP	Significance Determination Process
SECY	Secretary of the Commission (NRC)
SER	Safety Evaluation Report
SFPE	Society of Fire Protection Engineers
SLCS	Standby Liquid Control Systems
SNL	Sandia National Laboratories
SOV	Solenoid-Operated Valve
SRG	Special Review Group
SRP	Standard Review Plan
SRU	Signal Resistor Unit
SRV	Safety Relief Valve
SSA	Safe Shutdown Analysis
SSC	Structures, Systems, and Components
SSD	Safe Shutdown
SSEL	Safe Shutdown Equipment List
SWGR	Switchgear
TS	Technical Specification
TVA	Tennessee Valley Authority
TVAN	Tennessee Valley Authority Nuclear
V	Voltage
V&V	Verification and Validation
VCT	Volume Control Tank
XLPE	Cross-Linked Polyethylene

This page intentionally left blank.

CHAPTER 1. INTRODUCTION

1.1 Background

The fundamental safety objective of the regulatory program established by the U.S. Nuclear Regulatory Commission (NRC) is to ensure adequate protection of public health and safety. This means that the risk to the public from normal operation, anticipated transients, and accidents must be acceptably low, and the likelihood of accidents more severe than those postulated for design purposes must be extremely small. To achieve this goal, the NRC has promulgated regulations, staff positions, and clarification documents, which require that nuclear power plants (NPPs) must be conservatively designed, soundly constructed, and judiciously operated.

An NPP contains an extensive array of systems and components. To achieve a high level of safety, redundant (i.e., identical or diverse) safety systems are incorporated into the design of all NPPs that are currently operating in the United States. Redundancy provides assurance that failures affecting one system will not have a significant impact on plant safety because the plant design provides a "backup" system. The safety benefits of this important design feature could be negated, however, if the redundant systems were both susceptible to failure from a single cause. Fire is one example of such "common-mode" failure mechanisms. In the absence of suitable protection features and/or separation distances, a single fire could render redundant safety systems inoperable. In addition to a total loss of equipment function, lessons learned from actual fire events and cable fire test programs have shown that fire damage to power, control, and instrumentation circuits and cables may cause equipment to operate in undesired and frequently unexpected ways. Specific examples include spurious (unintended) equipment operations in the form of maloperations (failure to start/stop/actuate, inadvertent start/stop/actuation, etc.), false instrument signals, misleading indications, and loss of normal equipment control methods.

On March 22, 1975, the Brown's Ferry Nuclear Power Plant (BFN), operated by the Tennessee Valley Authority (TVA), experienced the worst fire (from a nuclear safety perspective), ever to occur in a commercial NPP operating in the United States. Although the licensee ultimately achieved safe-shutdown of the reactor, the event highlighted significant inadequacies in the fire protection programs (FPPs) established by the plants that were operating at that time. As a result of lessons learned from the Browns Ferry fire, the NRC issued new requirements and guidance, which significantly enhanced the FPPs (personnel, procedures, equipment, and plant design features) of NPPs. The revised program had three main objectives to (1) prevent significant fires, (2) ensure the capability to shut down the reactor and maintain it in a safe-shutdown condition, and (3) minimize radioactive releases to the environment in the event of a significant fire. Implementation of these three objectives satisfies the defense-in-depth (DID) approach as it applies to fire safety.

The NRC's regulatory framework for nuclear power plant FPPs is set forth in a number of regulations and supporting guidelines, including, but not limited to the following:
- Title 10, Section 50.48, of the *Code of Federal Regulations* (10 CFR 50.48)
- Appendix R to 10 CFR Part 50
- General Design Criterion 3 (GDC 3) of Appendix A to 10 CFR Part 50
- regulatory guides (RGs)
- generic communications [e.g., generic letters (GLs), bulletins (BLs), and information notices (INs)]
- NUREG-series technical reports, including NUREG-0800, "NRC Standard Review Plan" (SRP)
- associated branch technical positions (BTPs) and industry standards

The principal objective of this regulatory framework is to provide assurance that a fire will not significantly increase the risk of radioactive releases to the environment. The NRC's regulatory framework does not provide specific guidance for protection against economic or property loss.

The need to evaluate the effects of fire on circuits associated with the safe-shutdown systems was not explicitly stated in Appendix A to the Auxiliary and Power Conversion Systems Branch (APCSB) BTP 9.5-1. However, it is explicitly required in Appendix R to 10 CFR Part 50.[1] A commercial NPP contains a very large number of power, control, and instrument cables. A typical boiling-water reactor (BWR) requires approximately 60 miles of power cable, 50 miles of control cable and 250 miles of instrument cable. The fire at BFN damaged more than 1,600 cables, even though the fire was confined to a relatively small area of the plant (approximately 800 ft^2). While a single fire could affect a large number of cables, damage to many of these cables will have little or no impact on the operation of plant systems needed to achieve and maintain safe-shutdown conditions. Therefore, the NRC is concerned with those circuits and cables for which damage attributable to fire could impact the shutdown capability.

Specifically, these "circuits and cables of concern" to the NRC are as follows:

(1) circuits/cables needed to ensure the proper operation of *essential* shutdown systems and equipment (*"required circuits"*)

(2) circuits/cables associated with nonessential systems and equipment for which failure or maloperation resulting from a fire could impact the shutdown capability (*"associated circuits"*[2]).

Because circuits and cables of required shutdown systems (i.e., required circuits) frequently share certain physical or electrical configurations with cables of nonessential systems and equipment (i.e., associated circuits) fire damage to certain associated (nonsafety) circuits could impact the shutdown capability. Section III.G.2 of Appendix R to 10 CFR Part 50 provides various options for protecting circuits of concern (both required and associated) for post-fire safe-shutdown. Specifically, this section of the regulation reads as follows:

> *Where cables or equipment, including associated nonsafety circuits that could prevent operation or cause maloperation...of redundant trains of systems necessary to achieve and maintain hot shutdown conditions are located within the same fire area... one of the following means of ensuring that one of the redundant trains is free of fire damage shall be provided...*

Compliance with this requirement could be interpreted to mean, for example, that for each fire area, the specified fire protection features must (shall) be provided for *all* circuits and cables for which damage attributable to fire could impact the capability to achieve and maintain hot shutdown conditions. In its clarification of GL 81-12, the NRC defined "associated circuits" of concern. In addition, this clarification permitted the use of detailed circuit analyses as a means of demonstrating that fire damage to these nonessential circuits would not significantly impact the ability to achieve and maintain hot shutdown conditions. It should be noted that the use of circuit analysis in lieu of fire protection features is only permitted in the evaluation of fire damage to "associated circuits" (as defined in GL 81-12).

---

[1]   SECY-80-438A, "Commission Approval of the Final Rule on Fire Protection Program," September 30,1980.

[2]   For the purpose of this NUREG-series report, the term "associated circuits" is understood to be the "associated circuits of concern.

Circuits of equipment for which proper operation is needed to ensure the successful accomplishment of required hot shutdown functions (required circuits) must be provided with fire protection features sufficient to satisfy Section III.G.2 of Appendix R to 10 CFR Part 50 if damage to those circuits could adversely impact the desired operation of that equipment. The NRC has permitted the use of feasible manual actions under certain conditions, and rulemaking is underway to codify this alternative.

Therefore, it is not sufficient to consider only the effects of fire damage to cables of equipment needed to ensure operation of required shutdown systems and equipment (required circuits). Rather, the scope of the evaluation must also include consideration of the effects of fire damage to nonessential equipment and systems whose failure or inadvertent actuation could impact the shutdown capability (associated circuits of concern).

1.2 Purpose

To demonstrate compliance with the Fire Protection Rule (10 CFR 50.48) all operating plants have performed a deterministic evaluation of the capability to safely shut down the reactor in the event of fire. The purpose of this document is to facilitate understanding of this technically challenging process and the regulatory framework upon which it is based. This document assumes that the reader has had little or no involvement the development and/or implementation of fire protection criteria, post-fire safe-shutdown analysis, or related engineering disciplines. Explanatory text and/or graphic illustrations are used wherever practical.

It should be noted that this document describes only one possible approach for performing a deterministic assessment of the potential impact of fire on the ability to achieve and maintain safe-shutdown conditions. There are many acceptable methods of performing this type of analysis, and the NRC does not prescribe or endorse any one specific approach. This report does not discuss other important aspects of a comprehensive FPP, such as fire prevention measures, fire detection, or suppression systems. The criteria and assumptions described in this document are based on the NRC's regulatory framework for fire protection as it was in place at the time of this writing. This document only clarifies existing criteria. It does not contain any new or different staff positions or impose any new requirements. The information presented in this report must be used within the context of the licensing basis of each individual plant and with due consideration for the NRC's Backfit Rule, as specified in 10 CFR 50.109.

1.3 Document Summary

- Chapter 2, "Terminology," provides consistent interpretations of terms and phrases that may be encountered during the development or review of post-fire safe-shutdown analysis.

- Chapter 3, "Background Information and Experience Related to Fire-Induced Circuit Failures," provides fundamental design information to facilitate understanding of circuits and cables used in NPPs. In addition to describing the principal aspects of their design, construction, and application, this chapter discusses such topics as design and operational factors that influence cable selection, potential failure modes of circuits and cables, and the mechanisms (stressors) that can cause their failure. To illustrate the potential consequences that circuit and cable failures may have on plant operations, this technical discourse is followed by a chronicle of the impact that fire-induced cable and circuit failures had on the operation of redundant trains of safety systems during the Browns Ferry fire, as well as observations of various experts who participated in a recent cable fire test program sponsored by the Nuclear Energy Institute (NEI) and Electric Power Research Institute (EPRI).

- Chapter 4, "NRC Regulatory Requirements," provides a brief history of the development of fire protection regulations and guidelines governing the U.S. commercial nuclear power industry. In addition, this chapter presents a comprehensive discussion of requirements, guidelines, and staff positions that are specifically applicable to the performance of a deterministic analysis of post-fire safe-shutdown capability.

- Chapter 5, "Discussion of Post-Fire Safe-Shutdown Capability," describes the primary objectives of a comprehensive deterministic evaluation of the effects of fire damage, the qualitative hierarchy of fire damage limits, an overview of the evaluation process used to assess the potential effects of fire and its related perils on plant safety, the fundamental principles and assumptions that establish the "ground rules" for performing an appropriate evaluation, and a discussion of specific issues to be considered in the evaluation.

- Chapter 6, "Deterministic Analysis Process for Appendix R Compliance," describes the fundamental principals, assumptions, and criteria of a deterministic analysis for demonstrating compliance with regulatory requirements. In addition to defining the principal criteria and assumptions that form the basis of the analysis, this chapter presents a step-by-step description of a safe-shutdown analysis process.

- Chapter 7, "Maintaining Compliance," describes the impact that plant modifications may have on the plant's post-fire safe-shutdown capability and the administrative controls (procedures) that are typically needed to prevent future modifications from jeopardizing long-term compliance with the plant's fire protection licensing basis.

- Chapter 8, "Integration of Deterministic Criteria and Risk-Informed Information," provides risk-informed perspectives on post-fire safe-shutdown circuit analysis issues. This chapter was developed specifically to give staff responsible for plant inspection activities a general understanding of the fire risk analysis process and insights into the risk significance of fire-related circuit analysis issues.

- Appendix A, "Examples of Successful Implementation," describes various options available to resolve identified circuit/cable vulnerabilities. This discussion is supplemented by specific "real-world" examples to show how various licensees have successfully identified and appropriately resolved circuit/cable vulnerabilities.

- Appendix B, "Specific Circuit Analysis Issues," describes the NRC's expectations regarding certain specific circuit analysis issues that have been the subject of much recent debate. Specific topics discussed in this section include multiple spurious actuations, fire damage to nonessential systems, and multiple circuit faults. As in Appendix A, this appendix describes staff expectations in terms of "real world" examples of technical issues that the NRC staff identified during the reviews of SSAs developed by various licensees.

1-4

CHAPTER 2. TERMINOLOGY

The NRC developed the following definitions as an aid to ensuring consistent interpretations of terms that are commonly used in post-fire safe-shutdown analyses. To the extent practical, the staff derived these definitions from established fire protection guidance documents promulgated by the NRC (RGs, GLs, and INs) and industry-recognized standards including the Institute for Electrical and Electronics Engineers (IEEE) Standard 100, "IEEE Standard Dictionary of Electrical and Electronics Terms," and IEEE Standard 242, "IEEE Recommended Practice for Protection and Coordination of Industrial and Commercial Power Systems." In an effort to further minimize ambiguity, certain terms are supplemented with additional discussion, notes, graphic illustrations, and/or examples.

Actuated Equipment
The assembly of prime movers and driven equipment used to accomplish a protective action. (IEEE Std. 100-1988)

Actuation Device
A component or assembly of components that directly controls the motive power (electricity, compressed air, etc.) for actuated equipment. Examples of actuation devices include a circuit breaker, a relay, and a pilot valve used to control compressed air to the operator of a containment isolation valve. (IEEE Std. 100-1988)

Actuation
A change in position or operating state of a component. [See: Spurious Actuation/Operation.]

Adverse Effect
An undesired change in the operation or functional integrity of structures, systems, or components (SSCs). Adverse effects may occur as a result of exposure to the effects of fire (i.e., heat or smoke) and/or fire-suppression activities.

Affected Systems and Components
SSCs that may be adversely affected as a result of fire (including an exposure fire) or subsequent fire suppression activities in a single fire area.

Alternative Shutdown
The capability to safely shut down the reactor in the event of a fire using existing systems that have been rerouted, relocated, or modified. (RG 1.189)

Alternative Shutdown Capability
A defined and documented process (including equipment, personnel, and procedures) for accomplishing safe-shutdown conditions in the event of fire in areas where one train of the redundant systems (see note below) needed to achieve and maintain hot shutdown conditions has not been ensured to remain free of fire damage (i.e., not provided with fire protection features sufficient to satisfy applicable requirements. (Section III.G.2 of Appendix R to 10 CFR Part 50 or Position C.5.b of SRP Section 9.5.1)

Note: If the system is being used to provide its design function, it is generally considered to be *redundant.* If the system is being used *in lieu of* the preferred system because the redundant components of the preferred system do not meet the separation criteria of Section III.G.2, the system is considered to be an *alternative* shutdown capability. (GL 86-10, Question 5.8.3)

Ampacity
Current carrying capacity, expressed in amperes, of a wire or cable under stated thermal conditions. (IEEE Std. 100-1988)

Clarification: When current flows in a conductor, heat is produced because every conductor offers some resistance to the flow of current. The National Electrical Code® (NEC) (ANSI/NFPA 70) defines ampacity as "the current (in amperes) a conductor can carry continuously under the conditions of use without exceeding its temperature rating." The current-carrying capacity of a particular wire is dictated by its "ampacity" (that is, how many amps it can handle). Ampacity is a function of the cross-sectional area or diameter of the wire and its material type (e.g., copper or aluminum) and cable insulation condition for basic installation conditions. For more complex installation conditions, IEEE 835 provides more extensive and detailed tables. For installations involving cables in open cable trays, ICEA/NEMA P-54 should be consulted. Larger-diameter wires have larger cross-section areas and can safely carry more electrical current without overheating. The ampacity rating of a specific conductor may be obtained from tables in the NEC. These tables are based on the size of the wire, the maximum allowable operating temperature of the insulation material, and the installation conditions. The nominal ampacity values include a safety margin that is sufficient for most installations. However, there are instances where application of the NEC ampacity tables is insufficient. For example, although the addition of fire barrier wrap around cable trays and conduits will affect the ampacity of a conductor, the NEC tables do not address this problem. Several inches of fire barrier material can have a significant effect on the ampacity rating specified in the NEC tables. Since there are no derating tables in the NEC for this kind of situation, calculations must be performed to determine the current carrying capacity of the enclosed cables.

American Wire Gauge (AWG)
A standardized system used to designate the size or "gauge" of wire. As the diameter of wire decreases, the "AWG" number of the wire increases. The smallest AWG size is 40 and looks like a metal thread. "Four ought" (0000) is the largest AWG wire size designation. Wires larger than this size are designated by the "thousand circular mill" system or "KCMIL" sizes (known until recently as MCM).

Ampere
A standard unit of electric current flow (equal to a flow of 1 coulomb per second).

Any-and-All/One-at-a-Time
All potential spurious actuations that may occur as a result of fire in a single fire area must be addressed and prevented or their effects must be appropriately mitigated on a one-at-time basis. That is, in evaluating non-high/low-pressure interface components[3], the analyst must assume that "any and all" spurious actuations that could occur, will occur on a sequential, one-at-a-time, basis. For each fire area, the analyst should identify all potential spurious operations that may occur as a result of a postulated fire. While it is not assumed that all potential spurious actuations will occur instantaneously at the onset of fire, the analyst must consider the possibility that each spurious actuation will occur sequentially, as the fire progresses, on a one-at-a-time basis. If not appropriately prevented or mitigated, such sequential failures could result in concurrent failure of multiple devices.

Appendix R Cables
The set of cables that must remain free of fire damage to ensure that safe-shutdown conditions can be achieved within established criteria. [Synonym: required cables.]

Arcing Fault
See: High-impedance fault.

Associated Circuit Analysis
A documented, systematic evaluation of associated circuits of concern to post-fire safe-shutdown.

Associated Circuits (of Concern)
Those safety-related and nonsafety-related Class 1E and non-Class 1E cables that have a physical separation less than that specified in Section III.G.2 of Appendix R to 10 CFR Part 50 and have one of the following: (Reference GL 81-12 Clarification, Enclosure 2)

a.   *A power source that is shared with the shutdown equipment* (redundant or alternative) and is not electrically protected from the circuit of concern by coordinated breakers, fuses, or similar devices.

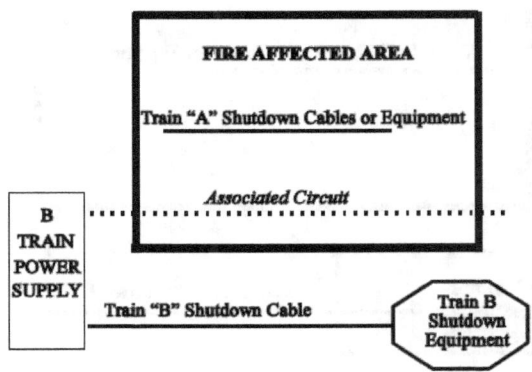

Associated Circuit Concern - Common Power Source

---

[3]   High/low-pressure interfaces are considered a special case to their severe consequence. (GL 86-10, Question 5.3.1)

b. *A connection to circuits of equipment of which spurious operation would adversely affect the shutdown capability* [e.g., residual heat removal (RHR)/reactor coolant system (RCS) isolation valves, automatic depressurization system (ADS) valves, power-operated relief valves (PORVs), steam generator atmospheric dump valves, instrumentation, steam bypass].

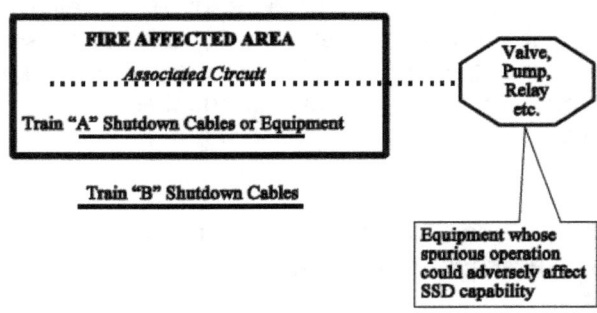

Associated Circuit of Concern - Spurious Operation

c. *An enclosure (e.g., raceway, panel, or junction box) that is shared with the shutdown cables* (redundant or alternative) and (1) is not electrically protected by circuit breakers, fuses, or similar devices, or (2) will allow propagation of the fire into the common enclosure.

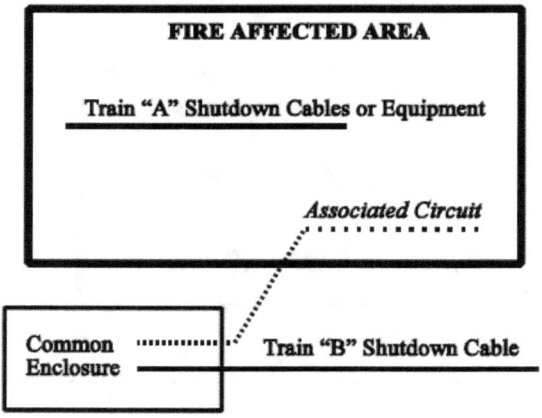

Associated Circuits of Concern - Common Enclosure

(SECY 80-438A, "Commission Approval of the Final Rule on Fire Protection Program," September 30, 1980)

Clarification: An *associated circuit of concern to post-fire safe-shutdown* may include any circuit or cable that is not needed to support the proper operation of required shutdown equipment (i.e., a nonessential circuit), but could adversely affect the plant's ability to achieve and maintain safe-shutdown conditions if it is damaged by fire. For example, while operation of the PORV in a pressurized-water reactor (PWR) may not be needed to ensure the operation of a defined shutdown system, its maloperation as a result of fire damage to connected cabling could have a significant impact on the plant's overall safe-shutdown capability.

Automatic
Self-acting; operating by its own mechanism when actuated by some monitored parameter, such as a change in current, pressure, temperature, or mechanical configuration. (RG 1.189)

Automatic Actuation Signal
A signal that is initiated in response to a previously defined variable or set of variables that will cause equipment to change position or operating mode, such as the undervoltage signal that causes an emergency power source [e.g., emergency diesel generator (EDG)] to automatically start and load in response to a low-voltage condition on safety-related switchgear (SWGR).

Bolted Fault
(1) A short circuit or electrical contact between two conductors at different potentials, in which the impedance or resistance between the conductors is essentially zero. (IEEE Std. 100-1988)
(2) A simplifying assumption used when calculating the value of short-circuit fault current to ensure that the short-circuit ratings of the equipment are adequate to handle the currents available at their locations. This assumption simplifies calculations, since the resulting values are a maximum and equipment selected on this basis will always have an adequate rating.
(ANSI/IEEE Std. 242-1986)

Cable
A conductor with insulation or a stranded conductor with or without insulation and other coverings (single-conductor cable) or a combination of conductors insulated from one another (multiple-conductor cable). (IEEE Std. 100-1988)

Cable Failure
A breakdown in the physical and/or chemical properties (e.g., electrical continuity, insulation integrity) of the cable conductor(s), such that the functional integrity of the electrical circuit cannot be ensured (e.g., interrupted or degraded).

Cable-Fire Break
Material, devices, or an assembly of parts, installed in a cable system, other than at a cable penetration of a fire-resistive barrier, to prevent the spread of fire along the cable system. (IEEE Std. 100-1988)

Cable Jacket
A protective covering over the insulation, core, or sheath of a cable. (IEEE Std.100-1988)

Cable and Raceway Database
A database, unique to the plant, which delineates the routing and location of cables and their associated raceways. (cable trays, conduits, pull-boxes, etc.)

Cable Penetration
An assembly or group of assemblies for electrical conductors to enter and continue through a fire-rated structural wall, floor, or floor-ceiling assembly. (IEEE Std. 100-1988)

Cable Routing
The pathway electrical wiring takes through the plant from power source or control point to component location.

Cable Size
See American Wire Gauge (AWG).

Cable-to-cable Fault
A fault condition of relatively low impedance between conductors of one cable and conductors of a different cable.

Circuit
(1) A conductor or system of conductors through which electrical current flows. (IEEE Std. 100-1988)
(2) Interconnection of components to provide an electrical path between two or more components.

Circuit Analysis
A systematic evaluation of the impact of fire-induced circuit/cable failure modes (e.g., hot shorts, open circuits, and shorts to ground) on the defined/credited shutdown capability.

Note: Performance of a detailed circuit analysis is not a requirement. Section III.G of Appendix R and Regulatory Position C.5.b of SRP Section 9.5.1 establish the fire protection design features that are necessary to ensure that SSCs important to safe-shutdown will remain free of fire damage. Where these fire protection features are provided, analysis is not necessary. When relied on in lieu of providing these features, circuit analyses must demonstrate a level of safety equivalent to that which would be achieved through compliance with applicable regulatory requirements.

Circuit Breaker
(1) A device designed to open and close a circuit by nonautomatic means, and to open the circuit automatically on a predetermined overload of current without injury to itself when properly applied within its rating. (IEEE Std. 100-1988)
(2) A mechanical switching device capable of making, carrying, and breaking currents under normal circuit conditions and also, making, carrying for a specified period of time, and breaking currents under specified abnormal circuit conditions such as those of a short circuit. (IEEE Std. 100-1988)

Circuit/Cable Fault
See: Fault, Fire-Induced Fault

Cold Shutdown Repair
Repair activities performed on equipment needed to bring the plant to cold shutdown conditions.

Note: Systems and equipment needed to achieve and maintain hot shutdown conditions must remain free of fire damage (i.e., repairs are not permitted).

Common Enclosure
An enclosure (e.g., cable tray, conduit, junction box) that contains circuits required for the operation of safe-shutdown components and circuits for non-safe-shutdown components. [See: Associated Circuits of Concern.] (RG 1.189)

Common-Mode Failure
Multiple failures that are attributable to a common cause, such as circuit faults resulting from the exposure of cables to the direct effects of fire (heat, smoke) and subsequent fire-suppression activities in a single fire area. (IEEE Std. 100-1988)

Common Power Supply/Source
A power supply that feeds safe-shutdown circuits and non-safe-shutdown circuits. [See: Associated Circuits of Concern.] (RG 1.189)

Conductor
(1) A substance or body that allows a current of electricity to pass continuously along it. (IEEE Std. 100-1988)
(2) A wire or combination of wires, not insulated from one another, suitable for carrying an electric current. (IEEE Std. 100-1988)

Clarification: For cables, the term "conductor" commonly refers to a single insulated wire located within a cable; for circuits, the term "conductor" may refer to a single wire, contact, wire termination, or other conductive pathway such as those used on printed circuit boards.

Conductor-to-Conductor Fault
(1) A circuit fault condition of relatively low impedance between two or more conductors of the same or different circuit.
(2) A cable failure mode of relatively low impedance between two or more conductors of the same multi-conductor cable (intra-cable fault) or between two or more separate cables (inter-cable fault).

Contact
A conducting part that co-acts with another conducting part to make or break a circuit. (IEEE Std. 100-1988)

Control Cable
Cable applied at relatively low current levels or used for intermittent operation to change the operating status of a utilization device of the plant's auxiliary system. (IEEE Std. 100-1988)

Control Circuit
The circuit that carries the electrical signals directing the performance of the controller, but does not carry the main power circuit. (IEEE Std. 100-1988)

Clarification: A control circuit is a low-voltage (typically 120-VAC or 125-VDC) circuit, consisting of switches, relays, and indicating devices, which direct the operation of remotely located plant equipment that is powered from a completely separate power supply.

Control Panel
An assembly of man/machine interface devices. (IEEE Std. 100-1988)

Control Power/Voltage
The voltage applied to the operating mechanism of a device to actuate it. (IEEE Std. 100-1988)

Clarification: Electrical power/voltage (typically 120-VAC or 125-VDC) used to power control circuit devices (e.g. relays, indicating lights).

Control-Power Transformer
A transformer that supplies power to motors, relays, and other devices used for control purposes. (IEEE Std. 100-1988)

Coordination (of Electrical Protection Devices)
The selection and/or setting of protective devices to sequentially isolate only that portion of the system where the abnormality occurs. To achieve this isolation, it is necessary to set protective devices so that only the device nearest the fault opens and isolates the faulted circuit from the system. It is obvious that such selectivity becomes more important with devices that are closer to the power source, as a greater portion of the system can be affected. Backup protective devices are set to operate at some predetermined time interval after the primary device fails to operate. A backup device is able to withstand the fault conditions for a longer period than the primary device. If a primary device fails to clear a fault and the backup device must clear it, the design of the protective system becomes suspect. To optimize the coordination of protective devices, good engineering practice require consideration of (1) the available maximum short-circuit currents, (2) the time interval between the coordination curves; and (3) load current. (IN 88-45)

Clarification: The design must ensure that electrical fault currents generated as a result of fire damage will not cause an interruption in the power being supplied to required shutdown equipment. To ensure this "continuity of service," cables and equipment fed from electrical power sources required for post-fire safe-shutdown must either be provided with suitable fire protection features (e.g. meet the criterial in Section III.G.2 of Appendix R to 10 CFR Part 50) or the fault-protection devices (relays, fuses, and/or circuit breakers) of the required power sources must be selectively coordinated. [See also: High-Impedance Fault.]

Coordination Study

The process of evaluating the performance of electrical distribution system protection devices (breakers, fuses, relays) to ensure that fault conditions caused by fire will be isolated and power outages to unaffected equipment will be minimized. A coordination study is based on a comparison of the time it takes individual overcurrent protection devices (circuit breakers, fuses, relays) to operate (trip) under abnormal (faulted) conditions. For post-fire safe-shutdown, this study must ensure that electrical power to shutdown equipment will not be interrupted as a result of fire-induced faults in nonessential loads (equipment or cables) of a required power supply [SWGR, load center, motor control center (MCC), fuse panel, etc.].

ANSI/IEEE Std. 242-1986, "IEEE Recommended Practices for Protection and Coordination of Industrial and Commercial Power Systems," provides detailed guidance on achieving proper coordination. (RG 1.189 and IN 88-45)

Credited Shutdown Equipment

The set of equipment that is relied on (credited in the SSA) for achieving post-fire safe-shutdown conditions in the event of fire in a specific fire area.

Current Carrying Capacity

See Ampacity.

Current Licensing Basis (CLB)

The set of NRC requirements applicable to a specific plant and a licensee's written commitments for ensuring compliance with and operation within applicable NRC requirements and the plant-specific design basis (including all modifications and additions to such commitments over the life of the license) that are docketed and in effect. The CLB includes the NRC regulations contained in 10 CFR Parts 2, 19, 20, 21, 26, 30, 40, 50, 51, 54, 55, 70, 72, 73, and 100, as well as the appendices thereto; orders; license conditions; exemptions; and technical specifications (TSs). The CLB also includes the plant-specific design basis information defined in 10 CFR 50.2, as documented in the most recent final safety analysis report (FSAR), as required by 10 CFR 50.71 and the licensee's commitments remaining in effect that were made in docketed licensing correspondence such as licensee responses to NRC BLs, GLs, and enforcement actions, as well as licensee commitments documented in NRC safety evaluations or licensee event reports. [See also RG 1.189.] (10 CFR 54.3)

Current Transformer
A device used to transform high currents used by a large equipment and SWGRs to lower levels that can safely be measured by standard metering equipment. The current reduction ratio of a current transformer (CT) is given on its nameplate. A CT with a current reduction ratio of 400:5 would reduce the current by a ratio of 400 divided by 5 or 80 times.

Note: The hazard of electric shock, burn, or explosion exists on an open-circuited CT. Death, severe personal injury, or equipment damage can result if the leads are touched when the CT is open-circuited. As much as 4,000 V has been measured on the secondary on large-core CTs with an open-circuited secondary. CTs must *always* be shorted *or* connected to a burden such as a meter or relay. Open-circuiting may also damage the CT insulation. Once a CT has been open-circuited, it must be demagnetized or accuracy may be reduced.
(Square D Application Bulletin No. 4200PD9203R8/95, April 1996)

Dedicated Shutdown
The ability to shut down the reactor and maintain shutdown conditions using SSCs dedicated to the purpose of accomplishing post-fire safe-shutdown functions. (RG 1.189)

Diagnostic Instrumentation
Attachment 1 to IN 84-09 lists the instruments that are necessary to achieve safe shutdown. Diagnostic instrumentation includes any additional instruments (beyond those listed in Attachment 1 to IN 84-09) that are needed to ensure proper actuation and functioning of safe-shutdown equipment and support equipment (e.g., flow rate, pump discharge pressure). The diagnostic instrumentation needed depends on the design of the alternative shutdown capability. (GL 86-10, Question 5.3.9)

Clarification: Section IX of IN 84-09 establishes the minium set of instrumentation that the NRC staff deems acceptable for meeting the alternative shutdown process monitoring function. Although this list includes "diagnostic instrumentation," it does not specifically define that term. Alternative shutdown strategies that rely on operator intervention (recovery actions) to mitigate equipment maloperations and/or failures that may occur as a result of fire must be supported by sufficient monitoring capability (diagnostic instrumentation) to ensure prompt detection of any failure(s) that may occur and confirm proper system response.

Emergency Control Station
The control stations located outside the main control room (MCR), where operations personnel take actions to manipulate plant systems and their controls to achieve safe-shutdown of the reactor.

Enclosure
An identifiable housing such as a cubicle, compartment, terminal box, panel, or raceway used for electrical equipment or cables. (IEEE Std. 100-1988)

Exposed (Circuits/Cables/Equipment/Structures)
(1) SSCs, that are subject to the effects of fire and/or fire-suppression activities.
(2) SSCs not provided with fire protection features sufficient to satisfy Section III.G.2 of Appendix R to 10 CFR Part 50 or Position C.5.b of SRP Section 9.5.1.

Exposure Fire
A fire in a given area that involves either in situ or transient combustibles and is external to any SSCs located in or adjacent to that same area. The effects of such fire (e.g., smoke, heat, or ignition) can adversely affect those SSCs that are important to safety. Thus, a fire involving one success path of safe-shutdown equipment may constitute an exposure fire for the redundant success path located in the same area, and a fire involving combustibles other than either redundant success path may constitute an exposure fire to both redundant trains located in the same area. (RG 1.189)

Failsafe Circuits
Circuits designed so that fire-induced faults will result in logic actuation(s) to a desired, safe mode that cannot be overridden by any subsequent circuit failures.

Fault
(1) Any undesired state of a component or system. A fault does not necessarily require failure. For example, a pump may not start when required because its feeder breaker was inadvertently left open. (IEEE Std. 100-1988)
(2) A partial or total local failure in the insulation or continuity of a conductor. (IEEE Std. 100-1988)
(3) A physical condition that causes a device, component, or element to fail to perform in a required manner (for example a short circuit, a broken wire, an intermittent connection). (IEEE Std.100-1988)

Fault Current
(1) A current that flows from one conductor to ground or another conductor owing to an abnormal connection (including an arc) between the two. (IEEE Std. 100-1988)
(2) A current that results from the loss of insulation between conductors or between a conductor and ground. (NEMA Std. ICS-1, 1988)

Clarification: Fault current is an abnormal level of current that is induced in an electrical circuit. Fault currents may be initiated by various mechanisms including insulation degradation, arcing, or physical contact between two conductors. Fault currents include short-circuit current (bolted fault), high-impedance (arcing) fault currents, and overload currents.

Feeder Breaker/Feeder
A general term used to describe a circuit breaker or fuse located upstream of an electrical load. Depending on usage, it may refer to a circuit breaker provided for a specific component (load breaker) or it may refer to a breaker located upstream of a SWGR, load center, or distribution panel. Opening a feeder breaker will cause a loss of power to all downstream loads.

Fire Area
The portion of a building or plant that is separated from other areas by rated fire barriers adequate for the fire hazard. (RG 1.189)

Fire Area Boundaries

As used in Appendix R to 10 CFR Part 50, the term "fire area" means an area that is sufficiently bounded to withstand the associated hazards and, as necessary, to protect important equipment within the area from a fire outside the area. In order to meet the regulation, fire area boundaries need not be completely sealed floor-to-ceiling, wall-to-wall boundaries. However, all unsealed openings should be identified and considered when evaluating the effectiveness of the overall barrier. Where fire area boundaries are not floor-to-ceiling, wall-to-wall boundaries with all penetrations sealed to the fire rating required of the boundaries, licensees must perform an evaluation to assess the adequacy of their plant's fire boundaries to determine whether the boundaries will withstand the hazards associated with the area. This analysis must be performed by at least a fire protection engineer and, if required, a systems engineer. (GL 86-10)

Fire-Induced Fault

An electrical failure mode (e.g., hot short, open circuit, or short to ground) that may result from circuit/cable exposure to the effects of fire (e.g., heat and smoke) and/or subsequent fire-suppression activities (e.g., water spray, hose streams).

Fire Suppression

Control and extinguishing of fires (firefighting). Manual fire suppression employs the use of hoses, portable fire extinguishers, or manually actuated fixed systems by plant personnel. Automatic fire suppression is the use of automatically actuated fixed systems such as water, Halon, or carbon dioxide (CO_2) fire suppression systems. (RG 1.189)

Fire-Suppression Impacts

The susceptibility of SSCs and operations response to suppressant damage (attributable to discharge or rupture). (NFPA 805)

Fire Zones

Subdivisions of fire areas (RG 1.189)

Note: Compliance with Section III.G.2 cannot be based on rooms or zones. (GL 86-10, Question 3.1.5)

Free of Fire Damage

In promulgating Appendix R to 10 CFR Part 50, the Commission provided acceptable methods for ensuring that necessary SSCs are free of fire damage (see Section III.G.2a, b and c); that is, the SSCs under consideration are capable of performing their intended functions during and after the postulated fire, as needed. Licensees seeking exemptions from Section III.G.2 must show that the proposed alternative provides reasonable assurance that this criterion is met. The term "damage by fire" also includes damage to equipment from the normal or inadvertent operation of fire-suppression systems. (GL 86-10)

Note: Section III.G.2 of Appendix R and Position C.5.b of SRP Section 9.5.1 establish the fire protection features that are necessary to ensure that systems needed to achieve and maintain hot shutdown conditions remain free of fire damage.

Fuse

(1) A device that protects a circuit by fusing open its current responsive element when an overcurrent or short-circuit current passes through it. (IEEE Std. 100-1988)

(2) A protective device that opens by the melting of a current-sensitive element during specified overcurrent conditions. (NEMA Std. FU-1 1986)

Fuse Current Rating

The AC or DC ampere rating that the fuse is capable of carrying continuously under specified conditions. (NEMA Std. FU-1 1986)

Fuse Voltage Rating

The maximum root-mean-square (RMS) AC voltage or the maximum DC voltage at which the fuse is designed to operate. (NEMA Std. FU-1 1986)

Ground

A conducting connection, whether intentional or accidental, by which an electric circuit or equipment is connected to the earth, or to some conducting body of relatively large extent that serves in place of the earth. (IEEE Std. 100-1988)

Grounded Circuit

A circuit in which one conductor or point (usually the neutral conductor or neutral point of transformer or generator windings) is intentionally grounded, either solidly or through a non-interrupting current-limiting grounding device. (IEEE Std. 100-1988)

High/Low-Pressure Interface

Reactor coolant boundary valves of which spurious operation as a result of a fire could (1) potentially rupture downstream piping on an interfacing system, or (2) result in a loss of reactor coolant inventory in excess of the available makeup capability.

High-Impedance Fault (HIF)

(1) An electrical fault of a value that is below the trip point of the breaker on each individual circuit. (GL 86-10, Question 5.3.8)

(2) A circuit fault condition resulting in a short to ground, or conductor-to-conductor hot short, where residual resistance in the faulted connection maintains the fault current level below the component's circuit breaker long-term setpoint. (RG 1.189)

Clarification: HIFs are typically initiated by damaged or degraded insulation and are characterized by low and erratic current flow. Unlike a short circuit (bolted fault), a HIF has an element of resistance between the affected power conductor and its return path (typically ground). This resistance limits the value of fault current. Because of these characteristics, HIFs may continue undetected by conventional circuit protective devices. (GL 86-10, Question 5.3.8) Should a sufficient number of these faults occur, the summation of fault currents may be sufficient to cause a trip of the upstream feeder breaker, resulting a loss of power to required shutdown loads connected to the affected power source. With regard to the analysis of their potential impact on post-fire safe-shutdown capability, HIFs should be postulated to occur simultaneously on all exposed cables located in the fire area and should be assumed to be of a magnitude that is just below the long-term trip point setting of the individual load breaker. (GL 86-10, Question 5.3.8)

Hot Short
Individual conductors of the same or different cables come in contact with each other
and may result in an impressed voltage or current on the circuit being analyzed. (RG 1.189)

Clarification: The term "hot short" is used to describe a specific type of short circuit fault
condition between energized and deenergized conductors. Should a deenergized conductor
come into electrical contact with an energized conductor (or other external source), the voltage,
current, or signal being carried by the energized conductor (or source) would be impressed
onto one or more of the deenergized conductors.

Important to Safety
NPP SSCs that are "important to safety" are those required to provide reasonable assurance
that the facility can be operated without undue risk to the health and safety of the public.
(RG 1.189)

Instrument Sensing Line
Small-diameter tubing (usually stainless steel, but sometimes copper) used to interconnect
plant process instrumentation.

Inter-Cable Fault
A fault between conductors of two or more separate cables.

Intra-Cable Fault
A fault between two or more conductors within a single multi-conductor cable.

Interlock
A device actuated by the operation of some other device with which it is directly associated
to govern succeeding operations of the same or allied devices.

Note: Interlocks may be either electrical or mechanical. (IEEE Std.100-1988)

Interrupting Device
A breaker, fuse, or similar device installed in an electrical circuit to isolate the circuit (or a portion
of the circuit) from the remainder of the system in the event of an overcurrent or fault
downstream of the interrupting device. (RG 1.189)

Isolating Device/Isolation Device
A device in a circuit which prevents malfunctions in one section of the circuit from causing
unacceptable influences in other sections of the circuit or other circuits. (IEEE Std. 100-1988;
RG 1.189)

Isolation Transfer Switch
A device used to provide electrical isolation from the fire-affected area and transfer control of
equipment from the main control room to the local control station (alternate shutdown panel).

Insulated Conductor
A conductor covered with a dielectric (other than air) having a rated insulating strength equal to
or greater than the voltage of the circuit in which it is used. (IEEE Std. 100-1988)

Insulation (Cable, Conductor)
That which is relied on to insulate the conductor or other conductors or conducting parts from ground. (IEEE Std. 100-1988)

Leakage Current (Insulation)
The current that flows through or across the surface of insulation and defines the insulation resistance at the specified direct current potential. (IEEE Std. 100-1988)

Local Control
Operation of shutdown equipment using remote controls (e.g., control switches) specifically designed for this purpose from a location other than the main control room (for example, Operating the EDG from controls provided at the remote/alternate shutdown panel).

Local Control Station
A control panel located in the plant which allows operation and monitoring of plant equipment from outside of the main control room. For post-fire safe-shutdown control functions and monitoring, variables on these panels must be independent (physically and electrically) from those in the MCR.

Local Operation
Manipulation of plant equipment from a location outside of the main control room (for example, manual operation of the circuit breakers or turning the handwheel on the valve to change its position).

Load Breaker
A circuit breaker that is located on the load side of a power source. [Synonym: branch breaker.]

Maloperation
The inability of a component to operate as desired or when expected.

Manual Action
Manual manipulation (operation) of equipment. These actions may be subdivided into the broad categories of "operator action" or "operator manual action."

Manual Valve
A valve that does not have the capability to be manipulated remotely.

Manually Operated Valve
A valve credited in the SSA or shutdown procedures for being manually manipulated.

Note: A manually operated valve may be a manual valve or a remotely motor-operated valve (e.g., MOV) that has its power and control capability disabled or removed.

Mitigating Action
A manual action (operator action or operator manual action) designed to stop the progression or reduce the severity of the unwanted condition.

Molded-Case Circuit Breaker
A circuit breaker that is assembled as an integral unit in a supporting and enclosing housing of molded insulating material. (IEEE Std. 100-1988)

Multi-Conductor Cable (Multiple-Conductor Cable)
A combination of two or more conductors cabled together and insulated from one another and from sheath or armor where used.

Note: Specific cables are referred to as 3-conductor cable, 7-conductor cable, 50-conductor cable, etc. (IEEE Std. 100-1988)

Nonessential (Conductor, Cable, Component, or System)
Class 1E, Non-Class 1E, safety-related, or nonsafety-related SSCs of which operation is not required to support the performance of systems credited in the SSA for accomplishing post-fire safe-shutdown functions.

Normally Closed or Normally Open
The status of a given component during the plant's normal operating modes. This terminology is usually applied to valve, circuit breaker, and relay operating positions.

Open Circuit
A failure condition that results when a circuit (either a cable or individual conductor within a cable) loses electrical continuity. (RG 1.189)

Clarification: A circuit fault condition where the electrical path has been interrupted or "opened" at some point so that current will not flow. Open circuits may be caused by a loss of conductor integrity as a result of heat or physical damage (break).

Operator Actions
Those actions taken by operators from inside the MCR to achieve and maintain post-fire safe-shutdown. These actions are typically performed by the operators controlling equipment that is located remote from the MCR.

Operator Manual Actions
Those actions taken by the operators to manipulate components and equipment from outside the MCR to achieve and maintain post-fire safe-shutdown. These actions are performed locally by operators typically at the equipment.

Overcurrent
Any current in excess of the rated current of equipment or the rated ampacity of a conductor. Overcurrent may result from overload, short-circuit, or ground-fault. A current in excess of rating may be accommodated by certain equipment and conductors for a given set of conditions. Hence, the rules for overcurrent protection are specific for particular situations.
(IEEE Std. 100-1988)

Overcurrent Protection
A form of protection that operates when current exceeds a predetermined value.
(IEEE Std. 100-1988)

Overcurrent Relay
A relay that operates when its input current exceeds a predetermined value.
(IEEE Std. 100-1988)

Overload
(1) Loading in excess of the normal rating of equipment. (IEEE Std. 100-1988)
(2) Generally used in referring to an overcurrent that is not of sufficient magnitude to be termed
   a short circuit. (IEEE Std. 100-1988)

Clarification: An overload is a circuit fault condition that occurs when the amount of current
flowing through the circuit (cable, wire) exceeds the rating of the protective devices (fuse, circuit
breaker, etc.). Without proper overload protection, wires can get hot or even melt the insulation
and start a fire. Overloads are most often between 1 and 6 times the normal current level.
Usually, they are caused by harmless temporary surge currents that occur when motors are
started or transformers are energized. Such overload currents, or transients, are normal
occurrences. Since they are of brief duration, any temperature rise is trivial and has no harmful
effect on the circuit components. (It is important that protective devices do not react to them).
A sustained overload current results in overheating of conductors and other components and
will cause deterioration of insulation, which may eventually result in severe damage and short-
circuits if not interrupted.

Paired Cable
A cable in which all of the conductors are arranged in the form of twisted pairs.
(IEEE Std. 100-1988)

Potential Transformer
A special class of transformer used to step down high distribution system-level voltages
(typically 480 V and above) to a level that can be safely measured by standard metering
equipment. Potential transformers (PTs) have a voltage reduction ratio given on their
nameplates. A PT with a voltage reduction ratio of 200:5 would reduce the voltage by a ratio
of 200 divided by 5 or 40 times.

Power Cable/Circuit
A circuit used to carry electricity that operates a load.

Pre-Fire Position/Operating Mode
Terminology used to indicate equipment status before a fire.

Protective Relay

A device used to detect defective lines or apparati or other power system conditions of an abnormal or dangerous nature and to initiate appropriate control action. A protective relay may be classified according to its input quantities, operating principal, or performance characteristics. (IEEE Std. 100-1988)

Clarification: Protective relays are small, fast-acting, automatic switches designed to protect an electrical system from faults and overloads. A single 4,160-V SWGR may have many relays, each with a specific purpose. Protective relays are classified by the variable they monitor or the function they perform. When a relay senses a problem (e.g., short circuit), it quickly sends a signal to one or many circuit breakers to open, or trip, thus protecting the remainder of the distribution system.

Raceway

An enclosed channel of metal or nonmetallic materials designed expressly for holding wires, cables, or busbars, with additional functions as permitted by code. Raceways include, but are not limited to, rigid metal conduit, rigid nonmetallic conduit, intermediate metal conduit, liquid-tight flexible conduit, flexible metallic tubing, flexible metal conduit, electrical nonmetallic tubing, electrical metallic tubing, underfloor raceways, cellular concrete floor raceways, cellular metal floor raceways, surface raceways, wireways, and busways. (RG 1.189; IEEE Std. 100-1988)

Rated Voltage

(1) The voltage at which operating and performance characteristics of apparatus and equipment are referred. (IEEE Std. 100-1988)
(2) For either single-conductor or multiple-conductor cables, the rated voltage is expressed in terms of phase-to-phase voltage of a three-phase system. For single-phase systems, a rated voltage of $\sqrt{3}$ * the voltage to ground should be assumed. (IEEE Std. 100-1988)

Recovery Action

Activities to achieve the nuclear safety performance criteria that takes place outside the MCR or outside of the primary control station(s) for the equipment being operated, including the replacement or modification of components. (NFPA 805, 2001 Edition)

Redundant Shutdown

(1) If the system is being used to provide its design function, it is generally considered to be redundant. If the system is being used in lieu of the preferred system because the redundant components of the preferred system do not meet the separation criteria of Section III.G.2, the system is considered to be an alternative shutdown capability. (GL 86-10, Question 5.8.3)
(2) For the purpose of analysis to Section III.G.2 criteria, the safe-shutdown capability is defined as one of the two normal safe-shutdown trains. If the criteria of Section III.G.2 are not met, an alternative shutdown capability is required. (GL 86-10, Question 5.1.2)

Note: For BWRs, the use of safety relief valves and low-pressure injection systems has been found to meet the requirements of a redundant means of post-fire safe-shutdown under Section III.G.2 of Appendix R to 10 CFR Part 50 (Letter from S. Richards, NRC, to J. Kenny, BWR Owners Group, dated December 12, 2000).

Relay

An electrically controlled, usually two-state, device that opens and closes electrical contacts to affect the operation of other devices in the same or another electric circuit. (IEEE Std. 100-1988)

Remote Control

Control of an operation from a distance; this involves a link, usually electrical, between the control device and the apparatus to be operated. (IEEE Std. 100-1988)

Note: Remote control may be accomplished from the control room or local control stations.

Remote Shutdown Location

A plant location external to the MCR that is used to manipulate or monitor plant equipment during the safe-shutdown process. Examples include the remote shutdown panel (RSP) or valves requiring manual operation.

Remote Shutdown Panel (RSP)

Depending on usage, the term "RSP" may refer to control and monitoring stations (panels) having significantly different design capabilities. For example, RSP may refer to either of the following:

(1) The control panel included in the plant design for the purpose of satisfying GDC 19 (shutdown attributable to loss of control room habitability).
   Note: The controls and instruments on this panel are not necessarily isolated from the effects of fire. For GDC 19, damage to the control room is not considered.
(2) The control panel included in the plant design for the purpose of controlling and monitoring alternative shutdown functions from outside the MCR.
   Note: Alternative shutdown systems need not be redundant, but must be both physically and electrically independent of the control room. (GL 86-10, Question 5.3.11)

Repair

To restore by replacing a part or putting together what is broken. (Webster's 9[h] New Collegiate Dictionary)

Clarification: In general, a repair may include any operator manual action involving (1) the use of a tool (screwdriver, pliers, wrench, etc.), (2) the installation of components (e.g., fuse, electrical/pneumatic jumpers), or (3) a modification of plant SSCs. Such repairs are only permitted on equipment needed to achieve and maintain cold shutdown conditions.

Note: (1) Tools do not include appropriately controlled equipment provided to facilitate the implementation of procedurally directed operator manual actions, such as ladders, flashlights, fuse pullers, extension bars/handles. (2) Removal of fuses (fuse pulling) is generally not considered a repair. However, this determination must be made on a case-by-case basis considering such factors as feasibility, time, adequacy of emergency lighting, potential for human error and personnel safety hazards. [See IN 84-09, Attachment I, XI for additional information.]

Required Circuits and Cables
Circuits and cables needed to support operation or prevent the maloperation of components identified as being necessary to achieve and maintain safe-shutdown for a particular fire area. In general, a circuit/cable is considered to be *required* for safe-shutdown if it is needed to ensure the operation of required equipment *and* fire-induced faults in the circuit (cable) can cause the required component(s) to fail and/or maloperate in an undesired condition for safe-shutdown.

Note: Required equipment designations may be found to vary between fire areas (e.g., a cable may be required for shutdown in the event of fire in one area, but not required in another).

Required Equipment List
See Safe-Shutdown Equipment List

Required Shutdown Equipment/Components
Equipment needed to ensure the capability to achieve and maintain post-fire safe-shutdown conditions may be accomplished within established criteria.

Required Shutdown System
The systems credited in the SSA for performing each nuclear safety function.

Resistance
Opposition of the flow of electricity through a material.

Clarification: A number of factors(such as wire diameter, wire length and any impurities in the makeup of the wire) determine the resistance to current flow. In general, smaller-diameter wires have more resistance than larger-diameter wires, and longer wires have more resistance than shorter wires. When electricity flows through any resistance, it dissipates energy in the form of heat.

Safe-Shutdown Analysis (Post-Fire Safe-Shutdown Analysis)
A documented evaluation of the potential effects of a postulated fire (including an exposure fire) and fire-suppression activities in any single area of the plant (fire area), on the ability to achieve and maintain safe-shutdown conditions in a manner that is consistent with established performance goals and safety objectives. (Sections III.G and III.L of Appendix R to 10 CFR Part 50 or Position C.5.b of SRP Section 9.5.1)

Safe-Shutdown Equipment List (SSEL)
A documented list of equipment and components that must operate or be prevented from maloperating to ensure the capability to achieve and maintain post-fire safe-shutdown conditions within established criteria. [Synonym: Required equipment list]

Safe-Shutdown System
All structures, equipment (components, cables, raceways, cable enclosures, etc.), and supporting systems [heating, ventilation, and air-conditioning (HVAC)] electrical distribution, station and instrument air, cooling water, etc.] needed to perform a shutdown function.

Selectivity

A general term describing the interrelated performance of relays and breakers, and other protective devices; complete selectivity is obtained when a minimum amount of equipment is removed from service for isolation of a fault or other abnormality. [See also: Coordination.] (IEEE Std. 100-1988)

Short Circuit

An abnormal connection (including an arc) of relatively low impedance, whether made accidentally or intentionally, between two points of different potential. (IEEE Std. 100-1988)

Short-Circuit Current (I_{sc})

Current that flows outside the normal conducting paths (e.g., conductor to ground).

Note: Unlike HIFs, this fault current is generally very large, since only the combined impedance of the object responsible for the short, the wire, and the transformer limit its magnitude. Short-circuit current is often 2 orders of magnitude greater than normal operating current. The symbol I_{sc} is frequently used to represent the value/magnitude of current flowing during a short-circuit fault condition.

Short to Ground

A short circuit between conductor(s) and a grounded reference point (e.g., grounded conductor, conduit, raceway, metal enclosure, shield wrap, or drain wire within a cable).

Solid Conductor

A conductor consisting of a single wire. (IPEEE Std. 100-1988)

Spurious Actuation/Operation

A full or partial change in the operating mode or position of equipment. These operations include, but are not limited to, (1) opening or closing normally closed or open valves, (2) starting or stopping of pumps or motors, (3) actuation of logic circuits, or (4) inaccurate instrument readings.

Spurious Indications

False indications (process monitoring, control, annunciator, alarm, etc.) that may occur as a result of fire and fire-suppression activities.

Spurious Signals

False control or instrument signals that may be initiated as a result of fire and fire-suppression activities.

Stranded Conductor

A conductor made from a number of smaller wire strands wrapped around each other.

Sub-Component

Components that are required to ensure the proper control and/or operation of main flowpath components (e.g., pumps, flowpath valves) and components such as flow switches, temperature switches, relays, transmitters, or signal conditioners that provide isolation or actuation signals to main components.

Tenability
The effects of smoke and heat on personnel actions. (NFPA 805)

Thermal/Hydraulic Timeline
A documented evaluation of the response of important reactor plant parameters to a postulated transient (thermal/hydraulic analysis) with respect to the time available to accomplish required shutdown functions. For example, the time available to establish auxiliary feedwater (AFW) following a reactor scram in a PWR would be determined by a thermal/hydraulic analysis. The objective of the thermal/hydraulic timeline is to compare this time to the time needed for operators to perform all system and equipment alignments necessary to establish a secure source of AFW.

Thermoplastic
A cable material that will soften, flow, or distort appreciably when subjected to sufficient heat and pressure. Examples include polyvinyl chloride (PVC) and polyethylene(PE).

Note: Cables using thermoplastic insulation *are not* usually qualified to survive the full environment qualification exposure condition of IEEE Std. 383. Many thermoplastic cables will, however, pass the limited flame spread test included in the IEEE Std. 383.

Thermoset
A cable material that will not soften, flow, or distort appreciably when subjected to heat and pressure. Examples include rubber and neoprene.

Note: Cables using thermoset insulation *are* usually qualified to IEEE Std. 383.

Time/Current Characteristic Curve (Trip Curves)
A graphic illustration of the operating characteristics of electrical protection devices (fuses, circuit breakers, or relays). The tripping characteristics of protective devices are represented by a characteristic tripping curve that plots tripping time versus current level. The curve shows the amount of time required for the protective device to trip at a given overcurrent level. The larger the overload or fault current, the faster the breaker/fuse will operate to clear the circuit (referred to as inverse time characteristics). A comparison of characteristic trip curves is necessary to determine whether proper coordination exists between devices.

Triplex Cable
A cable composed of three insulated single-conductor cables twisted together.
(IEEE Std. 100-1988)

Note: AC 3-phase (3Φ) power cables are commonly of triplex design.

Unprotected Circuit/Cable
A circuit/cable that is not provided with fire protection features sufficient to satisfy applicable requirements. (Section III.G.2 of Appendix R or Position C.5.b of SRP Section 9.5.1)

Voltage

The effective RMS potential between any two conductors or between a conductor and ground. Voltages are expressed in nominal values unless otherwise indicated. (IEEE Std. 100-1988)

Clarification: The electrical force that causes free electrons to move from one atom to another. Voltage is similar to pressure in a water pipe.

This page intentionally left blank.

CHAPTER 3. BACKGROUND INFORMATION AND EXPERIENCE RELATED TO FIRE-INDUCED CIRCUIT FAILURES

3.1 Background

Like any large industrial complex, an NPP contains an extensive array of systems and components, and nearly all of this equipment directly or indirectly depends on the continuous operation of one or more electrical circuits and cables. A typical BWR requires approximately 96 km (60 miles or 316,000 ft) of power cable, 77 km (50 miles or 254,000 ft) of control cable and 402 km (250 miles or 1,320,000 ft) of instrument cable. More than 1,600 km (1,000 miles or 5,280,000 ft) of cable went into the containment building of Waterford Steam Electric Generating Station, Unit 3, a pressurized-water reactor (PWR).[4] Because of their large quantity and the fact that much of the cable material is combustible (e.g., polymer insulation and outer jacket), cables frequently comprise a significant portion of the total combustible fire loading in many areas of a plant.

As evidenced by the Browns Ferry fire in 1975, electrical circuit failures resulting from fire-damaged cables can have a substantial impact on safe plant operations. Although the BFN fire was contained to a relatively small interior area of the plant, temperatures as high as 815.5 °C (1,500 °F) caused damage to more than 1,600 cables routed in 117 conduits and 26 cable trays. As described below, circuit failures resulting from damage to these cables caused equipment to operate in unexpected ways and significantly impeded the operators' ability to monitor and control reactor safety functions.

3.2 Circuit and Cable Primer

An electrical circuit is analogous to a circular path through which electrons flow. In the circuit illustrated in Figure 3-1, the flow path is from the negative battery terminal through the load (lamp) and back to the positive battery terminal. In more complex circuits, many paths may split off to various components, but they always form a line from one side (polarity) of the power source and return to the opposite side (polarity) of the power source.

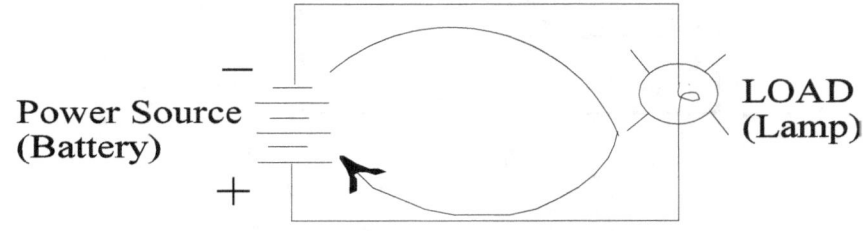

Figure 3-1 Circuit Illustration

---

[4]   NUREG/CR-6384, "Literature Review of Environmental Qualification of Safety-Related Electric Cables," Volume 1, April 1996.

For small, simple circuits, such as the one illustrated in Figure 3-1, the electrical path between components could be established by short lengths of individual wire conductors. However, in a large installation such as an NPP, the circuit components may be located at great distances from each other. For example, the power source in the above illustration may be a fuse panel that is located in the service water intake structure, while the load is a pump status indicator lamp that is located in the control room. For such applications, long lengths of cable containing one or more insulated wires or conductors are needed to establish the path for current to flow (i.e., complete the circuit). By contrast, in a complex facility, such as a power plant, many cables of various types, construction, and sizes are needed to distribute electric power, control signals, and process system information. As depicted in Figure 3-2, these cables are generally classified by the function they perform:

- *Power cables* distribute power from power supplies (SWGRs, MCCs, panel boards) to utilization equipment. Within the plant, power cables are classified by the level of voltage they carry. Medium-voltage power cables (4.16-kV, 6.9-kV) distribute power to auxiliary transformers, electrical SWGRs, and large motors. Low-voltage power cables (<1,000-V) supply power to MCCs, MOVs, pumps, and motors.

- *Control cables* allow remote control of a component or a permissive/interlock signal.

- *Instrument cables* transmit low-level signals from the instrument sensor to an indicator, controller, or recorder.

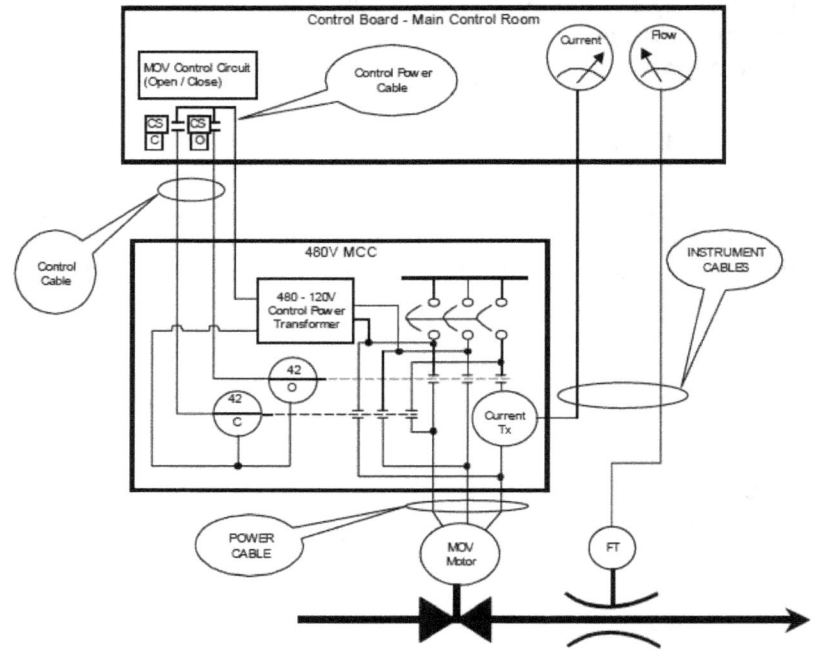

LEGEND
CS Control Switch
C Close
O Open
42O Open Relay
42C Close Relay
FT Flow Transmitter
MOV Motor Operated Valve
TX Transformer

Figure 3-2 General Cable Classifications

3-2

3.2.1 Cable Construction and Materials

As illustrated in Figure 3-3, most cables used in an NPP are composed of three parts, including a metallic conductor, insulation, and a protective polymer jacket.

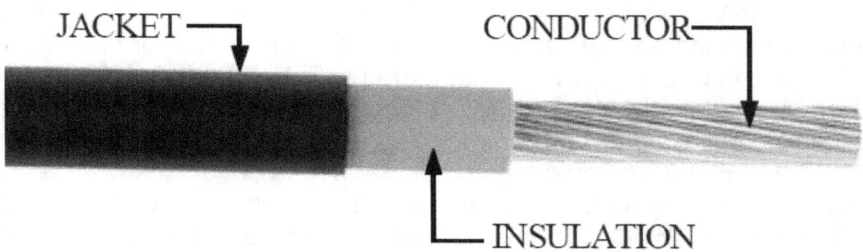

Figure 3-3 Cable Components

The conductor provides a low-resistance path for electrical current or signals. Copper and aluminum are popular conductor materials and may be either solid or stranded. As its name implies, a solid conductor is a single length of wire, whereas a stranded conductor is made by twisting individual strands of wire around each other until the desired conductor diameter or "gauge" is achieved. The conductor shown above in Figure 3-3 is a stranded conductor. While there is little difference in their electrical capabilities, stranded conductors are far more flexible than solid conductors of the same gauge, making them easier to install. The majority of cables found in plants contain stranded copper conductors. A cable may contain a single conductor (Figure 3-3), or a large number of separately insulated conductors. A cable containing more than one conductor is called a "multi-conductor cable." Figure 3-4 shows cross-sectional view of a multi-conductor cable containing seven conductors.

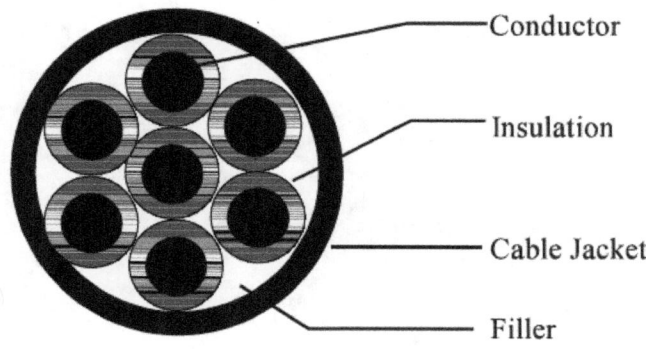

Figure 3-4 Multi-Conductor Cable (7-Conductor)

Insulation isolates the conductor from unwanted paths of current flow (e.g., grounded conduit or cable tray, other conductors, and personnel). Many different types of insulation materials are available to accommodate the specific application, environment, and service conditions of a cable. In most nuclear power applications, cables are insulated with either thermoset or thermoplastic materials. Thermoplastic materials are made from compounds that will re-soften and distort from their formed shapes by heating above a critical temperature peculiar to the material. Polyvinyl chloride (PVC) and polyethylene (PE) are examples of thermoplastic compounds. Thermoset insulation and jacket compounds will not re-soften or distort from their formed shapes by heating until a destructive temperature is reached. Insulation and cable outer jackets made from cross-linked polyethylene (XLPE), chlorosulfonated polyethylene (CSPE, commonly called Hypalon), and Neoprene are examples of thermoset materials. Cables that survive the full range of environment and flame spread conditions of IEEE 383 (IEEE 383-qualified cables) will likely have thermoset jackets and insulation with a failure temperature of approximately of 371 °C (700 °F). Cables that are not IEEE 383-qualified typically have thermoplastic jackets and insulation and may have a failure temperature as low as 218 °C (425 °F) depending on the melting or softening temperature of the specific thermoplastic polymer.

The voltage rating of a cable is the highest voltage that may be continuously applied and is generally a function of the type and amount (thickness) of insulation used. Cables used in low-voltage applications ($\leq$600 V) are generally rated at 600 V regardless of their actual application voltage. Cables in the low-voltage range include instrument circuits (50 V or less), control and control power circuits (120–250 V range) and certain power circuit applications (120, 480, and 600 V). Single- and multi-conductor cables used in medium-voltage applications (e.g., 4,160 V) are available with nominal voltage ratings of 5 kV, 8 kV, 15 kV, 25 kV, and 35 kV.

The cable jacket is usually a plastic cover that protects the cable from mechanical damage and chemical attack during installation and throughout its service life. Some of the more common jacket materials are PVC, Neoprene, and Hypalon. The jacket does not perform any electrical function. Where a high degree of physical protection is desired, cables may be furnished with a metallic outer sheath (or armor) made from interlocked aluminum or steel. Cables of this type are called "armored cables." Armoring protects the cable from penetration by sharp objects, crushing forces, and damage from gnawing animals or boring insects. Armored cables may be bare (i.e., exposed metal armor), or the armor may be covered with an additional layer of polymer jacket. Figure 3-5 illustrates an example of an armored power cable.

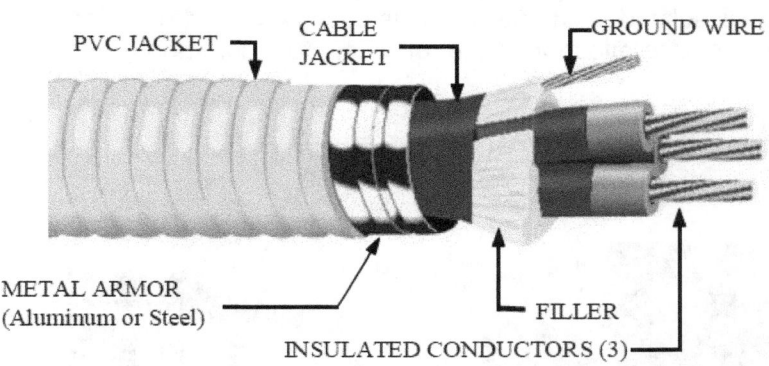

Figure 3-5 Armored Cable

Power cables may be of single- or multi-conductor design. Single-conductor cables are typically found inside electrical enclosures and cabinets, such as SWGRs and MCCs. A special type of multi-conductor cable called "Triplex cable" is commonly used in three-phase power applications, such as supplying power to an MOV from an MCC. A triplex cable contains three individually insulated conductors that are twisted around each other and contained within an outer jacket.

As shown in Figures 3-4 and 3-5, multi-conductor cables use a nonconductive "filler" material to occupy the openings (gaps) that are formed when a group of individual conductors are assembled. In addition to forming the shape (roundness) of the cable, fillers may also contribute to the flexibility and tensile strength of the cable.

Control and instrument cables are typically of multi-conductor design, as illustrated in Figure 3-7. Although the number of conductors that may be contained in a multi-conductor cable is theoretically unlimited, practical considerations such as the difficulty of installing long runs of very large-diameter cable tend to limit their size. Common control circuits employ multi-conductor cables having 3-, 7-, or 11-conductor configurations. Because of the need to block external sources of electrical "noise" generated by other plant equipment, instrument cables frequently use a number of "twisted/shielded pairs" of conductors contained within a protective outer jacket, as illustrated in Figure 3-7. The twisting of conductors reduces magnetic noise, while the shield and drain wire reduce electrostatic and radio-frequency interference. The shield consists of a conductive material (typically aluminum foil) that is wrapped around the twisted pairs of conductors. The uninsulated drain wire, which is in physical and electrical contact with the shield, provides for easier termination of the foil shield to a common ground point.

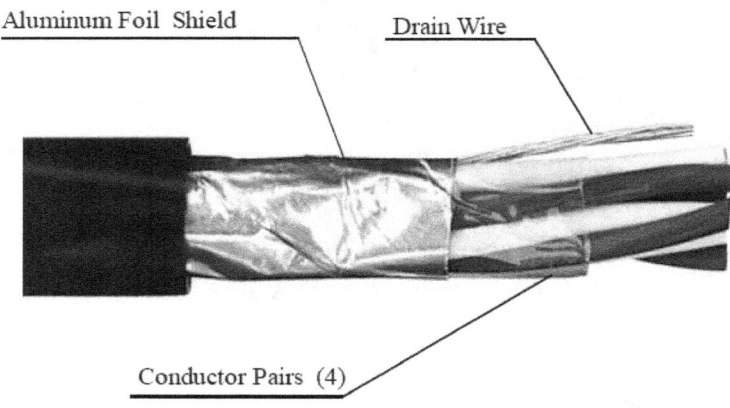

Figure 3-6 Illustration of Instrument Cable

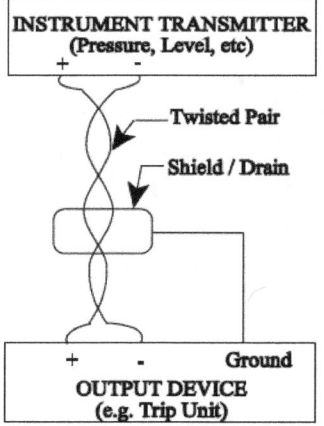

Figure 3-7 Twisted/Shielded Pair

3.2.2 Functional Considerations of Conductors and Cables

The fundamental purpose of a conductor is to provide a path for electrons to move from one location to another across a cable. The force or pressure that causes the electrons to move through the conductor is called *voltage*, which is measured in "volts." The quantity or flow rate of electrons moving through the conductor is called *current*. Current is measured in units called "amperes" or "amps," where 1 amp is equal to a flow of 1 coulomb per second through a wire (a coulomb is 6.28×10^{23} electrons). Simply stated, voltage causes current. Given a voltage and a complete path for the electrons (i.e., a complete circuit), current will flow. Given the path, but no voltage, or voltage without a path (e.g., an open circuit), there will be no current.

Resistance is a force that opposes the flow of electrons. Every material, including the most effective conductors (e.g., silver and gold), offers some resistance to current flow. Principal factors affecting the amount of resistance presented by a conductor include the following:
(1) Length: The longer the conductor, the higher the resistance.
(2) Diameter (gauge): The smaller the diameter of the conductor, the higher the resistance.
(3) Temperature: The higher the temperature, the higher the resistance.
(4) Material: Some materials are better conductors than others. Gold and silver are excellent conductors, but are also very expensive. Copper is widely used because it is a very good conductor and is not cost-prohibitive.

In the United States, cables are manufactured in accordance with the American Wire Gauge (AWG) standard, where "gauge" refers to the diameter of the metallic conductor (without insulation). The higher the gauge number, the smaller the diameter of the wire conductor. For example, wiring used to power receptacles in most U.S. households is AWG 12 or 14, while telephone wire is usually AWG 22 or 24. Because it has less electrical resistance over a given length, thick wire (i.e., small AWG number) can carry more current than thin wire (large AWG number). For example, a copper AWG 12 conductor is approximately 0.2 cm (0.08 inches) in diameter and can carry about 20 amperes of current. Conversely, an AWG 1 conductor has a diameter of approximately 0.76 cm (0.30 inches) and can carry about 150 amperes of current. Power cable conductors may range from 0.2 cm (0.08 inches) in diameter (AWG 12) to over 2.54 cm (1 inch). Because they carry less current, control cables commonly range from AWG 16 up through AWG 10, and instrumentation cables are generally AWG 16 or smaller.

The largest-diameter conductor specified in the AWG system is 0000 or 4/0 (pronounced "four ought"). Wire sizes larger than those covered in the AWG system are specified in "circular mills" (cmill). By definition, a circular mill is the area of a circle with a diameter of "1 mil" (1 one-thousandth of an inch). Because this unit is so small, the prefix "M" is normally used in denoting wire sizes. For example, a conductor that is 250,000 circular mils is normally denoted 250 MCM.

In most applications, the size of a cable is expressed in terms of the gauge (AWG) of its individual conductor(s) and the number of conductors it contains. For example, a cable that contains three AWG 12 conductors would be described as a "three conductor number 12" or "3/C, 12 AWG" cable.

Many people tend to think of a conductor's size (gauge) only in terms of its current-carrying capability (ampacity). For example, one general rule-of-thumb is that a cable containing AWG 12 copper conductors is sufficient to power loads supplied from a circuit breaker that has a 20-ampere trip point. While this rule-of-thumb may be sufficient for most home wiring applications, in large facilities such as NPPs, other factors such as cable length and ambient temperature may also have a significant impact on cable selection. As indicated above, the resistance of a conductor increases as its length increases, its diameter decreases, or the temperature of its surrounding environment (ambient temperature) increases. Heat is generated whenever a current flows through a conductor and, as the length of a cable increases, so does its resistance. This resistance, in turn, creates a voltage loss or "drop" in the cable. For example, if a cable is supplying power to a motor that is located some distance from its power source (e.g., MCC) and the gauge (diameter) of the cable conductors is not properly sized (increased) to accommodate for the additional resistance presented by the length of cable, the voltage measured at the motor will be less than that measured at the MCC and, in certain cases, may be insufficient for proper motor operation. Depending on the specific application, voltage drop and ambient temperature may be important considerations in the selection of cables.

The temperature rating of a cable/conductor is the maximum temperature at which its insulating material may be used in continuous operation without loss of its basic properties. The most common ratings are 60, 75, and 90 °C (or 140, 167, and 194 °F). Ampacity is the amount of current a cable/conductor can carry continuously under conditions of use without exceeding its temperature rating. This definition of ampacity recognizes that the maximum current that a conductor can carry continuously varies with the conditions of use as well as with the temperature rating of the conductor's insulation. For example, ambient temperature is a condition of use. A conductor with 60 °C (140 °F) insulation installed near a furnace so that the ambient temperature is 60 °C (140 °F) continuously has no current carrying capacity. Any current flowing through the conductor will raise its temperature beyond the 60 °C (140 °F) insulation rating. The ampacity of this conductor, regardless of its size, is therefore zero.[5]

The normal ambient temperature of a cable installation is the temperature the cable would assume at the installed location with no load being carried on the cable.[6] Ampacity limits for various combinations of cables and ambient temperatures are given in the NEC (ANSI/NFPA 70). In addition to ambient temperature, many other external factors can affect the ampacity of an electrical conductor. Examples include cable tray fill, heat generated by the conductor as a result of load current flow, heat generated by adjacent cables (e.g., within the same cable tray), and any insulating material that may surround the cable (e.g., fire-protective wrapping over a cable tray). Such factors must be taken into consideration when selecting conductors for a specific application.

---

[5]   *The National Electric Code Handbook*, P.J. Schram, Editor, National Fire Protection Association, Quincy, MA, 1997.

[6]   ANSI/IEEE Std. 141-1986.

3.3 Circuit Failure Modes and Mechanisms

As previously discussed, cables are composed of one or more electrical conductors. The individual conductors are electrically isolated from each other and from other possible diversion paths (e.g., ground) by a layer of electrical insulation material. When exposed to the effects of fire and its related perils (e.g., firefighting activities), insulation and other protective materials (e.g., jacket) may be subjected to a broad range of potentially damaging stressors or failure mechanisms. For example, heat could cause a significant reduction in the quality of electrical isolation provided by the conductor's insulation material or (in certain cases) cause it to completely melt away. Heat, combined with smoke and products of combustion, could initiate faults in electronic components and printed circuit boards. The addition of fire suppression agents could further exacerbate the effects of already damaged insulation. Mechanical forces, such as those that may be inflicted during firefighting activities (e.g., impact of a fire hose stream), could cause a further reduction in the physical and electrical integrity of circuits and cables. As a result, circuits and cables that are exposed to the effects of fire are expected to experience one or a combination of the following fault conditions or failure modes:

• *Open Circuit:* The loss of electrical continuity (i.e., the conductor is broken and the signal or power does not reach its destination), as illustrated in Figure 3-8.

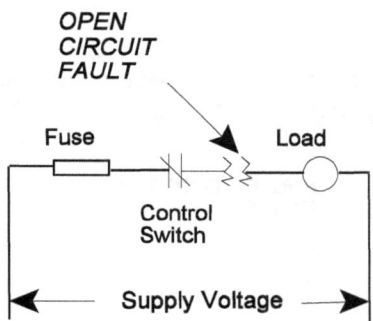

Figure 3-8 Open Circuit Fault

- *Short Circuit:* An abnormal connection of relatively low-impedance between two points of different potential , as illustrated in Figure 3-9.

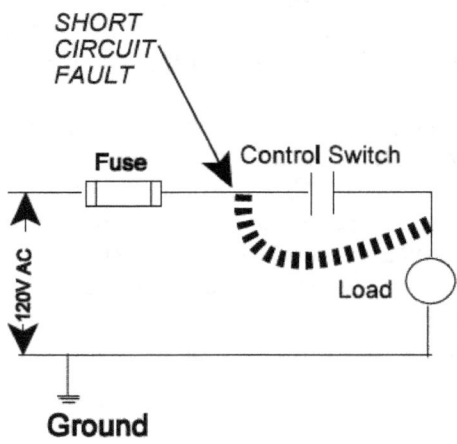

Figure 3-9 Short Circuit Fault

- *Shorts to Ground:* A conductor comes into electrical contact with a grounded conducting medium, such as a cable tray, conduit, or grounded conductor, as illustrated in Figure 3-10.

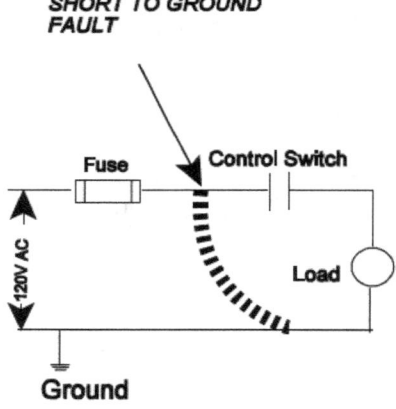

Figure 3-10 Short-to-Ground Fault

• *Hot Short:* A special type of short circuit condition that causes a previously un-energized conductor to become energized. As a result of this fault, the voltage, current, or instrument signal present in the energized conductor(s) is impressed on the previously un-energized conductor(s). As illustrated in Figure 3-11, a hot short could bypass circuit protective features and cause the unintentional actuation of equipment.

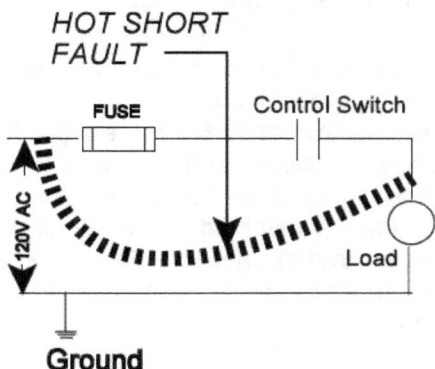

Figure 3-11 Hot Short Fault

• *High-Impedance Fault:* A special type of short-circuit condition in power cables where the fault contains some element of resistance to current flow. An arcing fault is a specific type of HIF. Rather than having direct contact offering minimal resistance to fault current (i.e., "bolted" fault condition), the arcing fault current must flow through or "arc over" a small air gap or water. "Because of the resistance of the arc and the impedance of the return path, current values are substantially reduced from the "bolted fault level."[7] For analytical purposes, HIF current is postulated to be a value that is just below the trip point of the individual circuit protective device (fuse or circuit breaker).

3.4 The Browns Ferry Fire

On March 22, 1975, a severe fire involving electrical cables occurred at BFN Unit 1, which is operated by TVA. The Browns Ferry plant consists of three BWRs, each of which is designed to produce 1,067 megawatts (MW) of electrical power. At the time of the fire, Units 1 and 2 were operating at 100-percent capacity, while Unit 3 was still under construction.

The fire began in a bank of cable trays in an area of the Unit 1 cable spreading room (CSR) where the trays passed through a penetration in a wall separating the CSR from the reactor building. At BFN, the reactor building functions as the secondary containment for the nuclear steam supply systems (NSSSs). To preclude uncontrolled and unmonitored releases of airborne radioactivity, the reactor building is designed (and required by license condition) to be maintained at a negative pressure of 62.3 Pa (0.25 inches of H_2O), in relation to the remainder of the plant and the outside environment. Each penetration through the reactor building wall was sealed with polyurethane (PU) foam to prevent leakage.

---

[7] ANSI IEEE Standard 242-1986, "IEEE Recommended Practice for Protection and Coordination of Industrial and Commercial Power Systems."

The penetration seal inspection process in place at the time of the fire used differential pressure as a means of identifying defective seals. If a penetration seal was defective, the flame would flicker and smoke from a candle would be drawn toward the seal. When workers used this method to test a modified seal, however, the candle flame was drawn into the penetration, igniting the sheet polyurethane (PU) foam that was used as a sealant material. The pressure differential between the CSR and the reactor building then fanned the fire, causing it to rapidly spread to a large number of cables located in trays on the opposite side of the wall.

The fire continued to burn for more than 7 hours as a result of a number of contributing factors, including the large amount of combustible cable insulation involved in the fire, the inaccessibility of fire in cable trays located approximately 6.5 meters (20 feet) above the floor, dense smoke, limited availability of breathing apparatuses and the operators' reluctance to use water to extinguish an electrical fire. Although the fire had a significant impact on plant operations, only a relatively small area of the plant was actually involved. In the CSR, damage was limited to a 2.32-m^2 (25-ft$^{2)}$ area adjacent to the penetration where the fire started. The major fire damage occurred on the opposite side of this penetration in an area of the reactor building measuring approximately 12.16 m x 6.08 m (40 ft x 20 ft).

Although damage was limited to a relatively small area of the plant, temperatures as high as 815.5 °C (1,500 °F) caused damage to more than 1,600 cables routed in 117 conduits and 26 cable trays. Of those, 628 cables were safety-related and their damage caused the loss of a significant number of plant safety systems, including redundant trains of emergency core cooling systems (ECCSs) and electric power and control systems. Fire-induced damage to cables located in the area, also impeded the functioning of normal cooling systems and degraded the capability to monitor the status of the plant.

As described in Section 3.3, when conductors of circuits and cables are exposed to the effects of fire and/or firefighting activities, their electrical integrity and, hence, their ability to properly function is compromised. The Browns Ferry fire demonstrated the impact that fire-induced cable faults can have on the operability of redundant plant safety systems. Table 3-1 depicts some of the more important consequences of the fire in Unit 1. Although not as severe, the fire also impacted Unit 2 operations for approximately 6 hours following initiation of the fire. Examples of abnormalities noted by Unit 2 operators include the loss of electrical power supplied from various 4-kV and 480-V shutdown boards, closure of the main steam isolation valves (MSIVs), loss of the manual actuation capability of all safety relief valves (SRVs), and loss of high-pressure coolant injection (HPCI) as a result of spurious closure of torus suction valves.

While certain operational consequences developed as the fire progressed, a review of the documented chronology of the event indicates that many abnormalities, including spurious ECCS alarms, false instrument indications, reductions in power level (resulting from a run-back of the reactor recirculation pumps for no apparent reason), and spurious starts and stops of RHR, core spray (CS), reactor core isolation cooling (RCIC), and HPCI pumps, were observed to occur rather quickly, during the first 20–30 minutes following the ignition of PU sealant material. The Browns Ferry fire was a clear demonstration of the impact that a fire involving redundant trains of electrical circuits and cables can have on operators' ability to monitor and control important plant parameters.

Table 3-1. Consequences of Cable Damage Attributable to Fire at Browns Ferry Unit 1[8]	
Consequence of Fire Damage	Attributed Cause
Loss of power supplied from 480-V shutdown boards 1A and 1B	• Fire-induced hot-short in circuit breaker trip indicator light caused voltage to be backfed to the breaker trip coil, thereby keeping it energized • Power cable faults
Spurious closure and inability to reopen MSIVs	Fire damage to MSIV control circuits
Spurious trip of reactor feedwater pump "A"	False high reactor water level signal to feedwater pump controller (Note remaining feed pumps B and C were manually tripped at the time of the scram)
Inoperability of HPCI	Fire-induced faults to cables associated with 250-VDC MOV board 1A (which powers HPCI valve controls), and cables associated with 480-V MOV board 1A (which powers the steam isolation valve)
Inoperability of redundant RHR systems (1A, 1B, 1C, and 1D)	Fire-induced failure of 480-V MOV boards 1A and 1B caused loss of power to valves. Also, fire-induced loss of power supplied from 4-kV shutdown board C caused a loss of RHR pump 1B.
Inoperability of redundant core spray (CS) systems (1A, 1B, 1C, and 1D)	Fire-induced failure of 480-V MOV board 1A and 1B caused loss of power to valves. Also, fire-induced loss of power supplied from 4-kV shutdown board C caused loss of CS pump 1B
Inoperability of redundant trains of standby liquid control systems (SLCSs) (1A, 1B)	Fire-induced loss of power from redundant 480-V shutdown boards 1A and 1B to pump motors and valves
Inoperability of RCIC	Inability to electrically operate steam isolation valve as a result of a cable fault and loss of power on 480-V MOV board 1B

---

[8] "Hearings Before the Joint Committee on Atomic Energy, Congress of the United States, First Session," September 16, 1975.

Table 3-1. Consequences of Cable Damage Attributable to Fire at Browns Ferry Unit 1 (continued)	
Consequence of Fire Damage	**Attributed Cause**
Loss of ability to operate all relief valves	Spurious closure and inoperability of 7 of 11 relief valves attributed to loss of power supplied from redundant 250-VDC boards 1A and 1B. Subsequent spurious closure of drywell air compressor flow control valve cut off air supply to remaining 4 relief valves, thereby rendering them inoperable for 4 hours
Abnormal behavior of instrumentation: • Observed ECCS alarms were contrary to system status • Random lights on ECCS panel began glowing alternately bright and dim	Fire damage to ECCS instrumentation circuits
Loss of operability of EDG "C" and loss of remote control capability of EDG "B" and EDG "D"	Fire damage to EDG control and instrumentation circuits

3.5 Insights and Observations Resulting from the Nuclear Energy Institute Fire Test Program

To further investigate the effects of fire conditions on circuit integrity and the potential for fire-induced spurious actuations, NEI and EPRI sponsored a series of 18 cable fire tests at Omega Point Laboratories (OPL), located in Elmendorf, Texas, during the period from January 8 through June 1, 2001. All tests were conducted within a steel enclosure that was 3.04 m x 3.04 m x 2.43 m (10 ft x 10 ft x 8 ft) with a single natural ventilation opening in one wall. Since the primary objective was to assess the potential for fire to cause undesired spurious actuations of equipment, the test included only control and control power (120-VAC) cables and did not include ungrounded DC circuits and power cables (480-VAC and 4,160-VAC). As a result, the tests did not fully evaluate the potential for fire to cause certain types of power circuit-fault conditions, such as HIFs.

In conducting the tests, OPL used three types of cables, including a specific type of multi-conductor armored cable having thermoset insulation, several types of thermoplastic cable, and several types of thermoset cable. OPL connected the tested cables to a single control circuit that had been selected as the object of study for spurious actuation. That control circuit was a NEMA-1 starter for an MOV. Important insights gained from this testing are highlighted by the following observations elicited from the experts responsible for reviewing the test results: [9]

---

[9] "Spurious Actuation of Electrical Circuits Due to Cable Fires: Results of Expert Elicitation," EPRI Technical Report 1006961, Final Report, EPRI Palo Alto, California, May 2002.

- *"Hot shorts leading to spurious actuations cannot be regarded as of negligible importance if the fire under consideration produces cable temperatures above the thresholds identified herein."*

- *"For the majority of the tests there was at least one devise actuation observed, and for several tests multiple actuations were observed... There was at least one spurious operation for almost every configuration tested... Overall, the likelihood of spurious actuation given failure was found to be somewhat higher than I might have assumed prior to conduct of the tests."*

- *"Thermoplastic cable is more likely to degrade to the point of allowing leakage currents large enough to cause device actuations or blown fuses than either armored cable or thermoset cable for the same exposure conditions."*

- *"It appears that 204.5 °C (400 °F) is the approximate degradation temperature of the thermoplastic cable used in these experiments and 371 °C (700 °F) is the approximate degradation temperature of the thermoset cable used in these experiments."*

- *"Water spray on damaged cables can cause spurious actuations to occur."*

- *"The available test data as a whole demonstrates that at least four factors are critical to the assessment of spurious actuation likelihood: armored versus non-armored cables; cables in trays versus cables in conduits; cable-to-circuit wiring configuration; and circuits without control power transformers (CPTs) versus circuits with CPTs."*

- *"The tested configuration used a 150 VA CPT on a nominal 120 V circuit. This application may bound, for example, NEMA size 1 starters which are limited to typically a maximum 7.5 HP motor. For circuits with a higher range, it is suggested to use the non-CPT values."*

- *"No open-circuit type of failures were observed which places an upper bound on such an end-point in the range of a 1-percent probability, given the number of possible open circuits."*

- *"Shorting to another conductor within the same cable is much more likely than shorting to a conductor in another cable."*

- *"The probability that a source conductor in a multi-conductor cable will short to an adjacent (different) single conductor cable (cable-to-cable short) is generally lower than the probability that a conductor-to-conductor short will occur within the multi-conductor cable."*

- *"For the cable configuration tested, the data indicate that a single-conductor cable will usually short to ground before shorting to another single conductor cable. For thermoset cables the probability is about 85–90-percent. For thermoplastic cables the probability is about 70–75-percent."*

- *"Undesired spurious actuations were caused by a single conductor cable shorting to an adjacent single conductor cable without grounding (cable-to-cable short). The probability for this case is estimated to fall between 0.05 and 0.30 with a best estimate point value near 0.20."*

- *"Undesired spurious actuations were caused by an interaction between an energized conductor within a multi-conductor cable having one grounded conductor and an adjacent single conductor cable. The probability for this case is estimated to fall between 0.05 and 0.20 with a best estimate point value near 0.10."*

- *"Several instances of multiple spurious actuations were observed in the same test, sometimes involving different conductors in the same multi-conductor cable."*

- *"For armored, multi-conductor, thermoset, cable having its armor shield maintained at ground potential, the probability of conductor-to-conductor shorts is estimated to be in the 20–30-percent range. This is significantly lower than the 70–80-percent range estimated for unarmored cable."*

- *"The opportunity for armored cable shorting to another cable (cable-to-cable short) was observed to be nil... this probability should be zero."*

This page intentionally left blank.

CHAPTER 4. NRC REGULATORY REQUIREMENTS

4.1   Safety Objective

The fundamental safety objective of the NRC's regulatory program is to ensure adequate protection of public health and safety. This means that the risk to the public from normal operation, anticipated transients, and accidents must be acceptably low, and the likelihood of accidents more severe than those postulated for design purposes must be extremely small. To achieve this high level of safety, redundant (i.e., identical or diverse) safety systems are incorporated into the design of all NPPs that are currently operating in the United States. Redundancy provides assurance that failures affecting one system will not have a significant impact on plant safety because the plant design provides a "backup" system.

To further increase that assurance, the safety equipment and cables of the redundant subsystems are typically segregated into divisions. The separate and redundant divisions of safety systems provide confidence that the failure of components or cables within one division will not adversely affect the plant's ability to accomplish required safety functions. In the absence of suitable protection features, such as separation distance or structural barriers, however, redundant trains of cables and equipment could be susceptible to a phenomenon known as "common-mode" failure, in which multiple failures in redundant systems may occur as a result of a common cause[10]. If a single event could induce failures in more than one of the redundant elements, the safety and reliability benefits afforded by this essential design feature could be negated. Flooding, earthquakes, and fire are three examples of events that have the potential to initiate common-mode failures in redundant safety systems.

As discussed in Chapter 3, a major fire at BFN Unit 1 on March 22, 1975, illustrated the impact that common-mode failures attributable to fire may have on the operation of a commercial NPP. Four days after that event, the NRC established a Special Review Group (SRG) to investigate the cause of the fire and evaluate the need to improve the FPPs at all NPPs. The SRG found serious design inadequacies regarding fire protection at Browns Ferry. In its report, entitled "Recommendations Related to the Browns Ferry Fire" (NUREG-0050, dated February 1976), the SRG provided more than 50 recommendations for improving fire prevention and control in existing facilities. The SRG specifically noted that the independence of redundant equipment at Browns Ferry was negated by not having a suitable degree of separation between cables associated with redundant trains of safety equipment. As a result, the SRG recommended that a suitable combination of electrical isolation, physical distance, barriers, and sprinkler systems should be applied to maintain the independence of redundant safety equipment and, therefore, the availability of safety functions despite postulated fires. In view of its findings, the SRG called for the development of specific guidance for implementing fire protection regulations, and for a comparison of that guidance with the FPP at each operating plant.

---

[10]   IEEE Std. 100, "The Authoritative Dictionary of IEEE Standards Terms" (IEEE Standard Dictionary of Electrical and Electronics Engineers), 1988.

The Browns Ferry fire was sufficiently significant to warrant major changes in the FPPs of NPPs operating in the United States. As discussed in this section, in the years following the Browns Ferry fire, the NRC and the nuclear industry expended considerable resources to develop and implement fire protection guidelines and regulatory requirements that would minimize both the probability of occurrence and the possible consequences of postulated fires. As a result, each operating plant currently has an approved FPP that is anchored in the long-established defense-in-depth (DID) safety principle of providing multiple protective barriers to prevent and mitigate accidents. This protection consists of administrative controls and personnel training to reduce the potential for fire to start, as well as plant design features to rapidly detect and promptly extinguish those fires that may occur. In addition, because of the potentially unacceptable consequences that an unmitigated fire may have on plant safety, each operating plant must demonstrate that in the event a fire were to initiate and continue to burn (despite prevention and mitigation features), the performance of essential shutdown functions will be preserved and radioactive releases to the environment will be minimized.

Recent studies have shown that the revised requirements for protecting SSCs that are important to safe-shutdown are beneficial to safety in the event of fire. Plant design changes required by the new regulatory framework (Appendix R to 10 CFR Part 50) have been effective in preventing a recurrence of a fire event of the severity experienced at Browns Ferry. In addition, according to a "Fire Risk Scoping Study" performed in 1989 by Sandia National Laboratories (SNL), plant modifications made in response to the new requirements have reduced the core damage frequencies (CDFs) at some plants by a factor of 10. The study also suggested that improper implementation of the regulatory requirements and degradation of fire protection DID could be risk-significant. The study concluded, for example, that weaknesses in either manual firefighting effectiveness or control system interactions could raise the estimated fire-induced CDF by an order of magnitude.

In GL 88-20, Supplement 4, the NRC asked each licensee to perform an individual plant examination of external events (IPEEE) for plant-specific severe accident vulnerabilities that are initiated by external events. Under the IPEEE program, the licensees systematically assessed the fire risk for each operating reactor and submitted the results to the NRC. The results of the IPEEE fire analyses provide important insights regarding reactor fire risk and confirm the results of the SNL "Fire Risk Scoping Study." For example, the IPEEE results show that fire events are important contributors to the reported CDF for a majority of plants, ranging on the order of $1 \times 10^{-9} - 1 \times 10^{-4}$ core damage events per reactor-year, with the majority of plants reporting a fire CDF in the range of $1 \times 10^{-6} - 1 \times 10^{-4}$ core damage events per reactor-year. In some cases, the reported CDF contribution from fire events can exceed that from internal events[11].

---

[11]   SECY 99-140, "Recommendation for Reactor Fire Protection Inspections," U.S. Nuclear Regulatory Commission, Washington, DC, May 20, 1999.

4.2   Background

Appendix A to 10 CFR Part 50, "General Design Criteria for Nuclear Power Plants," establishes the necessary design, fabrication, construction, testing, and performance requirements for SSCs that are important to safety. With regard to fire protection, GDC 3 states:

> *Structures, systems, and components important to safety shall be designed and located to minimize, consistent with other safety requirements, the probability and effect of fires and explosions. Noncombustible and heat-resistant materials shall be used wherever practical throughout the unit, particularly in locations such as the containment and control room. Fire detection and fighting systems of appropriate capacity and capability shall be provided and designed to minimize the adverse effects of fires on structures, systems, and components important to safety. Firefighting systems shall be designed to ensure that their rupture, or inadvertent operation does not significantly impair the safety capability of these structures, systems, and components.*

During the first decade or so of the U.S. nuclear reactor program, regulatory acceptance of FPPs at the Nation's NPPs was based on the broad performance objectives of GDC 3. Because of the lack of detailed implementation guidance at that time, however, the level of fire protection was generally found to be acceptable if the facility complied with local fire codes and received an acceptable rating from its fire insurance underwriter. Thus, the fire protection features in early U.S. NPPs were very similar to those of conventional, fossil-fueled, power generating stations.

The lessons learned from the Browns Ferry fire brought fundamental change to fire protection and its regulation in the U.S. nuclear power industry. As described in Section 3.4, the fire was started by plant workers who used a candle flame to test for air leakage through a penetration in a wall that separated the CSR from the reactor building. Although most of the fire damage was contained to a relatively small area of the reactor building [approximately 74.32 m² (800 ft²)], the fire affected more than 1,600 cables, routed in 117 conduits and 26 cable trays, of which 628 were important to safety. The resulting damage impeded the functioning of both normal and standby reactor cooling systems, significantly degraded the operators' ability to monitor important plant parameters, and forced operators to initiate emergency repairs in order to restore systems needed to place the reactor in a safe-shutdown condition.

The Browns Ferry fire demonstrated that the occupant safety and property protection concerns of the major fire insurance underwriters did not sufficiently encompass nuclear safety issues, particularly with regard to the potential for fire to cause the failure of systems and components that are important to safe-shutdown of the reactor. Investigations of the cause and possible consequences of this event revealed several significant fire protection vulnerabilities, including the following examples:
- apparent ease with which the fire started
- hours that elapsed before the fire was fully extinguished
- unavailability of redundant trains of plant safety equipment as a result of fire damage

On the basis of these findings, the NRC concluded that additional guidance and requirements beyond the existing fire protection regulation (GDC 3) were necessary. In recognition of the potential consequences of fire, and to ensure adequate fire safety in the overall design and operation of all NPPs operating in the United States, the NRC determined that established DID safety principles should be applied in the defense against fires.

DID is a fundamental safety philosophy that provides multiple layers of protection (i.e., barriers) to prevent and mitigate accidents. With regard to fire protection, the DID concept is aimed at achieving the following objectives:
- Prevent fires from starting.
- Rapidly detect, control, and extinguish those fires that do occur.
- Protect SSCs that are important to safety so that a fire that is not promptly extinguished by the fire protection activities will not prevent the safe-shutdown of the plant.[12]

The multiple levels of protection that are embodied in the DID philosophy ensure fire safety throughout the life of the plant by minimizing both the probability and the consequence of fires. While the NRC recognizes that no one level can be perfect or complete by itself, and strengthening any one level can compensate in some measure for known or unknown weaknesses in the others, each level of protection must meet certain minimum requirements.

Consistency with the DID philosophy is maintained if the plant meets the following criteria:
- Preserve a reasonable balance among prevention of core damage, prevention of containment failure, and mitigation of consequences.
- Avoid over-reliance on programmatic activities to compensate for weaknesses in plant design.
- Preserve system redundancy, independence, and diversity commensurate with the expected frequency and consequences of challenges to the system, as well as the associated uncertainties (e.g., no risk outliers).
- Preserve defenses against potential common-cause failures, and assess the potential for the introduction of new common-cause failure mechanisms.
- Prevent degradation of the independence of barriers.
- Preserve defenses against human errors.
- Maintain the intent of the the GDCs in Appendix A to 10 CFR Part 50.[13]

4.3   Development of Fire Protection Program Requirements

To assist licensees in enhancing their FPPs, the NRC staff incorporated the recommendations from the Browns Ferry SRG into a single technical guidance document identified as BTP APCSB 9.5-1, "Guidelines for Fire Protection for Nuclear Power Plants," dated May 1976. In so doing, the staff asked each licensee to submit an analysis that divided the plant into distinct fire areas and demonstrated that redundant trains of equipment required to achieve and maintain cold shutdown conditions were adequately protected from fire damage. However, the guidance contained in BTP APCSB 9.5-1 was only relevant to plants that filed an application for construction after July 1, 1979.

---

[12]   10 CFR Part 50, Appendix R , Section II, "General Requirements, Paragraph A, Fire Protection Program."

[13]   Regulatory Guide 1.174, "An Approach for Using Probabilistic Risk Assessment In Risk-Informed Decisions On Plant-Specific Changes to the Licensing Basis," U.S. Nuclear Regulatory Commission, Washington, DC, July 1998.

Consequently, the NRC staff sought to establish a suitable FPP without significantly affecting the design, construction, or operation of "older" plants that were either already operating or well past the design stage and into construction. Toward that end, in September 1976, the NRC issued Appendix A to BTP APCSB 9.5-1 "Guidelines for Fire Protection for Nuclear Plants Docketed Prior to July 1, 1976". This guidance provided acceptable alternatives in areas where strict compliance with BTP APCSB 9.5-1 would require significant modifications. Additionally, the NRC informed each plant that the staff would use the guidance in Appendix A to analyze the consequences of a postulated fire within each area of the plant, and asked licensees to provide results of the fire hazards analysis performed for each unit and TSs for the present fire protection systems.

Early in 1977, each pre-1979 licensee responded with an FPP evaluation that included a fire hazards analysis (FHA). The NRC staff reviewed these analyses using the guidelines of Appendix A to BTP APCSB 9.5-1. The staff also conducted inspections of operating reactors to examine the relationship between SSCs important to safety and the fire hazards, potential consequences of fires, and fire protection features. Based on the results of its reviews, the staff determined that additional guidance on the management and administration of FPPs was needed and, on August 29, 1977, the staff issued GL 77-02, "Nuclear Plant Fire Protection Functional Responsibilities, Administrative Controls, and Quality Assurance." This document provided the criteria used by the staff in reviewing specific elements of a licensee's FPP, including organization, training, combustible and ignition source controls, firefighting procedures, and quality assurance.

By the late 1970s, most operating plants had completed their analyses and implemented most of the FPP guidance of Appendix A to the BTP. Many fire protection issues were resolved during the BTP review process, and agreements were included in the NRC-issued safety evaluation reports (SERs). In certain instances, however, licensees refused to adopt some of the specified fire protection recommendations, such as the requirements for fire brigade size and training, water supplies for fire suppression systems, alternative or dedicated shutdown capability, emergency lighting, qualifications of penetration seals used to enclose places where cables penetrated fire barriers, and the prevention of reactor coolant pump oil system fires. Following deliberation, the Commission determined that, given the generic nature of some of the disputed issues, a rulemaking was needed to ensure proper implementation of the NRC's fire protection requirements. Accordingly, the Commission amended its regulations and, in November 1980, issued 10 CFR 50.48, "Fire Protection" (which specified broad performance requirements), and Appendix R, "Fire Protection Program for Nuclear Power Plants Operating Prior to January 1, 1979" (which specified detailed regulatory requirements for resolving the disputed issues).

As originally proposed (*Federal Register*, Vol. 45, No. 1&5, May 22, 1980), Appendix R would have applied to all plants that were licensed to operate before January 1, 1979, including those for which the staff had previously accepted the fire protection features as meeting the provisions of Appendix A to BTP APCSB 9.5-1. However, after analyzing comments on the proposed rule, the Commission determined that only 3 of the 15 items in Appendix R were of such safety significance that they should apply to all plants that were licensed before January 1, 1979. These three items are (1) fire protection of safe-shutdown capability, including alternative or dedicated shutdown systems; (2) emergency lighting; and (3) the reactor coolant pump oil system. The final rule required all reactors licensed to operate before January 1, 1979, to comply with these three items *even if the NRC had previously approved alternative fire protection features in these areas* (*Federal Register*, Vol. 45, November 19, 1980). In addition, the rule provided an exemption process that a licensee can request, provided that a required fire protection feature to be exempted would not enhance fire protection safety in the facility or that such modifications may be detrimental to overall safety.

By letter dated November 24, 1980, the Commission informed all power reactor licensees with plants licensed before January 1, 1979, of new fire protection regulations contained in 10 CFR 50.48 (to ensure that each plant had an FPP) and Appendix R to 10 CFR Part 50 (to ensure satisfactory resolution of disputed items). In its letter, the Commission stated that the provisions of Appendix R can be divided into two categories:

(4) Those provisions that the new rule requires licensees to backfit in their entirety, regardless of whether the NRC staff previously approved alternatives to the specific requirements of these sections. These requirements are set forth in Sections III.G, "Fire Protection of Safe-Shutdown Capability"; III.J, "Emergency Lighting"; and III.O, "Oil Collection Systems for Reactor Coolant Pump."

(5) Requirements concerning the "open items" of previous NRC staff fire protection reviews. (An "open item" is defined as a fire protection feature that the staff has not previously approved as satisfying the provisions of Appendix A to BTP APCSB 9.5-1, as reflected in a fire protection SER.)

The two enclosures to this letter included (1) a copy of the *Federal Register* Notice (45 FR 76602) and (2) a summary of open items that the staff identified during its evaluation of the plant's implementation of Appendix A to BTP APCSB 9.5-1.

4.3.1   NPPs Licensed Before January 1, 1979

With the exception of Sections III.G, J,and O (which were backfit by all plants operating before January 1, 1979, regardless of previous staff approvals of alternatives), those portions of Appendix A to the BTP APCSB 9.5-1 that were previously accepted by the staff remained valid. Therefore, Appendix R does not, by itself, define the FPP of any plant. For plants licensed before January 1, 1979 (pre-1979 plants), the FPP is defined by Appendix A to the BTP, the *applicable portions* of Appendix R to 10 CFR Part 50 (i.e., open issues from BTP APCSB 9.5-1 Appendix A reviews), and any additional commitments made by the licensee, as stated in the conditions of its operating license.

4.3.2   NPPs Licensed After January 1, 1979

As stated above, Appendix R is only required to be implemented by plants licensed to operate before January 1, 1979. FPPs at plants licensed after this date were typically reviewed by the staff during their initial licensing process. Certain plants in this category were required to implement specific sections of Appendix R (typically sections III.G., J, and O), as specified in their "Fire Protection" license condition. Consequently, there was no need to "backfit" Appendix R to plants licensed after January 1, 1979. Additionally, only two paragraphs of the Fire Protection Rule (10 CFR 50.48) apply to plants that were licensed after January 1, 1979. Specifically, those paragraphs are Paragraph A (requiring plants to have a fire protection plan that satisfies Criterion 3 of Appendix A to 10 CFR Part 50) and Paragraph B (requiring plants to complete all fire protection modifications needed to satisfy GDC 3 of Appendix A to 10 CFR Part 50 in accordance with the provisions of their operating licenses).

Guidelines acceptable to the staff for implementing GDC 3 at plants licensed after January 1, 1979, are presented in SRP Section 9.5.1, "Fire Protection Program." This document consolidates the guidance of BTP APCSB 9.5-1, Appendix A to BTP APCSB 9.5-1 (originally issued in August 1977), and the criteria of Appendix R to 10 CFR Part 50. Thus, SRP Section 9.5.1 may be considered a single-source reference that describes the features of an acceptable FPP.

4.4 Requirements, Guidelines, and Clarifications Related to Post-Fire Safe-Shutdown Capability

The NRC's regulatory framework for nuclear power plant FPPs is set forth in a number of regulations and supporting guidelines, including, but not limited to the following:
*   Title 10, Section 50.48, of the *Code of Federal Regulations* (10 CFR 50.48)
*   Appendix R to 10 CFR Part 50
*   General Design Criterion 3 (GDC 3) of Appendix A to 10 CFR Part 50
*   regulatory guides (RGs) and generic communications [e.g., generic letters (GLs), bulletins (BLs), and information notices (INs)]
*   NUREG-series technical reports, including NUREG-0800, "NRC Standard Review Plan" (SRP)
*   associated branch technical positions (BTPs) and industry standards

The comprehensive fire protection guidance and regulatory criteria described in these documents address the broad range of features that comprise an acceptable FPP. Consistent with the objectives of this report, however, this section discusses only those requirements, guidelines, and generic communications (clarification documents) that specifically relate to post-fire safe-shutdown capability and the performance of a safe-shutdown analysis.

Regulatory requirements of primary interest include GDCs 3, 5, 19, and 23 of Appendix A to 10 CFR Part 50; 10 CFR 50.48; and Sections III.G and III.L of Appendix R to 10 CFR Part 50. While the NRC recognizes that Appendix R is not applicable to plants that were licensed to operate after January 1, 1979, the technical requirements of Sections III.G and III.L were subsumed into the review guidance that the NRC staff developed for plants that were licensed to operate after that date (i.e., Position C.5.b of SRP Section 9.5-1). It is important to note that some of the regulations and guidelines described below are not applicable to each plant. Therefore, licensees and reviewers must refer to the plant-specific fire protection licensing bases when determining the applicability of regulations and guidelines for a specific NPP.

4.4.1 10 CFR Part 50, Appendix A, "General Design Criteria for Nuclear Power Plants"

For those plants to which its provisions apply, 10 CFR Part 50, Appendix A, "General Design Criteria for Nuclear Power Plants," establishes the necessary design, fabrication, construction, testing, and performance requirements for SSCs that are important to safety. Of these requirements, the following criteria have apply specifically to fire protection of NPPs.

- **GDC 3, "Fire Protection,"** requires that SSCs important to safety must be designed and located to minimize (consistent with other safety requirements) the probability and effect of fires and explosions. Noncombustible and heat-resistant materials are required to be used wherever practical, and particularly in locations such as the containment and control room. Fire detection and firefighting systems of appropriate capacity and capability are required to be provided and designed to minimize the adverse effects of fires on SSCs important to safety. GDC 3 also requires that firefighting systems must be designed to ensure that their failure, rupture, or inadvertent operation does not significantly impair the safety capability of these SSCs.

- **GDC 5, "Sharing of Structures, Systems, and Components,"** requires that SSCs important to safety must not be shared among nuclear power units unless it can be shown that such sharing will not significantly impair their ability to perform their safety functions, including, in the event of an accident in one unit, an orderly shutdown and cooldown of the remaining units.

- **GDC 19, "Control Room,"** requires that the design must provide a control room from which operators can take actions to operate the nuclear power unit under both normal and accident conditions, while limiting radiation exposure to control room personnel under accident conditions for the duration of the accident. GDC 19 also requires that equipment and locations outside the control room must be provided with the design capability to accomplish hot shutdown of the reactor, as well as a potential capability for subsequent cold shutdown of the reactor. It should be noted that the GDC 19 design criteria were largely based on environmental/habitability concerns within the control room. As a result, GDC 19 does not specifically consider the effect of equipment damage as a result of fire.

- **GDC 23, "Protection System Failure Modes,"** requires that the protection system must be designed to fail into a safe state or into a state demonstrated to be acceptable on some other defined basis if the plant experiences conditions such as disconnection of the system, loss of energy (e.g., electric power, instrument air), or postulated adverse environments (e.g., extreme heat or cold, fire, pressure, steam, water, radiation).

4.4.2   10 CFR 50.48, "Fire Protection"

Section 50.48(a) of 10 CFR Part 50 requires that each operating NPP must have a fire protection plan that satisfies GDC 3 of Appendix A to 10 CFR Part 50. It also specifies what such a plant should contain and lists the basic fire protection guidelines for the plan. Section 50.48(b) requires that all plants licensed before January 1, 1979, must satisfy the requirements of Sections III.G, J, and O, and other sections of Appendix R to 10 CFR Part 50, where approval of similar features had not been obtained prior to the effective date of Appendix R. Alternatively, plants licensed to operate after January 1, 1979, must meet the provisions of 10 CFR 50.48(a). The required schedules for licensees to comply with the provisions of Appendix R were established in 10 CFR 50.48(c). The rule also included provisions to allow licensees to file requests for exemptions from Appendix R requirements on the basis that the required modifications would not enhance the facility's fire protection safety or would be detrimental to overall facility safety. Upon approval by the staff, these exemptions become a part of the plant's fire protection licensing basis. The provisions of 10 CFR 50.48(c) have since expired and been deleted from the regulations.

In accordance with 10 CFR 50.48, each operating NPP must provide the means to limit fire damage to SSCs important to safety in order to ensure the capability to safely shut down the reactor. Licensees should develop an SSA that demonstrates the plant's capability to safely shut down for a fire in any given area. (See Chapter 6.)

4.4.3   10 CFR Part 50, Appendix R, "Fire Protection Program for Nuclear Power Facilities Operating Prior to January 1, 1979"

One of the principal goals of regulatory requirements and staff guidance issued since the Browns Ferry fire is to ensure that, in the event of fire in any area of the plant, one train of equipment needed to achieve and maintain safe-shutdown conditions in the reactor will remain free of fire damage. To achieve this objective, 10 CFR 50.48(b), which became effective on February 17, 1981, requires all NPP licensed before January 1, 1979, to meet the requirements of Section III.G, "Fire Protection of Safe-Shutdown Capability," of Appendix R to 10 CFR Part 50, regardless of any previous NRC approvals for alternative design features. Compliance with this criterion requires each licensee to reassess all areas of the plant and demonstrate for each area that suitable fire protection features (as specified in Section III.G.2 of Appendix R) are provided for redundant trains of cables and equipment necessary to achieve and maintain hot shutdown conditions. As part of this evaluation, the rule requires licensees to consider the potential effects of fire on associated nonsafety-related circuits and cables that could impact the shutdown capability. (See Chapters 3 and 6.) With regard to the fire protection of safe-shutdown capability, facilities that commenced operation on or after January 1, 1979, are subject to essentially the same criteria as those contained in Appendix R. These criteria have been imposed through license conditions or licensing commitments.

In developing the Fire Protection Rule, the Commission decided that the overall interest of public safety is best served by establishing some conservative level of protection and ensuring that level of compliance. The objective for fire protection of safe-shutdown capability is to ensure that at least one means of achieving and maintaining safe-shutdown conditions will remain available during and after any postulated fire in the plant. Because it is not possible to predict the specific conditions under which fire may occur and propagate, the design-basis protective features are specified rather than the design-basis fire. The fire protection features specified in Section III.G are not unique to the nuclear industry. Rather, they are based upon principles long accepted within that portion of U.S. industry that has been classified by their insurance carriers as "improved risk" or "highly protected risk."[14]

Section III.G.1 of Appendix R to 10 CFR Part 50 requires that fire protection features must be provided for SSCs that are important to safe-shutdown. These features must be capable of limiting fire damage so that the following conditions are maintained:

(a) One train of systems necessary to achieve and maintain hot shutdown conditions from either the control room or emergency control station(s) is maintained free of fire damage.

---

[14]   SECY 80-438A, "Rule on Fire Protection Program for Nuclear Power Plants Operating Prior to January 1, 1979," Enclosure A, U.S. Nuclear Regulatory Commission, Washington, DC, September 30, 1980.

(b) The extent of fire damage to redundant trains of systems and equipment necessary to achieve and maintain cold shutdown is limited so that at least one train can be repaired or made operable within 72 hours using onsite capabilities.

The fire areas falling under the requirements of III.G.1(b) are those for which an alternative or dedicated shutdown capability is not being provided. For those fire areas, Section III.G.1(b) requires only the capability to repair the systems necessary to achieve and maintain cold shutdown from either the control room or emergency control station(s) within 72 hours, not the capability to repair and achieve cold shutdown within 72 hours as required for the alternative or dedicated shutdown modes by Section III.L.[15]

Section III.G.2, provides various options for protecting the capability to achieve and maintain hot shutdown conditions, as follows:

> *Where cables or equipment, including associated nonsafety circuits that could prevent operation or cause maloperation due to hot shorts, open circuits or shorts to ground of redundant trains of systems necessary to achieve and maintain hot shutdown conditions are located within the same fire area outside of primary containment, one of the following means of ensuring that one of the redundant trains is free of fire damage shall be provided:*
> *(a) Separation of cables and equipment and associated nonsafety circuits of redundant trains by a fire barrier having a 3-hour rating. Structural steel forming a part of or supporting such fire barriers shall be protected to provide fire resistance equivalent to that required of the barrier; or*
> *(b). Separation of cables and equipment and associated nonsafety circuits of redundant trains by horizontal distance of more than 20 feet with no intervening combustibles or fire hazards. In addition, fire detectors and an automatic fire suppression system shall be installed in the fire area; or*
> *(c) Enclosure of cable and equipment and associated nonsafety circuits of one redundant train in a fire barrier having a 1-hour rating. In addition, fire detectors and an automatic fire suppression system shall be installed in the fire area.*
>
> *Inside non-inerted containments, one of the fire protection means specified above or one of the following fire protection means shall be provided:*
> *(d) Separation of cables and equipment and associated nonsafety circuits of redundant trains by horizontal distance of more than 20 feet with no intervening combustibles or fire hazards; or*
> *(e) Installation of fire detectors and an automatic fire suppression system in the fire area; or*
> *(f) Separation of cables and equipment and associated nonsafety circuits of redundant trains by a noncombustible radiant energy shield.*

Note: Since fire areas are frequently described in terms of the section of III.G that they meet, additional clarification is warranted with regard to the use of this terminology:
- For a fire area to "meet III.G.1," at least one train of shutdown systems and equipment must be completely independent (physically and electrically) of the fire area.

---

[15] GL 86-10, Enclosure 1, Paragraph 2, "Repair of Cold Shutdown Equipment," U.S. Nuclear Regulatory Commission, Washington, DC.

- A "III.G.2 Fire Area" contains redundant trains of shutdown equipment; however, one train has been ensured to remain free of fire damage (per the criteria contained in this section of the regulation).
- A "III.G.3 Fire Area" contains redundant trains of shutdown equipment or cables and one train has *not* been ensured to remain free of fire damage (per III.G.2 criteria) or redundant trains are vulnerable to damage as a result of fire suppression activities or the inadvertent actuation of fire suppression systems.

Interpretation 3 of GL 86-10 defines the term "free of fire damage" in Section III.G.1.a. The NRC provided this interpretation to clarify Section III.G.1.a, during the exemption process, for licensees who are attempting to justify the lack of III.G.2 separation features for redundant trains within a single fire area. For any fire area, an approved exemption is required where neither alternative safe-shutdown nor the separation features of Section III.G.2 are provided. [Reference: "Generic Guidance for Post-Fire Safe-Shutdown Analysis Assessment," Rev. G, Boiling-Water Reactor Owners Group (BWROG), p. 3-48, June 24, 1998.]

As indicated in the above text, Appendix R to 10 CFR Part 50 uses the term "free of fire damage." In promulgating Appendix R, the Commission provided acceptable methods for ensuring that necessary SSCs are free of fire damage. (See Appendix R, Section III.G.2a, b and c.) Specifically, the SSCs under consideration must be capable of performing their intended functions during and after the postulated fire, as needed.[16]

Where the protection of systems that are required to function properly for hot shutdown does not satisfy the requirement of Section III.G.2, or where redundant trains of systems required for hot shutdown located in the same fire area may be subject to damage from fire-suppression activities or from the rupture or inadvertent operation of fire suppression systems, Section III.G.3 requires that an alternative or dedicated shutdown capability must be provided and must be independent of cables, systems, or components in the area, room, or zone under consideration. In addition, Section III.G.3 further requires that fire detection and a fixed fire suppression system must be installed in the area, room, or zone under consideration. Specific criteria for implementing this capability are contained in Appendix R, Section III.L, "Alternative and Dedicated Shutdown Capability."

Although 10 CFR 50.48(b) does not specifically include Section III.L with Sections III.G, J, and O of Appendix R to 10 CFR Part 50 as a requirement applicable to all power reactors licensed before January 1, 1979, the appendix, read as a whole, and the Court of Appeals decision on the appendix, in the case of Connecticut Light and Power et al. vs. NRC, 673 F2d. 525 (D.C. Cir., 1982), demonstrate that Section III.L applies to the alternative safe-shutdown option under Section III.G if and where that option is chosen by the licensee[17].

Section III.G recognizes that the need for alternative or dedicated shutdown capability may have to be considered on the basis of a fire area, room, or fire zone. The alternative or

---

[16]   GL 86-10, Enclosure 1, Paragraph 3, "Fire Damage," U.S, Nuclear Regulatory Commission, Washington, DC.

[17]   GL 86-10, Enclosure 2, Question 5.1.3, U.S, Nuclear Regulatory Commission, Washington, DC.

dedicated capability should be independent of the fire area where it is possible to do so. When fire areas are not designated, or where it is not possible to have the alternative or dedicated capability independent of the fire area, careful consideration must be given to the selection and location of the alternative or dedicated shutdown capability to ensure that the performance requirement set forth in Section III.G.l is met. Where alternative or dedicated shutdown is provided for a room or zone, the capability must be physically and electrically independent of that room or zone. The vulnerability of the equipment and personnel required at the location of the alternative or dedicated shutdown capability to the environments produced at that location as a result of the fire or fire suppressants must be evaluated.

These environments may be due concerns such as the hot gas layer, smoke, drifting suppressants, common ventilation systems, common drain systems or flooding. In addition, other interactions between the locations may be possible in unique configurations. If alternative shutdown is provided on the basis of rooms or zones, the provision of fire detection and fixed suppression is only required in the room or zone under consideration. Compliance with Section III.G.2 cannot be based on rooms or zones[18]. While "independence" is clearly achieved where alternative shutdown equipment is outside the fire area under consideration, alternative shutdown equipment in the same fire area but independent of the room or the zone may also result in compliance with the regulation. The "room" concept must be justified by a detailed fire hazards analysis that demonstrates that a single fire will not disable both the normal shutdown equipment and the alternative shutdown capability.[19]

The remote shutdown systems recommended in Chapter 7 of the SRP are needed to meet GDC 19. These remote shutdown systems need to be redundant and physically independent of the control room in order to meet GDC 19. For GDC 19, damage to the control room is not considered. Alternative shutdown systems for Appendix R need not be redundant, but must be both physically and electrically independent of the control room.[20]

---

[18]   GL 86-10, Enclosure 2, Question 3.1.5, U.S. Nuclear Regulatory Commission, Washington, DC.

[19]   GL 86-10, Enclosure 1, Paragraph 6, "Alternative or Dedicated Shutdown," U.S. Nuclear Regulatory Commission, Washington, DC.

[20]   GL 86-10, Enclosure 2, Question 5.3.11, U.S. Nuclear Regulatory Commission, Washington, DC..

4.4.4 Generic Communications

To aid in developing a common understanding between licensees and NRC reviewers and inspectors, the staff has promulgated a number of clarification documents, principally in the form of GLs and INs. When considering guidance contained in generic communications, it is essential to note the following points:

(4) It is the Commission's position that regulatory guidance by itself cannot alter the specific regulatory requirements contained in the Commission's fire protection regulations.[21]

(5) NRC generic letters cannot legally create a new requirement for a specific course of action to resolve an issue. Generic communications have been used, however, to provide new or clarified interpretations of existing requirements.[22]

Table 4-1 summarizes the salient generic communications related to post-fire safe-shutdown capability.

---

[21] Letter from J. Hannon, NRC, to A. Marion, Nuclear Energy Institute; Subject: Adoption of NFPA Standard 805, dated April 6, 2001.

[22] Statement presented by Shirley Ann Jackson, Chairman, NRC, to the U.S. Senate Committee on Environment and Public Works, Subcommittee on Clean Air, Wetlands, Private Property, and Nuclear Safety, concerning NRC programs and nuclear safety regulatory issues, July 30, 1998.

Table 4-1. NRC Generic Communications	
Generic Communication	Description
GL 77-02	Provided guidance to supplement Appendix A BTP APCSB 9.5-1, regarding a licensee's fire protection organization, training of the fire brigade, control of combustibles and ignition sources, firefighting procedures, and quality assurance.
GL 81-12 and Clarification of GL 81-12	In these letters, the staff identified the information necessary to review licensee compliance with the alternative or dedicated shutdown requirements of Section III.G.3 of Appendix R to 10 CFR Part 50. These letters defined safe-shutdown objectives, reactor performance goals, necessary safe-shutdown systems and components, and associated circuit identification and analysis methods. GL 81-12 also asked licensees to develop TSs for safe-shutdown equipment that was not previously included in the existing plant-specific TSs.
GL 83-33	Provided clarification on the following requirements of Appendix R to 10 CFR Part 50: (a) detection and automatic suppression (b) fire areas (c) structural steel related to fire barriers (d) fixed suppression system (e) intervening combustibles (f) transient fire hazards It should be noted that certain licensees disagreed with, or found it difficult to implement, the interpretations provided in this GL. To pursue the matter with senior NRC management, the nuclear power industry formed the Nuclear Utility Fire Protection Group. To "...examine all licensing, inspection and technical issues and to make policy recommendations for expediting Appendix R implementation and for ensuring consistent levels of fire protection at all plants," by direction of the Executive Director for Operations (EDO), the staff formed the Steering Committee on Fire Protection Policy. Disagreements in the implementation of interpretations provided in GL 83-33 were ultimately resolved by issuance of GL 86-10, "Implementation of Fire Protection Requirements," on April 24, 1986.

Table 4-1. NRC Generic Communications	
Generic Communication	Description
IN 84-09	Provided guidance for conducting analyses and/or making modifications to implement requirements of Appendix R to 10 CFR Part 50, with respect to the following issues: (a) fire areas (b) fire barrier testing and configuration (c) protection of equipment necessary to achieve hot shutdown (d) licensee's reassessment for conformance with appendix r (e) identification of safe-shutdown systems and components (f) combustibility of electrical cable insulation (g) detection and automatic suppression (h) applicability of 10 CFR Part 50, Appendix R, Section III.L (i) instrumentation necessary for alternative shutdown (j) procedures for alternative shutdown capability (k) fire protection features for cold shutdown systems (l) RCP oil collection systems
IN 85-09	Alerted licensees to potential deficiencies in the electrical design of isolation/transfer switches, which do not provide redundant fuses upon transfer

Table 4-1. NRC Generic Communications	
Generic Communication	Description
GL 86-10	Provided additional guidance on acceptable methods of satisfying the NRC's regulatory requirements. Although the staff issued this document, it had the review and approval of the Commission. This letter addressed the following specific topics: (a) scheduler exemptions (b) documentation required to demonstrate compliance (c) applicable quality assurance requirements (d) NRC notification of deficiencies (e) incorporation of FPP into FSAR (f) standard fire protection license condition Through the implementation and adoption of a standard license condition, a licensee is allowed to make changes to its FPP without prior notification to the NRC in accordance with the provisions of 10 CFR 50.59, provided that the changes do not adversely affect the plant's ability to achieve and maintain post-fire safe-shutdown. Upon modification of the license to adopt the standard condition, the licensee could also amend the license to remove the fire protection TSs. (g) interpretations of Appendix R: • process monitoring instrumentation • repair of cold shutdown equipment • fire damage • fire area boundaries • automatic detection and suppression • alternative or dedicated shutdown capability (h) Appendix R questions and answers To assist the industry in understanding the NRC's requirements, and improve the staff's understanding of the industry's concerns, a series of workshops were conducted in each NRC region. This section presents the NRC's position as responses to the questions posed by the industry during these workshops.
GL 88-12	Provided additional guidance for implementation of the standard license condition and removal of the TSs associated with fire detection and suppression, fire barriers, and fire brigade staffing. The TSs associated with safe-shutdown equipment and the administrative controls related to fire protection audits were to be retained under the guidance of the GL.
IN 99-17	Alerted licensees to potential problems associated with post-fire safe-shutdown circuit analysis that could prevent the operation or lead to malfunction of equipment necessary to achieve and maintain post-fire safe-shutdown.

4.5 Fire Protection Licensing and Design Bases

With the issuance of the Fire Protection Rule (10 CFR 50.48, and Appendix R to 10 CFR Part 50), the NRC established the applicability of certain fire protection requirements, including those within the rule, on the basis of the licensing date for a given plant being before or after January 1, 1979. However, the progression of regulatory guidelines and requirements outlined above, coupled with a broad range of plant-specific attributes (design features, operating preferences, and exemptions to certain technical requirements), has created a unique set of circumstances for nearly every plant. Design and construction factors, such as plant type (PWR vs. BWR), age, size, NSSS supplier [Westinghouse Electric, Combustion Engineering (CE), Babcok and Wilcox (B&W), General Electric (GE)], architect/engineer, degree of separation provided for redundant shutdown systems in the initial plant design, type of cabling used (e.g., thermoset vs. thermoplastic insulation), and the individual preferences of a utility for system and equipment configurations can significantly influence the type and quantity of fire protection features needed to provide an acceptable level of protection. The influence that such factors have on the protection of safe-shutdown capability is considered by the staff and documented in plant-specific SERs (see below). As a result of these plant-specific differences, fire protection features imposed on one plant often differ considerably from those at another.

4.5.1 Plants Licensed Before January 1, 1979

The primary licensing basis for plants licensed to operate before January 1, 1979, comprises the plant's license conditions, Appendix R and any approved exemptions, and the staff's SERs of the FPP.

4.5.2 Plants Licensed After January 1, 1979

Plants licensed after January 1, 1979, are subject only to the requirements of 10 CFR 50.48(a) and, as such, must meet the provisions of GDC 3 as specified in their license conditions and as accepted by the NRC in their SERs. These plants are typically reviewed to the guidance of SRP Section 9.5-1. For these plants, where commitments to specific guidelines cannot be met, or alternative approaches are proposed, the differences between the licensee's program and the guidelines are documented in deviations.

4.5.3 Safety Evaluation Reports

Safety evaluation reports (SERs) document the staff acceptance of the plant's FPP or elements thereof. For plants licensed to operate prior to January 1, 1979, the staff's SERs also establish the extent to which the requirements of Appendix R to 10 CFR Part 50 apply. Plants for which the NRC previously accepted alternative fire protection features as satisfying the provisions of Appendix A to BTP APCSB 9.5-1, or accepted such alternatives in comprehensive SERs issued prior to publication of Appendix A to BTP APCSB 9.5-1 in August 1976, were only required to meet the provisions of Sections III.G (III.L), III.J, and III.O of Appendix R.

4.5.4 Exemptions and Deviations

When it promulgated Appendix R to 10 CFR Part 50, the Commission recognized that there would be plant conditions and configurations where strict compliance with specified fire protection design features would not significantly enhance the level of fire safety already provided by the licensee. Therefore, in cases where an FHA could adequately demonstrate that alternative fire protection features provided a level of fire safety equivalent to that required by the regulation, the licensee could apply for an exemption from the prescriptive requirements of Appendix R. Thus, the exemption process provided a means of allowing flexibility to meet the performance objectives of Appendix R through alternative means. For plants that began operation after January 1, 1979, guidance for the plants' FPPs is provided in BTP Chemical and Mechanical Engineering Branch (CMEB) 9.5-1. For these newer plants, the staff approved "deviations" from the guidance during the licensing process. Since Appendix R requirements are included in BTP CMEB 9.5-1, this report uses the term "exemptions" to refer to both BTP CMEB 9.5-1 deviations as well as Appendix R exemptions.

Through the performance of a detailed FHA of plant-specific conditions, a licensee may demonstrate that certain configurations, which do not meet the technical requirements of the regulation, will provide an adequate level of fire safety. For example, the evaluation of a fire area at a certain plant may find that although redundant shutdown components are adequately separated [>6.08 m (>20 ft) of horizontal separation distance)], the area between the components contains a small quantity of intervening combustibles in the form of cables routed in cable trays. Although this configuration does not satisfy the technical requirements of the rule (which specifies that the separation area must be free of intervening combustibles or fire hazards), when other protection features are considered (such as the use of armored sheathed cables, adequacy of installed fire detection systems, automatic and manual suppression capabilities, and the quantity and type of combustibles in the area), it may be shown that strict compliance with the technical requirements would not enhance fire safety. When such plant-specific conditions exist, licensees may request NRC approval of an exemption from the technical requirements of the regulation under 10 CFR 50.12. Under this provision, the Commission may grant exemptions from the requirements of the regulations in 10 CFR Part 50, which are authorized by law, will not present an undue risk to public health and safety, and are consistent with the common defense and security. The Commission will not consider granting an exemption unless special circumstances are present, as in the following cases:

- Application of the regulation in the particular circumstances conflicts with other rules or requirements of the Commission.

- Application of the regulation in the particular circumstances would not serve the underlying purpose of the rule or is not necessary to achieve the underlying purpose of the rule.

- Compliance would result in undue hardship or other costs that are significantly in excess of those contemplated when the regulation was adopted, or that are significantly in excess of those incurred by others similarly situated.

- The exemption would result in benefit to the public health and safety that compensates for any decrease in safety that may result from the grant of the exemption

- The exemption would provide only temporary relief from the applicable regulation and the licensee or applicant has made good faith efforts to comply with the regulation.

- There is present any other material circumstance not considered when the regulation was adopted for which it would be in the public interest to grant an exemption. If such condition is relied on exclusively for compelling the Commission to grant the exemption, the exemption may not be granted until the EDO has consulted with the Commission.

As previously stated, plants licensed after January 1, 1979, have FPPs that were typically reviewed and approved under the guidance contained in SRP Section 9.5.1 and, therefore, are not subject to the specific regulatory requirements of 10 CFR 50.48 and Appendix R. For these plants, a license amendment or NRC staff approval of a deviation from a specific NRC guideline is necessary when an alternative approach is used to satisfy the requirements of GDC 3. As with an exemption, the licensee must submit a sound technical justification for the alternative approach for NRC review and approval, along with its license amendment or deviation request.

4.5.5 Standard Plant License Condition

Most operating plant licenses contain a section on fire protection. License conditions for plants licensed prior to January 1, 1979, typically contain a condition requiring implementation of modifications to which the licensee committed as a result of the FPP review with respect to the BTP. These license conditions were added by amendments issued between 1977 and February 17, 1981, the effective date of 10 CFR 50.48 and Appendix R. As a result of numerous compliance, inspection, and enforcement issues associated with the various plant license conditions, the staff developed a "standard licensing condition (see below), which the staff transmitted to licensees in GL 86-10, along with the NRC's recommendation that licensees should adopt the standard condition. The staff also issued GL 88-12 to provide additional guidance regarding removal of the fire protection requirements from the plant-specific TSs. The staff promulgated these changes specifically to give licensees greater flexibility in managing and implementing their FPPs and to clarify the fire protection licensing basis for each facility.

If the licensee has adopted the standard license condition and incorporated the FPP in its FSAR, the licensee may make changes to the approved FPP without prior Commission approval only if those changes would not adversely affect the ability to achieve and maintain safe-shutdown in the event of a fire, as documented in a safety evaluation. In addition to planned changes, a safety evaluation may be required for nonconforming conditions. GL 86-10 recommended that licensees incorporate the FPP by reference in the facility's FSAR. Incorporating the FPP and major commitments (including the FHA) by reference in the FSAR places the FPP (including the systems, administrative and technical controls, organization, and other plant features associated with fire protection) on a consistent status with other plant features described in the FSAR. GL 86-10 further recommended adopting the standard license condition, requiring licensees to comply with the provisions of the approved FPP as described in the FSAR and establishing when NRC approval is required for changes to the program. The licensee should maintain, *in auditable form*, a current record of all such changes, including an analysis of the effects of the changes on the FPP, and should make such records available to NRC inspectors upon request. All changes to the approved program should be reported, along with the FSAR revisions required by 10 CFR 50.71(e).

If the FPP committed to by the licensee is required by a specific license condition and is not part of the FSAR for the facility, licensees may be required to submit amendment requests even for relatively minor changes to the FPP.

The NRC transmitted to licensees the following standard license condition for fire protection in April 1986 as part of GL 86-10 with information on its applicability to specific plants:

> **Fire Protection**
> *[Name of Licensee] shall implement and maintain in effect all provisions of the approved fire protection program as described in the Final Safety Analysis Report for the facility (or as described in submittals dated ----------) and as approved in the SER dated ----------- (and supplements dated ---------) subject to the following provision:*
>
> *The licensee may make changes to the approved fire protection program without prior approval of the Commission only if those changes would not adversely affect the ability to achieve and maintain safe-shutdown in the event of a fire.*

The adoption of the standard license condition in conjunction with the incorporation of the FPP in the facility's FSAR provides a more consistent approach to evaluating changes to the facility, including those associated with the FPP.

Within the context of the standard fire protection license condition, the phrase "not adversely affect the ability to achieve and maintain safe-shutdown in the event of a fire," means to maintain sufficient safety margins. (See RG 1.174 for additional information.)

If a proposed change involves a change to a license condition or technical specification that was used to satisfy NRC requirements, a license amendment request should be submitted. When a change that falls within the scope of the changes allowed under the standard fire protection license condition is planned, an evaluation is made to determine whether the change would adversely affect the ability to achieve and maintain safe-shutdown. The evaluation should include the effect on the FHA and the consideration of whether circuits or components, including associated circuits, for a success path of equipment needed for safe-shutdown are being affected or a new element introduced in the area. If this evaluation concludes that there is no adverse effect, this conclusion and its basis should be documented and be available for future inspection and reference. If the evaluation finds that there is an adverse effect, or that it is outside the basis for an exemption (or deviation) that was granted (or approved) for the area involved, the licensee should make modifications to achieve conformance, justify and request an exemption, or deviation from the NRC. (See GL 86-10, Questions 8.19, 8.20, and 8.21 for additional information.)

CHAPTER 5. DISCUSSION OF POST-FIRE SAFE-SHUTDOWN CAPABILITY

5.1 Fire Protection Program Objectives

The primary objective of FPPs at U.S. nuclear power reactors is to minimize both the probability of occurrence and the consequences of fire. As discussed in Chapter 4, to achieve this goal, FPPs are based on a "DID" safety concept that is aimed at achieving the following objectives::
- Prevent fires from starting.
- Rapidly detect, control, and extinguish those fires that do occur.
- Protect SSCs that are important to safety so that a fire that is not promptly extinguished by the fire protection activities will not prevent the safe-shutdown of the plant.

This section focuses on the final element of the DID concept—ensuring that in the event a fire were to occur (despite prevention efforts) and continue to develop (despite features provided for its rapid detection and prompt extinguishment), the SSCs important to safe-shutdown would remain free of fire damage.

Redundancy is a fundamental safety feature incorporated into the design of all commercial NPPs operating in the United States. In essence, redundancy provides assurance that failures affecting one system will not have a significant impact on plant safety because the plant design provides a "backup" system. To further increase that assurance, the safety equipment and cables of the redundant subsystems are typically segregated into divisions. The separate and redundant divisions of safety systems provide confidence that the failure of components or cables within one division will not adversely affect the plant's ability to accomplish required safety functions.

To a certain extent, this design feature also provides a measure of safety against the possible consequences of fire. The level of confidence achieved through redundancy, however, is highly dependent on the degree of separation and independence provided for the redundant elements. In the absence of suitable protection features, such as separation distance or structural barriers, however, redundant trains of cables and equipment could be susceptible to a phenomenon known as "common-mode" failure, in which multiple failures in redundant systems may occur as a result of a common cause[23]. If a single event could induce failures in more than one of the redundant elements, the safety and reliability benefits afforded by this essential design feature could be negated. As demonstrated by the Browns Ferry fire, common-mode failures attributable to fire may cause equipment to fail and/or interact in ways that are not readily predictable.

The need to fully consider the potential consequences of fire damage to redundant divisions of safety equipment was emphasized by the SRG established by the NRC to investigate the Browns Ferry fire event:

> *The chronicle of the Browns Ferry fire includes many examples of unavailability of redundant equipment. Evidently, the independence provided between redundant subsystems and equipment was not sufficient to protect against common-mode failures.*

---

[23] IEEE Std. 100, "The Authoritative Dictionary of IEEE Standards Terms" (IEEE Standard Dictionary of Electrical and Electronics Engineers), 1988.

Minimizing the potential for fire to cause common-mode failures in redundant divisions of shutdown equipment, is an essential element of the "DID" philosophy for fire protection. Achieving this objective requires that plant safety systems must be designed so that in the event that a fire should start (despite the fire prevention program) and continue to burn for a considerable time, it will not preclude the capability to achieve safe-shutdown functions.

5.2   Fire Damage Limits

Achieving safe-shutdown conditions is a sequential process that relies on the operation of various plant systems to achieve and maintain both hot and cold shutdown conditions. While certain shutdown functions, such as initial reactivity control must be immediately available, other functions, such as decay heat removal (DHR) may not be needed for some time after a reactor trip. The longer after reactor trip that a function is required, the more time the operators have to analyze the situation and take the necessary steps in order to effectively operate the systems that are needed to provide the function. Thus, fire damage to systems that are needed to achieve and maintain hot shutdown conditions poses a greater threat to safety than damage to equipment that is only needed to achieve and maintain cold shutdown. The need to ensure an adequate level of fire protection for systems and equipment needed to perform hot shutdown functions was underscored in the Commission's comments on Appendix R to 10 CFR Part 50. In its Statements of Considerations on the Fire Protection Rule (SECY 80-438A), the Commission included the following statement:

> When considering the consequences of a fire in a given fire area, in evaluating the safe-shutdown capabilities of the plant, we must be able to conclude that one train of equipment that can be used immediately to bring the reactor to hot shutdown conditions remains unaffected by that fire.

The regulation clearly specifies the relationship between the specific shutdown functions performed (i.e., hot or cold shutdown) and the level of fire damage permitted to plant systems. Specifically, Appendix R, Section I, "Introduction and Scope," establishes the fire damage limits based on the safety function of the SSCs, as summarized in Table 5-1.

Table 5-1. Fire Damage Limits Based on the Safety Function of the SSCs	
Safety Function	Fire Damage Limit
Hot Shutdown	One train of equipment necessary to achieve hot shutdown from the control room or emergency control station(s) must be maintained free of damage by a single fire, including an exposure fire
Cold Shutdown	Both trains of equipment necessary to achieve cold shutdown may be damaged by a single fire, but damage must be limited so that at least one train can be repaired or made operable within 72 hours using onsite capabilities
Design-Basis Accident	Both trains of equipment necessary for mitigation of consequences following design-basis accidents may be damaged by a single exposure fire

Additionally, 10 CFR 50.48(b) requires that all licensed NPPs operating prior to January 1, 1979, must meet the requirements of Section III.G, "Fire Protection of Safe Shutdown Capability," of Appendix R to 10 CFR Part 50, regardless of any previous approvals by the NRC for other design features. Compliance with this criterion requires that each licensee must demonstrate that, in the event of an exposure fire in any single area of the plant, one of the redundant trains of cables and equipment necessary to achieve and maintain hot shutdown conditions will remain free of fire damage. Although hot shutdown equipment must remain free of fire damage, equipment required to achieve and maintain cold shutdown may be damaged, provided that the necessary repairs can be completed within the time restrictions established in the regulation. (Note: Facilities that began operation on or after January 1, 1979, are subject to essentially the same criteria as those contained in Appendix R, which have been imposed through license conditions or licensing commitments).

It should also be noted that not all safety-class equipment requires the same level of protection from fire. SSCs that are only used to mitigate the consequences of design-basis accidents do not require the same level of fire protection as those needed to accomplish post-fire safe-shutdown. The basis for this position is provided in Section I, "Introduction and Scope," of Appendix R to 10 CFR Part 50:

> *Because fire may affect safe-shutdown systems and because the loss of function of systems used to mitigate the consequences of design-basis accidents under post-fire conditions does not per se impact public safety, the need to limit fire damage to systems required to achieve and maintain safe-shutdown conditions is greater than the need to limit fire damage to those systems required to mitigate the consequences of design-basis accidents.*

5.3 Evaluation Process Overview

To ensure the ability of achieve and maintain safe-shutdown conditions in the event of fire, licensees perform a comprehensive assessment of the potential effects of fire and its related perils (direct flame impingement, hot gases, smoke migration, firefighting water damage, etc.) in each fire area. The overall objective of this deterministic evaluation, which is frequently referred to as an "SSA," is to identify potential fire vulnerabilities and develop protective measures that are consistent with established requirements.(e.g., Section III.G of Appendix R to 10 CFR Part 50). This is a technically complex process, involving personnel who have expertise in fire protection, plant operations, electrical engineering, and mechanical systems engineering disciplines.

Information developed during performance of the FHA provides the initial input for the SSA. For example, in addition to identifying the plant fire areas, the FHA will contain important information related to fire barrier ratings, equipment locations, fire detection and suppression capabilities, etc. This information is then supplemented by facility design and engineering data, additional analyses and studies, and data developed by direct observation or walkdown of facility spaces and systems.

The NRC neither prescribes nor endorses a specific approach for performing a deterministic assessment of fire damage on the ability to achieve safe-shutdown conditions (i.e., SSA). Differences in plant design, construction, equipment layout and operating preferences have resulted in many variations in plant-specific approaches. However, the overall process of performing an SSA remains fairly consistent between plants.

As illustrated in Figure 5-1, the determination of post-fire safe-shutdown capability typically includes two principal assessments, namely a "systems analysis" and a "fire area analysis." As part of the systems analysis, the licensee defines required shutdown functions and identifies redundant trains or "paths"of plant systems capable of accomplishing each of these functions. The licensee then identifies equipment, cables, and circuits that are needed to ensure the operation of these systems or that may adversely affect the shutdown capability if they are damaged as a result of fire. After identifying the equipment and cabling needed to ensure safe-shutdown, the licensee may determine their physical location (by fire area). The licensee then performs a "fire area analysis" to assess the potential consequences that a postulated fire in each area may have on the plant's ability to achieve and maintain safe-shutdown conditions. Figure 5-1 provides an overview of this process, and Chapter 6 presents a more detailed discussion of the SSA process.

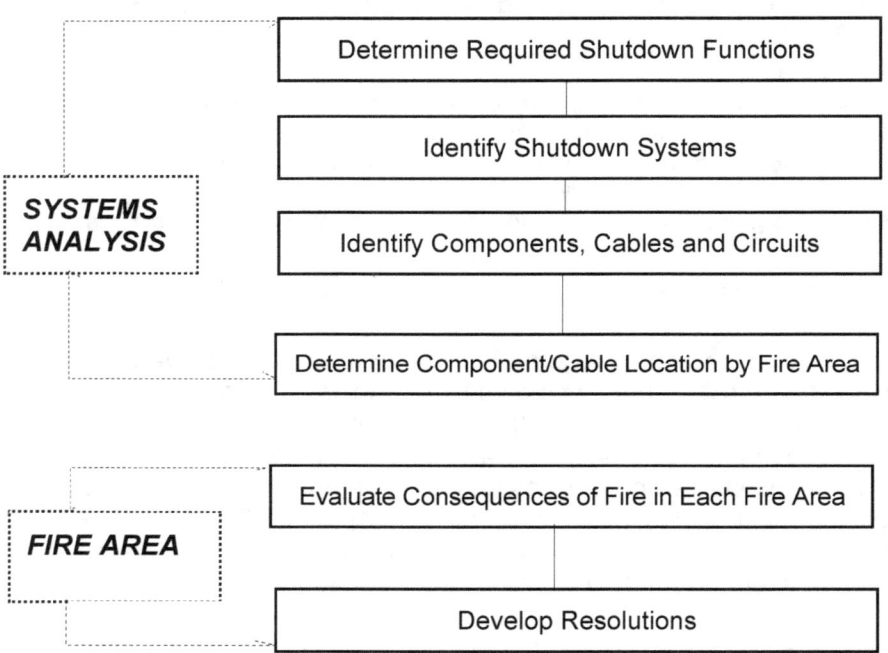

Figure 5-1 Overview of the Safe-Shutdown Evaluation Process

Because the SSA is based on large quantities of information and data, computer programs are frequently used to sort, manage, and analyze the data needed develop a safe-shutdown capability for the facility.

Conducting an SSA is an iterative process. As changes to the SSA database are implemented and facility modifications are installed, additional analysis must be performed to demonstrate that the changes have not compromised the previous analysis.

5.4   Analysis Assumptions

The following fundamental principles and assumptions establish the "ground rules" for performing an acceptable SSA:

Fire Hazards Analysis

An FHA, performed by qualified individuals, divides the plant into distinct fire areas and identifies fire hazards and major equipment located within each of those areas.

Shutdown Functions, Systems and Equipment

The systems and equipment needed for post-fire safe-shutdown are those systems necessary to perform the shutdown functions defined in Section III.L of Appendix R to 10 CFR Part 50. These functions are reactivity control, reactor coolant makeup, reactor heat removal, process monitoring, and associated support functions. Section III.L also defines the acceptance criteria for systems performing these functions:

> *During the post-fire shutdown, the reactor coolant system process variables shall be maintained within those predicted for a loss of normal a.c. power, and the fission product boundary integrity shall not be affected (i.e., there shall be no fuel clad damage, rupture of any primary coolant boundary, or rupture of the containment boundary).*

Except for BWR shutdown methodologies that rely on the use of low-pressure injection systems (see below), these criteria apply to the systems needed to satisfy both Section III.G and III.L of Appendix R to 10 CFR Part 50.[24]

Exposure Fire

The evaluation of safe-shutdown capability is based on the occurrence of a single *exposure fire* in an area containing (or presenting a fire hazard to) components, equipment, or cabling relied on for post-fire safe-shutdown. An exposure fire is defined as a fire in a given area that involves either in situ (permanently installed) or transient combustibles, but is external to any SSCs located in (or adjacent to) that same area. The effects of such fire (e.g., heat, smoke, or ignition) can adversely affect SSCs important to safety. Thus, a fire involving one train of safe-shutdown equipment may constitute an exposure fire for the redundant train located in the same fire area. Also, a fire involving combustibles other than either redundant train may constitute an exposure fire to both redundant trains located in the same fire area. Each fire area must be analyzed for the effects of an exposure fire.

---

[24]   IN 84-09, Section V, p. 4, U.S. Nuclear Regulatory Commission, Washington, DC, February 13, 1984.

Damage Expectations

In general, all cables and equipment that are exposed to the effects of fire (i.e., do not meet protection criteria of Appendix R, Section III.G.2) should be assumed to experience damage unless the staff has reviewed and approved a plant-specific exemption to these requirements. Licensees cannot take credit for fire to cause a loss of function if such a loss would simplify the shutdown scenario. For example, assuming that fire causes a loss of offsite power may be nonconservative.

Cause of Failures

The only failures considered are those that are directly attributable to the fire and/or fire-suppression activities. No other failures or independent events are assumed to occur concurrently with the fire.

Availability of Shutdown Systems

At the onset of the postulated fire, all safe-shutdown systems (including applicable redundant trains) are assumed to be operable and available for post-fire safe-shutdown. Systems are assumed to be operational with no repairs, maintenance, testing limiting conditions of operation (LCOs), etc., in progress. The unit is assumed to be operating at full (100-percent) power under normal conditions and normal lineups with a 3-month 100-percent power history.

Use of Low-Pressure Injection Systems at BWRs

The use of SRVs in conjunction with low-pressure injection (LPI) systems meets the requirements of a redundant means of post-fire safe-shutdown under Section III.G.2 of Appendix R to 10 CFR Part 50. When this methodology (SRV/LPI) is employed, the shutdown performance criteria identified in Section III.L do not apply. Rather, licensees who designate SRV/LPI as a redundant means of post-fire safe-shutdown must show that SRV/LPI can achieve and maintain hot shutdown in accordance with Sections III.G.1 and III.G.2 of Appendix R.[25]

Availability of Offsite and Onsite Power Sources

For the case of redundant shutdown, licensees may credit offsite power if it can be demonstrated to be free of fire damage. For fires not requiring implementation of an alternative or dedicated shutdown capability, offsite power is assumed to remain available unless fire can result in its loss. In the absence of an evaluation of the impact of fire on the availability of the offsite power sources, the analysis should demonstrate the capability of achieving shutdown conditions where offsite power is available *and* where offsite power is not available for up to 72 hours. For fire areas requiring an alternative or dedicated shutdown capability, the analysis should demonstrate the capability of achieving shutdown conditions where offsite power is available *and* where offsite power is not available for up to 72 hours. After 72 hours, offsite power can be assumed to be restored.

---

[25] Letter from S. Richards, NRC, to J. Kenny, BWROG, dated December 12, 2000.

Multiple-Unit Sites

Unrelated fires in two or more units are not postulated to occur simultaneously. However, where a single fire can impact more than one unit of a multi-unit site, the licensee must demonstrate the ability to achieve and maintain safe-shutdown conditions in each of the affected units.

Automatic Equipment Operation

Automatic equipment operation may or may not occur during a fire. Licensees cannot take credit for fire to cause a loss of automatic functions if such a loss would simplify the alternative shutdown scenario. For fire areas requiring alternative shutdown capability, licensees should consider the "worst case" scenario. For other fire areas, licensees may credit automatic operation of components and logic circuits in the analysis if they demonstrate that the circuitry associated with the automatic operation will remain unaffected by the postulated fire (i.e., satisfies established fire protection/separation criteria).

Relay/Switch Contact Positions

All relay, position switch, and control switch contacts in control circuits are in the position or status that correspond to the normal operation of the device. Test and transfer switches in control circuits are in their normal position.

Repair Activities

Repair activities (e.g., wiring changes, fuse replacement, use of pneumatic or electric jumpers, or other modifications) are not permitted for systems that are required to achieve and maintain hot shutdown conditions. Modifications and repair activities are permitted for cold shutdown systems provided that (1) for areas *not requiring* an alternative shutdown capability, the licensee can demonstrate that all repair activities can be accomplished within 72 hours or, (2) for areas requiring an alternative shutdown capability, all needed repairs can be performed and cold shutdown achieved within 72 hours.

Cable and Circuit Failure Modes

It is not deemed possible to accurately predict the manner in which damaged cables or circuits may fail. Various types of electrical failure modes (e.g., hot shorts, open circuits, or shorts to ground) must be assumed to occur as a result of fire damage.

Single-Failure Criterion

Because it is only one of several levels of defense, the shutdown capability does not have to meet the single-failure criterion.

Redundant vs. Alternative Shutdown Systems and Equipment

For the purpose of analysis of compliance with Section III.G.2 criteria (i.e., redundant train shutdown capability), the safe-shutdown capability is defined as one of the two normal safe-shutdown trains. If the system is being used to provide its design function, it is generally is considered to be redundant. If the system is being used in lieu of the preferred system because the redundant components of the preferred system do not meet the separation criteria of Section III.G.2, the system is considered an alternative shutdown capability. (Reference GL 86-10.)

Post-Fire Operating Procedures

The only requirement for post-fire operating procedures is for those areas where alternative shutdown is required. For other areas of the plant, shutdown would be achieved utilizing one of the two normal trains of shutdown systems. Shutdown in degraded modes (one train unavailable) should be covered by present operator training and abnormal and emergency operating procedures (EOPs). If the degraded modes of operation are not presently covered, the operations staff should assess the need for additional training or procedures. (Reference GL 86-10.)

5.5   Redundant Shutdown Capability

As experienced during the Browns Ferry fire, SSCs that are exposed to the effects of fire may be damaged, and this damage may lead to unexpected consequences in the operation of plant safety systems. On February 20, 1981, the NRC forwarded GL 81-12, which restated the regulatory requirement for each licensee to reassess areas of the plant containing cables or equipment, including associated nonsafety circuits, of redundant trains of systems necessary to achieve and maintain hot shutdown conditions.

Failing to adequately identify circuits, components, and systems required to achieve and maintain safe-shutdown and protect them from the effects of fire could result in damage to redundant trains of shutdown systems and significantly impair the ability to safely shutdown the plant in the event of fire. Consequently, one of the key outcomes of the SSA evaluation process is the identification of plant locations (fire areas) that contain redundant trains of SSCs important to safe-shutdown. As described in Section 4.3 above, when redundant trains of cables or equipment, including associated nonsafety circuits necessary to achieve and maintain hot shutdown are found to be located in the same fire area, the fire protection requirements of Section III.G.2 of Appendix R must be satisfied. If not, the licensee must provide an alternative or dedicated shutdown capability or request an exemption.

Areas of the plant that meet the separation requirements of Section III.G.2 are frequently referred to as "redundant shutdown" fire areas.

5.6   Alternative Shutdown Capability (10 CFR Part 50, Appendix R, Section III.G.3)

In certain areas of the plant, redundant trains of equipment required for hot shutdown may be located in close proximity. Typical examples include the MCR and CSR, where redundant trains of shutdown equipment may be separated by only a few inches. In such cases, compliance with fire protection features specified in Section III.G.2 of Appendix R cannot be readily achieved. When areas such as these are identified, an alternative or dedicated shutdown capability must be provided that is both physically and electrically independent of the area under consideration.

Alternative shutdown capability is provided by rerouting, relocating, or modifying existing systems. An example of an alternative shutdown capability would be the installation of isolation switches to isolate safety-related circuits from fire damage. Alternative shutdown capability can also be provided by implementing procedures specifying "alternative" methods of operation, such as manual operations and/or evacuation of the normal control station(s) such as the control room.

Dedicated shutdown capability is provided by installing new structures and systems for the sole function of post-fire safe-shutdown. Examples of dedicated shutdown capability include installation of emergency generators, process instrumentation, or other equipment which is intended to be used only for safe-shutdown purposes (i.e., dedicated to safe-shutdown).

The alternative or dedicated shutdown capability may be unique for each area, or it may be one unique combination of systems for all fire areas requiring this capability. For those areas requiring alternative or dedicated shutdown capability, fire detection and a fixed fire-suppression system must also be installed in the fire area of concern.

The design-basis event for considering the need for alternative or dedicated shutdown capability is a postulated fire in a specific fire area containing redundant safe-shutdown cables/equipment in close proximity where it has been determined that fire protection means cannot ensure that safe shutdown capability will be preserved. Licensees should consider two cases in which (1) offsite power is available; and (2) offsite power is not available. (Reference GL 86-10.)

The SSA must demonstrate that, during a post-fire safe-shutdown, the reactor coolant process variables will be maintained within those predicted for a loss of normal AC power and the integrity of the fission product boundary will not be affected. Integrity of the fission product boundary includes (1) no fuel clad damage, (2) no rupture of any primary coolant boundary, and (3) no rupture of the containment boundary.

The alternative or dedicated shutdown capability shall be able to achieve and maintain sub-critical conditions in the rector, maintain the reactor coolant inventory, achieve and maintain hot standby conditions (hot shutdown for a BWR) for an extended period of time, achieve cold shutdown conditions within 72 hours, and maintain cold shutdown conditions thereafter.

Performance goals for the shutdown functions identified in the SSA are as follows:

- The reactivity control function should be capable of achieving and maintaining cold shutdown reactivity conditions.

- The reactor coolant makeup function should be capable of maintaining the reactor coolant level above the top of the core for BWRs and within the level indication of the pressurizer for PWRs.

- The reactor heat removal function should be capable of achieving and maintaining DHR.

- The process monitoring function should be capable of providing direct readings of the process variables necessary to perform and control the above functions.

The systems used for alternative or dedicated shutdown need not be designed to (1) seismic Category I criteria, (2) single-failure criteria, or (3) other design-basis accident criteria, except for the portions of these systems that interface with or impact existing safety systems.

It should be noted that safe-shutdown performance goals and functions to be performed are specified in the regulation (Appendix R, Section III.L). However, specific methods for achieving these objectives are left to the individual plants to determine and demonstrate.

Implementation of an alternative or dedicated shutdown capability will require operators to perform many activities at local control stations outside the MCR. All operator activities should be prescribed in abnormal operating procedures that have been integrated into the overall plant operator training and qualification program. As alternative/dedicated shutdown procedures are developed, timely performance of all manual operator actions in the process must be ensured. Verification that time-dependent actions are satisfied in the written procedures is accomplished by performing a thermal-hydraulic timeline analysis, where various types of transients are analyzed to determine how much time the operating crew has to implement each of the safe-shutdown functions before exceeding the established performance criteria. These transients may involve a fire-induced spurious equipment operation or the generation of a false signal, with an assumed concurrent loss of offsite power. Typical examples include the loss of main feedwater in a PWR or inadvertent opening of the turbine bypass valves in a BWR that could cause over-pressurization of the main condenser as a result of the loss of circulating water (resulting from the concurrent loss of offsite power).

In summary, this analysis and verification will include the following confirmations:

- The procedural steps or operator manual actions can be performed (by verifying that operators will have access to required equipment).

- The analysis criteria are satisfied. For example, the performance of time-sensitive steps within allotted times (derived from the results of the plant's thermal-hydraulic analysis).

- Required support equipment (such as ladders and valve handles) are available (pre-positioned and administratively controlled) for use when needed.

Other alternative/dedicated shutdown implementation considerations include the following:

- Confirmation that the minimum shift complement of operators, exclusive of operators who are part of the fire brigade, is adequate to properly implement the safe-shutdown procedures.

- Job performance measures covering the major tasks in the post-fire safe-shutdown procedures have been integrated into the overall plant operator training and qualification program.

- Confirmation of the availability and adequacy of emergency lighting (this is necessary because alternative/dedicated shutdown procedures frequently require the performance of operator manual actions throughout the plant). Section III.J of Appendix R requires that fixed emergency lighting units must be provided for locations in the plant associated with post-fire safe-shutdown implementation, including the ingress and egress routes of the operators to those locations.

- Confirmation of the availability and adequacy of communication systems. Most alternative and dedicated shutdown strategies rely heavily on the operators' ability to confirm or verify the operation of plant equipment and then report this information back to another operator stationed at central location (typically the RSP). The communication system relied on to ensure this capability provides a vital shutdown support function. In addition to remaining free of fire damage, the designated method of communications should not (1) be affected by a loss of offsite power, (2) interfere with any in-plant instrumentation, or (3) have dead zones in areas where communication is vital to the shutdown process.

5.7 Specific Considerations

5.7.1 Operator Manual Actions

In the early 1990s, the NRC identified significant performance deficiencies with Thermo-lag fire barrier material. At that time, the industry used this material extensively to meet the fire protection requirements specified in Section III.G.2 of Appendix R for cable trays, conduits, and other enclosures containing circuits required to achieve and maintain hot shutdown conditions. During the subsequent Thermo-lag resolution process, many licensees attempted to minimize the use of this material by re-analyzing their plants and developing alternative protection strategies. While many approaches, such as cable rerouting, use of different equipment, or use of rated fire barriers of different materials are clearly acceptable, some licensees replaced the fire barriers with the use of operator manual actions. In some cases, this may not provide the level of fire protection required by the regulation.

In general, reliance on operator manual actions does not satisfy the specific technical requirements of Section III.G.2 of Appendix R to 10 CFR Part 50. However, in certain cases, the staff has reviewed and approved the use of operator manual actions on a plant-specific basis. These approvals are documented in plant-specific safety evaluations and incorporated into the plants' fire protection licensing bases. One example is an exemption granted to Alabama Power Company for the Joseph M. Farley Nuclear Plant, dated November 19, 1985 (NUDOCS Accession No. 8512060395). The staff has developed a rulemaking plan, identified in SECY 03–100, to allow use of feasible operator manual actions in Section III.G.2 areas without prior staff approval. The Commission approved the proposed rulemaking plan in September 2003, as well as the staff's proposal to provide enforcement discretion for feasible manual actions without prior staff approval.

As discussed in Section 5.6, operator manual actions are permitted to accomplish alternative shutdown in accordance with Appendix R, Section III.G.3, provided that the required operator manual actions are incorporated into post-fire operating procedures, verified to be physically possible, and capable of being performed within the time constraints defined by a thermal-hydraulic analysis developed for the specific shutdown scenario (e.g., fire in control room with one worst-case spurious actuation), *and* provided that sufficient staffing, communications, and emergency lighting are ensured to remain available.

Where operator manual actions are relied on to ensure the successful accomplishment of required shutdown functions, it is expected that they can be safely and effectively performed in a sufficiently timely manner. The following factors should be considered when determining the acceptability of operator manual actions:

- Available indications: If credited to support operator manual actions, diagnostic indications shall have the following capabilities:
    - Show the need for the action.
    - Operate effectively, given the postulated fire.
    - Verify that the intended safety function has been accomplished.

- Environmental considerations: Environmental considerations encountered while accessing and performing operator manual actions shall be demonstrated to be consistent with the human factor considerations for visibility, habitability, and accessibility, including the following:
  - Emergency lighting shall be provided as required in Appendix R, Section III.J, or by the licensee's approved fire protection program.
  - Radiation shall not exceed the limits specified in 10 CFR 20.1201.
  - Temperature and humidity conditions shall not adversely affect the capability to perform the operator manual actions (e.g., see NUREG/CR-5680, "The Impact of Environmental Conditions on Human Performance"), or the licensee shall provide an acceptable rationale for why temperature and/or humidity do not adversely affect the ability to perform operator manual actions.
  - Smoke and toxic gases from the fire shall not adversely affect the capability to access the required equipment to perform the operator manual actions.
  - All locations where operator manual actions are performed, including the pathways to those locations, shall be accessible.

- Staffing and training: All plant operators, under all staffing levels, shall be capable of performing all required actions in the times required for a given fire scenario. The use of operators shall be independent from any collateral fire brigade or control room duties that they may need to perform as a result of the fire. Operators required to perform the manual actions shall have been appropriately trained and shall be continuously available to perform the actions required to achieve and maintain safe shutdown.

- Communications: To achieve and maintain safe shutdown, communications capability shall be adequate for performance of the operator manual actions that must be coordinated with other plant operations, with this communications capability continuously available.

- Equipment: Any equipment required to support operator manual actions, including keys, self-contained breathing apparatuses (SCBAs), and personnel protective equipment, shall be readily available, easily accessible, and functional. Credit shall not be taken for the use of non-functional equipment or equipment for which functionality may have been adversely affected by the fire as a result of smoke, heat, water, combustion products, or spurious actuation effects (e.g., over-torquing an MOV as a result of a spurious signal, as discussed in IN 92-18).

- Procedures: Procedural guidance on the use of required operator manual actions shall be readily available and easily accessible.

- Demonstration: The capability to successfully accomplish required operator manual actions within the time allowable using the required procedures and equipment shall be demonstrated using the same personnel/crews who will be required to perform the actions during the fire. Documentation of the demonstration, as well as any training periodically provided to the operators, shall be provided.

- Complexity and number: The degree of complexity and total number of operator manual actions required to effect safe shutdown shall be limited, such that their successful accomplishment under realistically severe conditions is ensured for a given fire scenario. The need to perform operator manual actions in different locations shall be considered when sequential actions are required. Analyses of the postulated fire time line shall demonstrate that there is sufficient time to travel to each action location and perform each action required to support the associated shutdown function(s), such that an irrecoverable condition does not occur.

These factors represent an expansion of the "Inspection Criteria for FP Manual Actions," which the NRC issued in March 2003 as Enclosure 2 to Attachment 71111.05 of the FP Inspection Procedure. Specifically, the March 2003 inspection criteria, gave significant latitude, as follows:

> *For an interim period, while rulemaking is in progress... acceptance criteria can be developed which would facilitate evaluations of certain manual actions.*

The March 2003 inspection criteria were based on the NRC's inspection experience and addressed the following factors:
- diagnostic instrumentation
- environmental considerations
- staffing and training
- communications and accessibility
- procedures
- verification and validation (V&V)

In addition to these factors, manual operator actions may not include repair activities that are needed to achieve and maintain hot shutdown conditions. Appendix A to this document provides additional guidance on the use of operator manual actions. In addition, consult the following reference sources for the FP inspection criteria:
- NRC Inspection Manual Chapter 0609, "Significance Determination Process"
- Input from FP risk-related studies sponsored by the NRC's Office of Nuclear Regulatory Research (RES)
- Feedback from a meeting with the Advisory Committee on Reactor Safeguards (ACRS), Subcommittee on Fire Protection
- Performance-shaping factors used in human reliability analysis techniques

5.7.2   Repairs

Section III.G.1 of Appendix R to 10 CFR Part 50 states that one train of systems needed to achieve and maintain hot shutdown conditions must be free of fire damage. Thus, one train of systems needed for hot shutdown must be ensured to remain operable both during and after a fire. Operability of the hot shutdown systems must exist without repairs. In general, fuse removal for the purpose of preventing the maloperation of equipment is not considered a repair, provided that the fuse removal is routine and can be performed in a manner that does not subject the operator to an undue safety hazard (e.g., reaching into an energized 4 kV SWGR). However, the replacement of fuses is considered a repair.

Repairs are allowed for cold shutdown systems. However, the time requirements for completing repairs is dependent on the shutdown method employed. For areas provided with an alternative or dedicated shutdown capability, Appendix R, Section III.L.5, states that, *"equipment and systems comprising the means to achieve and maintain cold shutdown conditions shall not be damaged by fire; or the fire damage to such equipment and systems shall be limited so that the systems can be made operable and cold shutdown can be achieved within 72 hours."*

This time limit should not to be confused with the requirements for completing repairs for areas that do not require an alternative shutdown capability. For these areas, Section III.G.1.b requires only the capability to repair the systems necessary to achieve and maintain cold shutdown from either the control room or emergency control station(s) within 72 hours, not the capability to repair and achieve cold shutdown within 72 hours as required for the alternative or dedicated shutdown modes by Section III.L.

Procedures for repairing damaged cold-shutdown equipment should be prepared in advance with replacement equipment stored on site. All repairs should be of sufficient quality to ensure safe operation until the plant is restored to an operating condition.

5.7.3   Diagnostic Instrumentation

Certain post-fire safe-shutdown strategies rely on the operators to take mitigating actions in response to equipment perturbations that may be caused by fire. For example, a shutdown strategy for one fire area may rely on operators to manually close a tank discharge valve in the event that it spuriously changes position (i.e., opens) as a result of fire damage to its control cabling. To ensure this capability, sufficient "diagnostic instrumentation" (such as tank level indicator and/or level alarm annunciators) must be available to enable the operators to promptly detect the undesired change in valve position and take corrective actions necessary to defeat it (i.e., manually close the valve).

As stated in GL 86-10, "diagnostic instrumentation" is instrumentation, beyond that identified in Attachment 1 to IN 84-09, which is needed to ensure proper actuation and functioning of safe- shutdown and support equipment (e.g., flow rate, pump discharge pressure). The specific diagnostic instrumentation needed depends on the design of the shutdown capability. When the shutdown strategy relies on the use of procedures to direct operator actions or operator manual actions in response to equipment upsets that may occur as a result of fire, sufficient diagnostic instrumentation must be ensured to remain available (i.e., free of fire damage) so that the success of operator activities can be readily confirmed.

CHAPTER 6. DETERMINISTIC ANALYSIS PROCESS
FOR APPENDIX R COMPLIANCE

The Browns Ferry event was of sufficient significance to warrant major changes in fire protection design features of NPPs in the United States. Consequently, the NRC issued its new regulation as 10 CFR 50.48 and Appendix R to 10 CFR Part 50, which became effective on February 17, 1981. One of the key requirements of this regulation was to backfit Section III.G, "Fire Protection of Safe Shutdown Capability," to all NPPs that were licensed to operate before January 1, 1979. This section establishes the minimum acceptable fire protection design features that are necessary to ensure that licensees can achieve safe-shutdown in the event of fire in any area of the plant. The fundamental objective of Section III.G is to extend the DID concept to fire safety by obtaining reasonable assurance that, in the event a fire were to start (despite the fire prevention program) and continue to propagate (despite fire protection activities), one train of SSCs needed to achieve and maintain safe-shutdown conditions will remain available. Figure 6-1 illustrates how fire damage to circuits and cables may adversely affect the shutdown capability.

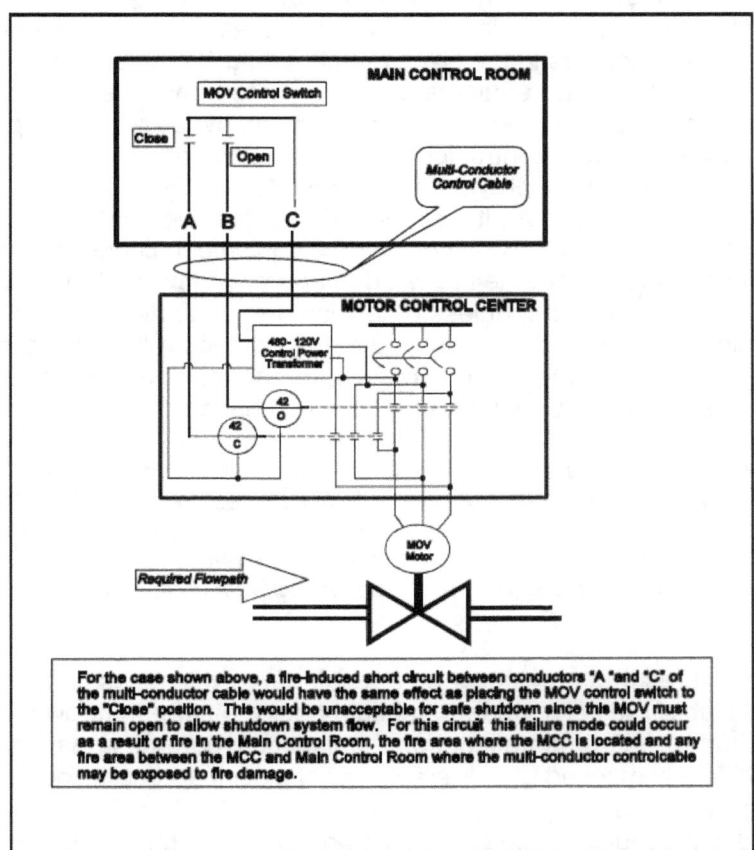

Figure 6-1 Potential Effect of a Fire-Induced Circuit Failure

Section III.G required each licensee to perform a comprehensive evaluation of each fire area and demonstrate, through the performance of a deterministic assessment of potential fire damage, that SSCs of which failure (or maloperation) could impact the ability to achieve and maintain safe-shutdown conditions are provided with suitable fire protection features (i.e., as required by Section III.G.2 of Appendix R, or justified in a staff-approved exemption). For locations of the plant where compliance with the the fire protection design features specified in Section III.G.2 may not be feasible because redundant trains of cables and/or equipment are located in close proximity (such as the control room or CSR), Section III.G.3 requires licensees to provide an alternative or dedicated shutdown capability that is independent (both physically and electrically) from the fire area under consideration. In either case, the evaluation of a fire in any area must conclusively demonstrate that one train of equipment that can be used to immediately bring the reactor to hot shutdown conditions remains unaffected by fire.

6.1 Principles of a Deterministic Evaluation of Post-Fire Safe-Shutdown Capability

The SSA for each plant must specifically identify all systems and equipment upon which the licensee will depend to perform essential shutdown functions. It must also include an evaluation of any circuits or cables in the fire area that could (1) adversely affect the operability of identified shutdown systems and equipment or (2) initiate transients that could preclude the successful accomplishment of required shutdown functions by feeding back potentially disabling fault conditions to power supplies, control logic or instrumentation circuits. In addition, the SSA must describe how the licensee will prevent or appropriately mitigate such disabling conditions. Otherwise, the licensee cannot ensure its reliance on the identified safe-shutdown equipment. Because it is not possible to predict the manner in which equipment (cables, circuits or components) may fail, the SSA must assume that the fire will damage any unprotected equipment located in the fire area under evaluation and, unless otherwise demonstrated through the performance of more detailed evaluations, it must be assumed that this damage will fail the affected equipment in a mode that adversely impacts safe-shutdown. In summary, the NRC expects that such evaluations will be based on the following deterministic premise:

> *Cables and components that are exposed to the effects of fire and its related perils (i.e., not provided with fire protection features sufficient to meet Section III.G of Appendix R) will be damaged, and, unless demonstrated otherwise through the performance of suitably comprehensive and conservative engineering evaluations, it is assumed this damage will cause connected equipment to fail or malfunction in an undesired manner for shutdown.*

Not all circuit/cable failures that may occur as a result of fire will necessarily have an adverse impact on the plant's ability to achieve and maintain post-fire safe-shutdown conditions. The electrical distribution, instrumentation, communications, control, and process systems of a commercial NPP are composed of a diverse array of electrical circuits/cables, and fire damage to many (if not most) of these circuits will have no adverse effect on the ability to achieve and maintain safe-shutdown conditions. In certain instances, it may be possible to demonstrate, through the performance of a detailed analysis of the potential effects of fire damage, that even if a fire were to damage certain circuits of required shutdown components, the damage would be acceptable because it will not have any effect on the ability of the component to perform its intended shutdown function. For example, consider the circuit illustrated in Figure 6-2.

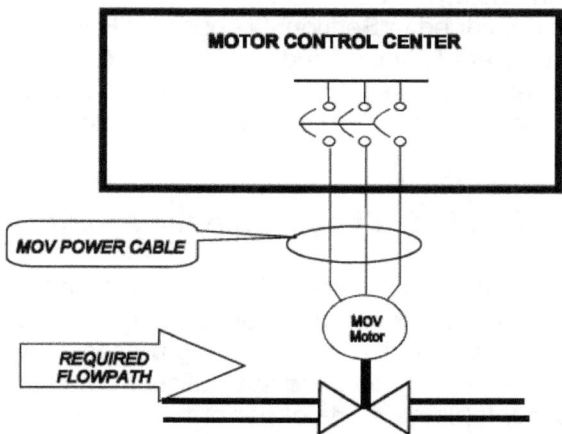

MOTOR CONTROL CENTER

MOV POWER CABLE

MOV Motor

REQUIRED FLOWPATH

For this case, the MOV is "OPEN" during normal plant operations and must be assured to remain "OPEN" for post-fire safe shutdown (to allow fluid to flow in the required shutdown system). Since the MOV will fail in the "as is" position on loss of motive power, fire damage which results in a loss of power to the MOV would not impact the post-fire safe shutdown capability.

Figure 6-2   Fire Damage to Certain Circuits of Required Shutdown Equipment May Not Pose a Threat to the Shutdown Capability

In this case, an MOV is open during normal plant operations. To ensure successful achievement of safe-shutdown conditions, the MOV must remain open to allow fluid to flow through the required flowpath. For this case, the required shutdown component is an MOV which, by design, will fail in the "as-is" (open) position upon a loss of motive power. Therefore, if it can be shown that fire damage to the power cable would only result in a loss of motive power to the MOV, the analysis has demonstrated a level of safety equivalent to that which would be achieved through compliance with Section III.G.2, and the power cable would not require any additional fire protection features. As stated in the staff's clarification of GL 81-12, *"Our interest is only with those circuits (cables) whose fire-induced failure could affect shutdown."*

6.2 Use of "Appendix R" Terminology

Throughout this document, and particularly in this section, reference is made to post-fire safe-shutdown criteria contained in Sections III.G and III.L of Appendix R to 10 CFR Part 50. Since its inception, reference to "III.G.2" has become synonymous with redundant train shutdown capability, and "III.L" is commonly referred to when discussing alternative shutdown capability irrespective of the actual requirements specified in the plant's fire protection licensing basis.

In addition to simplifying the discussion, the use of such "Appendix R terminology" is generally acceptable because the guidelines contained in SRP Section 9.5.1 include the acceptance criteria listed in Appendix R to 10 CFR Part 50 and 10 CFR Part 50.48. It should be noted, however, that the use of this terminology is not intended to imply that Appendix R requirements are applicable to all plants. As described in Section 4, Appendix R is *only* specifically applicable to a limited number of plants that were fully licensed and operating before January 1, 1979. The staff typically reviewed the post-fire safe-shutdown capabilities of plants licensed after this date during the initial licensing process for conformance to guidelines contained in Position C.5.b of SRP Section 9.5.1.

6.3   Overview of the Post-Fire Safe-Shutdown Analysis Process

A comprehensive evaluation of the potential impact of fire damage on the ability to achieve and maintain safe-shutdown conditions within the performance goals and criteria specified in Appendix R to 10 CFR Part 50 is a technically challenging process, involving the expertise of personnel knowledgeable in plant operations and specialists from various engineering disciplines. There are many acceptable methods of performing a fire SSA, and the NRC neither prescribes nor endorses any one approach. The SSA should be a bounding analysis that identifies the range of possible fire impacts within each fire area and ensures that appropriate measures are in place to prevent this damage from affecting the ability to safely shut down the plant. For each fire area, the SSA will define the set of systems necessary to accomplish required shutdown functions in accordance with established performance criteria. The selected systems form the basis for the selection of individual components and cables needed to ensure that each system will be capable of accomplishing its intended shutdown function.

The detailed methods used by a particular plant operating organization will vary with plant-specific conditions, such as design, construction, cable configuration, equipment layout, and operating preferences. Therefore, it is not possible to develop a "one-size-fits-all" procedural process for performing a deterministic analysis sufficient to satisfy Appendix R concerns. However, the *overall approach* for ensuring the availability of at least one shutdown *"success path"* (i.e., the minimum set of SSCs necessary to achieve and maintain safe-shutdown in the event of a fire) for each fire area, is fairly consistent among plants regardless of plant design or vintage. Figure 6-3 illustrates an overview of this approach and provides references to the specific subsection of this chapter that describes each of the steps.

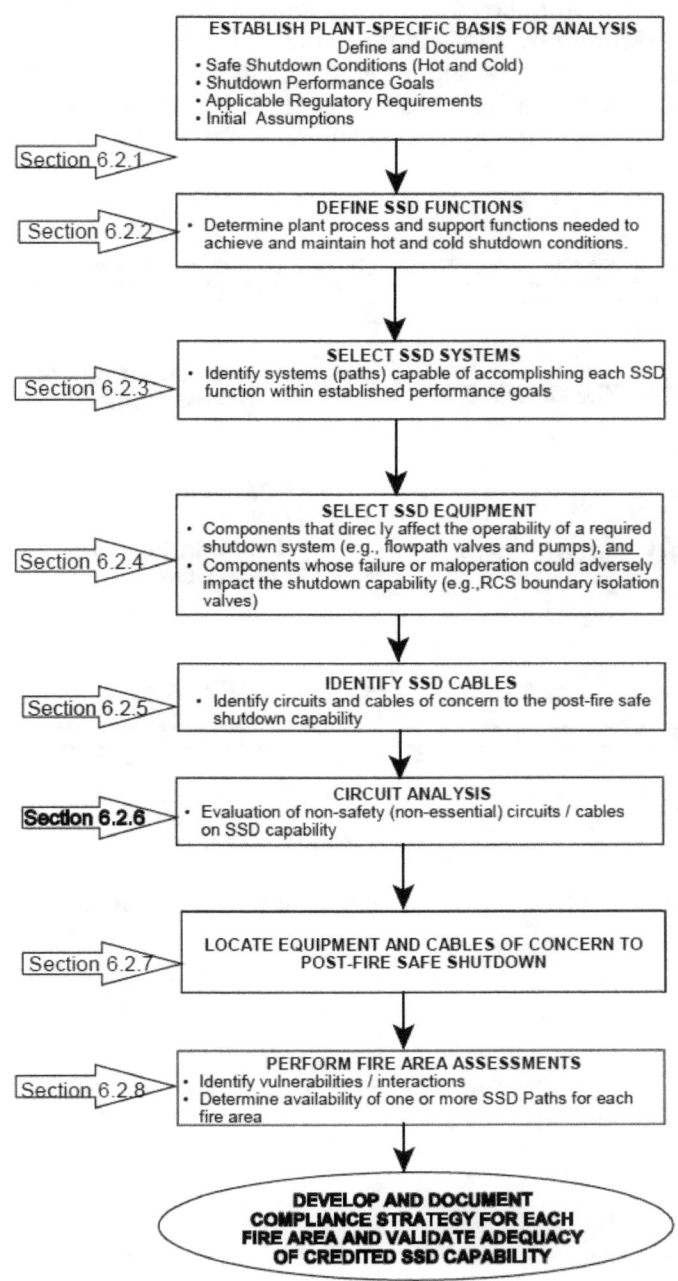

Figure 6- 3 Overview of Post-Fire Safe Shutdown Analysis Process

It should be noted that, for the purpose of this discussion, it is assumed that a comprehensive FHA has already been performed by qualified fire protection engineers to divide the plant into separate and distinct fire areas that are separated from other fire areas by rated fire barriers that are adequate for the anticipated fire hazard. As depicted in Figure 6-4, the fire area boundaries represent the extent of fire spread assumed in the SSA.

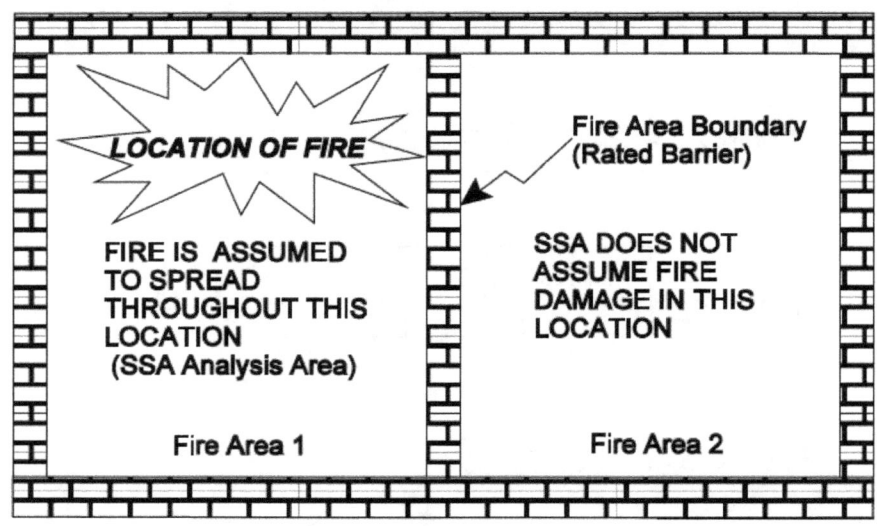

Figure 6-4 Fire-Rated Boundaries Determine Extent of Fire Spread Assumed in SSA

6.4   Methodology

In demonstrating the plant's safe-shutdown capability, the SSA integrates the following evaluations:

(1) **Safe-Shutdown System Selection/Path Development** identifies systems that are capable of accomplishing shutdown safety functions (e.g., reactivity control, reactor coolant makeup, DHR, etc.).

(2) **Plant Configuration** compares equipment locations and cable routing with the fire area boundary information established in the FHA.

(3) **Safe-Shutdown System Performance** demonstrates that, following a fire, sufficient equipment of adequate capacity and capability will remain available to achieve and maintain the reactor n a safe-shutdown condition.

(4) **Associated Circuits Effects** demonstrates that a fire cannot, through its effects on nonessential/nonsafety electrical circuits, prevent safe-shutdown systems and equipment from accomplishing their intended functions or initiate an event that is beyond the capability of the safe-shutdown systems.

6.4.1   Establish the Plant-Specific Technical and Licensing Bases for the Safe-Shutdown Analysis

6.4.1.1   Assemble Plant-Specific Information

The first step in the SSA process is to review available documentation to obtain an understanding of the available plant systems and functions required to achieve and maintain safe-shutdown. The following documentation is typically needed to perform the SSA:

- *Fire Protection Licensing Basis Documents* include the FSAR, plant operating license conditions, TSs, applicable regulatory requirements (Appendix R or SRP Section 9.5.1), and fire protection safety evaluations issued by the staff.

- *Fire Hazards Analysis* identifies the fire areas, characterizes the hazards, and describes the fire protection features within each fire area.

- *Plant System Descriptions* are the detailed descriptions of the functions and capabilities of each plant system, including those systems capable of accomplishing the safe-shutdown functions. They should include both front line and support systems necessary for operation of the system. Support systems do not directly provide safety functions, but are required to ensure that the front line systems can perform the safety functions as required. Examples of support systems include cooling water, electrical power distribution, instrument air, and HVAC.

- *Plant System Design Drawings,* also known as piping and instrumentation diagrams (P&IDs), identify the components that make up a system and the flowpath of that system, and identify any interconnections to other systems that could degrade the system under certain fire damage conditions. Electrical drawings needed for review typically include electrical distribution one-line diagrams, cable block diagrams, logic diagrams, cable and raceway layout drawings, and instrument loop diagrams.

- *Applicable Operating Procedures* document the plant's normal, emergency, and abnormal operating procedures.

6.4.1.2 Define and Document Safe-Shutdown Conditions for the Plant

In order to develop an effective strategy for achieving and maintaining the reactor in a "safe-shutdown" condition, it is first necessary to define the plant-specific parameters that must be satisfied in order to declare that a "safe-shutdown" condition has been achieved. For fire events, safe-shutdown includes both hot shutdown and cold shutdown conditions. The plant's TSs document the plant-specific parameters for each of these conditions.

6.4.1.3 Define and Document the Safe-Shutdown Performance Goals

Guidance for determining the functional and performance requirements of systems upon which the plant relies to accomplish both redundant (III.G.2) and alternative (III.L) shutdown was initially provided by the NRC in IN 84-09, "Lessons Learned from NRC Inspections of Fire Protection Safe Shutdown Systems." Specifically, IN 84-09 states that the systems and equipment needed for post-fire safe-shutdown (*both* redundant and alternative) are those systems necessary to perform the safe-shutdown functions defined in Section III.L of Appendix R. Section III.L defines the acceptance criteria for such systems as follows:

- During post-fire safe-shutdown, the reactor coolant system process variables shall be maintained within those predicted for a loss of normal AC power

- The fission product boundary integrity shall not be affected (i.e., there shall be no fuel clad damage, rupture of any primary coolant boundary, or rupture of the containment boundary).

By letter dated December 12, 2000 (Reference: Richards Letter) the staff documented its evaluation of a BWR Owners Group (BWROG) Fire Protection Committee position regarding the use of low-pressure injection systems as "redundant" shutdown systems under Appendix R. The staff position documented in this evaluation clarified the information initially provided in IN 84-09. Specifically, in its review of the applicability of Section III.L requirements, the staff concluded that *"Section III.L performance criteria are applicable only to alternative or dedicated shutdown capability, and need not be met for redundant post-fire safe-shutdown capability."* As a result of the staff's clarification of IN 84-09, RG 1.189 now defines performance criteria for shutdown systems as follows:

> **Regulatory Position 5.1 Safe-Shutdown Performance Goals for Redundant Systems**
> *"Ensure that fuel integrity is maintained and that there are no adverse consequences on the reactor pressure vessel integrity or the attached piping. Fuel integrity is maintained provided the fuel design limits are not exceeded."*

> **Regulatory Position 5.2 *Alternative or Dedicated Shutdown Design and Performance Goals***
> ***5.2.1 Alternative or Dedicated Safe-Shutdown System Design Goals***
> *During the post-fire safe-shutdown, the reactor coolant system process variables should be maintained within those predicted for a loss of normal ac power, and the fission product boundary integrity should not be affected (i.e., there should be no fuel clad damage, rupture of any primary coolant boundary, or rupture of the containment boundary).*
>
> *The systems used for alternative or dedicated shutdown need not be designed to (1) seismic Category I criteria, (2) single-failure criteria, or (3) other design-basis accident criteria, except the portions of these systems that interface with or impact existing safety systems.*

5.2.2 Safe-Shutdown Performance Goals for Alternative or Dedicated Systems
The performance goals for the safe-shutdown functions should be:
- *The reactivity control function should be capable of achieving and maintaining cold shutdown reactivity conditions.*
- *The reactor coolant makeup function should be capable of maintaining the reactor coolant level above the top of the core for BWRs and within the level indication of the pressurizer for PWRs.*
- *The reactor heat removal function should be capable of achieving and maintaining DHR.*
- *The process monitoring function should be capable of providing direct readings of the process variables necessary to perform and control the above functions.*

RG 1.189 further states that the capability of the required shutdown functions should be based on a previous analysis, if possible (e.g., those analyses in the FSAR). The equipment required for alternative or dedicated shutdown should have the same or equivalent capability as that relied on in the above referenced analysis. It should be noted that specific methods for achieving these objectives are left to the individual plants to determine and demonstrate.

6.4.1.4   Define and Document Initial Assumptions

In order to proceed with the analysis, it is necessary to establish a set of initial assumptions or "ground rules" that define the fundamental criteria and conditions under which the evaluation process is to be performed. For compliance with Appendix R, the SSA must be based on the following considerations:

- *Exposure Fire:* The analysis must assume that a single *"exposure fire"* will occur in any fire area. An exposure fire is defined as a fire in a given fire area that involves either in-situ (permanently installed) or transient combustibles, but is external to any SSCs located in (or adjacent to) that same fire area. The effects of such fire (e.g., smoke, heat or ignition) can adversely affect those SSCs important to safety. Thus, a fire involving one train of safe-shutdown equipment may constitute an exposure fire for the redundant train located in the same fire area. Also, a fire involving combustibles other than the redundant train may constitute an exposure fire to both redundant trains located in the same area

- *Extent of Fire Damage:* For analysis purposes, it is assumed that only a single exposure fire will occur in any fire area at a given time. Since it is not deemed possible to accurately predict the manner in which equipment (cables, circuits or components) may fail, this analysis must assume that the fire will damage any unprotected equipment located in the fire area under evaluation, and, unless demonstrated otherwise through the performance of more detailed evaluations, it must be assumed this damage will cause the affected equipment to fail in an undesired manner for safe-shutdown. The fire area boundaries represent the extent of fire spread assumed in analysis. During the performance of a comprehensive SSA, all areas of the plant will be individually analyzed for an exposure fire.

- *Failures:* The only failures considered are those that are directly attributable to the fire. No other failures or independent events are assumed to occur concurrently with the fire. No other design-basis events or failure consequences need be postulated in conjunction with the exposure fire, except for those caused by the fire itself.

- *Equipment Availability:* At the onset of fire, all safe-shutdown systems are assumed to be operable and available for post-fire safe-shutdown.

- *Availability of Offsite Power:* For fires not requiring implementation of an alternative or dedicated shutdown capability, offsite power is assumed to remain available unless fire can result in its loss. In the absence of an evaluation of the impact of fire on the availability of the offsite power sources, the analysis should demonstrate the capability of achieving shutdown conditions where offsite power is available and where offsite power is not available for up to 72 hours. For fire areas requiring an alternative or dedicated shutdown capability, the analysis should demonstrate the capability of achieving shutdown conditions where offsite power is available and where offsite power is not available for up to 72 hours. After 72 hours, offsite power can be assumed restored.

- *Automatic Equipment Operation:* Automatic equipment operation may or may not occur during a fire. For fire in areas requiring an alternative or dedicated shutdown capability, a loss of automatic functions must be assumed. For example, in the event of a loss of offsite power (LOOP) the EDGs will normally start automatically on undervoltage. However, in developing the alternative shutdown capability operation of this automatic start feature cannot be assumed. For other fire areas, automatic operation of components and logic circuits may be credited in the analysis, but only if the circuitry associated with the automatic operation is known to be unaffected by the postulated fire (i.e., satisfy separation requirements of Section III.G.2 of Appendix R). If the automatic actuation of equipment will be lost as a result of fire in these areas, manual initiation of systems required to achieve and maintain safe-shutdown, via manipulation of controls located in the main control room, is acceptable if it can be demonstrated that reliance on such operator actions will provide an equivalent level of safety to that which would be achieved by performance of the automatic functions.

- *Plant Status:* The plant is operating at 100-percent power upon the occurrence of the fire.

- *Equipment Status:* Components are in their normal operating position or status at the time of the fire. All relay, position switch, an control switch contacts are in the position or status that corresponds to the normal operation of the device. Test and transfer switches in control circuits are in their normal position.

- *Use of Repair Activities:* Repair activities, (which are generally defined as any activity requiring the use of tools such as wiring changes, installation of electrical or pneumatic jumpers, and fuse replacements) are not permitted for systems required to achieve and maintain hot shutdown conditions. Modifications and repairs are permitted for cold shutdown systems as described below.

- *Multi-Unit Sites:* Where a single fire can impact more than one unit, the ability to achieve and maintain safe-shutdown for each affected unit must be demonstrated.

- *Passive Components:* The operation of passive components that are not electrically controlled or operated, such as manually actuated valves and check valves, is not assumed to be affected by fire damage.

- *Time Constraints and Limitations of Fire Damage*

Hot Shutdown Systems (All Areas)
When considering the consequences of fire in a given fire area, it must be conclusively demonstrated that one success path of equipment, that can be used immediately to bring the reactor to *hot shutdown* conditions, remains unaffected by fire.

Cold Shutdown Systems (Areas not Requiring an Alternative Shutdown Capability)
For areas of the plant not requiring an alternative or dedicated shutdown capability, it must be demonstrated that fire damage to one success path of equipment needed for achieving cold shutdown will be limited so that equipment can be returned to an operating condition within 72 hours.

Cold Shutdown (Areas Requiring an Alternative or Dedicated Shutdown Capability)
For areas requiring an alternative or dedicated shutdown capability, it must be demonstrated that cold shutdown capability can be restored *and cold shutdown conditions achieved* within 72 hours.

6.4.2   Define Required Safe-Shutdown (SSD) Functions

Required shutdown functions are those plant process and support functions that must be accomplished and controlled to ensure that the reactor is brought to and maintained in a safe-shutdown condition without exceeding the shutdown performance goals described above. (See Section 6.4.1.3.) Successful accomplishment of each of the following shutdown functions is necessary to preclude the occurrence of an unrecoverable plant condition (e.g., uncontrolled primary depressurization, loss of DHR capability or breach of the RCS boundaries):

- *Reactivity Control*
  This function is necessary to decrease the power output of the reactor core to the decay heat level. The reactivity control function must be capable of achieving and maintaining reactor shutdown from the initial scram shutdown to cold shutdown conditions. This function must be capable of compensating for any positive reactivity increases as a result of Xenon-135 decay, reactor coolant temperature decreases occurring during cooldown, and RCS dilution. The safe-shutdown performance and design requirements for the reactivity control function can be met without automatic scram/trip capability. The SSA must only provide the capability to manually scram/trip the reactor. For PWR the analysis must demonstrate that a method for ensuring that adequate shutdown margin is maintained. This is typically accomplished by ensuring an adequate concentration of borated water is utilized during RCS makeup/charging.

- *Reactor Coolant Makeup Control*
  The reactor coolant makeup control function must be capable of ensuring that sufficient makeup inventory is provided to compensate for reactor coolant system fluid shrinkage during cooldown and to replace any coolant that may leak from the system. Maintenance of adequate inventory prevents overheating of the reactor fuel, which could lead to core damage.

Systems performing this function must be capable of maintaining reactor coolant level above the top of the core for BWRs[25] and within the level indication of the pressurizer for PWRs.

- *Reactor Coolant Pressure Control*
  Pressure control is required to ensure that the RCS is operated within prescribed pressure-temperature limits, to prevent RCS peak pressure limitations from being exceeded and (for PWRs) to minimize void formation within the reactor vessel during natural circulation cooldown.

- *Decay Heat Removal*
  The DHR function must be capable of removing both decay and latent energy from the reactor core and primary systems at a rate such that overall system temperatures can be maintained within acceptable limits. This function shall also be capable of achieving cold shutdown conditions and maintaining cold shutdown thereafter.

- *Process Monitoring*
  To adequately change system alignments, control safe-shutdown equipment, and ensure the shutdown process remains within acceptable performance criteria, operators must be provided with sufficient instrumentation to monitor the status of process system variables. Direct readings of the variables used to control the shutdown process are required.

In GL 81-12, "Fire Protection Rule," and IN 84-09, "Lessons Learned form NRC Inspections of Fire Protection Safe Shutdown Systems" the NRC provides guidance regarding the minimum set of instrumentation deemed necessary for *alternative* or *dedicated* shutdown capabilities. The minimum process monitoring capability described in these documents includes the following instruments:

Instrumentation Needed for Alternative or Dedicated Shutdown of a BWR:
a. Reactor water level and pressure.
b. Suppression pool level and temperature.
c. Emergency or isolation condenser level.
d. Diagnostic instrumentation for shutdown systems. (See Note 1.)
e. Level indication for all tanks used.

Instrumentation Needed for Alternative or Dedicated Shutdown of a PWR:
a. Pressurizer pressure and level.
b. Reactor coolant hot leg temperature or core exit thermocouples, and cold leg temperature.
c. Steam generator pressure and level (wide range).
d. Source range neutron flux. (See Note 2.)
e. Diagnostic instrumentation for shutdown systems. (See Note 1.)
f. Level indication for all tanks used [e.g., condensate storage tank (CST)].

Note 1   Diagnostic instrumentation is instrumentation, beyond that identified above, that is needed to ensure the proper actuation and functioning of safe-shutdown equipment and associated support equipment (e.g., flow rate, pump discharge

---

[25]   Short-term core uncovery may be permissible when using low-pressure injection systems at BWRs (see Richards Letter to BWROG, December 2000).

pressure). The diagnostic instrumentation needed is plant-specific and should be based on the design of the alternative shutdown capability (GL 86-10). Sufficient instrumentation must be ensured to remain available (unaffected by fire) to allow operators to detect malfunctions that may occur, take appropriate corrective actions without resorting to potentially complex troubleshooting activities, and ensure activity was successfully accomplished.

Note 2   In a letter dated September 10, 1985, the NRC Committee for Review Generic Requirements (CRGR) instructed the Office of Nuclear Reactor Regulation (NRR) to eliminate the staff position for a source range neutron flux monitor as part of the Appendix R alternative shutdown instrumentation in PWRs.

Enclosure 1 of GL 86-10, "Implementation of Fire Protection Requirements" states that the instrumentation listed above provides an acceptable method for compliance with the *alternative shutdown* requirements of the regulation (i.e., Section III.L.2.d of Appendix R). This list, however, does not exclude other alternative methods of compliance. A licensee may propose to the staff alternative instrumentation to comply with the regulation (e.g., boron concentration indication). While such a submittal is not an exemption request, it must be justified based on a technical evaluation.

Instrumentation Needed for Redundant Shutdown Capabilities:

For redundant shutdown capabilities, where shutdown activities are controlled from within the main control room, one train of systems needed to achieve and maintain hot shutdown conditions must remain free of fire damage. (Section III.G of Appendix R) As a result, additional specific guidance, such as that discussed above for alternative shutdown capabilities is not necessary. For these areas, the determination of required process and diagnostic instrumentation should to be based on the plant-specific operating procedures (including normal, abnormal, and EOPs) that would be used to shutdown the reactor in the event of an unmitigated fire. Since the same shutdown functions are generally required to be performed for both alternative and redundant shutdown, this monitoring capability is expected to be fairly consistent with the instrumentation listed above.

Sufficient instrumentation must be ensured to remain available to implement the shutdown methodology described in the SSA and applicable procedures. For shutdown strategies that rely on operator actions as a means of mitigating equipment maloperations that may occur as a result of fire damage, sufficient diagnostic instrumentation must be available for operators to detect the maloperations and initiate appropriate responses in a timely manner, without resorting to complex and potentially hazardous troubleshooting activities.

When sufficient diagnostic instrumentation is not ensured to remain unaffected by fire, reliance on the operators ability to detect fire-induced maloperations that may occur and perform activities needed to defeat them before an unrecoverable condition is achieved cannot be ensured. For example, during a fire an operator observes the ensured method of monitoring pressurizer level to be decreasing. Since many possible maloperations of plant equipment are capable of causing this indication (e.g., spurious closure of a makeup flowpath valve, loss of a makeup pump, open bypass valve, open PORV or open head vent) without the benefit of additional diagnostic instrumentation, the operator's ability to determine the cause of this indication (pressurizer level decreasing) may be significantly compromised.

It should be noted that the use of operator actions as an immediate response to a confirmed fire has been shown to reduce the need for diagnostic instrumentation. An example of this approach would be a shutdown procedure that, immediately upon confirmation of fire, directs operators to close the MSIVs in the control room as a means of preventing their undesired operation (failure to close) as a result of potential fire damage to their control circuits.

For this case, an immediate action is taken to prevent a possible undesired outcome. Since no reliance is placed on the operators ability to detect a possible failure, the need for some diagnostic instrumentation may be eliminated.

- *Supporting Functions*
  To ensure the successful accomplishment of the above shutdown functions, several support systems and equipment are necessary. The supporting functions shall be capable of providing the process cooling, lubrication, etc., necessary to permit the operation of equipment used to accomplish the above shutdown functions. The specific support systems needed will vary with the shutdown methodology developed by the plant. Typical examples include electrical distribution systems, HVAC and essential room cooling, component cooling water, essential service water, and communications capability (e.g., portable radios, sound powered phones).

6.4.3 Select Shutdown Systems

The next step in the process is to identify a system or combination of systems capable of accomplishing each of the required shutdown functions described above (Section 6.4.2). This may be accomplished by a review the design documentation, such as system descriptions, system drawings, and plant procedures, described in Section 6.4.1.1. Once identified, these systems can be combined into safe-shutdown success paths and given a unique designation (e.g., SSD Path 1, SSD Path 2, etc.). A description of each path should then be documented. For example, shutdown paths for a BWR may be described as follows:

Path 1   Control Rod Drive System; Division I train of ADS, Division I CS in Alternative Shutdown Cooling and Division I of RHR in Suppression Pool Cooling Mode

Path 2   Control Rod Drive System; Division II train of ADS, Division II CS in Alternative Shutdown Cooling and Division II RHR in Suppression Pool Cooling Mode

In addition, systems necessary to support the operation of the above "front line" systems should also be identified as safe-shutdown systems (e.g., electrical distribution systems, instrumentation, cooling water systems and HVAC). A summary of the post-fire safe-shutdown analysis process to this point is illustrated in Figure 6-4a.

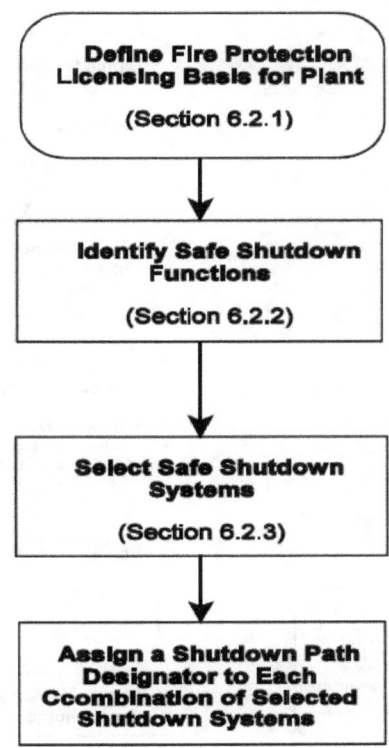

Figure 6.4a   Safe Shutdown System Selection and Path Development

6.4.4   Select and Locate Required Shutdown Equipment

The systems identified above form the basis for the selection of safe-shutdown components. The next step in the process is to identify the specific equipment necessary for the identified systems to perform their shutdown function. This process is illustrated in Figure 6-4b.

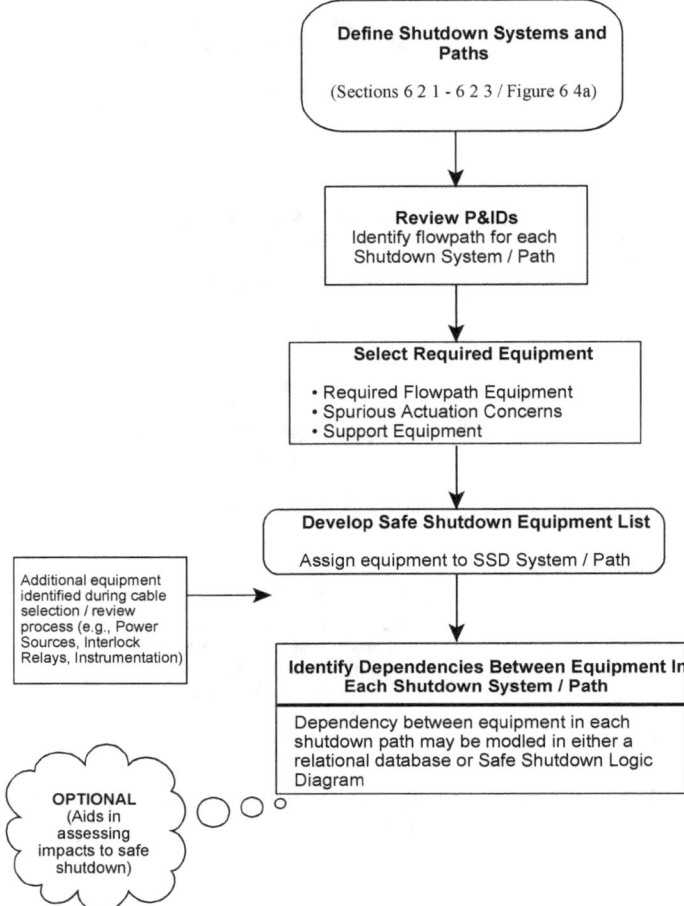

Figure 6.4b Safe Shutdown Equipment Selection

Using piping and instrumentation drawings (P&IDs) for the systems comprising each safe-shutdown path, the mechanical equipment required for the operation of each system may be identified. The selected equipment should be related back to the safe-shutdown systems it supports and be assigned to the same safe-shutdown path as that system. Equipment that could spuriously operate and impact the safe-shutdown capability may also be identified during the review of the P&IDs. This equipment should be related to the particular safe-shutdown path that it can affect. Equipment that can result in a loss of reactor inventory in excess of the available make up capability (i.e., initiate a fire-induced LOCA) should also be identified by a review of P&IDs for systems physically connected to the reactor vessel. The following criteria and assumptions are applicable to the selection of safe-shutdown equipment:

- Exposure fire damage to manual valves and piping is not assumed to adversely impact their ability to perform their safe-shutdown function.

- Manual valves are assumed to be in their normal position as shown on P&IDs.

- A check valve that closes in the direction of potential flow diversion is assumed to seat properly with sufficient leak tightness to prevent flow diversion.

- The effects of fire on instrument tubing must be considered. Heat generated by the fire may cause subsequent effects on instrument readings and/or signals. The fire area location of the instrument tubing should be determined and the effects of fire damage to it should be considered when evaluating the effects of a postulated fire in the area. In addition, the effects of fire on heat sensitive components such as copper sweated fittings should also be considered.

As a result of this review process, a list of "safe-shutdown components" also called "required components" will be generated for each system. This list should include: (1) components that are required to operate in order to ensure the proper operation of systems credited in the analysis (e.g., SSA) for achieving and maintaining post-fire safe-shutdown conditions; *and* (2) components of which inadvertent actuation or maloperation could significantly degrade the capability of these credited systems to perform their intended shutdown function. The following examples represent the typical required components:

(1) Components that must start and/or continue to operate on demand such as required pumps, fans, air compressors and motors.

(2) Electrically actuated or controlled components that must change operating status or position, such as a normally closed valves located in a required flowpath.

(3) Electrically actuated or controlled components that must *not* change position or operating mode. Examples include a normally closed valves that constitute a system boundary or diversion flowpath and normally open valves located in a required flowpath.

(4) Components needed to ensure the proper operation of shutdown equipment and systems. Examples include: Power supplies (EDGs, battery banks, inverters, battery chargers, SWGRs, MCCs, load centers, and distribution panels) room coolers and air bottles.

(5) Components that can cause equipment and systems to automatically actuate and/or change operating state in an undesired manner for safe-shutdown. Examples include interlock circuits, pressure switches, temperature switches, solid-state control systems, and various instrumentation devices.

The resulting list of equipment SSEL establishes the basis for identifying *required circuits and cables* (i.e., circuits and cables needed to support operation of the identified shutdown paths) *as well as associated nonsafety circuits* for which damage attributable to fire could impact (adversely affect) the achievement of safe-shutdown conditions.

Illustration of Equipment Selection Process

The following are *guidelines* regarding which components to include on the SSEL for each system evaluated. (Refer to Figure 6-5.)

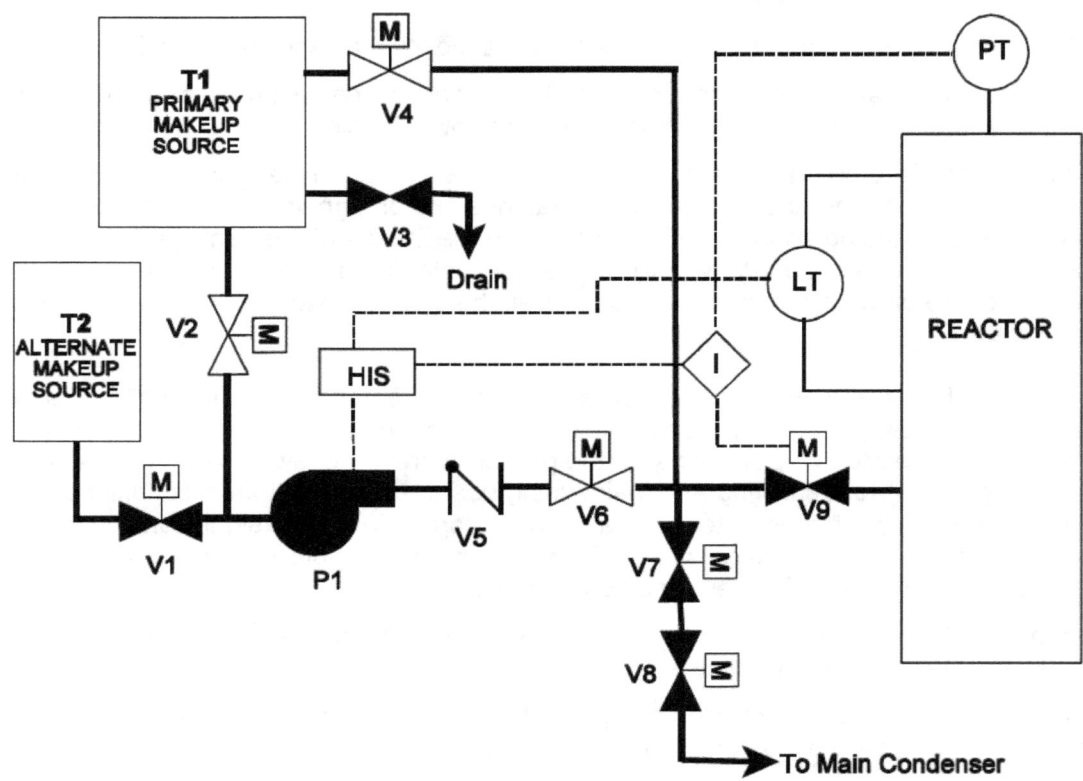

Figure 6-5 Example System

Components of interest are those needed to ensure the successful accomplishment of a required shutdown functions. This includes components that are needed to ensure the proper functioning of required shutdown systems (e.g., pumps and valves located in a required flowpath) as well as components of which maloperation attributable to fire could impact the shutdown capability (e.g., pressurizer PORVs, ADS valves, instrumentation). The following examples represent the typical components to be included in the SSEL:

- *Valves and HVAC dampers that constitute system boundaries* should be included if fire-induced faults could cause them to change position *and* their maloperation (e.g., inadvertent/spurious opening) would significantly impact the capability of the system to perform its intended shutdown function (e.g., by creating a flow leakage/diversion path that cannot be adequately compensated for by the system). Valves V7 and V8 in Figure 6-5 fall into this category.

- *Valves and dampers (e.g., HVAC dampers) in the flow path* that are power operated should be included. Associated valve operators should also be included as part of the valve/damper. These components should be included whether or not they are required to change position during shutdown if a fire-induced fault could cause them to change position. These components ensure that the process flow path is maintained. [Valves V1, V2, P1, V4 and V9 in Figure 6-5) fall into this category.]

- *For tanks*, all inlets and outlet lines should be evaluated for their functional requirements and isolation. For lines that are not required to be functional, a means of isolation should be included when necessary to prevent unnecessary drawdown of the tank. Tank inventory must be evaluated to ensure that it is always sufficient to support the system requirements. Tanks T1 and T2 in Figure 6-5 are an example of this category.

- *Interlock circuitry* between safe-shutdown components and safe-shutdown/non-safe shutdown components should be reviewed to determine if additional components require inclusion. This is to ensure that a failure of a non-safe-shutdown component would not prevent the safe-shutdown system from operating as required. (In Figure 6-5, the interlocks between the reactor level and the pump and the reactor pressure and valve V9 and the pump are examples of this category.)

- *All necessary process and diagnostic instrumentation* (e.g., process flow, pressure, temperature, level, indicators and recorders)

- *Power supplies* or other electrical components that support operation of required shutdown components should be included (SWGR, EDGs, MCCs, load centers, inverters, batteries, relays, control switches, flow switches, pressure switches, level switches, transmitters, controllers, transducers, and signal conditioners).

6.4.5 Identify Required Circuits and Cables

As discussed above, to achieve safe-shutdown conditions certain *shutdown functions* (e.g., reactivity control, DHR, reactor coolant inventory and pressure control, etc.) must be accomplished and controlled to ensure that the reactor is brought to and maintained in a safe-shutdown condition within design parameters established in the applicable licensing basis documents. The systems identified as being needed to accomplish these functions are classified as *"required shutdown systems."* Similarly, the equipment that must operate or be prevented from mal-operating in order for the required shutdown systems to accomplish their intended shutdown functions, are considered *"required equipment" or "required components."* Once identified, required components are listed on the SSEL. The SSEL establishes a starting point for identifying *required circuits and cables* (i.e., circuits and cables needed to support operation of the identified/credited shutdown paths) **as well as** the *associated nonsafety circuits* for which damage attributable to fire could impact (adversely affect) the achievement of safe-shutdown conditions by initiating an event that exceeds the design capability of the credited/required shutdown systems.

After the listing of required components is developed SSEL for each shutdown path, the circuits/cables needed to support the operation of this equipment are identified and evaluated. For each required component, all circuits (cables) that: (a) are needed to ensure proper operation, or (b) could cause maloperation/undesired actuation must be identified. A circuit/cable is considered to be *required for safe-shutdown* if it is connected to or associated with the operation of a required shutdown component *and* fire damage to the circuit/cable can cause the component to fail in an undesired manner for post-fire safe-shutdown. In addition to the set of circuits/cables needed to ensure the acceptable operation of required shutdown components, *associated circuits of concern to post-fire safe-shutdown* must also be identified and analyzed. As discussed below, these circuits have one of the following:

(1) A *common power source* with the shutdown equipment and the power source is not electrically protected from the circuit of concern by coordinated breakers, fuses, or similar devices

(2) A *common enclosure* (e.g., raceway, panel, junction box) with shutdown cables and a) are not electrically protected by suitably sized circuit breakers, fuses or similar devices, or b) will allow fire to propagate into the common enclosure

(3) *A connection to equipment of which spurious operation or maloperation may adversely affect the shutdown capability* (Note: As discussed in Section 6.4.4 above, the identification of "spurious operation components" is typically performed as part of the review of P&IDs to identify required shutdown equipment)

The following paragraphs provide criteria and guidance for selecting safe-shutdown cables and determining their potential impact to equipment required for achieving and maintaining safe shutdown for the condition of an exposure fire. The objective of the cable selection criteria is to ensure that circuits and cables of required shutdown equipment are identified and that these cables are properly related to equipment with functionality they could affect. Through this cable-to-equipment relationship, cables become associated with the same safe-shutdown path as the equipment affected by the cable.

6.4.5.1 Cable Identification

- *Scope:* The list of cables of which failure could impact the operation of a piece of safe shutdown equipment includes more than those cables that are directly connected to the equipment. The relationship between cable and affected equipment should be based on a review of electrical or elementary wiring diagrams. In addition to the cables that are physically connected to the equipment, the list of required cables will include any cables interlocked to the primary electrical schematic through secondary schematics. To ensure that all cables that could affect the operation of the safe shutdown equipment are identified, the power, control, instrumentation, interlock, and equipment status indications should be investigated. Schematic diagrams should be reviewed to identify additional circuits and cables for interlocked circuits that also need to be considered for their impact on the ability of the equipment to operate as required in support of post-fire safe-shutdown.

- *Cable/Component Associations:* Each cable should be related back to the same shutdown path as the equipment it supports. In cases where the failure of a single cable could impact more than one piece of shutdown equipment, the cable should be associated with each piece of shutdown equipment.

- *Isolation Devices:* Electrical devices such as relays, switches, and signal resistor units (SRUs) are considered to be acceptable isolation devices. In the case of instrument loops, the isolation capabilities of the devices in the loop should be evaluated to determine that an acceptable isolation device has been installed at each point where the loop must be isolated so that a fault would not impact the performance of the instrument function.

- *Screening:* Circuits that do not impact the desired safe-shutdown performance or expected operation of a component, such as those illustrated in Figure 6-2 above, may be screened from further evaluation unless some reliance on these circuits is necessary. However, these circuits must be ensured to be isolated from the component's control scheme in such a way that a cable fault would not impact the performance of the circuit.

- *Power Cables:* Electrical distribution system (EDS) equipment needed to provide power to shutdown equipment may be identified from a review of the electrical schematics associated with the shutdown equipment. For each component requiring electric power to perform its safe-shutdown function, the cable that supplies power to the component should be identified. Initially, only the power cables from the immediate upstream power source are identified for these interlocked circuits and components. A further review of the electrical distribution system is needed to capture the remaining equipment from the electrical power distribution system necessary to support delivery of power from either the offsite power source or the EDGs to the safe-shutdown equipment. This equipment should then be added to the SSEL. This information will be needed to support the *Associated Circuits — Common Power Source Analysis* described in Section 6.4.5.2.

- *Automatic Initiation Logic:* The automatic initiation logic for the credited post-fire safe-shutdown systems is not required to support safe-shutdown; each system can be controlled manually by operator actuation. However, if not protected from the effects of fire, the fire-induced failure of automatic initiation logic circuits must be verified to not adversely affect any post-fire safe-shutdown system function. Otherwise it would need to be included in the SSEL.

6.4.5.2   Identification of Associated Circuits

The overall objective of the SSA, is to demonstrate that in the event of an exposure fire in any single area of the plant, SSCs important to safe-shutdown will remain available to accomplish required shutdown functions (e.g., reactivity control, reactor coolant makeup, and pressure control, DHR) as needed. Because circuits and cables of the required shutdown systems frequently share certain physical or electrical configurations with cables of nonessential systems and equipment (i.e., not required for post-fire safe-shutdown) it is not sufficient to only consider the effects of fire damage to cables of required components. For example, consider the cable configuration shown in Figure 6-6. In this case, the cable that supplies power to a nonessential load is powered from the same power supply as equipment relied on for safe-shutdown. While a fire that causes a loss of the nonessential load may not directly impact the shutdown capability, a fire that damages the power cable of the nonessential load could significantly impact the shutdown capability if damage to this cable resulted in a loss of the required (Train B) power supply. Because fire damage to certain nonessential equipment and cables may adversely affect the operability of required shutdown systems, in performing the SSA the analyst must consider the effect of fire on both the primary, or "front-line" shutdown equipment and any nonessential equipment and cabling that may affect the ability of required shutdown systems to accomplish their intended shutdown function if they are damaged by fire. That is, the scope of the evaluation must extend beyond the limited set of equipment that comprises the defined shutdown paths. A suitably comprehensive evaluation will address the potential impact of fire damage to any circuit/cable located within the fire area that could adversely affect the post-fire safe-shutdown capability.

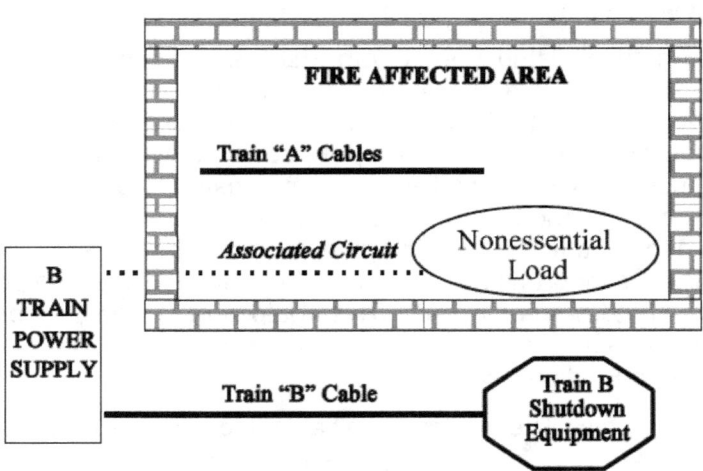

Figure 6-6 Associated Circuit

6-22

6.4.5.2.1 Associated Circuit Configurations of Concern to Post-Fire Safe Shutdown

Section III.G.2 of Appendix R to 10 CFR Part 50 requires that separation features be provided for equipment and cables, including associated nonsafety circuits that could prevent the operation or cause the maloperation (attributable to hot shorts, open circuits, or shorts to ground) of redundant trains of systems necessary to achieve and maintain hot shutdown conditions. An *associated circuit of concern* to post-fire safe-shutdown may include any circuit or cable that, while not needed to support the proper operation of required shutdown equipment (i.e., a nonessential/nonsafety circuit), could adversely affect the plant's ability to achieve and maintain safe-shutdown conditions. Associated circuits of concern may be found to be associated with circuits of required systems through any of the following configurations:

- Circuits that share a **common power source** (e.g., SWGR, MCCs, fuse panel) with circuits of equipment required to achieve safe-shutdown

- Circuits that share a **common enclosure**, (e.g., raceway, conduit, junction box, etc.) with cables of equipment required to achieve safe-shutdown

- Circuits of equipment of which **spurious operation** or maloperation may adversely affect the shutdown capability

Methods for identifying each type of associated circuit defined above are discussed in the following sections.

Circuits Associated by Common Power Source

The electrical distribution system is one of the most important support systems of any installation. Electrical power supplies (e.g., SWGRs, MCCs, fuse and circuit breaker panels) required to power shutdown equipment in the event of fire are identified during the selection of required shutdown equipment (Section 6.4.4). Once identified, the analyst must then ensure that in the event of fire, the required power supplies will remain available, as needed to ensure the continuity of service to essential shutdown loads. In the event of an electrical fault condition, a properly engineered system will allow only the protective device nearest the fault to open while not disturbing the remainder of the system.

In many cases, relatively few of the components that are normally powered from a specific power supply are needed to accomplish required shutdown functions. While providing power to the remaining "nonessential" loads (equipment) may not be necessary to accomplish safe-shutdown, it must be ensured that fire initiated faults on the power cables to this equipment will not affect the shutdown capability by causing a trip of the protective devices (e.g., circuit breaker, fuse, or relay) located upstream of the required supply. To address this concern, the SSA must be extended to consider the effects of fire-induced faults on all circuits of required power supplies identified in Section 6.4.4. To ensure that fire-induced faults on these circuits will not affect the capability of achieving safe-shutdown conditions, this analysis must ensure that circuits which share a common power source with circuits of required equipment are provided with: (a) fire protection features sufficient to satisfy Section III.G.2, or (b) suitably coordinated electrical protective devices.

The common power source associated circuit concern is illustrated in Figure 6-7a. In this case, in the event of fire in Fire Area II, Train A safe-shutdown equipment (Pump A) located in Fire

Area I and powered by safe-shutdown Bus A, is relied on to accomplish safe-shutdown. Although the Train A pump is located in a separate fire area, it may be vulnerable to loss as a result of fire in Fire Area II. This is because, as shown in Figure 6-7a, a Train A *associated circuit power cable* is also located in Fire Area II. Although this cable and its connected load (Pump X) are not needed to perform a shutdown function, the absence of suitable *coordination of electrical protective devices*, a fire-induced electrical fault on the this cable could cause the upstream feeder breaker of Bus A (Breaker 1) to trip before the individual branch breaker (Breaker 2). Because this would result in a loss of electrical power to all shutdown equipment powered from safe-shutdown Bus A, failures of this type are unacceptable.

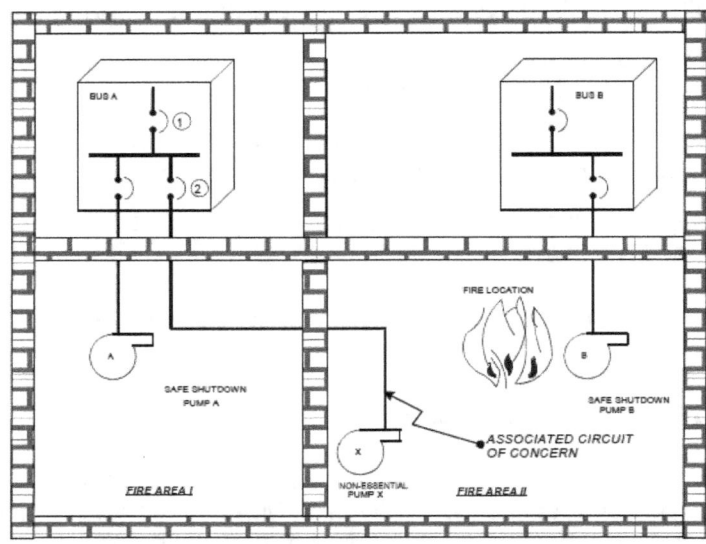

Figure 6-7a Common Power Source Associated Circuit

The common power source associated circuit concern consists of two items:

(1) coordination of electrical protective devices (circuit breakers, relays, fuses, etc.)
(2) multiple high impedance faults (MHIFs)

Coordination of Electrical Protective Devices

To minimize the effect of an electrical fault on system operation, the tripping characteristics of electrical protective devices (fuses, circuit breakers and relays) should be sufficiently coordinated so that electrical faults will be rapidly isolated by the protective device located nearest the fault. Although the term "coordination" is often used, "selectivity" or "selective tripping" more precisely describes post-fire safe-shutdown concerns. Selectivity means positive coordination over the entire range of possible fault currents, ensuring that the faulted circuit is cleared and that other parts of the system are not affected. Examples of both a non-selective system and a system that is provided with fully selective protective devices are illustrated in Figures 6-7b and 6-7c. In the non-selective system shown in Figure 6-7b, a branch circuit fault would cause fuses D, C and B to open, resulting in a loss of power to all loads supplied from the system. In the fully selective system shown in Figure 6-7c, the fault is isolated by fuse D and the remainder of the system remains undisturbed.

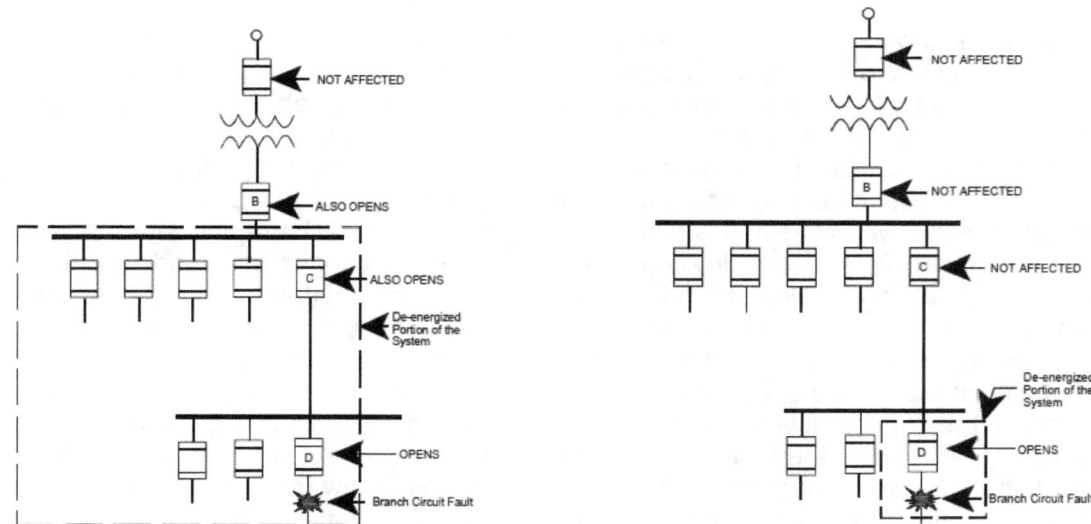

Figure 6.7b - Non-Selective Coordination   **Figure 6.7c - Selective Coordination**

Depending on their design and/or individual trip settings, fault protective devices of the same type (e.g., fuse or circuit breaker) and rating (20 amp, 30 amp etc.) may have significantly different tripping characteristics. A coordination study consists of the selection or setting of all series protective devices from the load upstream to the power supply. In selecting or setting these protective devices, a comparison is made of the operating times of all the devices in response to various levels of overcurrent. The objective, of course, is to design a selectively coordinated electrical power system. The operating response of a specific protective device is graphically represented by time-current characteristic curves. Time-current characteristic curves are presented on a log-log graph where the ordinate (y-axis) represents a time range from 0.01 to 1,000 seconds, and the abscissa (x-axis) represents the current level. By overlaying the time-current curves of two protective devices or comparing them in some other manner, their selectivity may be quickly determined. If the curves of the two devices intersect, for example, the intersection area indicates conditions under which both devices may trip. If such a pair of circuit breakers were used in an electrical distribution system, those conditions could result in both devices tripping. On the other hand, if the curves of the circuit breakers are distinctly separate and do not intersect, the circuit breakers are said to be coordinated.

A new or revised coordination study should be made when the available short-circuit current from the power supply is increased; when new large loads are added or existing equipment is replaced with larger equipment; or when protective devices are upgraded.

Multiple High-Impedance Faults

In the previous paragraphs, the need for circuit protective device "selectivity" was discussed. The evaluation of selectivity typically considers "worst-case" fault conditions initiated by "bolted faults." A "bolted fault" develops when the conductor of a faulted cable is in firm contact with a conductor that is at a different potential, such as a cable tray (phase to ground fault). Since this fault condition offers little, if any, impedance (resistance) to the flow of fault current, it will result in a maximum value of fault current being drawn from the affected power source. In a properly coordinated (selective) system, this high value of fault current will be rapidly interrupted and cleared by the circuit protective device closest to the fault. Under certain conditions, however, insulation degradation resulting from fire damage may cause a different kind of fault condition known as an "HIF." In almost every case, this type of fault occurs between one phase and ground. However, instead of establishing direct contact to ground potential (as for a "bolted fault" condition) the faulted conductor is not mechanically firm or is erratic. As a result an arc develops in the air gap between the faulted conductor and ground. This arc introduces an element of resistance to the flow of fault current that is not present in a bolted fault. As a result, the magnitude of high-impedance fault currents are relatively low (in comparison to a bolted fault) and in many cases, the arcing fault will be of such a low value that it is less than the continuous current rating of the overcurrent protective for the circuit involved.

In the majority of cases, an arcing fault starts as a small breakdown in insulation. Ionization of the atmosphere and destruction of insulation cause the fault to develop into a self-sustaining arcing fault. In a 480-V system, tests and calculations have indicated that this sustained current can be as low as 20-percent of the available bolted three-phase current[26]. Although the individual faults are not of sufficient magnitude to cause a trip of the individual load breakers, a coordination problem could exist if the cumulative effect of these faults were to cause the upstream feeder breaker of a required power source to trip. To fully demonstrate that a required power source will not be impacted by fire damage to its connected cabling, the potential impact of HIFs should be considered. This evaluation involves determining the effect of such faults on all cables of a required power supply that may exposed to fire damage. (See Figure 6-8.)

For the purpose of performing this analysis, the following assumptions are applicable:

• The HIF current of each cable that may be exposed to fire damage is postulated to be a value that is just below the trip point setting of the individual protective device for the load.

• All unprotected load cables of the power supply being evaluated, that are located within the zone of influence of the fire (e.g., located in the same fire area/zone), are assumed to simultaneously fault to the HIF condition.

• The total load current to be considered is the sum of all high-impedance faults *plus* the normal operating load current on the bus.

---

[26] "Good Design Prevents High-Impedance Fault," *Actual Specifying Engineer*, Vol. 17, No.4, Medalist Publications, Inc., Chicago, IL, 1967.

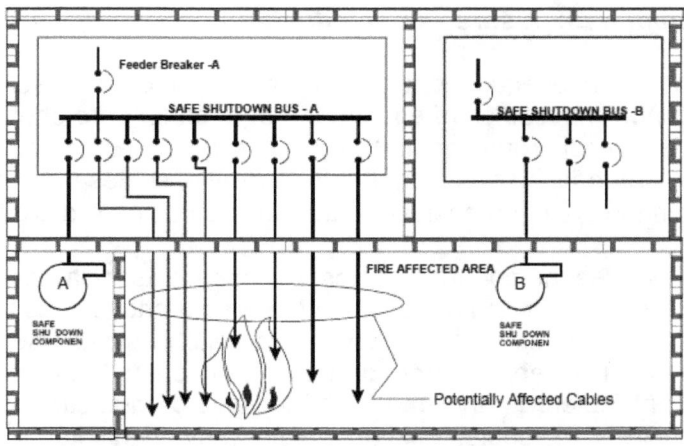

Figure 6-8 Illustration of Multiple HIF Concern

Circuits Associated by Common Enclosure

Cables that are not needed to perform a shutdown function (nonessential cables) frequently share a common enclosure (e.g., cable tray, conduit, junction box, panel, etc) with cables of shutdown equipment. Since the routing of these nonessential cables is generally unknown, they may be damaged by fire in any area of the plant. In the absence of suitably sized electrical protection devices (fuse or circuit breaker) and/or fire protection features, damage to these cables could also damage the required cables located within the common enclosure. In addition, if fire were to spread along these cables into an adjacent fire area due to inadequate cable penetration seals, the safe shutdown equipment or cables located in the adjacent fire area could also be impacted. This condition would exceed the criteria and assumptions of this methodology (i.e., multiple fires and fire spread beyond area under consideration).

Circuits that share enclosures with safe-shutdown circuits must be analyzed to determine the potential effect that fire damage to these circuits (cables) may have on the safe-shutdown capability. This concern consists of two issues:

(1)   *Cable ignition*: Fire-initiated electrical faults on inadequately protected cables could cause an over current condition, resulting in secondary ignition.

(2)   *Fire propagation*: The effects of the fire may extend outside of the immediate area or into the common enclosure by means of fire propagation.

As described in the following paragraphs, either of these cases could result in damage that could disable redundant trains of required shutdown equipment.

Case 1: Common Enclosure — Cable Ignition

Cables of nonessential equipment may share a common enclosure (e.g., raceway, conduit, or panel) with cables of equipment required for safe-shutdown. In the absence of adequate electrical protection (i.e., properly sized fuses and circuit breakers), heat generated by fire-induced faults on the nonessential cables may cause a secondary fire to occur within the common enclosure, thereby damaging required cables.

Figure 6-9 illustrates the common enclosure concern associated with "cable ignition." As shown in this diagram, a fire occurs in fire area II and causes a fault on an associated circuit cable that is not properly protected by a suitably sized fuse. As a result of this condition, the fault current will propagate along the entire length of the affected cable, into an adjacent fire area (Fire Area I). If the value of fault current exceeds the current carrying capacity of the cable, a secondary fire may be initiated in Fire Area I, resulting in the loss of redundant trains of shutdown equipment (Instruments A and B).

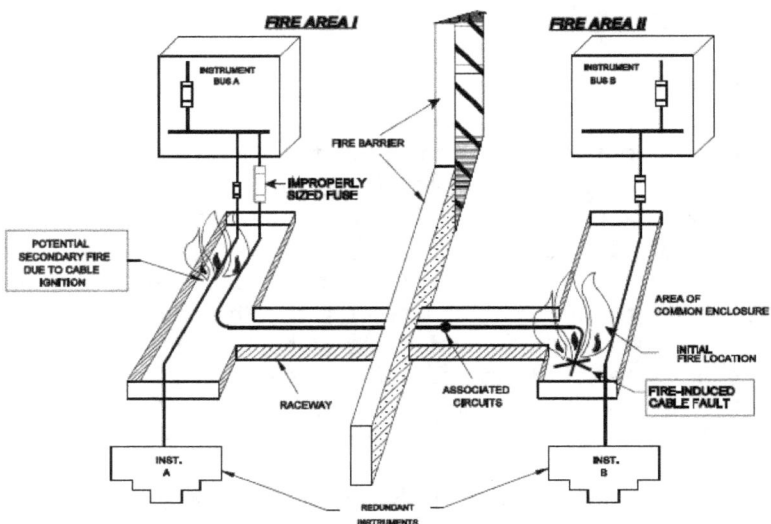

Figure 6-9 Common Enclosure - Case 1: Cable Ignition

Case 2 Common Enclosure — Fire Propagation

Cables of equipment that is not needed for safe-shutdown may traverse fire areas containing redundant trains of shutdown equipment. When fire protection features, such as fire stops and penetration seals are not provided, there is a potential for a cable to serve as a pathway for fire to propagate (travel) into adjacent fire areas. This concern is illustrated in Figure 6-10. In the example shown, the initial fire in Fire Area II will render instrument "B" inoperable. Since the cable tray is not provided with suitable protection features (e.g., penetration seals or fire stops), a fire that affects Instrument "B"cables could also propagate along the associated circuit cables and impact the redundant Instrument "A"cables located in the adjacent fire area.

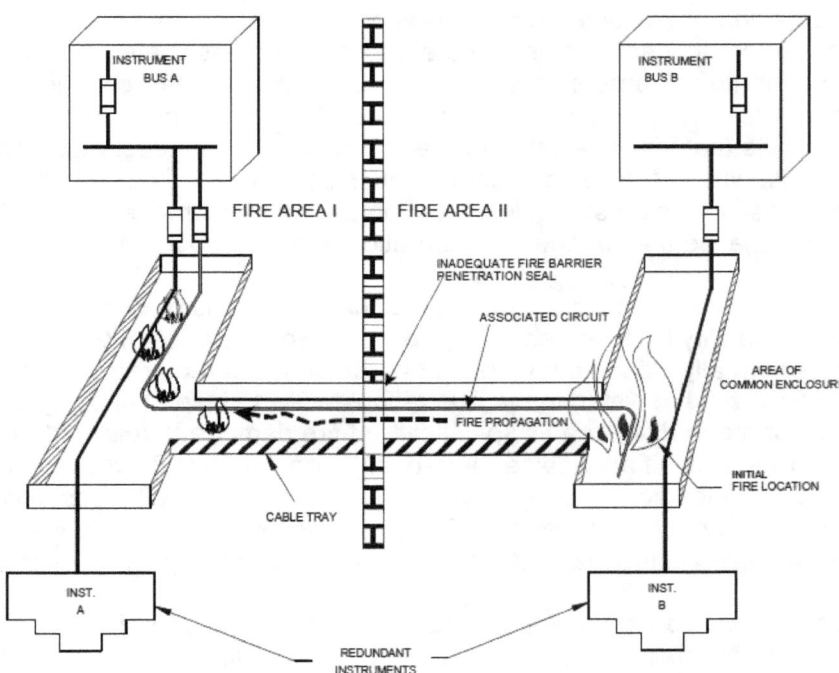

Figure 6-10 Common Enclosure Associated Circuit Case 2: Fire Propagation

Spurious Actuations and Signals

Cable damage attributable to fire or its related perils (e.g., firefighting and fire-suppression activities) can cause connected equipment to operate in an undesirable and/or unexpected manner. For example, a fire-induced short circuit on control wiring of a normally open MOV, could cause the valve to inadvertently close, thereby blocking a required flow path. Conversely, the spurious opening of a normally closed valve could divert flow from a required flow path. Additional examples include false instrument indications, the spurious starting or stopping of electrically powered equipment, such as pumps and motors, and the initiation of false control and interlock signals.

The achievement of safe-shutdown is dependent on the active control of some components and preventing the maloperation of other components. The circuits of both categories of components have the potential for being associated circuits of concern by spurious operation. Components which must actively operate (change position or operating status) at some point in the safe-shutdown sequence must be analyzed to identify circuits (cables) which if damaged could prevent the desired component operation; likewise, passive components, such as a normally closed MOV that is required to remain closed for safe shutdown, must be analyzed to ensure that fire-induced cable faults cannot cause the spurious maloperation of the component.

An example of how fire-initiated spurious actuations of equipment may impact the shutdown capability is illustrated in Figure 6-11. For this case, MOV-1, located in Fire Area IV, is normally closed during plant operation and is required to remain closed for safe-shutdown. As depicted in the illustration, MOV-1 could spuriously actuate (open) as a result of fire in Fire Area I. Specifically, if fire damage to relay "R" control circuits in this area were to initiate a false "auto-open" signal, relay "R" would actuate, closing contact RC1. Since actuation of contact RC1 has the same effect as closing the "open" contact of the MOV control switch (CS-O), motor-contactor solenoid 42-O would energize, resulting in the inadvertent actuation (undesired opening) of MOV-1.

Circuits that could cause undesirable spurious equipment operations must be identified and evaluated for their effect on safe-shutdown capability. The specific method used to prevent or control spurious equipment operations must be consistent with the potential severity of the spurious actuation. For example, since their inadvertent operation may place the plant in a potentially unrecoverable condition [loss of coolant accident (LOCA)], the spurious opening of valves which form a high/low pressure interface boundary would have a high consequence on the shutdown capability. As discussed in Section 6.3, given the severe consequences associated with this event, high/low-pressure interface boundaries are subject to more stringent analysis criteria. For example, the analysis must consider multiple, simultaneous, hot shorts of the required polarity and sequence as a credible event.

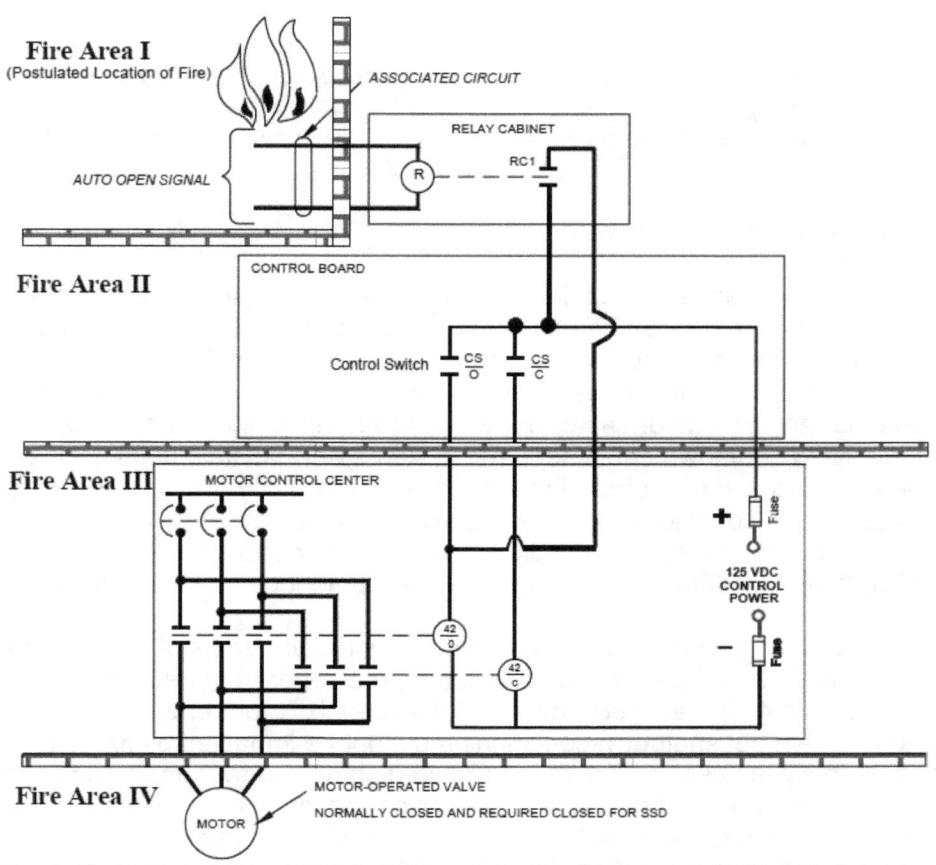

Figure 6-11 Example of the Spurious Actuation Associated Circuit Concern

While the spurious actuation of components having a high consequence on the ability to achieve safe-shutdown conditions must be precluded, other spurious equipment operations may not require this level of protection, provided it can be demonstrated that their inadvertent or "spurious" actuation would not impact on the safe-shutdown capability of the plant. A specific example of this case is a spurious actuation which causes the loss of ventilation in an area containing safe-shutdown equipment. If it can be demonstrated that the required equipment will remain operable (i.e., capable of performing its intended function) for a sufficient length of time without ventilation, plant modifications necessary to preclude the spurious operation may not be necessary.

As described in Section 6.4.4, potential spurious components of concern may be identified from a review of system design documents (e.g., flow diagrams, electrical schematics, etc.). During this review, components of which inadvertent operation could prevent the system from performing its intended shutdown function are identified and included in the SSEL. This list should include components of nonessential systems of which spurious operation could affect the shutdown capability. Once identified, appropriate methods of control can be planned. However, it is imperative that the safe-shutdown analysis include a thorough evaluation of all plant systems so that potential spurious equipment operations of concern can be properly identified for each fire area.

6-31

6.4.6   Circuit Analysis

"The need to evaluate the effects of fire on circuits associated with the safe-shutdown systems was not explicitly stated in Appendix A to BTP APCSB 9.5-1. It is explicitly required in Appendix R." (Reference: SECY-80-438A, "Commission Approval of the Final Rule on Fire Protection Program," September 30, 1980.)

6.4.6.1   Background/Objective

The evaluation of the consequences of fire in a given fire area must conclusively demonstrate that one train of equipment that can be used immediately to bring the reactor to hot shutdown conditions remains unaffected by fire. The systems and equipment that will be depended upon to perform essential shutdown functions must be identified in the FHA and/or the SSA for the plant. It follows that any circuits or cables in the fire area that could (1) adversely affect the operability of identified shutdown equipment and systems or (2) initiate plant transients that could preclude the successful accomplishment of required shutdown functions, by feeding back potentially disabling fault conditions to power supplies, control logic or instrumentation circuits, must be evaluated and such disabling conditions prevented or appropriately mitigated. Otherwise, reliance on the identified safe-shutdown equipment cannot be ensured.

In addition to establishing protection requirements for redundant trains of systems necessary to achieve and maintain hot-shutdown conditions (i.e., the set of "required" shutdown equipment identified in Section 6.4.4 above), Section III.G.2 of Appendix R further specifies that the ability to achieve and maintain hot shutdown conditions must not be impacted by fire which damages nonsafety circuits that are associated with the required shutdown systems. Additionally, with regard to alternative or dedicated shutdown capabilities, Sections III.L.3 and III.L.7 of Appendix R require the shutdown capability to be independent (physically and electrically) of the specific fire area(s) under consideration and isolated from associated nonsafety circuits such that a postulated fire involving associated circuits will not prevent safe-shutdown.

Associated circuits of concern are defined as cables (circuits) that may affect the safe-shutdown capability and/or prevent the achievement of post-fire safe-shutdown conditions if they are damaged by fire. Associated circuits may be safety-related or nonsafety-related. These circuits are a concern as long as their failure could impact the defined method of achieving and maintaining post-fire safe-shutdown conditions (i.e., the method credited in the plant's SSA). Specific associated circuit configurations of concern to post-fire safe-shutdown include circuits that share a common enclosure or power source with shutdown circuits and circuits that could cause equipment to spuriously actuate in an undesired manner for safe-shutdown. Each of these configurations is described above in Section 6.4.5.

6.4.6.2   Circuit Analysis Criteria and Assumptions

The fire protection design options delineated in Section III.G.2 of Appendix R provide assurance that cables and equipment located in a specific fire area under consideration will remain free of fire damage. It is not deemed possible to accurately predict the manner in which cables or circuits which lack such protection may fail when subjected to fire and its related perils (e.g., fire-suppression system actuation and physical insults resulting from fire-damaged equipment and firefighting activities). Therefore, analytical approaches used to demonstrate an equivalent level of fire safety to that which would be achieved through compliance with the regulation, are expected to assume that the exposed cables (circuits) will be damaged and then evaluate the possible consequences of this damage on the ability to achieve and maintain safe-shutdown conditions. Such an evaluation would require consideration of one or more (i.e., combination) of the following failure modes:

(1) Open circuits resulting in a loss of electrical continuity (see Section 6.4.6)

(2) Short circuits between individual conductors of a multi-conductor cable (see Section 6.4.6).

(3) Short circuits between conductors of different cables (see Section 6.4.6)

(4) "Hot shorts" where un-energized circuits are inadvertently energized by fire damage which causes conductors of different potential to establish electrical contact (short). A "hot-short" may be compared to the actuation of a light switch. Prior to actuation of the switch, the light is off because its conductors are not energized. Following actuation of the switch (or in our case, development of a hot short) a pathway for current flow is completed between the energized conductors and the formally de-energized conductors and the light illuminates (see Section 6.4.6)

(5) Short circuits between conductors of logic circuits located in equipment and cabinets that are exposed to fire damage (e.g., MCCs, control boards, instrument panels).

(6) Direct or "bolted" low-impedance short circuits of energized conductors to grounded reference potentials. (see Section 6.4.5)

(7) Arcing (high impedance) short circuits of energized conductors. (see Section 6.4.5)

Criteria/Assumptions

For the purpose of performing an evaluation of fire-induced circuit failures, the following criteria and assumptions are applicable:

- The fire is assumed to occur anywhere in the fire area and to extend throughout the fire area under consideration. Unless provided with suitable fire protection features (per Section III.G.2 of Appendix R) the fire must be assumed to impact the performance of all equipment and cables located in the fire area.

- Credit cannot be taken for the proper function of any electrical circuit that has not been fully analyzed.

- Credit may be taken for automatic actuation signals to position equipment to the desired shutdown condition but only if it can be demonstrated that the fire will not affect the proper operation of the circuits and equipment that generate the automatic signals. Credit cannot be taken for automatic signals if the equipment or circuits that generate the automatic signals are exposed to fire damage.

- It cannot be assumed that fire will affect any electrical circuit in such a way as to cause equipment to fail in its desired safe-shutdown position.

- There is no limit on the number of circuit/cable faults that may occur as a result of fire damage in a given fire area. Any circuit/cable located in the fire area of consideration that lacks suitable fire protection features (per Section III.G.2) must be assumed to be damaged by the effects of fire and/or its related perils.

• In determining the potential for fire to cause undesired spurious equipment actuations, components other than high/low pressure interface valves, need only consider the effect of a single hot short. However, this single fault (hot short) must be considered to occur in combination with other possible circuit failure modes (open circuits, shorts to ground).

• If it is determined that more than one hot short is required to cause a component to spuriously actuate and the component is not a high/low pressure interface valve *and* the conductors of concern are not located in a single (multi-conductor) cable, then spurious operation of the component is not considered credible. (See Figure 6-11a.)

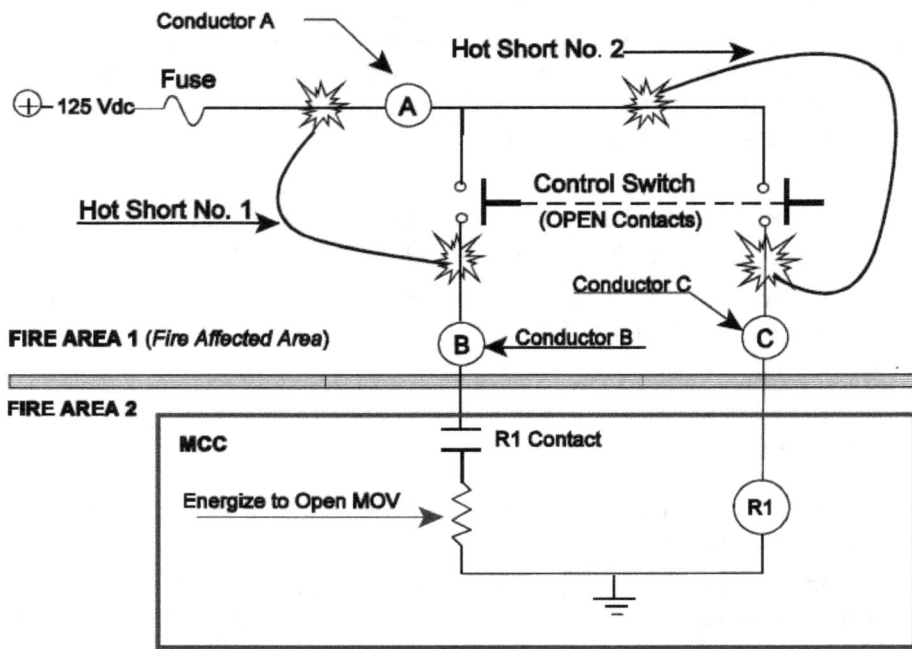

Figure 6.11a Consideration of Multiple Hot-Shorts

For fire in Fire Area 1 Both Hot Short No. 1 [Conductor A to B] _and_ Hot Short No. 2 [Conductor A t o C must occur to cause the spurious opening of the MOV. With the following two exceptions, multiple hot shorts of this nature are not considered credible:

1. For High / Low Pressure interface valves, each valve having exposed circuits in the fire affected area would need to consider the occurrence of multiple hot shorts as a credible event. OR

2. IF: Conductors necessary to cause a spurious actuation (e.g., A, B, and C)are all located in same multi-conductor cable and the cable is not adequately protected from fire damage (per III.G.2)
    THEN :   Spurious operation of the component must be considered as a credible event , whether or not the component is part of a High / Low Pressure Interface.

- The evaluation of high/low pressure interface components must consider the occurrence of multiple, simultaneous, hot shorts of the required polarity and sequence as a credible event. Given the unacceptable consequences associated with this event, the analysis must consider the occurrence of hot shorts on all three phases of the components power cable in the proper sequence (i.e., Phase A to Phase A; Phase B to Phase B and Phase C to Phase C) as a credible event.

- Multiple conductor-to-conductor hot shorts in cables containing more than a single conductor (i.e., multi-conductor cables) are credible and must be evaluated. It is not sufficient to only consider the effect of a single fault on each conductor on a one at a time basis (see Figure 6-11a).

- "Hot shorts" may result from a fire-induced insulation breakdown between conductors of the same cable (circuit), a different cable (circuit), or from some other external source resulting in an undesired impressed voltage or signal on specific conductors.

- Circuit failures resulting in spurious actuations of equipment must be assumed to exist until action is taken to isolate the affected circuit from the fire area or other actions are taken, as appropriate, to negate the effects of the faulted condition that is causing the spurious actuation. It cannot be assumed that the fire would eventually clear the circuit faults.

- "Open circuits" may result from a fire-induced break in conductors resulting in the loss of circuit continuity.

- "Shorts to ground" may result from a fire-induced breakdown of cable (circuit) insulation, resulting in the conductor being applied to ground potential.

- Where a single fire can impact cables that can cause the spurious opening of high/low pressure interface isolation valves, it must be assumed that all of the affected valves will spuriously actuate *simultaneously*.

- For each fire area all potential spurious operations that may occur as a result of a postulated fire should be identified and evaluated for their impact on the safe shutdown capability. With the exception of components comprising a high/low pressure interface boundary, spurious actuations having the potential to impact the shutdown capability must either be prevented or the effects of each actuation must be appropriately mitigated on a one-at-time basis. That is, the analyst must assume that "any and all" spurious actuations that could occur, will occur, but on a sequential, one-at-a-time, basis. It is not assumed that all spurious actuations that could occur as a result of fire damage will occur instantaneously at the onset of the fire. However, the analyst must consider the possibility for each spurious actuation to occur sequentially, as the fire progresses, on a one-at-a-time basis. In the absence of suitable fire protection features, the potential for such sequential failures to result in the concurrent failure of two or more devices must be considered. Analysis approaches that arbitrarily limit the number of spurious actuations that may occur (such as assuming that only one spurious actuation will occur for each fire event) as a result of fire damage are inconsistent with regulatory requirements. GL 86-10 Question and Answer Section 5.3 provides additional guidance.

- Analysis methodologies that attempt to predict the number of circuit faults and/or spurious equipment actuations that may occur as a result of fire damage to exposed circuits and cables may lack sufficient technical basis may not be valid. For example, without additional justification it is not acceptable to assume that only one spurious actuation or one hot short would occur as a result of fire in any fire area, unless it has been reviewed by the staff for a specific licensee's application.

- All cables, regardless of type or manufacture, including IEEE-383 qualified cables, will support combustion. No credit may be taken for the ability of cables to "self-extinguish".

- For fires requiring implementation of an alternative or dedicated shutdown capability, it is necessary to identify all potential spurious operations that may result from the fire and evaluate the impact of each on the ability to achieve and maintain safe shutdown. Spurious operations could occur on a circuit that is isolated from the fire area under consideration during the time it takes the operator to evacuate the MCR and assume control of the plant at a remote location (e.g., RSP, therefore, spurious operations must be postulated on circuits that can be isolated as well as circuits that cannot be isolated from the fire area under consideration. That is, the potential for spurious operations of equipment to occur prior to actuation of isolation devices (e.g., isolation/transfer switches) must be considered. If the actuation can be appropriately controlled or mitigated by actuation of the isolation/transfer switch, actuation of the transfer switch is considered to be an adequate mitigating action. For those circuits that are not capable of being isolated from the fire area under consideration, it must be assumed that they will spuriously actuate as a result of fire damage on a one-at-a-time, sequential basis.

- A "hot short" between conductors of different cables does not need to be postulated to occur on a safe-shutdown cable that is routed individually (by itself) in a metallic conduit or in a metallic conduit that does not contain other energized circuits (conductors). If this justification is used provisions must be made to ensure that future circuit changes or cable routing modifications do not alter this condition.

- Fire is not expected to damage cables that are routed in "embedded" conduits (i.e., conduits that are located within the confines of a structural concrete floor, wall or ceiling).

- All components are assumed to be in their normal position as shown on the P&IDs.

- Circuit contacts are assumed to be positioned (i.e., open or closed) consistent with the normal mode of the component as shown on the schematic drawings.

- Unless demonstrated otherwise, the effect of fire damage to instrumentation circuits cannot be predicted. That is, the instrument may fail full scale high, full scale low or at some intermediate point. It cannot be assumed that fire damage would always cause an instrument to fail at some pre-determined point (e.g., full downscale, mid-range or full upscale).

- The evaluation of the potential impact of fire-induced spurious actuations on safe shutdown capability must consider all possible failure modes of the equipment or components under consideration. This includes, for example, the potential for fire-induced circuit/cable damage to cause mechanical failure of MOVs as described in IN 92-18.

6.4.6.3   Types of Circuit Failures

Sections III.G.2 and III.L.7 of Appendix R delineate the cable and circuit failure modes that must be considered in the evaluation of post-fire safe-shutdown capability as open circuits, shorts to ground and hot shorts. This section provides specific examples of each of these types of circuit failure conditions.

6.4.6.3.1   Open Circuits

An open circuit is a fire-induced break in a conductor resulting in a loss of circuit continuity. An open circuit will prevent the ability to control or power the affected equipment. Deterioration of fiber optic cables leads to a loss of signal and has a similar effect.

Potential consequences of open circuits on the safe-shutdown capability include, but are not limited to the following:

· a loss of power to required shutdown equipment

· an inability to control essential shutdown equipment

· a loss of power to an interlocked relay or other device that may change the state of the equipment (e.g., a solenoid that is required to remain energized for safe shutdown becomes de-energized)

· an open circuit on the secondary winding of certain types of current transformers may result in initiation of secondary fires at the location of the current transformer. The potential for this occurrence is largely dependent on the rating, type and design of current transformers used and, therefore, must be evaluated on a case-by-case basis

The condition of an open circuit on a grounded control circuit is illustrated in Figure 6-12. In the circuit illustrated an open circuit at location No. 1 is equivalent to a blown fuse — equipment operation will not be possible. An open circuit at location No. 2 will prevent opening or starting of the equipment but will not impact the ability to close or stop the equipment.

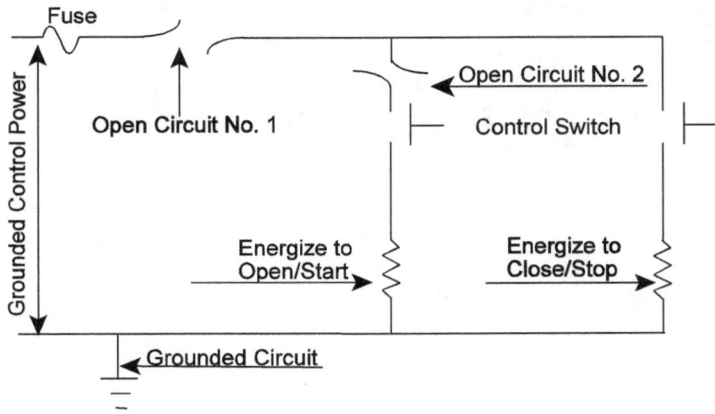

Figure 6-12 Open Circuit Example

6.4.6.3.2   Shorts to Ground: Grounded Circuits

A short to ground results from a degradation (breakdown) of cable/conductor insulation. This fault condition results in a ground potential on the affected conductor. A short to ground can have all of the same effects as an open circuit and, in addition, a short to ground can also impact the control circuit or power train of which it is a part. In the case of a grounded circuit illustrated in Figure 6-13, a short on any part of the circuit would present a concern for tripping the isolation device (i.e., fuse) thereby causing a loss of control power. For the circuit illustrated a short to ground at location No. 1 will result in the control power fuse blowing and a loss of power to the control circuit. This will result in an inability to operate the equipment using the control switch. As discussed in Section 6.4.5.2.1, depending on the coordination characteristics (selectivity) between the fuse and its upstream protective devices (fuses, circuit breakers that provide power to the fuse in this circuit) the power to other circuits could also be affected. This failure mechanism should be evaluated as part of the associated circuits common power source analysis. A short to ground at location No.2 will have no effect on equipment operation until the close/stop control switch is closed. Should this occur the effect will be identical to the short to ground at location No.1. A short to ground at this location would not affect the ability to open/start the equipment until the close/stop control switch is placed in the closed position.

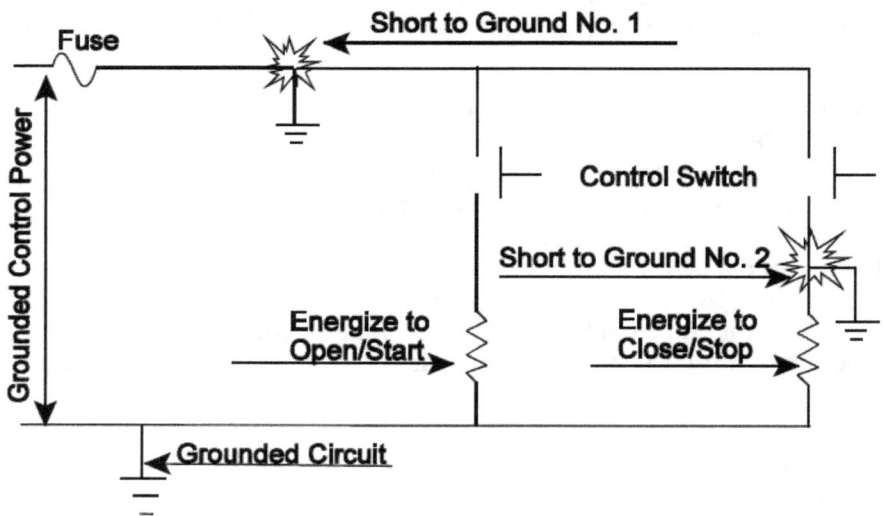

Figure 6-13 Shorts to Ground (Gounded Circuit)

6.4.6.3.3   Shorts to Ground: Ungrounded Circuits

In the case of an ungrounded circuit (such as most 125 VDC control power schemes) a single short to an external ground reference (e.g., cable tray, conduit or metallic enclosure) on any part of the circuit may not cause the circuit isolation device to trip. To illustrate this concept consider the simple light circuit illustrated in Figure 6-14. In this case a battery is being used to supply power to the lamp and there is no reference to any external grounded reference potential, such as a metal cable tray. This is a simple example of an ungrounded circuit. For a circuit such as this, connecting a single wire (to simulate a short) from the positive (+) side of the battery to a grounded cable tray will not have any effect on the operation of the lamp since there is no complete path for fault current to flow back to the battery. However, the occurrence of an additional (second) short on the negative (-) side of the circuit will provide a complete path for current to flow, causing the fuse to blow and resulting in an inability to illuminate the lamp. It should be noted that the second ground fault may occur as a result of fire damage to this circuit or any other circuit that is also fed from the same ungrounded power source (e.g., battery). Therefore, the potential for an ungrounded circuit to become grounded as a result of fire damage must be considered.

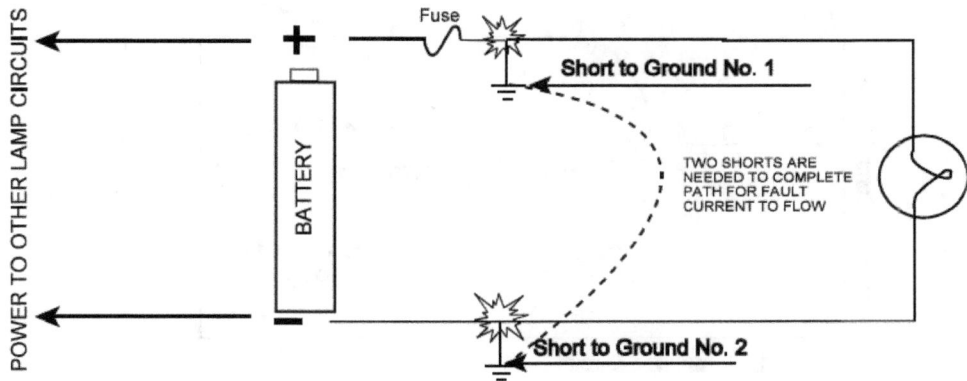

Figure 6-14 Ungrounded Circuit Illustration

6.4.6.3.4   Hot Shorts

In a "hot short" fault condition an energized conductor comes in electrical contact with other un-energized conductors. As a result of this fault, (short circuit between conductors) an undesired voltage or signal is impressed on conductors that were previously un-energized. A hot short fault condition may occur between conductors of the same cable, a different cable, or some other external source. An example of a hot short fault condition is illustrated in Figure 6-15. For the circuit illustrated in Figure 6-15, a hot short at location no.1 would energize the open/start relay and result in the undesired (spurious) opening or starting of the equipment being controlled by this circuit. This condition would be unacceptable for safe-shutdown if the desired operating mode of the affected equipment were closed or stop. A hot short at location No.2 would energize the close/stop relay and result in the undesired (spurious) closure or stopping of the equipment being controlled by this circuit. This condition would be unacceptable for safe-shutdown if the desired operating mode of the affected equipment were open/start.

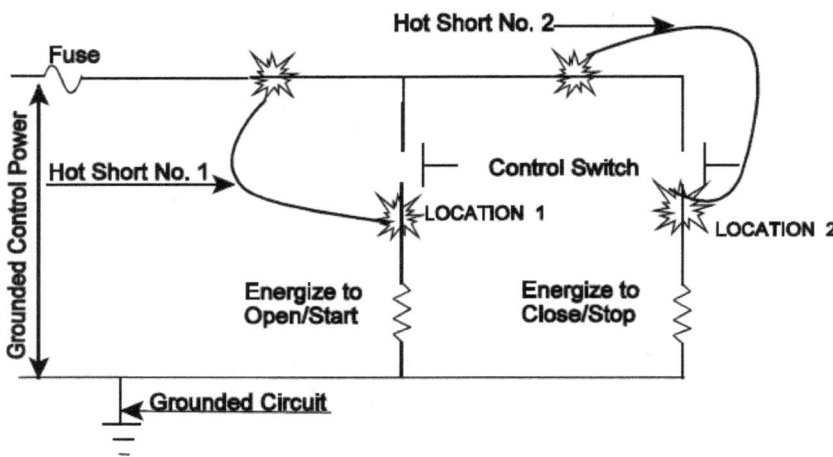

Figure 6-15 Hot Short Example

The hot shorts illustrated in Figure 6-15 are derived from energized conductors in the same circuit. However, it should be noted that the same hot short fault conditions could also be established as a result of electrical contact (short) between locations 1 and 2 and conductors connected to any other energized source, including those that may be external to this circuit.

In the case of an ungrounded circuit, a single hot short may be sufficient to cause a spurious actuation. A single hot short can cause a spurious actuation if the hot short comes from a circuit from the positive leg of the same ungrounded source as the affected circuit. There are also additional cases where a hot short on an ungrounded circuit, in combination with a short to ground can cause a spurious actuation. In reviewing these cases, the "common denominator" is that in every case, the conductor in the circuit between the control switch and the control coils (open/start or close/stop) must be involved.

6-40

Given the possibility of a short to ground being caused by the fire, it should be assumed that a spurious operation will result whenever the fire affects the conductor between the control switch and the control coils. Since a hot short from the same source or grounding of ungrounded circuits cannot be ruled out, it should be assumed that ungrounded circuits will behave the same as grounded circuits in their response to hot shorts.

6.4.7   Locate Equipment, Cables, and Circuits of Concern to Post-Fire Safe Shutdown

At this point in the analysis process, plant process and support *functions* that must be accomplished to achieve and maintain hot and cold shutdown conditions have been defined (Section 6.4.2), *shutdown systems* (redundant and/or alternative) capable of accomplishing each of the required shutdown functions have been determined and assigned a unique safe-shutdown path designation (Section 6.4.3). With the shutdown paths defined, the equipment needed to ensure the proper operation of each path is identified and documented in the SSEL (See Sections 6.4.4). The SSEL establishes a starting point for identifying *required circuits and cables* (i.e., circuits and cables needed to support operation of the identified shutdown paths) **as well as** *associated nonsafety circuits* that could impact (adversely affect) the achievement of safe-shutdown conditions if they are damaged by fire. (See Section 6.4.5.) Following their identification, associated circuits of concern are then evaluated to assess the potential impact of fire and related perils (e.g., fire suppression activities) on the shutdown capability of the plant. (See Section 6.4.6.)

As discussed in Section 6.1, the post-fire safe-shutdown analysis is performed on a fire area basis. With the equipment, circuits and cables of concern to post-fire safe-shutdown identified, their physical location in the plant is then determined. The specific fire area where each piece of shutdown equipment is located may be determined from a comparison of plant design documents (e.g., equipment layout drawings) to the fire area delineations identified in the FHA. The location of this equipment (i.e., fire area) should then be verified as necessary by field walkdowns and entered into the SSEL.

The routing of cables, including all raceway and cable endpoints, may be determined from a review of plant design drawings (e.g., conduit and cable raceway drawings) and/or cable installation data (e.g., cable pull tags). In certain cases, cable routing information may be obtained by joining the list of safe-shutdown cables with an existing cable and raceway database. For either case, field walkdowns should be performed as necessary to confirm the accuracy of the design information used in the evaluation.

To understand the potential impact of an exposure fire within each fire area, the results of the preceding evaluations should be tabulated in a report that includes such information as:
- fire area designation, location, and description
- shutdown path/systems relied on to achieve SSD [required path(s)]
- potentially affected unit(s)
- potentially affected shutdown path/system
- potentially affected cables [identify function (power, control, instrument) and whether damage can result in a spurious actuation, SSD path/system, affected equipment]
- potentially affected equipment (ID, type, description, SSD path, location, normal operating mode, required operating mode/position for SSD, etc.)

6.4.8   Perform Fire Area Assessments

For each fire area the evaluation of the consequences of fire must conclusively demonstrate that one train of equipment that can be used immediately to bring the reactor to hot shutdown conditions remains unaffected by fire. Systems needed to achieve and maintain cold shutdown may be damaged by fire but the extent of damage to these systems must be limited so that any necessary repairs can be implemented and shutdown conditions achieved within the time constraints described in Section 6.4.1.4.

There are many acceptable approaches to achieve the above objectives and the NRC does not prescribe or endorse any one specific approach. The approach presented in this document starts by defining safe-shutdown success paths (See Section 6.4.1–Section 6.4.5) and then each fire area is evaluated to determine the affected equipment in each fire area. From the resulting list of affected equipment, the impact of fire on the ability to achieve and maintain safe-shutdown conditions can be determined for each area. The various steps involved in this approach are illustrated in Figure 6-16. Another approach may start with the fire area and identify the redundant divisions (trains) of equipment and cables that are located in the fire area. From this information a shutdown success path that relies on the use of equipment associated with the "least affected" division could be developed. With the shutdown success path determined for the area, the impact of any interactions between cables and equipment in the area can then be assessed.

Regardless of the approach used, the SSA should be a bounding analysis which identifies the range of possible fire impacts within each fire area and ensures that appropriate measures are in place to prevent this damage from affecting the ability to safely shutdown the plant. For each fire area, the SSA must define a set of systems and equipment that are capable of accomplishing the required shutdown functions in accordance with established performance criteria.

The degree of physical separation provided for redundant trains of shutdown systems may vary widely among plants. Later generation plants that were designed and/or constructed after the Browns Ferry fire, tend to have a greater amount of physical separation inherent in their design. Older plants, however, (typically those receiving an operating license prior to the promulgation of Appendix R) typically were not designed with this concept in mind. Regardless of plant vintage however, the evaluation of a specific fire area may find at least one shutdown success path to be completely independent (both physically and electrically) of the fire area under evaluation. For these cases, the method(s) [e.g., SSD success path(s)] available to achieve safe-shutdown in the event of fire in the area is documented and no further evaluation is necessary. In other cases, however, an adequate level of separation may not already exist (i.e., at least one train of shutdown equipment/shutdown path is *not* independent of the fire area). For these cases, at least one shutdown success path must be identified and provided with suitable fire protection features as described below.

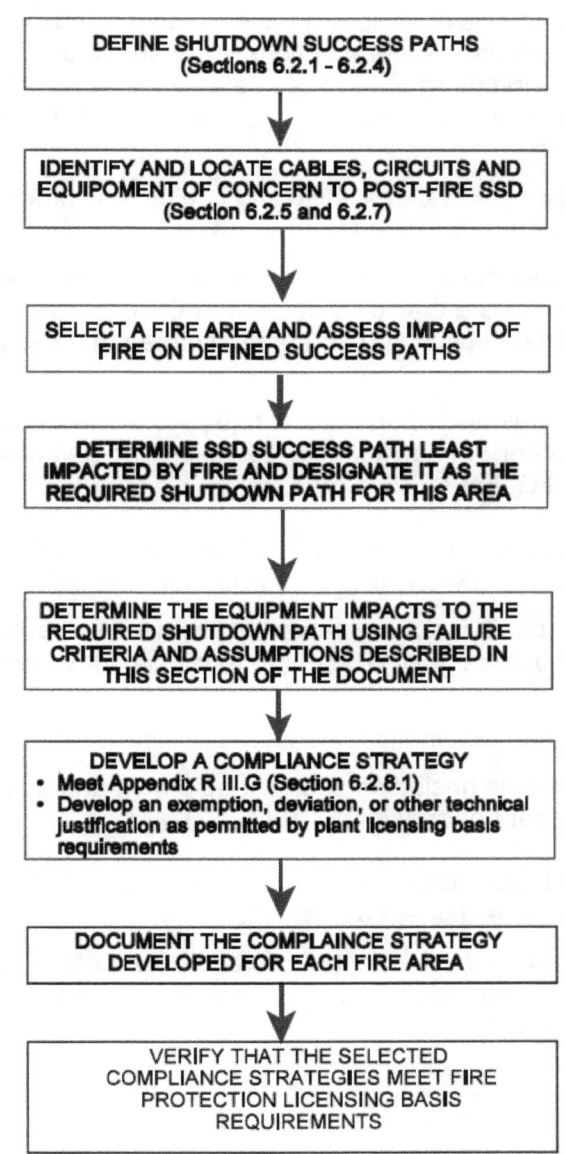

```
┌─────────────────────────────────────┐
│   DEFINE SHUTDOWN SUCCESS PATHS      │
│       (Sections 6.2.1 - 6.2.4)       │
└─────────────────────────────────────┘
                  │
                  ▼
┌─────────────────────────────────────┐
│ IDENTIFY AND LOCATE CABLES, CIRCUITS AND │
│ EQUIPOMENT OF CONCERN TO POST-FIRE SSD   │
│       (Section 6.2.5 and 6.2.7)          │
└─────────────────────────────────────┘
                  │
                  ▼
┌─────────────────────────────────────┐
│ SELECT A FIRE AREA AND ASSESS IMPACT OF │
│    FIRE ON DEFINED SUCCESS PATHS        │
└─────────────────────────────────────┘
                  │
                  ▼
┌─────────────────────────────────────┐
│  DETERMINE SSD SUCCESS PATH LEAST     │
│ IMPACTED BY FIRE AND DESIGNATE IT AS THE │
│ REQUIRED SHUTDOWN PATH FOR THIS AREA   │
└─────────────────────────────────────┘
                  │
                  ▼
┌─────────────────────────────────────┐
│ DETERMINE THE EQUIPMENT IMPACTS TO THE │
│ REQUIRED SHUTDOWN PATH USING FAILURE   │
│ CRITERIA AND ASSUMPTIONS DESCRIBED IN  │
│    THIS SECTION OF THE DOCUMENT        │
└─────────────────────────────────────┘
                  │
                  ▼
┌─────────────────────────────────────┐
│     DEVELOP A COMPLIANCE STRATEGY     │
│ • Meet Appendix R III.G (Section 6.2.8.1) │
│ • Develop an exemption, deviation, or other technical │
│   justification as permitted by plant licensing basis │
│   requirements                        │
└─────────────────────────────────────┘
                  │
                  ▼
┌─────────────────────────────────────┐
│  DOCUMENT THE COMPLAINCE STRATEGY     │
│    DEVELOPED FOR EACH FIRE AREA       │
└─────────────────────────────────────┘
                  │
                  ▼
┌─────────────────────────────────────┐
│      VERIFY THAT THE SELECTED         │
│ COMPLIANCE STRATEGIES MEET FIRE       │
│ PROTECTION LICENSING BASIS            │
│        REQUIREMENTS                   │
└─────────────────────────────────────┘
```

Figure 6.16   Fire Area Assessment Flowchart

One train of systems necessary to achieve and maintain hot shutdown conditions must be free of fire damage (Appendix R Section III.G.1.a). For cases where adequate fire area separation does not exist (i.e., redundant trains of shutdown systems are located in the same fire area), Section III.G of Appendix R provides several options for ensuring that the hot shutdown capability is protected from fires. The first three options, as defined in Section III.G.2, provide the following methods for protecting redundant trains of equipment located in fire areas that are *outside* of non-inerted containments:

- Enclose one of the redundant systems, including cables, equipment and associated nonsafety circuits, in a 3-hour fire-rated barrier. (III.G.2.a)

- Separate redundant systems, including cables, equipment and associated nonsafety circuits, by a horizontal distance of more than 6.08 m (20 ft) with no intervening combustibles or fire hazards. In addition, fire detection and an automatic fire-suppression system are required. (III.G.2.b)

- Enclose redundant systems (including cables, equipment and associated nonsafety circuits) in a 1-hour fire-rated barrier. In addition, fire detection and an automatic fire-suppression system are required. (III.G.2.c).

The next three options, as defined in Section III.G.2, provide methods for protecting redundant trains of equipment located in fire areas that are *inside* non-inerted containments:

- Separate redundant systems, including cables, equipment and associated nonsafety circuits, by a horizontal distance of more than 6.08 m (20 ft) with no intervening combustibles or fire hazards. (III.G.2.d)

- Install fire detection and an automatic fire suppression systems. (III.G.2.e)

- Separate redundant systems (including cables, equipment and associated nonsafety circuits) by a noncombustible radiant energy shield. (III.G.2.f)

The last option, as defined by Section III.G.3, provides an alternative or dedicated shutdown capability to the redundant trains damaged by a fire:

- Ensure that alternative (or dedicated) shutdown equipment are independent (both physically and electrically) of the cables, equipment, and associated nonsafety circuits of the redundant systems damaged by the fire.

CHAPTER 7. MAINTAINING POST-FIRE SAFE-SHUTDOWN:
Configuration Management for Post-Fire Safe-Shutdown Analysis

A post-fire safe-shutdown analysis is based on a "snapshot" of the configuration of plant SSCs and cable routing information that existed at the time the analysis was performed. However, the plant design features and operating practices that form the basis of the analysis are rarely static. Over its operating life, a plant may make modifications to improve its safety, reliability, and efficiency. If not properly evaluated, plant modifications can significantly compromise the results presented in the SSA and, in certain instances, may threaten the plant's ability to achieve and maintain safe-shutdown conditions in the event of fire. Effective maintenance of the plant's post-fire safe-shutdown capability, as described in the SSA and its supporting calculations and procedures, requires that all proposed changes to the plant design and operations, whether permanent or temporary, must be evaluated for their impact on the shutdown capability.

Figure 7-1 illustrates how even a seemingly straightforward modification involving the installation of nonsafety-related equipment can impact the shutdown capability. In this case, the licensee is making a modification to provide a more efficient means of transferring water between two nonsafety-related tanks. Key components being added as part of this modification include a pump (Pump X), piping, a nonsafety-related SWGR (SWGR 1A-1), motor-operated pump suction and discharge valves, and related controls and instrumentation. The pump is to be located in a fire area where the SSA credits the use of Division B equipment and is to be powered from a new Division A power source (SWGR 1A-1).

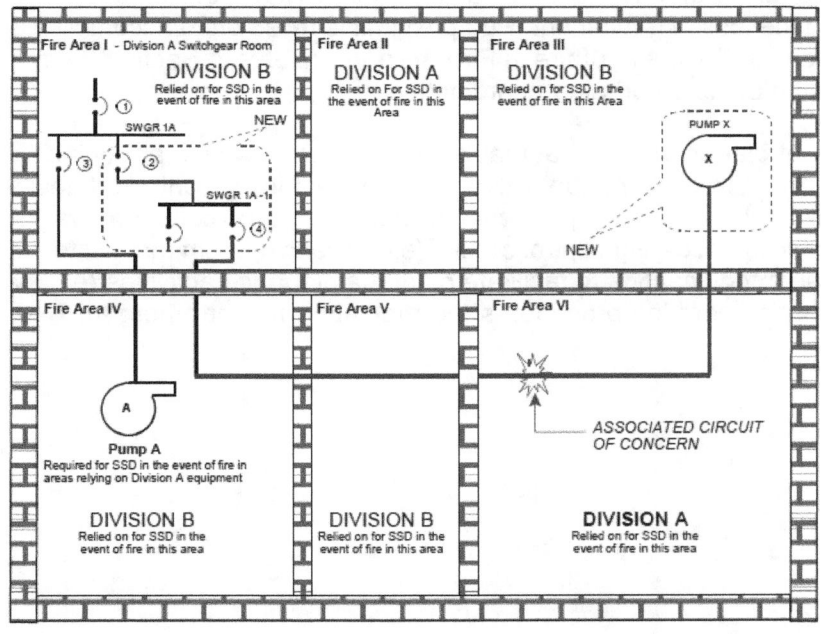

Figure 7-1 Modification Impacting the SSD Capability

The proposed design also includes the following attributes:

- The pump and the entire system in which it is located will not need to remain operational to accomplish required shutdown functions.

- Maloperation of the pump (e.g., an unintended start or stop) would have no impact on the shutdown capability of the plant.

- The pump is powered from a nonsafety-related power source (SWGR 1-1A) that does not power any safe-shutdown components. A fire-induced loss of SWGR 1-1A would not impact safe-shutdown.

- The power source (SWGR 1-1A) is physically located in a fire area where equipment from the redundant division (Division B) is relied on to accomplish post-fire safe-shutdown.

- Before installing this modification, the SSA demonstrated an acceptable level of coordination between load and feeder breakers of SWGR 1A.

While these considerations may suggest that the planned modification would not impact the plant's post-fire safe-shutdown capability, a potential vulnerability still exists. Specifically, as shown in Figure 7-1, the cable that provides power to Pump X traverses several fire areas. Note that this routing includes an area (Fire Area VI) where Division A equipment (including required SSD Pump A powered from SWGR 1A) is relied on for post-fire safe-shutdown. Because the cable has not been provided with fire protection features (e.g., rated barrier wrap), it is susceptible to fire damage. If this modification were to be installed without ensuring that the new circuit breakers installed in SWGR1A (Breaker 4) and SWGR 1A-1 (Breaker 2) properly coordinate with the upstream feeder breaker (Breaker 1), a fire in Fire Area VI could cause Breaker 1 to trip and, thereby, result in the loss of equipment (Pump A) that is relied on to accomplish essential shutdown functions in the event of fire in Fire Area VI. (Refer to Section 6.4.5.2.1 for a more detailed discussion of circuit breaker coordination.)

Other examples of plant changes that may affect the shutdown capability include replacing a passive component (e.g., a manual valve) with an electrically controlled device (e.g., an MOV), rerouting cables, replacingf circuit protective devices (fuses, circuit breakers, relays), installing or removing interlocks, modifying control circuits (e.g., changing from manual to automatic control), making temporary modifications to facilitate plant maintenance activities (e.g., welding), and changing the plant's operating procedures (normal, abnormal or emergency).

Requirements governing the fire protection of safe-shutdown capability must be maintained over the life of the plant. This capability is provided through the establishment of administrative control procedures, which specify that changes in plant design and operations (both permanent and temporary) must be subjected to an appropriate level of review. This assessment must be performed by qualified personnel knowledgeable of the plant's post-fire safe-shutdown analysis. These procedures must address the following specific configuration control issues:

- **Modifications:** All modifications (i.e., permanent or temporary additions, deletions, or changes) to plant SSCs must be reviewed for their potential impact on the plant's post-fire safe-shutdown capability (as documented in the SSA, its supporting calculations, and procedures).

- **Fuse Replacement and Changes in Circuit Breaker or Relay Settings:** Ensure that fuses, circuit breakers, and relays having ratings or settings other than those selected to ensure proper coordination for post-fire safe-shutdown are not accidentally used. To ensure that future plant changes will not compromise circuit breaker and fuse coordination studies referenced in the SSA, the replacement of fuses in power sources required for post-fire safe-shutdown should be performed in accordance with approved procedures. In addition, the coordination study should be maintained current with the most recent modification.

- **Procedure Changes:** The review of permanent and/or temporary procedure changes should consider the following factors:
  (a) the effects of the change on the plant's capability to achieve and maintain post-fire safe-shutdown
  (b) changes to responsibilities and tasks assigned to fire brigade members
  (c) changes to responsibilities and tasks assigned to operations staff members who are responsible for achieving and maintaining safe-shutdown from inside the control room and from alternative shutdown location(s)

This page intentionally left blank.

CHAPTER 8. INTEGRATION OF DETERMINISTIC CRITERIA AND RISK-INFORMED INFORMATION

8.1 Overview of a Risk-Informed Approach

It is NRC policy to increase the use of risk information in the regulatory decisionmaking process (Reference Final Policy Statement, 1995). Risk combines two factors, including (1) the likelihood (or frequency) that an event will occur and lead to undesired consequences and (2) the severity of those undesired consequences. In this chapter, the event of interest is a fire that challenges nuclear safety. The potential undesired consequence of such an event is an offsite release of radioactive materials. For the commercial nuclear power industry, the severity of the release consequences is measured by the potential impact on public health.

In practice, risk is usually quantified using probabilistic risk assessment (PRA). PRA results are most often expressed using two intermediate risk measures; namely, CDF and large-early release frequency (LERF)[27]. CDF reflects the frequency (in events per reactor year) with which a given plant might expect to experience an accident leading to core damage. LERF reflects the frequency with which one might expect an accident to occur and lead to a large release of radioactive materials relatively early in the accident sequence. In this context, the term "early" is measured in relation to population evacuation times. Both CDF and LERF are considered indirect measures of risk because they do not directly quantify the public health consequences of potential plant accidents. CDF and LERF are used as risk measures because they are generally indicative of the potential that public health consequences might occur.

The NRC's risk-informed policy, as embodied in RG 1.174, weighs regulatory compliance issues against both CDF and LERF criteria. To date, the NRC has not formally risk-informed the fire protection portions of the regulatory requirements (e.g., Appendix R to 10 CFR Part 50). However, aspects of the fire protection regulatory process have begun to incorporate risk information. For example, the NRC staff is currently engaged in a rulemaking activity related to the recently adopted 2001 Edition of the National Fire Protection Association Standard 805 (NFPA 805), "Performance-Based Standard for Fire Protection for Light-Water Reactor Electric Generating Plant." NFPA 805 utilizes risk information in evaluating the acceptability of proposed plant changes that impact fire protection. A second example is the significance determination process (SDP) and, in particular, the fire protection SDP, which assesses fire-related inspection findings based on risk measures.

The current discussion is intended to provide risk-informed perspectives on the post-fire safe shutdown circuit analysis issues for use by the NRC staff and, in particular, those staff members responsible for plant inspection activities. The discussion does not establish any new requirements or regulatory compliance criteria. Rather, the discussion is intended to assist the NRC staff in understanding, and potentially assessing, the risk-significance of the fire-related safe-shutdown circuit analysis issue.

---

[27] Note that the calculation of CDF is also known as the Level 1 analysis. Level 2 refers to the containment performance analysis (e.g., LERF), and Level 3 refers to the analysis of offsite release and public health consequences.

It should be noted that fire-induced circuit failure modes and effects risk analysis is an area of ongoing technical discussion and development. Some aspects of the problem are in their first stages of application, and quantification methods have not yet been fully developed or demonstrated. Hence, this discussion is preliminary and subject to change as additional insights develop. Efforts to further develop risk analysis methods for fire-induced circuit faults is ongoing through the NRC's Office of Nuclear Regulatory Research (RES). More detailed discussions on this topic can, for example, be found in LaChance, et al., 2000.

8.2 Fire Risk Analysis Overview

For the purposes of this chapter, risk insights will be discussed in the context of CDF as the primary risk measure of interest. It should be recognized that the success criteria assumed in a typical fire PRA are not the same as those applied in a regulatory fire protection framework. In the regulatory context, the fire safe-shutdown analysis considers the ability to achieve both hot and cold shutdown. Hot shutdown must be possible within 24 hours, and the regulations do not allow for hot shutdown repair actions. Cold shutdown has a longer mission time, 72 hours, and, within certain limits, repair actions are allowed. In contrast, a typical PRA considers success to be achieving and maintaining a stable hot shutdown condition such that core damage is prevented. In a typical PRA, scenarios are analyzed until a safe and stable plant condition is achieved, but generally only out to 24 hours. (Note that the potential for core damage accidents that occur beyond this period should not be dismissed out of hand.) PRAs do not generally consider cold shutdown. Hence, the PRA/CDF success criteria align most closely with the regulatory hot shutdown requirements. The only correspondence to the regulatory cold shutdown requirements would be found in a low power and shutdown fire risk analysis and very few of these have been performed to date. This section provides an overview of current fire CDF quantification practice. This overview provides a convenient framework for our discussion of risk perspectives on post-fire safe-shutdown circuit analysis.

Both regulatory requirements and fire PRAs focus first and foremost on the fire hazard, or risk, associated with fires that impact some bounded region of the plant. However, how these bounded regions are defined in the regulatory context often differs from the definitions used in a fire PRA. In the regulatory context plants are partitioned into fire areas; that is, physical regions that are bounded on all sides by fire-rated boundary elements sufficient to contain the fire hazards(RG 1.189). Fire PRAs are generally based on fire compartments[28], a less rigorously defined subdivision of the plant. In a fire PRA, a given fire area may be retained in whole as a fire PRA compartment, or the fire area may be partitioned into two or more fire compartments. Defining fire PRA compartments involves the application of analyst judgment. Fire PRA compartments may credit features that would not be credited in defining fire areas in the regulatory context. This may include non-fire-rated partitions, partitions with unsealed penetrations, active partitions (e.g., heat activated roll-up doors), water curtains, and even extended spatial separation. In most fire PRAs, each fire compartment represents a region wherein, based on the judgement of the analyst, the damaging effects of the majority of fires are expected to be confined.

---

[28] Note that the terminology applied varies between analyses. Some analysts may refer to fire zones, analysis zones, rooms, or other designations to describe the physical analysis boundaries drawn to support the fire PRA. The concept remains the same.

The CDF analysis systematically considers risk contributions arising from fires in each fire compartment. The risk quantification results may be reported at a fire scenario level (e.g., for a given fire ignition source), but are more typically reported at a compartment level. Hence, PRAs will often cite a CDF contribution for each individual fire compartment (e.g., the Cable Spreading Room). An explicit analysis is also conducted to assess the risk contribution of fires that might impact multiple fire compartments. The results for the multi-compartment (or room-to-room) fire scenarios are often reported separately. It is also common to report a total fire-induced CDF for the plant as a whole (i.e., the sum of the individual compartment and multi-compartment contributors).

In the most general terms, the likelihood that a fire might initiate a core damage accident is assessed on the basis of the following three-factor formula:

$$CDF = \sum_i f_i \left(\sum_j P_{cd,j|i} \left(\sum_k P_{CD:k|i,j} \right) \right)$$

The first term (f_i) on the right-hand side represents the fire occurrence frequency. The summation over the index "I" implies that the plant-wide fire-induced CDF is based on the sum of contributions from many individual fires involving a number of fire compartments. Fire frequency includes consideration of both fixed (e.g., fixed electrical and mechanical equipment; fixed components that might experience a leak of lubricating oil or flammable gases including hydrogen; semi-permanent storage items; etc.) and transient fire ignition sources (e.g., maintenance materials staged in anticipation of an outage, inservice maintenance support materials, welding and cutting operations, refuse, etc.).

Consistent with the fire PRA plant partitioning practices as described previously, fire frequency may be quantified at a compartment level reflecting all possible fire ignition sources in a given fire compartment (e.g., a battery room). However, fire frequency may also be quantified at a more detailed level. For example, fire frequency may be expressed for a particular fire ignition source (e.g., a motor or pump), or for a specific group of fire ignition sources (e.g., a bank of SWGR or general transient fuel sources). Fire frequency is usually based on statistical analysis of the evidence provided by past fire events; i.e., a fire event database. A number of such databases are available from both public and private sources. (Reference NUREG/CR-4586 and EPRI TR-1000894.)

The second term ($P_{cd,j|I}$) reflects the conditional probability that, given a particular fire (I), a particular physical damage state (j) will be induced. The physical damage state is defined by the plant equipment, components, and/or electrical cables damaged by the fire. Note that the nomenclature P_{cd} implies the probability of either "component damage" or "critical damage," depending on the analysts' use of terminology. The second summation implies that a given fire might lead to more than one physical damage state depending, for example, on the duration of the fire and, by implication, the physical extent of fire damage. Calculation of the component damage term typically involves the analysis of fire growth behavior, component thermal response and damage, and fire detection and suppression. It is in this part of the analysis that fire models, for example, are applied.

8-3

Given the first and second terms, the analyst is postulating that a fire has occurred and has damaged some set of plant components. The loss of some plant components implies that some subset of the plant systems and/or functions are damaged and/or rendered inoperable. The third and final term ($P_{CD,k|i,j}$) reflects the conditional probability that given the physical damage state (j) resulting from the fire (I), operators will fail to achieve safe-shutdown and core damage will result. Summation over the index (k) implies that for a given physical damage state, the plant will still have available various options (or paths) for achieving safe-shutdown. Each safe-shutdown path will have a unique likelihood of success/failure. The calculation of P_{CD} includes consideration of system faulting behaviors given the fire damage, operator performance, and random equipment failures independent of the fire. The summation of the contributions from each failure path leading to core damage is often referred to as the conditional core damage probability (CCDP) associated with a given physical plant damage state.

The risk importance of any given fire compartment can be weighed in terms of the absolute CDF contribution and based on the relative contribution of a given fire compartment to the overall fire CDF. For example, even if a plant has a total fire CDF that is considered low, the fire PRA will still typically identify and analyze in detail the risk-dominant fire compartments; that is, those compartments that contribute most to fire risk. Typically, on the order of two-to-ten fire compartments are found to dominate the plant fire risk estimates. The risk-dominant fire compartments often include areas such as the main control room, cable spreading room, auxiliary electrical equipment or relay rooms, and emergency SWGR areas. Other compartments may be risk important on a plant-specific basis.

One of the most significant factors in determining which compartments are fire risk dominant is the routing of important power, control, and instrument cables through the plant. Fire risk is often dominated by fires leading to the failure of electrical cables. Hence, fire compartments through which important electrical cables pass tend to be fire risk-dominant. A second significant factor is the presence, or absence, of significant fire ignition sources in a fire compartment. For example, even a fire compartment such as the cable spreading room may be found to have a relatively low fire risk if it lacks significant fire ignition sources.

Many fire compartments will ultimately be found to contribute little to plant risk. In a fire PRA, a formal process is used to 'screen out' such compartments. The first screening step is usually based on qualitative arguments. For example, compartments that contain no safety-related equipment or electrical cables, and where fires cannot induce a plant transient (e.g., manual trip), are often qualitatively screened as insignificant risk contributors. A second stage of screening is typically conducted based on conservative quantification of the three-factor formula cited previously. Quantitative screening generally focuses on the potential severity of fire damage and the likelihood of core damage given fire damage (i.e., the second and third terms). Compartments will rarely screen on fire frequency alone because virtually all compartments have a non-trivial fire frequency (generally no less than 1×10^{-4} fires per reactor year, or 1×10^{-4}/ry and often higher).

It is important to note that fire PRAs usually credit components, systems, and functions that are not credited in the post-fire safe-shutdown analysis[29]. The post-fire safe-shutdown analysis is, first and foremost, intended to ensure that one train of equipment necessary to achieve and maintain safe shutdown will remain free of fire damage. However, other plant systems not credited in the post-fire safe-shutdown analysis will likely survive any given fire event and, in reality, could be used as available to support the post-fire plant recovery efforts. This fact presents a sometimes difficult challenge to fire risk analysis. Fire is a very spatially-oriented phenomena. Even given a rather severe fire, fire-induced component and electrical cable failures will likely occur only in a specific and limited physical region of the plant. Hence, accurate information on component and cable locations is often critical to the fire damage analysis. The more accurate the available information is, the more accurate the risk estimates can be made.

Because the post-fire safe-shutdown analysis is, in essence, a success-path analysis, it credits a limited subset of the plant systems. The electrical cables and components required to support these credited systems are traced within the plant and their locations are generally well known, at the least to the level of their presence in, or absence from, each fire area. However, for systems not credited in the safe-shutdown analysis, the associated components and electrical cables may not be traced and their locations may not be known. To avoid undue optimism, the analyst must verify that a fire cannot cause damage to a system's components and electrical cables before credit for the system's function can be taken in the risk analysis. For those systems not credited in the safe-shutdown analysis, this can require tedious and time consuming efforts, in particular, to trace electrical cables through the plant. An approach that is often taken is to assume failure of a system unless the lack of a fire threat to the system's components and electrical cables can be verified for a given fire scenario. If the failure assumption is found to be critical to the quantification, then additional verification and cable tracing efforts may be undertaken.

8.3   Circuit Analysis and the Risk Analysis Framework

It is now possible to express the issues of circuit analysis in the context of the computational framework described previously. The first term in the CDF equation, the fire frequency, has essentially no interaction with the circuit analysis issues. Similarly, the second term, the likelihood that the fire will lead to some level of physical damage, is also not directly relevant to the circuit analysis issues. Circuit analysis comes into play through the third term, the likelihood that the fire-induced equipment failures will lead to core damage. In this context, we are especially interested in the fire-induced failure of electrical cables.

A fire may cause failures in power, control/indication, and/or instrument cables associated with various plant systems and functions. The response of the impacted systems, the circuit or system fault mode, will depend on the mode of electrical cable failure observed. The process of examining the various electrical cable failure modes in order to identify the potential circuit or system fault modes is referred to here as the process of circuit analysis. More formally, this is referred to as the electrical cable failure modes and effects circuit analysis.

---

[29]   Note that some methods used in the Individual Plant Examination External Events (IPEEE) studies credited only the Appendix R systems (NUREG/CR-1742). When rigorously applied, such approaches typically yield conservative estimates of fire risk.

Circuit analysis is complicated in part because electrical cables may experience one or more of several failure modes. Furthermore, the failure behavior may be dynamic, changing throughout the course of the fire event. Each unique combination of electrical cable failures can potentially induce a unique circuit fault mode. Circuit fault modes of potential interest include loss of function, loss of control, loss of indication, corrupted indications or signals, and spurious actuation. Since the electrical cable failure behavior may be dynamic, the circuit's faulting behavior may also be dynamic. To illustrate, consider that two of the possible cable failure modes of particular importance are hot shorts and shorts to ground. Conductor-to-conductor short circuit cable failure modes, including hot shorts, are likely to transition to shorts to ground given an enduring fire exposure. Therefore, in some cases it may be important to assess both the initial cable failure mode, the anticipated duration of a specific failure modes, and the impact of an ultimate short to ground. As the fire scenario develops, multiple cables may fail at discrete points in time, and multiple circuit faults may come into play. This introduces the further question of concurrent behavior involving multiple circuits; for example, how likely is it that two or more circuits might experience concurrent spurious actuations.

It is not possible to exhaustively explore all of the potential electrical cable failure modes in a fully dynamic context for any but the most simplistic of fire damage state scenarios. Hence, it is widely recognized that some optimization of the circuit analysis process is both necessary and desirable. The specific optimization framework being discussed here is fire-induced core damage risk. That is, the process of circuit analysis is optimized to focus attention on those electrical cables, cable failure modes, and circuit fault modes that may be risk significant.

Typically, a given circuit or system will have a limited set of specific fault modes that will be unique in the context of fire risk. Depending on the circuit, some fault modes may be benign while others might challenge the safe-shutdown process. For example, loss of function in a valve may have little risk impact if operation of the valve is not required to mitigate the accident scenario. However, spurious actuation of that same valve might challenge safe-shutdown by opening an undesired coolant flow diversion path, or by closing a desired coolant flow path.

Circuit faulting behavior influences the likelihood of successful shutdown in three primary ways:

- Circuit faulting can lead to the unavailability of one or more desired plant systems.

- Circuit faulting might cause the maloperation of one or more plant systems (e.g., a spurious actuation or change of operational state).

- Circuit faulting may compromise instrument and control signals that operators depend on in their response to the event (e.g., the loss of control and instrument signals, or transmission of corrupted signals).

Each of these circuit faulting effects can have unique implications for fire risk. The practical objective of PRA circuit analysis is to identify the risk-important circuits and circuit fault modes, and to then quantify the likelihood that such faults might be observed during a given fire. Insights gained to date related to this objective are discussed below.

8.4 A Mechanistic View of the Problem

A mechanistic view of the circuit analysis problem is being developed in support of broader fire PRA development activities (LaChance, et al., 2000). As an entry condition to the fire PRA circuit analysis task, it is assumed that fire modeling tools of some type (potentially including expert judgement) have been applied separately and have predicted the failure of one or more electrical cables. Under the mechanistic view of circuit analysis, the problem is first split into two major pieces; namely, the electrical cable failure mode analysis and the circuit fault mode analysis. The discussions provided in this chapter are organized based on this mechanistic view.

The cable failure mode analysis addresses the short circuiting behavior of the damaged electrical cables. That is, given electrical cable failure, an analysis is performed to determine the relative likelihood that a particular mode of cable failure will occur. The circuit fault mode analysis considers the potential responses of the circuit to various cable failure modes. For example, the circuit fault mode analysis determines whether or not spurious actuation is possible given failures involving a particular electrical cable, and if so, what combination(s) of conductor shorting behaviors could lead to a spurious actuation fault mode.

There is a degree of iteration between the cable failure mode and circuit fault mode analyses. The circuit fault mode analysis will likely identify a unique combination of conductors that, if they short together, would cause a spurious actuation. Furthermore, the circuit fault mode analysis might also find that if one particular conductor were to become involved in the short circuit (e.g., a grounded conductor), the spurious actuation would be self-mitigated. Based on these insights, the cable failure mode analysis would be asked to estimate the likelihood that a combination of conductors leading to spurious actuation, and not involving the grounded conductor, will short together given electrical cable failure.

In practice, the iterative nature of the problem is addressed by dividing the cable failure mode analysis into two further steps. The first step is to consider the electrical cable failure behavior independent of the circuit. That is, the electrical cable in and of itself will experience some combination of conductor short circuits or failure modes. Conductors may short to other conductors in the same cable, they may short to the conductors of another electrical cable, or they may short to an external ground. This behavior should, at least to some extent, be relatively independent of the nature of the circuit to which the cable is connected. However, cable failures must also be considered in the context of the circuit to which the cable is connected, and this is the second step in the cable failure mode analysis.

A circuit utilizes each conductor in a given cable in a particular way. In a control circuit, for example, some conductors are energized to supply control power to the circuit, some conductors are normally de-energized and carry control power to an actuating device when the circuit is exercised (e.g., a control action is taken), other conductors will typically carry control indication signals back to the control station, one or more conductors may be grounded, and finally, some conductors may not be used in the circuit at all (spare conductors). The circuit fault behavior (i.e., how the circuit responds to the cable failures) will typically depend on the types of short circuits (the failure modes) experienced by key conductors among those conductors servicing the circuit. For example, short circuits between the normally energized (or source) conductors, and those normally de-energized conductors that feed power to actuating devices (the potential spurious actuation target conductors) can lead to spurious actuation of the circuit (the so-called hot short induced spurious actuation).

In the consideration of circuit faulting behavior, the initial cable failure behavior is often of paramount importance. In particular, the relative likelihood of conductor-to-external ground versus conductor-to-conductor short circuits is critical. Shorts to ground will generally trip circuit protective features leading to a loss of either control or motive power (see Section 8.6 below). In contrast, conductor-to-conductor short circuits carry the potential to cause spurious actuation of circuits and components. Hence, if a short to ground is the first failure mode observed, other potential failure modes may be rendered essentially moot. That is, if circuit protection is tripped open by a short to ground, then it may not be possible for a subsequent hot short to energize or spuriously actuate that system. However, the importance of subsequent failures and failure mode transitions must be viewed in the context of the circuit under analysis. For example, multiple shorts to ground on an ungrounded DC circuit may have more significant risk implications than only the first such short to ground.

It should also be noted that the cable failure and circuit fault mode discussions which follow are based largely on information gathered during fire experiments involving the failure of electrical cables. Virtually all of the available data is based on small-to-medium scale tests. Small-scale tests in particular cannot fully simulate actual plant installation and fire exposure conditions. Medium scale tests come closer to an actual application, but still cannot, or do not, capture potentially important features and variations of actual plant installations, fire exposure conditions, and conditions during fire suppression. This is true even of the most recent NEI/EPRI electrical cable fire tests (EPRI TR-1003326), even though these tests represent one of the most relevant data sources currently available. Hence, the data and insights derived from such data must be viewed in the context of how those data were gathered.

The available data are both limited and uncertain. The direct extrapolation any given test result to a particular application may be inappropriate. This is especially true in the circuit analysis context given that the data available have illustrated, but not fully investigated, the importance of various factors to the cable failure and circuit response behavior. In the discussion which follows, the author has tried to stress the uncertainties associated with our current understanding of cable failure behavior while at the same time providing as many numerical probability insights as possible. Both the qualitative and quantitative insights described here must be considered preliminary.

8.5   Electrical Cable Failure Modes

In both the regulatory and risk contexts, the failure of an electrical cable implies that the cable is no longer free of fire damage; that is, it is no longer "capable of performing its intended function during and after the postulated fire, as needed" (GL 86-10). From an electrical perspective, the function of an electrical cable is to provide a medium for the transmission of electrical energy (power and/or signals) between two points in a common electrical circuit while simultaneously maintaining the electrical isolation of the transmission path from other elements of the same circuit and from other co-located circuits. Failure, therefore, implies loss of continuity in the energy transmission path or diversion of a sufficient fraction of the available electrical energy to an unintended circuit destination such that proper function of the circuit is no longer ensured.

As discussed previously in Chapter 3, electrical cables are manufactured in a wide range of configurations. The primary configuration features that define a given electrical cable are the size of the individual conductors (expressed using the AWG), the number of conductors, shielding and/or armoring features, and the insulation/jacket materials used in the construction.

There are four primary modes of cable failure of potential interest. These failure modes relate to the electrical behavior of the conductors associated with a given electrical cable, as follows:

- A *conductor to external ground short circuit* results in the diversion of electrical energy to ground.

- A *conductor to conductor short circuit* may result in the diversion of electrical energy from one conductor (the source conductor) to one or more unintended conductors [the target conductor(s)]. One special case of the conductor to conductor short circuit is the *hot short*, that is, the shorting of an energized conductor to a non-energized and non-grounded conductor.

- *Conductor insulation resistance degradation* may result in the partial diversion of the available electrical energy to an unintended conductor path.

- A *loss-of-conductor continuity* (or open circuit conductor failure) is a physical break in the conductor that will result in electrical energy being unable to reach the intended circuit destination.

Before proceeding, two points related to cable failure behavior should be observed. First, the likelihood estimates discussed here are all conditional values given that an electrical cable has been damaged. That is, the likelihood that a particular fire might cause electrical cable damage is not included, only the likelihood that certain failure modes might be observed given that one or more electrical cables have been damaged by a fire.

Second, the discussion focuses on the initial failure mode (that is, the first failure mode that might be observed given failure). As noted previously in Section 8.4, cable failure behavior may be dynamic, but the initial failure mode is of paramount importance. Some limited discussion of this dynamic behavior is provided, primarily in the context of the duration of hot shorts. Given a fire exposure of sufficient duration and intensity, the available experimental evidence indicates that all of the conductors in the damaged electrical cables will ultimately short to the grounded raceway. However, in the context of a real fire event, fires do not burn forever, and fires do not always create intensely damaging exposures. Hence, the shorting behavior of a given electrical cable could, for example, involve sustained hot shorts, shorts to ground, or hot shorts that later transition to shorts to ground.

8.5.1 Conductor-to-Conductor Short Circuits

Conductor-to-conductor short circuits are broadly categorized as either intra- or inter-cable. Intra-cable conductor-to-conductor shorting implies that the short circuit involves conductors within a single multi-conductor electrical cable. Inter-cable conductor-to-conductor shorting implies that the short circuit involves the conductors of two or more separate electrical cables (single and/or multi-conductor). Note that it is possible to have both intra- and inter-cable conductor-to-conductor short circuits active concurrently.

Conductor-to-conductor short circuit electrical cable failures have the potential to induce a range of circuit faulting behaviors. Such failures can lead to loss of circuit function, corrupted indications, loss of control, and spurious actuations. The actual circuit fault observed is entirely dependent on which conductors actually short together. However, the relative likelihood of conductor-to-conductor short circuits is of critical interest to the risk quantification.

In this context, we are primarily interested in initial cable failures that are manifested as a conductor-to-conductor short circuit that does not simultaneously involve a short to an external ground. As discussed below, one or more of the shorting conductors may be grounded, in which case, the conductor-to-conductor short circuit may have the same circuit fault effect as a conductor-to-external ground short. However, from a mechanistic view of cable failure, the first question to ask is the likelihood that the initial short circuit involves only conductors and not an external ground. One can then consider the nature of the conductors present and potential combinations of conductors, each of which may have unique circuit faulting effects.

There is currently little data available on cable failure modes and effects. A recent review sponsored by the RES identified a small number of experiments providing relevant data but also concluded that most electrical cable fire experiments provided little or no information on cable failure modes and effects (LaChance, et al., 2000). Hence, of particular note is a recently completed set of tests performed by NEI and EPRI with the participation of the NRC (NUREG/CR-6776 and EPRI TR-1003326). These tests provide the most relevant data on cable failure modes and effects currently available and will be discussed in some detail.

A total of 18 fire tests were conducted, each involving a cable tray and four to five monitored cable bundles. The tests explored a limited range of fire exposure conditions, cable types, and routing conditions. The data have provided many interesting insights into cable failure modes and effects behavior. However, the data are subject to substantial limitations, and caution must be exercised in extrapolating the results to any specific application.

First, the data were gathered in an atypical room. The test room was a steel plate box of limited dimension [3.04 m x 3.04 m x 2.43 m (10 ft x 10 ft x 8 ft)]. Given the steel room construction, heat losses from the walls and ceiling of the room were much greater than would be anticipated given a wall material such as concrete or gypsum wallboard. Hence, the relationship between the fire intensity and the room temperature was somewhat distorted in comparison to other room fire tests. In many of the tests, the room temperatures hovered very near the anticipated failure threshold temperatures for, in particular, the thermoset cables being tested. Hence, consistent with past experiments, the fire damage times were often prolonged (in some cases in excess of one hour). At higher exposure temperatures, the damage times would have certainly been shorter. It was also observed that some of the larger fires burned in an under-ventilated condition (as evidenced by an increase in room temperature when the size of the ventilation opening was increased during a given test). Hence, fires may not have reached the full burning intensity cited as the nominal fire intensity in the test reports.

Second, the circuit tests conducted by NEI used a surrogate MOV control circuit. The same circuit, with some variations, was used in all tests. The characteristics of this circuit may not be typical of other types of control circuits. Further, quantification of the circuit fault mode results is in part dependent on the circuit design, in particular, the number and placement of fuses, the number of energized conductors, the number of target conductors, and the presence of a ground conductor in the control cable. For another circuit with a different combination of conductors, the results could be quite different. For example, the presence of a grounded conductor in each multi-conductor electrical cable almost certainly contributed to a higher incidence of shorts to ground and a lower likelihood of spurious actuation.

Finally, the tests used primarily AC power sources. The NRC portions of the tests did involve some DC testing, but experimental problems caused much of the DC data to be compromised. The data did result in some conflicting information, hence, the applicability of AC circuit test results to DC circuits remains uncertain.

The results for the NEI MOV circuits were expressed primarily in the context of two circuit fault modes; namely, fuse blows (indicating an energized conductor shorting to ground or to a grounded conductor) versus spurious actuations. Overall, a substantial fraction of the cable failures resulted in a spurious actuation circuit fault mode.

The NRC sponsored portions of the tests focused on monitoring conductor shorting behavior through measurements of the conductor insulation resistance (IR) values during the fire tests. As the electrical cables are heated, the electrical insulation value of the insulation material is degraded. This degradation was monitored for both conductor-to-conductor and conductor-to-external ground. As a result, the actual shorting patterns between various conductors and between each conductor and ground could be determined. The initial cable failures were dominated by intra-cable conductor-to-conductor short circuits. The conditional probability of this mode of cable failure was estimated as 80-percent or higher based on these and other tests (conditional on electrical cable failure attributable to fire).

One possible explanation for the high likelihood of intra-cable conductor-to-conductor short circuits revolves around manufacturing practices associated with multi-conductor electrical cables. When multi-conductor electrical cables (with more than two conductors) are constructed, the individual conductors are first formed and insulated. The various insulated conductors are then brought together and the filler[30] and jacket materials are applied. In the jacketing process, the insulated conductors are generally twisted around each other to form a tight arrangement. If, for example, a length of a multi-conductor electrical cable is laid out along the floor, one typically observes a spiral pattern in the outer ring of conductors. This spiraling may leave a residual tension between the conductors. As the insulation materials are heated and lose their physical integrity (i.e., either melting or charring) this residual tension may draw the conductors together.

8.5.2 Combinatorial Models for Conductor-to-Conductor Shorting

Section 8.5.1 has discussed conductor-to-conductor short circuit failures in a very broad context that is essentially independent of the circuit to which the electrical cable is attached. There is, however, an interest in more specific modes of conductor-to-conductor shorting that would be relevant to a given circuit. Some analysts have proposed the application of combinatorial models to address this problem. To date, such models have not been assessed for validity, hence, their application to risk analysis remains unproven.

---

[30] Filler materials fill voids between the individual conductors within a multi-conductor electrical cable and may include materials such as paper, natural fibers, or polymeric (e.g., nylon) fibers.

The most obvious example where such a model might be applied is in estimating the likelihood of hot shorts leading to spurious actuation. To illustrate the combinatorial model approach, consider a circuit where there is one specific conductor (one target conductor) within a seven-conductor electrical cable that, if energized, would cause a spurious actuation. Further assume that there is one other conductor in the same electrical cable that can provide the energizing source for the hot short (act as the source conductor). The analyst concludes that intra-cable shorting is the mode of cable failure most likely to cause a spurious actuation. The spurious actuation analysis then needs to estimate the likelihood that a cable failure will create a hot short between the one source conductor and the one target conductor of interest. The analyst might then consider the total number of conductor pair shorting combinations available. For a seven-conductor electrical cable there are 21 such combinations possible. Only one of these pair combinations leads to spurious actuation. Hence, the analyst might conclude that the likelihood of the spurious actuation is 1 in 21. This is a very simplistic example intended only to illustrate the approach; however, it is not a recommended approach. Indeed, the available experimental evidence would indicate a much higher likelihood of spurious actuation for this configuration illustrating that this simplistic model fails to capture the important behaviors adequately.

Potential problems with such approaches have not yet been resolved. First, the shorting behavior of multi-conductor cables is complex and often involves more than two conductors in a shorting group. Second, the conductor shorting behavior is not totally random, but rather, tends to involve adjacent conductors within the electrical cable. Hence, the likelihood that any two conductors might short together is dependent in large part on their relative proximity to each other within the electrical cable. In most cases the analyst will not know the exact orientation of circuit functions and individual conductors in an electrical cable. The conductor-to-circuit wiring configuration may need to be treated as an aleatory uncertainty, and that uncertainty could be substantial. Third, many circuits will contain a "mitigating conductor" (e.g., a grounded conductor) that if involved in the shorting could mitigate a hot short (e.g., by tripping the circuit protection features). Again, the combinatorial models need to address this aspect as well.

Combinatorial models represent a potentially valuable approach that will likely see further development in the near future. For example, one participant in the recent EPRI expert panel proposed a more complex combinatorial model that incorporates an advanced view of cable failure behavior (see Appendix B of EPRI TR-1006961). The model appeared to work well in comparison to the experimental data available to the expert panel, but remains unproven in a more general context.

8.5.3 Conductor-to-External Ground Short Circuits

For all electrical cables, there is a potential that the insulated conductors will short to an external ground source. In particular, the raceways in which electrical cables are routed (trays and/or conduits) are generally metal (often galvanized steel and less commonly aluminum) and are typically grounded. Hence, most electrical cables have more or less ready access to an external ground plane once the cable insulation breaks down.

Note that a conductor-to-conductor short circuit that happens to involve a grounded conductor will have the same circuit faulting effect as a conductor-to-external ground short circuit. However, in the mechanistic view of cable failure modes and effects, the relative likelihood of a conductor-to-conductor short involving a grounded conductor is treated separately. The current discussion focuses on the role of the external ground sources in cable failure modes and effects behavior.

The conductor to external ground failure mode can introduce unique circuit consequences. For most AC circuits, shorts to ground will render a control or power circuit non-functional, but will also have a mitigating effect on, in particular, the possibility of spurious actuation circuit faults. Shorts to ground on an energized electrical cable of a grounded circuit will generally cause circuit protection devices to trip deactivating the impacted circuit. This could impact either the control or motive power of a circuit depending on which electrical cables are impacted (see Section 8.5). Also note that if a conductor-to-conductor short circuit does form, and if any one of the involved conductors shorts to an external ground (or is itself grounded), then all of the involved conductors will also short to ground. In a risk context, the primary interest is the conditional likelihood that a short to ground will be observed before a hot short that might lead to a spurious actuation failure. Note that ungrounded DC circuits are unique with regard to shorts to an external ground. A single short to ground on an ungrounded DC circuit has essentially no impact on circuit performance. However, multiple shorts to ground may adversely impact the circuit. In effect, for an ungrounded DC circuit, the external ground acts as an external conduit for the formation of conductor-to-conductor shorts.

For multi-conductor electrical cables, 20-percent or less of the observed cable failures are likely to involve an initial short to external ground. For single conductor electrical cables the likelihood of a short to external ground failure is estimated to be substantially higher (perhaps 50-percent or higher) but there is little experimental data available to support this contention.

Experiments show that given a sustained damaging fire, all of the conductors in the damaged electrical cables will ultimately short to ground. Hence, another potentially important consideration in the context of fire risk is the transition time associated with this behavior (e.g., transitions from conductor-to-conductor to conductor-to-external ground short circuits). This transition behavior is important because it may, for example, determine whether a valve might fully reposition, or for how long a PORV might remain open, or how long an operator might have to recover a spurious actuation before the control function is lost.

Experimental evidence indicates that, given a sustained damaging fire, initial cable failures will likely transition to shorts to external ground over a wide range of times. In the recent NEI tests (EPRI TR-100326), for example, some of the spurious actuation circuit faults were of momentary duration (e.g., less than one second) while others were maintained for in excess of 11 minutes. The average duration of a spurious actuation signal was 1-3 minutes depending on the cable type. It should also be noted that in the NEI tests, one of the conductors in the multi-conductor control cable was grounded, and short circuits to this grounded cable would mitigate the actuation signal.

Overall, the test data available do suggest that sustained conductor-to-conductor shorts are possible. It should also be noted that suppression of the fire could "lock in" conductor-to-conductor electrical cable failures such that the short to external ground transition might not be observed in all cases. Hence, it would not be appropriate to assume that shorts to an external ground would mitigate all potential spurious actuation failure within any given time period. Statistically this is certainly a non-trivial possibility that increases in likelihood the longer a fire lasts. However, it is far from certain that this transition will occur, especially given aggressive firefighting activities.

Overall, short to external ground cable failures are high likelihood events given fire-induced cable failures and should be considered in a risk-informed analysis. Recall also that conductor-to-conductor short circuits may have the exact same impact as a conductor to external ground short circuit if one (or more) of the involved conductors happens to be grounded.

8.5.4   Loss of Conductor Insulation Resistance (IR)

Polymeric insulation materials, thermoset and thermoplastic, dominate the current electrical cable applications in the U.S. nuclear power industry. When these materials are heated, they lose their electrical insulation value. Based on available equipment qualification test results (NUREG/CR-4537), the degradation in resistance is logarithmic with linear increases in temperature. An example of this behavior is illustrated in Figure 8-1 (reproduced from NUREG/CR-6681). This same mechanism can lead to a loss of insulation resistance failure mode when electrical cables are heated in a fire.

In general terms, this mode of failure is associated with a degradation of the electrical cable that is less severe than an actual short circuit condition. This mode would be active at temperatures below the melting point of a thermoplastic material, and below the nominal gross failure threshold of thermoset materials. For some circuits, a significant degradation in the insulation resistance between individual conductors or between conductors and ground could compromise the performance of the circuit.

This mode of failure is particularly relevant to instrumentation circuits. A typical instrumentation circuit operates at 4-20 mA. Given the nature of the instrument loop circuit, a breakdown in the instrument cable insulation could cause all or part of the intended current signal to be diverted, bypassing the instrument display device. This would bias, or corrupt, the instrumentation reading. Note that the direction of the bias will be predictable because while one can divert some of the intended signal, one cannot increase the current flow to the indication device. The direction of the bias will always be towards the low-current indication, although whether low current corresponds to high or low on the process variable scale must be determined for each specific case. For other types of circuits (i.e., those with more robust electrical energy), this mode of failure is unlikely to compromise circuit function. Rather, for higher-energy circuits, actual short-circuit conditions will be the failure modes of interest.

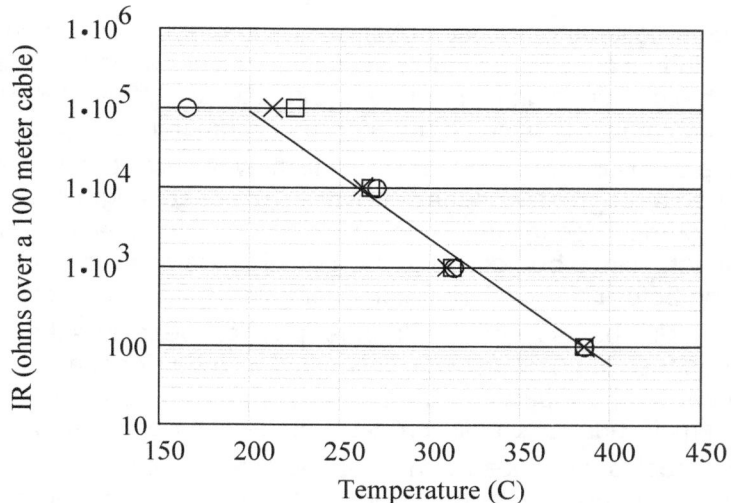

Figure 8.1 IR versus Temperature Behavior of a Typical Electrical Cable Insulation Material

This plot shows test data and a linear regression curve fit for a Brand Rex cross-linked polyethylene (XLPE) insulated 12 AWG 3-conductor electrical cable. The data are from Table 4 of NUREG/CR-5655, "Submergence and High-Temperature Steam Testing of Class 1E Electrical Cables" (NUREG/CR-5655). Similar plots can be generated for any given cable type, size and voltage rating given test data that reports IR as a function of temperature.

A recent test series examined this behavior for instrument circuits (NUREG/CR-6776). In general, a pronounced difference was noted between the behavior of thermoset and thermoplastic insulated instrument cables. Thermoplastic insulated electrical cables tended to fail abruptly and catastrophically with little or no indication of degraded signals prior to loss of signal. Thermoset insulated electrical cables illustrated a prolonged period of corrupted signal transmission before a complete loss of signal was observed. Hence, the use of thermoset insulated electrical cables appears to increase the potential that operators might be misled by a corrupted signal. An offsetting observation was that the thermoplastic insulated electrical cables failed far more quickly than did the thermoset insulated electrical cables. This is also consistent with the observation that thermoset insulated electrical cables are generally more resistant to fire-induced failure than are thermoplastic insulated electrical cables.

8.5.5 Loss of Conductor Continuity

As noted previously, a loss of conductor continuity implies that the physical and electrical integrity of the conductor itself is lost. That is, the conductor breaks. Note that this mode of failure may also be referred to as an open circuit cable failure, although this may lead to confusion with use of the term open circuit in the context of a mode of circuit faulting. An open circuit as a fault mode generally implies the opening of circuit protection devices (fuses or breakers). A loss-of-conductor continuity cable failure can have similar effects on a circuit, especially if the failure is associated with an energized power supply conductor.

Loss-of-continuity conductor failures have been observed both in actual fires and during tests. However, this mode of failure is considered highly unlikely to occur as the initial failure mode. Evidence taken from both experience and experiments indicates that fire-induced loss-of-conductor continuity failures may be observed under three circumstances as follows:

- During a prolonged fire exposure, the conductor material may melt causing a loss of conductor continuity. This is often a progressive behavior over an exposed length of electrical cable, rather than an abrupt or localized failure. In most cases, all of the electrical cable insulation materials would have long since burned away; hence, all of the conductors would have shorted to ground long before a loss-of-conductor continuity failure were observed.

- Loss-of-continuity may also be associated with other physical behaviors that could place an undue physical load on the electrical cables. This might include, for example, the collapse of cable supports or raceways, the impact of a hose stream on a badly damaged electrical cable, or physical stressors that may cause electrical cables to come loose from a terminal connection.

- High-energy electrical cables (i.e., those with a high voltage and/or current potential) may experience repeated, short duration, high-intensity arcing shorts (either phase-to-ground or phase-to-phase). These shorts are typically of such high energy that the conductor material is melted and/or vaporized at the location of the short causing the short to self-mitigate. Circuit protection devices (fuses and breakers) have a finite current/time response behavior, and conventional circuit protection devices are not designed to detect arcing faults (arcing fault circuit interrupters are available but are not widely used in the U.S. nuclear power industry). Hence, the circuit protective features may not be activated/tripped by these short duration arcing short circuits. If this behavior is repeated a sufficient number of times, the conductor continuity may eventually be lost.

The risk implications of a loss-of-continuity cable failure must be viewed in the context of the circuit under analysis. No concise risk analysis of this question has yet been conducted. Loss-of-conductor continuity failures are not expected to be risk-significant, in part because of their low likelihood of occurrence and in part because they are not expected to introduce unique risk scenarios or insights. The rationale for the second part of this conclusion depends on the type of circuit considered, as follows:

- **For control and instrument cables** the available power is not sufficient to induced high-energy arcing conditions. Hence, loss-of-conductor continuity failures will only be observed in long duration fires, and then only after all conductors have shorted to ground. This implies that other modes of cable failure (i.e., conductor-to-conductor and conductor-to-external ground short circuits) will determine the circuit faulting behavior.

- **For power cables** it is possible that a loss-of-conductor continuity might occur as a result of high-energy arcing. However, for power circuits, the loss-of-conductor continuity cable failure will mimic an open circuit fault associated with tripping of circuit protection features; namely, power will be unable to reach its intended destination. This same mode of circuit faulting is observed given a sustained short-to-external ground or phase-to-phase conductor shorting behaviors. Hence, in terms of the impact on the power electrical cables' own circuit, no unique fault modes are introduced. The only difference given a loss-of-conductor continuity failure is that the side of the broken conductor(s) leading back to the power supply source might remain energized; hence, these conductors might be available as a hot short source for other electrical cables. In the hot short analysis, the existence of an appropriate source is generally assumed unless the lack of such a source can be confirmed. Hence, again, the loss-of-conductor continuity failure should introduce no unique risk scenarios or insights.

8.5.6 Summary of Electrical Cable Failure Mode Insights

For multi-conductor electrical cables the dominant mode of cable failure anticipated is intra-cable conductor-to-conductor short circuits. Evidence in this area is strong and indicates that 80% or more of all fire-induced multi-conductor cable failures will initially involve intra-cable conductor-to-conductor short circuits. This appears to apply to both thermoset and thermoplastic insulated electrical cables. (Recall that not all intra-cable conductor-to-conductor shorts involve hot shorts leading to spurious actuation as discussed further below.)

The available data indicate that inter-cable conductor-to-conductor shorting is possible, but is less likely to occur than is intra-cable conductor-to-conductor shorting. The data also indicate that inter-cable shorting is more likely given thermoplastic insulated electrical cables than it is given thermoset insulated electrical cables. The available data on inter-cable shorting is not sufficient to provide firm estimates of conditional likelihoods. However, for thermoplastic insulated electrical cables, the likelihood of inter-cable conductor-to-conductor short circuits is probably 0.5 or less. For thermoset insulated electrical cables the likelihood of inter-cable shorting is probably 0.1 or less. For both electrical cable types the likelihood of inter-cable shorting may be much lower depending on the cable raceway configuration and fire exposure conditions.

For both electrical cable types, thermoplastic and thermoset, the likelihood of a hot short versus a short to ground will depend on a number of configuration factors that are currently not well characterized. While some of these factors may have little influence on the intra-cable shorting behavior, they likely have a stronger influence on the likelihood of inter-cable shorting. That is, for some configurations inter-cable shorts cannot be considered a rare event while for others, the likelihood may be very low. Factors that are believed to have a significant impact on the likelihood of inter-cable shorting include the following (NUREG/CR-6776):

- The nature of the fire exposure: Direct flame/plume exposures that heat the cables from below may be more prone to cause shorts to ground than would radiant heating that heats the cables from above.

- The loading of the raceway: A tray with many electrical cables would be more likely to experience inter-cable shorting than a sparsely loaded cable tray.

- Trays with maintained spacing of the electrical cables: For such configurations (generally used only for larger power cables), inter-cable shorting independent of the grounded raceway appears to be highly unlikely.

- The position of the critical electrical cables within the raceway: Electrical cables located at the bottom of a tray would be more likely to short to ground than electrical cables located on top of a cable load.

- Cable tray type: Cable tray type (e.g., ladder back versus solid bottom) impacts the cable support loading and may impact the failure behavior, but this parameter has not been investigated.

- Use of conduits: Electrical cables in conduits appear to have a higher likelihood of shorts to ground and a lower likelihood of hot-short induced spurious actuation in comparison to electrical cables in cable trays. This appears to apply to both intra- and inter-cable shorting behaviors.

It also appears that loss-of-conductor continuity failures are unlikely to occur as an initial failure mode. Such failures are likely to occur, but only after extended fire exposures or after repeated arcing faults for higher energy electrical cables. This failure mode is not expected to contribute significantly to fire risk.

Combinatorial models may be used in the future as a tool to estimate the likelihood of specific cable failure modes, and in particular the likelihood of hot shorts leading to spurious actuation. However, these models have not been fully developed and remain unproven.

8.6   Circuit Fault Modes

The risk implications of cable failure induced circuit faults will be discussed in the context of the three primary circuit types or functions; namely, power, indication/control, and instrumentation. For each circuit type, the cable failure modes and circuit fault modes of potential interest are somewhat unique. Fault modes of interest for each circuit type are as follows:
- Power circuit fault modes
    a.   Loss of primary or motive power to a system or component
    b.   Hot shorts leading to spurious actuation
    c.   Multiple high impedance faults

- Control and indication circuit fault modes
    d.   Loss of control function or power
    e.   Spurious actuation in control circuits
    f.   Loss of control indications
    g.   False control indications

- Instrumentation circuit fault modes
    h.   Loss of permissive signals
    i.   False permissive signals
    j.   Corrupted instrument gage readings

Each of these circuit types is discussed in detail in the subsections that follow.

8.6.1 Power Circuit Fault Modes

Loss of Primary Motive Power

For power circuits, many electrical cable failures will lead to a loss of primary motive power to plant devices[31]. A loss of primary motive power implies that the faulted system stops operating. Continuously operated devices such as pumps, fans, and motors will stop and/or will be unable to start. Intermittent operating devices such as MOVs would cease movement, if movement were in progress at the time of the cable failure, and would be unable to move through normal control functions (in some cases manual repositioning would still be possible, e.g., using a handwheel). Devices that require continuous power to maintain position, such as a solenoid operated valve, would cease to be operable and would stay in, or reposition to, their de-energized condition.

Loss of primary motive power could result from the following power cable failures:
- phase-to-ground short circuits involving an energized conductor
- phase-to-phase short circuits involving two or more energized conductors
- hot shorts to a power circuit of higher voltage potential

In each case, the cable failures would lead to opening of circuit protective features (e.g., breakers and/or fuses) — an open circuit fault mode for the power supply circuit.

Given the many ways that power cable failures might lead to an open circuit fault condition, the loss of motive power will be the predominant fault mode given the failure of power cables. It can nominally be assumed that 99-percent or more of the power cable failures would lead to this mode of circuit faulting.

Power Cable Hot Shorts Leading to Spurious Actuation

The likelihood of power cable failure induced spurious operations depends in large part on the nature of the power supply system. Single-phase AC power systems may be somewhat vulnerable to spurious actuation faults, whereas three-phase AC and ungrounded DC systems appear to have a far lower likelihood of spurious operation.

For the ungrounded DC and three-phase AC systems, multiple concurrent inter-cable hot shorts of the proper polarity are required to induce spurious actuation of plant components as a result of failures in power cables. However, the conditions leading to this fault mode are quite specific and are considered highly unlikely to occur. In general, a spurious actuation induced by power cable failures for these two types of systems requires either two or three (depending on whether the system is DC or three-phase AC) concurrent hot shorts of the proper polarity such that the attached device is appropriately powered.

---

[31] Motive power is distinguished from control power. Motive power is the source of energy that runs a primary electrical device such as a motor, while control power is a separate, although potentially dependent, light power circuit used to energize secondary control devices such as relays which in turn control the flow of motive power to the primary component.

For a single-phase AC system, the neutral power leg is typically tied to ground. Hence, a single hot short from the 'hot' leg of the AC system to a hot leg power lead for another device can cause spurious operation. Return power can be transmitted through the common ground, bypassing the neutral conductor (hence, a neutral-to-neutral short circuit may not be required). General practice for NPPs in the United States is to use separate electrical cables for each power supply circuit. Hence, spurious actuation would generally require an inter-cable hot short. Given that only one inter-cable conductor-to-conductor hot short is required, the likelihood is higher in comparison to the DC and three-phase AC cases. In all three cases, the voltage and current characteristics of the source conductors must be compatible with the target device. Application of an excessive voltage may damage the target device rather than cause it's actuation. Similarly, application of a DC source to an AC device (or vise-versa) would likely damage rather than activate the target device. Finally, if the current available to the source conductors is not sufficient to power the target device, then an overcurrent condition will likely trip the source conductors' protective circuit features mitigating the fault.

Nominally, the probability of such spurious actuation faults given the failure of power cables is judged to be low for all three cases, although no specific investigation of this potential has yet been undertaken. The conditions required depend on the nature of the power source involved:

- For three-phase AC power circuits (typical of large motors and MOVs), a spurious actuation would require three concurrent hot shorts, each provided by a source of compatible power (voltage and current). Shorts to an incompatible power source (wrong voltage or inadequate current) would likely either damage the target component or trip circuit protection on the source bus. All three source conductors must also be powered from the same electrical bus. Reversal of two phases of the source/target configuration could cause the target device to operate in reverse, and could well damage the target device. Spurious actuations for this configuration are considered highly unlikely and are estimated to have a 0.001 conditional likelihood or less. Note that if a ground conductor is routed with the energized conductors (e.g., a triplex cable with ground), the likelihood of a spurious actuation will be further reduced.

- For single-phase AC power circuits (typical of smaller motors and MOVs), the neutral is generally tied to ground so only one hot short from a power source of proper voltage and current would be required. Again, shorts to a source bus of improper voltage or inadequate current would likely either damage the target component or trip circuit protection for the source conductors. This is also considered an unlikely occurrence, but the conditional probability of occurrence given cable failure may be as high as 0.1 depending on the nature of the power cables and grounding provisions. For most cases the likelihood should be lower. For example, if an explicit ground conductor is routed with the high and low potential power cables (e.g., a two-conductor electrical cable with ground or three-conductor electrical cable), then the likelihood of a spurious actuation will be lower. Use of armored electrical cables could essentially eliminate this possibility because there is virtually no possibility of inter-cable shorts independent of the grounded armor. Routing of electrical cables in conduits would also reduce the likelihood even if the conduit contains more than one power cable such that inter-cable shorting remains a possibility.

- For an ungrounded DC power system, two concurrent hot shorts of the proper polarity are required to induce a spurious actuation. Alternatively, one of the two polarity hot shorts might be provided through the effects of multiple shorts to ground, however, one side of the power supply system must remain isolated from ground or circuit protection would be tripped.

If the DC voltage is not appropriate to the target device, the device would either fail to operate or might be damaged. Adequate current is also needed.

Overall, spurious actuations that are induced by failures in those electrical cables that provide motive power to a device are considered unlikely. The highest likelihood case is single-phase power systems, and while unlikely, this type of circuit fault might still have some non-trivial contribution to risk and should be considered. For the ungrounded DC and three-phase AC power systems, the occurrence of a power cable failure induced spurious actuation appears unlikely. Hence, consideration of such fault modes for other than high consequence applications (e.g., high-low pressure interfaces) does not appear to be warranted. The conditions required to cause such faults are simply too specific and too restrictive to be considered likely, and the potential for such faults will likely have little risk significance.

Multiple High-Impedance Faults

There is a potential that concurrent failures involving several power cables could introduce a unique failure mode for plant power distributions systems. In particular, if multiple power cables fed from a common bus experience low quality or high impedance shorts, each electrical cable could experience current leakage beyond that expected as a result of the normal operation of the powered component. Enough faults of this type could create a demand on a higher level circuit protection device that exceeds the protection level of the higher level bus, without exceeding the protection level of the individual circuits. The physics of such behaviors remain poorly understood, and cannot be dismissed out of hand. However, based on the knowledge we have regarding cable failure behavior, this mode of failure is considered to be unlikely in practice. Several factors work against such an occurrence.

One such factor is the precise quality of the faults required to create such a situation. The multiple high impedance fault scenario postulates that several electrical cables are leaking current at levels just below the trip point of the nearest up-stream circuit protection device. This would require a sustained fault with a rather precise resistance, and indeed a resistance that is relatively low.

However, the shorting behavior of energized electrical cables does not favor the formation of such shorts. Experiments do show that electrical cables will tend to degrade progressively over time (NUREG/CR-6776, NUREG/CR-5655, and NUREG/CR-5546). The data show that electrical cables energized to a non-trivial level (i.e., greater than approximately 50 V) display an abrupt shorting behavior beyond a certain level of degradation. It appears that once the degradation reaches the point where the insulation is providing about 1,000–10,000 ohms IR, there is an abrupt transition to a low-impedance or dead-short fault.

A second factor working against this scenario is timing. The multiple high impedance fault scenario requires that several faults be active concurrently. This is certainly possible, but experimental evidence suggests that even electrical cables located in a common tray will fail at discrete times rather than all at once. The issue of timing combined with the need for a sustained fault of a rather precise resistance value would appear to indicate that a multiple high-impedance fault, leading to the tripping of a higher level power distribution bus, while possible, is of low likelihood.

Finally, the risk implications of the multiple high impedance fault issue are mitigated to some extent given that operator response could potentially recover the undamaged circuits The scenario does not postulate that the higher level bus is damaged beyond recovery, simply that the circuit protection trips at a level higher in the distribution system than the level at which the actual cable failures occurred. Hence, isolating the damaged circuits would allow for re-setting of the tripped breaker/fuse and recovery of the higher level bus. The timing of such recovery actions, and the likelihood of success, would need to be considered in a risk assessment.

In the context of a fire PRA, the loss of a higher level bus, when treated, would typically be assumed to occur as a result of a random failure of the nearest circuit protection feature to trip on demand. In this scenario, a single electrical cable failure might fail to be isolated by the first upstream circuit protection feature, and would therefor cascade to the next level bus. The risk implications of the multiple high impedance issue could be estimated using a similar approach by increasing the random failure probability of the local circuit protection device to reflect the likelihood of the multiple high impedance fault scenario. The effect on the plant systems would be similar, although the multiple high impedance fault scenario would require that more failed circuits be isolated before the higher level bus can be recovered. Such an exercise has not yet been conducted.

8.6.2 Control and Indication Circuit Fault Modes

In U.S. NPPs, the control and indication functions tend to be combined in a common circuit for a given device. For example, the open/close/in-motion indicator lights for an MOV tend to be a part of the overall control circuit and the conductors associated with the indication functions are often routed in the same electrical cable as those associated with the control functions. Hence, circuit fault modes for both control and indication circuits are treated as a common subject.

Loss of Control Function or Power

One likely mode of circuit faulting for control and indication circuits is a loss of control function. For continuously operating systems, a loss of control function may leave the system components running. For some devices, such as solenoid operated valves, a loss of control power can lead to repositioning of the device to the fail-safe condition. For other devices, such as an MOV, the loss of control function would leave the device in its prior state and render the normal controls ineffective at changing that state. Loss of control function fault modes are of potential risk importance if the system or function lost must be manipulated to support hot shutdown. This would include both front line and support systems. Loss of control function failures impacting only cold shutdown functions are not likely to be risk significant provided that hot shutdown can be achieved. Loss of control function failures for containment isolation functions are also of low risk significance unless the ability to achieve hot shutdown is also compromised.

A loss of control function would typically be associated with failures in the control system electrical cables, and in particular, either a loss of control power or other short circuit conditions that will divert the control power in the event that a control operation is attempted. In most cases, a loss of control function will be associated with a loss of the control power source. If the conductors that supply control power to the control circuit short to ground (or across polarities for DC circuits), then circuit protection for the control power circuit would likely trip. In some cases, a control cable failure can leave a control circuit nominally intact. However, upon any attempt to manipulate the control circuit various faults can occur that would render the control system inoperable (e.g., see MOV circuit analysis examples in LaChance, et al., 2000).

Spurious Actuation in Control Circuits

The issue of spurious actuations (or spurious operations) has received much attention. Spurious actuation is one specific type of "maloperation" fault as identified in the NRC fire regulations. Spurious actuation involves activation of a functional mode of a system or component caused by fire-induced electrical cable failures. Based on current understanding of the circuit analysis issues, the most likely source of spurious actuations will be control circuit electrical cable failures. Because the shorting behavior of the electrical cable conductors is complex, the analysis of spurious actuations is also complex.

A spurious actuation is generally caused by hot shorts, but not all hot shorts will lead to a spurious actuation, so care must be taken in estimating the likelihood of a spurious actuation. The short circuit must involve the right set of conductors. For many circuits, a specific pair of conductors must be involved in a common short. For grounded circuits, the short must not involve an external ground or grounded conductor. For ungrounded DC circuits, a pair of correct-polarity hot shorts is required. The exact configuration of shorts that could cause spurious actuation is potentially unique for each circuit in the plant; however, in practice many circuits will share common configurations and common failure/fault modes. The number of unique configurations that might need to be considered has not been determined.

A detailed analysis of spurious actuation is a tedious undertaking for most circuits. For the purposes of regulatory compliance, simplified methods of analysis are often employed. One common approach is the "hot probe" analysis. Under this approach the analyst assumes that a source conductor of proper voltage and current will be available. Each conductor in a circuit is then systematically energized by this "hot probe" source conductor to determine if a spurious actuation is possible. For the purposes of risk assessment, the regulatory analysis results can be applied, but generally only with some considerable uncertainty in quantification of the results. A more rigorous quantification requires a more rigorous analysis.

Time may also be a factor for some cases. Time may be important from two primary perspectives. Specifically, time may be important from the perspective of when the spurious actuation occurs and how long it persists. For example, a spurious actuation may open an solenoid-operated valve (SOV), but if the actuation is mitigated within a short period of time, the fault may have minimal risk implications. Similarly, a hot short may initiate a spurious actuation of an MOV, and the duration of the hot short may determine whether the valve fully repositions or only partially repositions. For some systems, a hot short might start the system (e.g., a pump), but mitigation of the hot short might cause the system to stop.

The questions of timing are also important when the issue of multiple spurious actuations is considered. In some cases, spurious actuations may only be risk significant if they are postulated in combination with other spurious actuations (or potentially other specific system faults). Hence, the timing of onset and the duration of the faults will influence the likelihood that any two or more spurious actuations might be active simultaneously.

The only experimental study that has directly assessed electrical cable failures leading to spurious actuation are the recent joint NEI/NRC electrical cable failure modes and effects tests described previously. In particular, the NEI MOV circuit tests provided many insights into spurious actuations.

As previously noted, a number of spurious actuations were observed, and the likelihood of spurious actuation given electrical cable failure was found to depend on a number of factors. Overall, the likelihood of spurious actuation given cable failure cannot be considered small. For most configurations a screening value ranging from 0.1 to 1.0 would be appropriate. A recent EPRI expert panel estimated the spurious actuation likelihood for the "base case" configuration[32] of this circuit ranges from 0.1 to 0.5 due only to intra-cable hot shorts (Reference EPRI TR-1006961). Variations from the base case led to other likelihood estimates, including the following general effects:

- Armored electrical cables showed a somewhat lower likelihood of intra-cable hot shorting, presumably due to the prevalent ground plane represented by the grounded armor.

- Electrical cables in conduits appeared less susceptible to hot-short induced spurious actuations, again presumably due to the prevalent ground plane represented by the grounded conduit.

- The lack of a CPT in the circuit increased the likelihood of a hot-short induced spurious actuation (by a factor of approximately 2). Note that CPTs are common in MOV control circuits.

- Inter-cable conductor-to-conductor short circuits are substantially less likely than intra-cable conductor-to-conductor short circuits. One explanation for the lower likelihood of inter-cable shorting is that there is no inherent residual tension between the conductors of two separate electrical cables as there is between the conductors of a multi-conductor electrical cable (see previous description).

- As compared to thermoset insulated electrical cables, the thermoplastic insulated electrical cables showed a similar likelihood of intra-cable hot shorts leading to spurious actuation, but an increased likelihood of inter-cable hot shorts leading to spurious actuation.

Multiple Spurious Actuations

A particular aspect of the spurious actuation question is the likelihood that multiple spurious actuations might be observed during a given fire. The evidence both from testing an actual fire experience clearly indicates that multiple spurious actuations are possible. However, it is appropriate to consider multiple spurious actuations in a more structured context.

There are several potential aspects to the multiple spurious actuation question, each of which may have unique risk implications. One of the most critical questions relates to timing. Specific issues related to timing include the following:

- Simultaneous behaviors: Simultaneous implies that events occur at essentially the same moment in time. To date no specific applications where simultaneity has been a critical factor to risk have been identified. Based on our understanding of electrical cable failure behavior, the onset of multiple cable failures simultaneously is possible, but appears unlikely. The most likely case leading to simultaneous spurious actuation faults would be where multiple faults might be created by the failure of a single cable. If the multiple faults require the failure of multiple cables, simultaneity appears unlikely. Fire testing indicates that even within a given raceway cable failures tend to be somewhat distributed over some time period, usually

---

[32] The base case involved a thermo-set insulated electrical cable in a cable tray with a control power transformer (CPT) in the circuit to limit the available total circuit power.

measured in minutes. Several factors likely account for this observation. For example, the heating from a fire is generally nonuniform; variations in electrical cable size lead to variations in their thermal response; variations in cable placement within a raceway lead to variations in the heating rate. Overall, it would appear that simultaneous spurious actuation faults are not of substantial concern in the risk context unless they can be induced by the failure of a single electrical cable.

- Concurrent behaviors: Concurrent implies that multiple faults occur at discrete points in time, but that they endure for a sufficient period of time that they overlap. Note that in this context we are referring to circuit faults, not cable failure. Note in particular that a self-mitigating cable hot short (e.g., a hot short that subsequently shorts to ground) may not mitigate the fault condition. For example, a repositioned MOV may not return to its original position when the hot short self-mitigates. Rather, some active intervention by plant operators may be required to mitigate the fault.

- Sequential behaviors: Sequential faulting implies that one fault is mitigated before being followed by another fault at a later time. Clearly, sequential behaviors are possible if not likely. For example, it appears that the 1975 Browns Ferry fire involved primarily a sequential series of spurious actuations (see discussion below) that were either self-mitigated or mitigated through operator actions during the event. However, even with sequential faults, some risk important scenarios may arise.

The test data and experience clearly indicate that concurrent hot shorts are possible. Hence, concurrent spurious operations are also possible. During the NEI MOV circuit tests, for example, some tests experienced concurrent hot shorts on two separate control circuits given the exposure of just four control circuits to potential actuation. This would tend to indicate a high potential for concurrent hot shorts and spurious actuation faults. One factor in this behavior was likely the co-location of the cables in a common raceway. The failure behavior for electrical cables located in separate raceways has not been explored extensively, although some data is available. The intensity of the fire exposure will be the primary factor in determining the timing of electrical cable failures, especially when multiple raceways are exposed.

An example where concurrent spurious actuation faults would be important is a case with two normally-closed SOVs in series in a significant diversion path. For the diversion path to open both valves must open and be held open concurrently. Self-mitigation of either hot short (e.g., by a subsequent short to ground) would return that valve to the normally closed position closing the diversion path.

A similar situation involving two MOVs, rather than SOVs, presents some interesting insights. Even given sequential self-mitigating hot short cable failures, both valves may be left open concurrently. That is, once each MOV repositions, mitigation of the hot short may not return the valve to a closed position. Rather, it is likely that mitigation of the hot short will cause a loss of control power and a loss of the normal control function while leaving the valve in the open position. Similar behaviors could be observed in circuits with latching or locking relays where even a momentary hot short might lock in a spurious actuation circuit fault. Again, the existence of concurrent spurious actuation faults is distinct from the existence of concurrent hot short cable failures for certain cases.

The assumption of sequential faults is, in essence, the basis most commonly used for current fire safe-shutdown analyses, and is the so called "any and all, one at a time" approach (a detailed discussion of the application of this approach is provided in Appendix B of this document). Two additional considerations related to multiple spurious actuations are as follows:

- Multiple actuations of a single system: It appears likely that a system that experiences one spurious actuation signal will experience two or more such signals. This was observed in both the 1975 Browns Ferry fire and during the NEI MOV circuit tests. It is also nominally consistent with the NRC/RES insulation resistance measurements made during the NEI tests as well. In the NRC/RES measurements, groups of conductors were observed to form dynamic conductor shorting groups, a behavior that could lead to multiple actuations of a circuit as a result of the failure of a single control cable.

- Actuations involving multiple systems: Both experience and testing demonstrate the potential for the actuation of multiple systems. During the NEI MOV circuit tests, for example, as many as three of the four exposed circuits experienced spurious actuations during a given test.

Overall, one cannot dismiss the possibility of multiple spurious actuations, either concurrently or sequentially. Further, one cannot dismiss either multiple actuations of a single system, or the spurious actuation of multiple systems. The obvious question is how likely are such events and how many spurious actuations are reasonable to postulate? Given the NEI MOV circuit tests in particular, the likelihood of spurious actuation of a circuit (given damage to a susceptible control cable[33]) was relatively high. The likelihood was found to be dependent on a number of factors, and varied over a fairly wide range. Important factors explored in the tests were discussed previously.

Given the identification of several important factors, it is not possible to cite a single value that would be characteristic of a "typical" control circuit. In broad terms, the mean likelihood of actuation (given failure of a susceptible control cable as observed in the NEI MOV circuit tests) ranged from about 0.1 to about 0.6, depending on how the tests are parsed. For at least one configuration, the EPRI expert panel cited an upper bound estimate of the spurious actuation likelihood of 1.0. This range represents a significant variation even given that a limited set of potential factors of importance were varied, that only one basic control circuit configuration was tested, and that the factors varied were only explored over a limited range. Overall, there is still at least one order of magnitude uncertainty in the likelihood of spurious actuation for any given circuit (assuming some level of susceptibility).

Given spurious actuation likelihoods of this order, the possibility of multiple spurious actuations cannot be dismissed. Given the data, the number of spurious actuations may be limited only by the number of susceptible cables damaged by the fire. This still, however, leaves open the questions of likelihood (how likely is it that two or more actuations would be experienced) and timing (sequential versus concurrent faults). Neither question, unfortunately, has a clear cut answer. One can, for estimation purposes, assume nominal likelihoods based on the NEI tests for a given circuit. If the conditions of the associated electrical cables are well characterized, then the estimates can be refined. If one assumes circuits with the highest level of susceptibility (e.g., a mean value of 0.6 given cable damage), and assuming independence between failures, then as many as four spurious actuations would still have a likelihood of $(0.6)^4$, which equals 0.13.

---

[33] By susceptible control cable we mean a control cable configuration wherein intra-cable shorts do hold the potential to cause a spurious actuation.

It is likely that more risk consequence mitigation will be achieved by considering the likelihood of damage to multiple control cables than from consideration of the likelihood of spurious actuation given control cable failure. In particular, most electrical cables used by the U.S. nuclear industry are fairly robust and resistant to fire damage (thermoset insulated electrical cables in particular). Experience illustrates that most fires are small, causing damage to few, if any, exposed electrical cables. These observations substantially reduce the likelihood that fires leading to multiple spurious actuations will occur. Nonetheless, given a severe fire and damage to many electrical cables, it appears that one or more spurious actuations are likely.

Lost or Misleading Control Indications

As noted previously, the indication functions are generally carried by conductors that reside in the same electrical cable with the control functions for the same circuit. (Note that instrument signals are discussed in Section 8.6.3 below.) There are various circuit fault modes of potential interest to these indication functions. Fault modes of potential interest include the following:

*   Hot shorts can illuminate indicators inconsistent with the actual system status (e.g., a valve open light might illuminate even though the valve is actually closed).

*   A short to ground can fail an indication (e.g., an indication lamp may go out).

*   Some indication faults may not be manifested until an attempt is made to operate circuit (e.g., given an attempt to operate a valve, both the open and closed indicator lights might be illuminated).

The importance of such faults to risk is primarily driven by the operator's response. Operators take control actions based on the signals presented to them. False indications may lead to unsafe actions. The importance of such faults may be mitigated by redundancy in the signals available to operators. Further, inconsistency between corrupted and intact signals may lead operators to diagnose control circuit problems. For example, if an operator sees both open and closed indicators illuminated for a single valve, they may conclude a circuit fault has occurred and will not place much faith in that circuit. Indeed, experience includes cases where operators have diagnosed the existence of a fire based on faults in their control circuits.

The risk importance of indication circuit faults has not yet been assessed. No fire risk analysis to date has explicitly considered this issue.

8.6.3 Instrumentation Circuit Fault Modes

Instrument circuits present potentially unique circuit analysis concerns. Instrument circuits provide critical information regarding the status of the plant to operators. As opposed to status indicators (discussed previously in Section 8.6.2), instrument circuits provide a variable output reading that is proportional to some process variable (e.g., temperature, pressure, level, flow rate, current draw on an electrical circuit, etc.). Instruments are important to post fire safe-shutdown for several reasons:

*   Instruments provide operators with needed information on the status of the plant The degradation of instrument reading (e.g., transmission of a corrupted reading) might mislead operators into taking improper response. A complete loss of an instrument reading might be more obvious, but deprives the operator of potentially important information.

- Instruments are often associated with permissive interlocks. A loss of an instrument signal might cause a loss of the permissive signal. This could in turn cause the shutdown, or prevent the startup, of a desired system. (An example of this is cited below where the fire-induced failure of an oil pressure signal cable caused a false low oil pressure signal and prevented the operators from starting the associated pump.)

- Some instrument signals are tied to automatic control systems or functions. Degradation in these instrument readings could lead to the undesired actuation of automated control functions.

Note that to date no fire PRA has systematically evaluated the implications of fire-induced failures in instrument circuits. Hence, the available insights in this area are limited.

Instrument Loop Fire Damage Testing

During the joint NEI/NRC electrical cable fire tests described previously, several instrument cables were tested (NUREG/CR-6776). These tests utilized a simulated 4–20 A instrument loop, a common instrument circuit configuration. With respect to instrumentation cable failures, the following insights were observed:

- The instrument cables failed earlier in the test than did the co-located control cables. The instrument cables tested were all rather small, and this result generally reflects the thermal mass effect. That is, smaller cables heat more quickly, and hence fail more quickly, than do larger cables.

- Thermoplastic insulated instrument cables failed early in the fire tests, and the signal was lost quite abruptly. The instrument readings in such cases would abruptly change from normal to full loss of signal (off-scale low). Such behavior would likely be an obvious indicator to plant operators of a problem in the circuit.

- Thermoset insulated cables experienced degradation and failure later in the exposures, and over a more extended time period, typically of several minute duration. The initial degradation was manifested as an unsteady drop in the simulated process variable value. The degradation in some cases became progressively worse over a period of some minutes. Eventually, a sudden loss of signal was observed in each case. Such behavior may not be as obviously indicative of instrument circuit degradation.

- The behavior of an instrument circuit given cable degradation (e.g., the signal bias direction) can be predicted based on fairly simple circuit analysis.

Loss of Permissive Signal

The loss of a permissive instrument reading may induce a loss of function for the associated system. In some cases, multiple signal losses may be required to cause a loss of function (e.g., given a two out of three polling scheme). Loss of function faults might be recoverable, but only if operators can bypass the permissive signal and re-start the system. Such recovery actions are probably not covered by the operator's procedures, and hence, may be unlikely. Success would require 'on-the-fly' circuit diagnosis and modification. Such operations would not typically be credited in a fire PRA.

It appears that few fire PRAs have explicitly considered the implications of loss of permissive signals. The extent to which such failures are captured would depend on the approach taken. If the regulatory compliance safe-shutdown equipment list included those electrical cables that carry the permissive signal for safe-shutdown systems, then loss of those electrical cables was likely assumed to cause loss of the system. However, particularly for systems not credited in the safe shutdown analysis but credited in the fire PRA, permissive signals may or may not have been identified as a part of the plant shutdown model, and as a part of the electrical cable tracing efforts.

False Permissive Signal

There is a potential that certain types of corrupted or lost signals could cause a spurious actuation signal to be generated through automatic control systems. This potential would depend on the control logic. For example, multiple sensor line polling might make such spurious control signals unlikely. Some advanced circuits may also be designed to detect and reject corrupted signals. Again, the potential risk significance of such faults has not been addressed in any PRA known to the authors of this chapter.

Corrupted Instrument Gage Readings

As noted previously, the instrument signals are of critical importance to operators and are used to guide the operator actions or operator manual actions. A complete loss of several control signals may mean that operators would not know the actual reactor status. This, of course, depends on number of independent or redundant sensors available. It is also important to note that an instrument reading that is completely lost is likely to be readily apparent to operators as a damaged circuit. A more difficult question arises if one postulates that corrupted signals are transmitted to operators.

If corrupted signals are transmitted to operators, they may be misled as to reactor status and may take improper response. For example, a false low water level signal could lead operators to activate additional water sources leading to overcooling of the reactor vessel. A false high level reading could lead operators to shut down or throttle coolant injection systems potentially leading to voiding of the core. To date, no fire PRA known to the authors has systematically addressed such issues.

As noted previously, a pronounced difference between thermoplastic and thermoset insulated cables has been observed which is directly relevant to the potential for transmission of corrupted signals. Thermoplastic insulated cables experienced a sudden failure with no appreciable pre-failure degradation of the transmitted signal. In contrast, thermoset insulated cables degraded over a period of minutes before ultimate loss of signal. Hence, it would appear that the potential for corrupted signals is primarily a factor for plants that utilize thermoset insulated instrument wires. While thermoset insulated cables are known to be predominant in control and power cable applications in the United States, the proportion of plants using thermoplastic versus thermoset insulated instrument cables is not known.

8.6.4 Summary of Circuit Fault Insights

Circuit faults have been discussed in the context of three primary circuit functions; namely, power, control/indication, and instrument circuits. Insights have been derived from both testing, and as discussed in Section 8.7 below, experience.

For power circuits, it is anticipated that most electrical cable failures will lead to a loss of motive power to the related components. Such losses will generally not be recoverable without some repair to replace or bypass the damaged electrical cables. Spurious actuations attributable to hot shorts in power cables are considered unlikely, but the actual likelihood depends on the nature of the power supply system.

The highest likelihood case involves single-phase AC power systems where only a single hot short is needed to cause a spurious operation. In general, an inter-cable hot short is required because of the common practice of utilizing separate electrical cables for each power circuit. A nominal upper bound conditional probability for these cases is estimated at 0.1, although a number of factors could reduce this probability substantially. For these systems some consideration of the risk implications of power cable failure induced spurious actuations would appear appropriate.

The likelihood of spurious actuation for ungrounded DC and three-phase AC power systems is far lower because multiple concurrent correct-polarity, correct voltage inter-cable hot shorts are required. Given the configuration of most power cables, and the apparently low likelihood of inter-cable hot shorts, such concurrent faults appear of low likelihood. Furthermore inter-cable hot shorts in power cables are unlikely to be sustained for any substantial period of time; hence, they are not likely to be risk significant.

One unique aspect of power cables discussed is the issue of multiple high-impedance faults. These scenarios postulate the concurrent existence of several electrical cable short circuits. Furthermore, the short circuit fault paths must each be of a very specific quality (i.e., fault resistance) in order for the postulated scenario to come about. For a number of reasons discussed previously, this would appear to be an unlikely scenario. In a risk context, loss of a higher-level bus attributable to failures in lower-level supply cables can be addressed based on random failure of the first line of circuit protection. In order to further assess the potential risk significance of such scenarios, these random failure probabilities could be adjusted to account for multiple high impedance fault scenarios, but no analysis of this type has yet been undertaken.

For control/indication circuits, many potential failure modes involving both the control and indication functions were discussed. The indication function circuit faults are primarily of interest to risk analysis in relation to their potential impact on operator actions or operator manual actions. No fire PRA to date has considered these issues; hence, their importance to risk is not known. The control functions, on the other hand, have broad-ranging implications. The one control circuit fault mode given the most attention has been spurious actuations. Both experience and experiments indicate that spurious operations are of relatively high likelihood given the failure of electrical cables that are susceptible to inducing such faults. Although a number of factors have been identified that substantially impact this behavior. Spurious actuation probabilities conditional on cable damage vary by at least one order of magnitude given variations in the identified factors.

A particular aspect of the spurious actuation fault mode discussed at some length was the question of multiple spurious actuations. Based on the existing evidence, multiple spurious actuations are both possible and potentially likely given the failure of multiple control cables susceptible to inducing such faults. Using the current estimates of the conditional probability of spurious actuation (given electrical cable failure), it is difficult to justify the screening of any given number of spurious actuation faults based on low likelihood and on a generic basis. For some special cases, such screening might be justified (e.g., cases involving armored electrical cables, cases involving electrical cables in conduits, and cases that require inter-cable hot shorts rather than intra-cable hot shorts). However, no firm basis for such screening has yet been established.

In the case of instrument circuits, the importance of circuit faults was discussed in the context of permissive signals and their impact on operator actions or operator manual actions. Again, no fire PRA to date has included a rigorous treatment of instrument circuit failure; hence, risk insights in this area are lacking.

8.7   Experience-Based Spurious Actuation Insights

As a closing discussion, this section provides a brief summary of insights related to spurious actuation circuit faults that derive from actual fire experience. In the experience base there are several fire incidents, both in the US and abroad, that illustrate spurious actuations. Chapter 3 of this report has already discussed the occurrence of multiple spurious actuations during the 1975 Browns Ferry electrical cable fire. The following additional spurious actuation examples are cited in NUREG/CR-6738:

*   During a 1982 fire at the Armenian NPP, three reported spurious actuations and other control and indication problems are reported, all apparently caused by fire-induced electrical cable failures.

*   The main generator breakers were closed inadvertently as a result of fire damage to the associated control cables. This led to the non-operating generators being connected to the grid and in turn caused secondary fires in one of the turbine-generators and in the startup transformer.

*   One of the diesel generators spuriously disconnected from its emergency loads apparently as a result of control cable damage. Attempts to correct the failure during the fire were not successful.

*   One feedwater pump spuriously started following damage to an electrical cable, apparently in the control circuits. In this last case, the fault that actuated the pump by-passed the normal start logic allowing the pump to start without first starting the lube-oil pumps. Hence, the pump ran for some period without proper lubrication. The fault also by-passed or defeated the normal control room start/stop functions and operator attempts to shut down the pump from the main control room failed. The pump was ultimately secured by electrica technicians who isolated the pump from the power bus manually.

*   Neutron flux and other reactor related instrumentation indicated conditions that may not have been the actual conditions of the reactor. This was likely because many of the instrument

cables were degraded and/or failed by the fire. These indications led to the actuation of various emergency signals. This incident is one of the few incidents where there is specific information indicating that multiple spurious actuations actually occurred during a fire.

- During a 1988 fire at the Ignalinan NPP, there were a number of cases where equipment was lost as a result of spurious trip signals caused by the failure of instrument and control cables. These included the following events:

- The control room received oil level alarms for one of the main coolant pumps and the pump tripped automatically. Failures in the oil level indicator and alarm circuit electrical cables are suspected to be the cause of the trip (rather than an actual drop in oil inventory).

- Instrumentation and control cable failures led to the opening of supply breakers for two normal 6 kV buses and two essential (nonsafety) buses.

- Control cable damage tripped Transformer 5 and prevented it from taking up the loads for these buses.

- A 1991 fire at Chernobyl Unit 2 was attributed to cable damage that resulted from poor cable pulling practices during plant construction. In this instance, a conductor-to-conductor short in a multi-conductor electrical cable led to spurious closure of a generator breaker, grid back-feed into the generator, generator rotor failure, turbine oil and generator hydrogen release and a large fire. In this case, an electrical cable failure caused spurious component actuations that in turn caused the fire.

- During a 1995 fire at Waterford, the event sequence log and the control room operator observations indicate erratic behavior in the position indication of a breaker or a pump. There is no verification in the incident report regarding the behavior of these items in the field. Hence, it is not clear if these are spurious indications only or are, in fact, spurious actuations.

Based on this experience, it is reasonable to conclude that given fire-induced electrical cable failures, spurious actuations are possible, if not likely. Event reports are not sufficiently detailed, however, to allow for a reliable statistical estimate of the likelihood of a spurious actuation given a fire and/or given fire damage. Fire event descriptions do not, in general, provide a sufficient level of detail regarding component/electrical cable damage and systems performance during a fire to support such an analysis with confidence.

The data also show that multiple spurious actuations involving either a single system (i.e., a system that actuates repeatedly during an event) or multiple systems are also possible. Again, data limitations prevent us from providing reliable estimates of the likelihood that any given number of actuations might occur in a fire. The cases noted previously show spurious actuations impacting up to three independent systems during a single fire event.

CHAPTER 9. REFERENCES

U.S. Nuclear Regulatory Commission Documents

Regulations

10 CFR Part 50, "Domestic Licensing of Production and Utilization Facilities."

10 CFR 50.48, "Fire Protection."

10 CFR Part 50, Appendix A, "General Design Criteria for Nuclear Power Plants."
    GDC 3, "Fire Protection."
    GDC 5, "Sharing of Structures, Systems, and Components."
    GDC 19, "Control Room."
    GDC 23, "Protection System Failure Modes."

10 CFR Part 50, Appendix R, "Fire Protection Program for Nuclear Power Facilities Operating Prior to January 1, 1979."

Final Policy Statement, "Use of Probabilistic Risk Assessment Methods in Nuclear Activities," U.S. Nuclear Regulatory Commission, *Federal Register*, V60, p. 42622, August 16, 1995.

Regulatory Guides

RG 1.6, "Independence Between Redundant Standby (Onsite) Power Sources and Between Their Distribution Systems," March 1971.

RG 1.32, "Criteria for Safety-Related Electric Power Systems for Nuclear Power Plants," Revision 2, February 1977.

RG 1.75, "Physical Independence of Electrical Systems," Revision 2, September 1978.

RG 1.174, "An Approach for Using Probabilistic Risk Assessment in Risk-Informed Decisions on Plant-Specific Changes to the Licensing Basis," July 1998.

RG 1.189, "Fire Protection for Operating Nuclear Power Plants," April 2001.

Branch Technical Positions

BTP APCSB 9.5-1, "Guidelines for Fire Protection for Nuclear Power Plants," May 1, 1976.

BTP APCSB 9.5-1, Appendix A, "Guidelines for Fire Protection for Nuclear Power Plants Docketed Prior to July 1, 1976," February 24, 1977.

BTP ASB 9.5-I, "Guidelines for Fire Protection for Nuclear Power Plants," Revision 1, March 1979.

BTP CMEB 9.5-1 (Formerly BTP ASB 9.5-1), "Guidelines for Fire Protection for Nuclear Power Plants," Revision 2, July 1981.

Generic Letters

GL 77-02, "Nuclear Plant Fire Protection Functional Responsibilities, Administrative Controls, and Quality Assurance," August 29, 1977.

GL 80-100, "Resolution of Fire Protection Open Items," November 24, 1980.

GL 81-12, "Fire Protection Rule (45 FR 76602, November 19, 1980)," February 20, 1981, and Clarification Letter, March 1982.

GL 83-33, "NRC Positions on Certain Requirements of Appendix R to 10 CFR Part 50," October 19, 1983.

GL 85-01, "Fire Protection Policy Steering Committee Report," January 9, 1985.

GL 86-10, "Implementation of Fire Protection Requirements," April 24, 1986.

GL 86-10, Supplement 1, "Fire Endurance Test Acceptance Criteria for Fire Barrier Systems Used to Separate Redundant Safe Shutdown Trains Within the Same Fire Area to Implementation of Fire Protection Requirements," March 25, 1994.

GL 88-12, "Removal of Fire Protection Requirements from Technical Specifications," August 2,1988.

GL 91-18,"Information to Licensees Regarding Two NRC Inspection Manual Sections on Resolution of Degraded and Nonconforming Conditions and on Operability," Revision 1, October 8, 1997.

Bulletins

BL 75-04, "Cable Fire at Browns Ferry Nuclear Power Station," March 24, 1975.

BL 75-04A, "Cable Fire at Browns Ferry Nuclear Plant," April 3, 1975.

BL 75-04B, "Cable Fire at Browns Ferry Nuclear Power Station," November 3, 1975.

Information Notices

IN 84-09, "Lessons Learned From NRC Inspections of Fire Protection Safe-Shutdown Systems (10 CFR Part 50, Appendix R)," February 13, 1984.

IN 85-09, "Isolation Transfer Switches and Post-Fire Shutdown Capability,"January 31, 1985.

IN 87-50, "Potential LOCA at High- and Low-Pressure Interfaces from Fire Damage," October 9, 1987.

IN 88-45, "Problems in Protective Relay and Circuit Breaker Coordination," July 7, 1988.

IN 90-69, "Adequacy of Emergency and Essential Lighting," October 31, 1990.

IN 91-17, "Fire Safety of Temporary Installations," March 11, 1991.

IN 91-77, "Shift Staffing at Nuclear Power Plants," November 26, 1991.

IN 92-18, "Loss of Remote Shutdown Capability During a Fire," February 28, 1992.

IN 93-71, "Fire at Chernobyl Unit 2," September 13, 1993.

IN 95-33, "Switchgear Fire and Partial Loss of Offsite Power at Waterford Unit 3," August 23, 1995

IN 95-36, "Potential Problems with Post-Fire Emergency Lighting," August 29, 1995.

IN 95-48, "Results of Shift Staffing Study," October 10, 1995.

IN 97-37, "Main Transformer Fault With Ensuing Oil Spill Into Turbine Building," June 20, 1997.

IN 98-31, "Fire Protection System Design Deficiencies and Common-Mode Flooding of Emergency Core Cooling System Rooms at Washington Nuclear Project Unit 2," August 18, 1998.

IN 99-17, "Problems Associated With Post-Fire Safe-Shutdown Circuit Analyses," June 3, 1999.

NUREG-Series Reports

Bennett, P.R., A.M. Kolaczkowski, and G.T., Medford, "Summary Report: Electrical Equipment Performance Under Severe Accident Conditions (BWR/Mark I Plant Analysis), NUREG/CR-4537, U.S. Nuclear Regulatory Commission, Washington, DC, September 1986.

Jacobus, M. J., and G.F. Fuehrer, "Submergence and High Temperature Steam Testing of Class 1E Electrical Cables," NUREG/CR-5655, U.S. Nuclear Regulatory Commission, Washington, DC, May 1991.

Kazarians, M., and G. Apostolakis, "Fire Risk Analysis for Nuclear Power Plants," NUREG/CR-2258, U.S. Nuclear Regulatory Commission, Washington, DC, September 1981.

Nowlen, S.P., M. Kazarians, and F. J. Wyant, "Risk Methods Insights Gained from Fire Incidents," NUREG/CR-6738, U.S. Nuclear Regulatory Commission, Washington, DC, September 2001.

Nowlen, S.P., "Ampacity Derating and Cable Functionality for Raceway Fire Barriers," NUREG/CR-6681, U.S. Nuclear Regulatory Commission, Washington, DC, August, 2000.

Nowlen, S. P., "An Investigation of the Effects of Thermal Aging on the Fire Damageability of Electric Cables," NUREG/CR-5546, U.S. Nuclear Regulatory Commission, Washington, DC, May 1991.

NUREG-0050, "Recommendations Related to Browns Ferry Fire," Report by Special Review Group, U.S. Nuclear Regulatory Commission, Washington, DC, February 1976.

NUREG/CR-1742, "Perspectives Gained From the Individual Plant Examination of External Events (IPEEE) Program," U.S. Nuclear Regulatory Commission, Washington, DC, April 2002.

NUREG-0800, "Standard Review Plan for the Review of Safety Analysis Reports for Nuclear Power Plants, Section 19.0, Use of Probabilistic Risk Assessment in Plant-Specific Risk-Informed Decision Making: General Guidance," U.S. Nuclear Regulatory Commission, Washington, DC, November 2002.

NUREG-0800, "Standard Review Plan for the Review of Safety Analysis Reports for Nuclear Power Plants, LWR Edition," Section 9.5.1, "Fire Protection System," U.S. Nuclear Regulatory Commission, Washington, DC, July 1996.

Subudhi, M., "Literature Review of Environmental Qualification of Safety Related Electric Cables," NUREG/CR-6384, Volume 1, U.S. Nuclear Regulatory Commission, Washington, DC, April 1996.

Wheelis, W. T., "Users Guide for a Personal-Computer-Based Nuclear Power Plant Fire Data Base," NUREG/CR-4586, U.S. Nuclear Regulatory Commission, Washington, DC, August 1986.

Wyant, F. J., and S. P. Nowlen, "Cable Insulation Resistance Measurements Made During Cable Fire Tests, NUREG/CR-6776, U.S. Nuclear Regulatory Commission, Washington, DC, June 2002.

Commission Papers

SECY 80-438A, "Rule on the Fire Protection Program for Nuclear Power Plants Operating Prior to January 1, 1979," September 30, 1980.

SECY 83-269, "Fire Protection Rule for Future Plants," July 5, 1983.

SECY 93-143, "NRC Staff Actions to Address the Recommendations in the Report on the Assessment of the NRC Fire Protection Program," May 21, 1993.

SECY 95-034, "Status of Recommendations Resulting from the Reassessment of the NRC Fire Protection Program," February 13, 1995.

SECY 96-267, "Fire Protection Functional Inspection Program," December 24, 1996.

SECY 99-040, "Second Interim Status Report–Fire Protection Functional inspection Program," February 5, 1999.

SECY 99-140, "Recommendations for Reactor Fire Protection Inspections," May 20, 1999.

SECY 99-182, "Assessment of the Impact of Appendix R Fire Protection Exemptions on Fire Risk," July 9, 1999.

Inspection Program Documents

IMC 0609 Appendix F, "Fire Protection Significance Determination Process," 2000.

IP-64100, NRC Inspection Manual, IM-64100, "Post-Fire Safe-Shutdown, Emergency Lighting and Oil Collection Capability at Operating and Near-term Operating Reactor Facilities."

IP-64704, "Fire Protection Program," June 24, 1998.

IP-71111.05, "Triennial Fire Protection Inspection Procedure," March 2003.

NRC Inspection Reports

IR 50-254/98-011 and 50-265/98-011, "Fire Protection Inspection Quad Cities Nuclear Generating Station Units 1 and 2."

IR 50-259/00-08, 50-260/00-08, and 50-296/00-08, "Fire Protection Baseline Inspection Browns Ferry Units 1, 2, and 3."

IR 50-282/98-016 and 50-306/98-016, "Inspection of Prairie Island Nuclear Generating Station Fire Protection Functional Inspection Self Assessment."

IR 50-313/01-06 and 50-368/01-06, "Triennial Fire Protection Baseline Inspection of Arkansas Nuclear One."

IR 50-335/98-201 and 50-389/98-201, "Fire Protection Functional Inspection of St. Lucie Plant."

IR 50-387/97-201 and 50-388/97-201, "Fire Protection Functional Inspection of Susquehanna Steam Electric Station."

IR 50-458/97-201, "Fire Protection Functional Inspection River Bend Station Unit 1."

Letters and Memoranda

BWROG Letter 1999 (BWROG-99-079), W.G. Warren to J. Hannon, "BWR Owners Group Appendix R Fire Protection Committee Generic Guidance for BWR Post-Fire Safe-Shutdown Analysis," November 15, 1999.

Collins Letter 1997, S.J. Collins to R.E. Beedle (NEI), "Assessment of NEI Concerns Regarding NRC Information Notice 92-18, Potential for Loss of Remote Shutdown Capability During a Control Room Fire," March 11, 1997.

Dembek Memo 1999, S. Dembek to S.A. Richards, "Summary of Meeting with Boiling-Water Reactor Owners Group (BWROG) Appendix R Committee on Post-Fire Safe-Shutdown Circuit Analysis Issues (Fire-Induced Circuit Failures)."

Denton Letter, Harold R. Denton, NRC, Letter to S.A. Bernsen, Bechtel Power Corporation (No subject), April 30, 1982.

Hannon Letter 2001, J. Hannon to D. Modeen (NEI), "Nuclear Energy Institute/Electric Power Research Institute Fire Testing: Comprehensiveness With Respect to Outstanding Circuit Analysis Issues (TAC No. MA4745)," February 1, 2001.

Holahan Memo, Gary M. Holahan, Memo to Dennis Crutchfield, Subject: "Request for Assistance: Determine Whether Two Hot Shorts in a Multiconductor Cable Associated with a Non-High/Low-Pressure Interface Should Be Analyzed for Fire-Induced Spurious Actuation (GL 86-10, Section 5.3.1, Non-High/Low-Pressure Interfaces in Ungrounded AC and DC Circuits) (AITS 205-89)," December 4, 1990.

Mattson Memo 1982, Roger J. Mattson, Memo to Richard H. Vollmer. Subject: "Position Statement on Allowable Repairs for Alternative Shutdown and on the Appendix R Requirement for Time Required To Achieve Cold Shutdown," July 2, 1982.

Mattson Memo 1983, Roger J. Mattson, Memo to D. Eisenhut, Subject: "Task Interface Agreement #8 3-53, 'Physical Independence of Electrical Systems,' TAC No. 51567," July 22, 1983.

Richards Letter 2000, S.A. Richards, Letter to J.M. Kenny, BWR Owners Group, "BWROG Appendix R Fire Protection Committee Position on SRVs and Low-Pressure Systems Used as Redundant Shutdown Systems Under Appendix R," December 12, 2000.

Rubenstein Memo 1982, L.S. Rubenstein, Memo to Roger J. Mattson, Subject: "Use of the Automatic Depressurization System (ADS) and Low-Pressure Coolant Injection (LPCI) To Meet Appendix R, Alternate Shutdown Goals," December 3, 1982.

Rubenstein Memo 1983, L.S. Rubenstein, Memo to Roger J. Mattson, Subject: "Statement of Staff Position Regarding Source Range Flux, Reactor Coolant Temperature, and Steam Generator Pressure Indication to Meet Appendix R, Alternate Shutdown Capability," January 7, 1983.

Stello Letter, Victor Stello, Jr., Letter to David Bixel, Consumers Power Company, Subject: "Manpower Requirements for Operating Reactors," Docket No. 50-255," June 8, 1978.

Thadani Memo 1993, A.C. Thadani to T.E. Murley, Subject: "Report on the Reassessment of the NRC Fire Protection Program," February 27, 1993.

Vollmer Memo 1983, R.H. Vollmer, Memo to Darrel G. Eisenhut, Subject: "Emergency Lighting Requirements," (TIA 83-87; TAC 52308)," December 21, 1983.

Licensee Event Reports

LER 219/92-011, "Design Deficiency Causes Noncompliance with Appendix R Criteria," Oyster Creek, September 15, 1992.

LER 247/96-007-00, "Potential Challenge of High/Low Pressure Interface," Indian Point Unit 2, April 29, 1996.

LER 247/96-014-00, "Loss of Process Monitoring Function During Postulated Fires (Appendix R)," Indian Point Unit 2, August 26, 1996.

LER 266/97-020-01, "Conditions Outside Appendix R Safe-Shutdown Analysis," Point Beach Nuclear Plant Unit 1, October 14, 1997.

LER 266/97-032-00, "Inadequately Rated Electrical Buses Could Disable Switchgear and Cause Secondary Fires That Prevent Shutdown Per Appendix R," Point Beach Nuclear Plant Unit 1, July 30, 1997.

LER 266/99-008-00, "Postulated Fire Could Lead to Loss of Redundant Trains of Charging Capacity," Point Beach Nuclear Plant Unit 1, November 3, 1999.

LER 266/00-008-00, "Inadequate Procedural Guidance for Spurious Opening of RHR to Containment Sump Valves SI-851A/B During Appendix R Alternate Shutdown," Point Beach Nuclear Power Plant Unit 1, October 19, 2000.

LER 266/01-006-00, "Appendix R Requirements Not Satisfied for Unanalyzed Fire-Induced Damage to the Auxiliary Feedwater System," Point Beach Nuclear Plant Units 1 and 2, February 4, 2002.

LER 272/99-011-00, "125 VDC Control Power Circuits for 4 kV Breakers Do Not Meet Requirements of 10 CFR Part 50 Appendix R," Salem Generating Station Unit 1, November 12, 1999.

LER 280/99-003-00, "Potential Loss of Charging Pumps Due to Main Control Room Fire," Surry Power Station Unit 1, April 28, 1999.

LER 298/96-009-00, "Appendix R Safe-Shutdown Analysis Vulnerabilities," Cooper Nuclear Station, June 1, 1998.

LER 298/00-002-00, "Appendix R Safe Shutdown Analysis Vulnerability Due to Potential Conductor-to-Conductor Hot Shorts," Cooper Nuclear Station, February 10, 2000.

Other Documents

ANSI/IEEE C.2, "National Electrical Safety Code."

"Design Basis Document for Appendix R, Susquehanna Steam Electric Station Units 1 and 2," DBD076, Pennsylvania Power and Light LLC, July 12, 2001.

Engineering Calculation EC-013-0843, "SSES 10 CFR Part 50 Appendix R Compliance Manual," Susquehanna Steam Electric Station Units 1 and 2," Pennsylvania Power and Light LLC, April 22, 2002.

Engineering Calculation G13.18.3.6*07, "Safe Shutdown Common Enclosure Associated Circuit Analysis, Gulf States Utilities," September 27, 1994.

EPRI TR-1000894, "Fire Events Database for U.S. Nuclear Power Plants: Update Through 1999," Electric Power Research Institute, Palo Alto, California 2000.

EPRI TR-1003326, "Characterization of Fire-Induced Circuit Faults: Results of Cable Fire Testing," Electric Power Research Institute, Palo Alto, California 2002.

EPRI TR-1006961, "Spurious Actuation of Electrical Circuits Due to Cable Fires: Results of an Expert Elicitation," Electric Power Research Institute, Palo Alto, California 2002.

"Generic Guidance for BWR Post-Fire Safe-Shutdown Analysis," Revision 0, GE-NE-T43-00002-00-02, November 1999.

"Good Design Prevents High-Impedance Fault," *Actual Specifying Engineer*, Volume 17, No. 4, Medalist Publications, Inc., Chicago, IL, 1967.

IEEE Std. 100-1998, "IEEE Standard Dictionary of Electrical and Electronics Terms," 4th Edition, Institute of Electrical and Electronics Engineers.

IEEE Std. 242-1986, "IEEE Recommended Practices for Protection and Coordination of Industrial and Commercial Power Systems (Buff Book)," Institute of Electrical and Electronics Engineers.

IEEE Std. 141-1986, "IEEE Recommended Practices for Electric Power Distribution for Industrial Plants (Red Book)," Institute of Electrical and Electronics Engineers.

IEEE Std. 383, "IEEE Standard for Type Test of Class IE Electric Cables, Field Splices, and Connections for Nuclear Power Generating Stations," Institute of Electrical and Electronics Engineers.

IEEE Std. 690-1984, "IEEE Standard for the Design and Installation of Cable Systems for Class 1E Circuits in Nuclear Power Generating Stations," Institute of Electrical and Electronics Engineers.

IEEE Std. 835, "Standard Power Cable Ampacity Tables," Institute of Electrical and Electronics Engineers.

LaChance, J., et al., "Circuit Analysis — Failure Mode and Likelihood Analysis: A Letter Report to the USNRC," Sandia National Laboratories, Albuquerque, New Mexico, May 8, 2000.

NEI-00-01, Draft Revision C, "Guidance for Post-Fire Safe-Shutdown Analysis," Nuclear Energy Institute, October 2001.

NEI-00-01, Draft Revision D, "Guidance for Post-Fire Safe-Shutdown Analysis," Nuclear Energy Institute, October 2002.

NFPA Fire Protection Handbook, Section 6, 18th Edition, National Fire Protection Association, Quincy, Massachusetts.

NFPA 805, "Performance-Based Standard for Fire Protection for Light-Water Reactor Electric Generating Plants," 2001 Edition, National Fire Protection Association.

Ramsey, C., et al., "United States Department of Energy Reactor Core Protection Evaluation Methodology for Fires at RBMK and VVER Nuclear Power Plants, DOE/NE-0113 Revision 1, U.S. Department of Energy (DOE), June 1997.

Sullivan, K., et al., "A Historical Fire Protection Licensing Document Describing Requirements for Commercial Nuclear Power Plants Operating in the United States," USNRC Technical Report R7017/U7010, Brookhaven National Laboratory (BNL), Upton, New York, March 1995.

Sullivan, K., "Electrical Post-Fire Safe Shutdown Assistance for FPFI Procedure," Technical Letter Report to NRC Office of Nuclear Reactor Regulation, Brookhaven National Laboratory (BNL), Upton, New York, September 23, 1996.

Sullivan, K., and R.E. Deem, "Baseline Tri-Annual Fire Protection Inspection — Braidwood Nuclear Power Station," Technical Letter Report Input to NRC Region III, Brookhaven National Laboratory (BNL), Upton, New York, April 9, 2003.

Sullivan, K., "U.S. Commercial Nuclear Reactor Plant Post-Fire Safe-Shutdown Circuit Analysis History and Safety Significance/Discussion of Potential Severity of Fire-induced Reactor Plant Transients," Technical Letter Report to the USNRC Office of Nuclear Reactor Regulation (JCN J-2427), Brookhaven National Laboratory (BNL), Upton, New York, July 20, 1998.

This page intentionally left blank.

APPENDIX A.
SUCCESSFUL IMPLEMENTATION OF APPENDIX R CIRCUIT ANALYSIS

APPENDIX A.
SUCCESSFUL IMPLEMENTATION OF APPENDIX R CIRCUIT ANALYSIS

A.1 Circuits of Concern to Post-Fire Safe-Shutdown

As described in Chapter 6, circuits of concern to post-fire safe-shutdown, fall into one of two broad categories:

(1) Circuits/cables of equipment needed to ensure the proper operation of the *systems* credited in the SSA for performing essential shutdown functions ("required" or "safety" circuits)

(2) Circuits/cables of equipment that, if damaged by fire, could impact the shutdown capability ("associated, "nonessential" or "nonsafety" circuits of concern)

Required Circuits

Because a cable or circuit is related to the operation of a required shutdown component does not necessarily mean it is of concern to post-fire safe-shutdown. As discussed below in Section A.2 below, the determination of whether or not a specific cable is required for safe-shutdown may depend on such factors as the specific function of the cable, the position/status (open, closed, running, stopped, etc.) of the component at the onset of fire, and the desired position/status of the component for shutdown. In general, a circuit/cable is considered to be required for safe-shutdown if it has the following characteristics:

(1) It is related to the operation of a required shutdown component

(2) Fire-induced faults in the circuit/cable can prevent the operation or cause a maloperation of the shutdown system in which the component is located

Power and control cables of Pump P-1 in Figure A-1 and control cables of valve V6 are typical examples of "required cables." In contrast,"associated nonsafety cables" are not directly related with the operation of any of the credited shutdown systems. Cable/circuits related to the operation of Valves V-9 and V-10 in Figure A-1 are examples. Although not needed to ensure operation of the credited shutdown systems, fire damage to circuits such as these could significantly impact the shutdown capability.

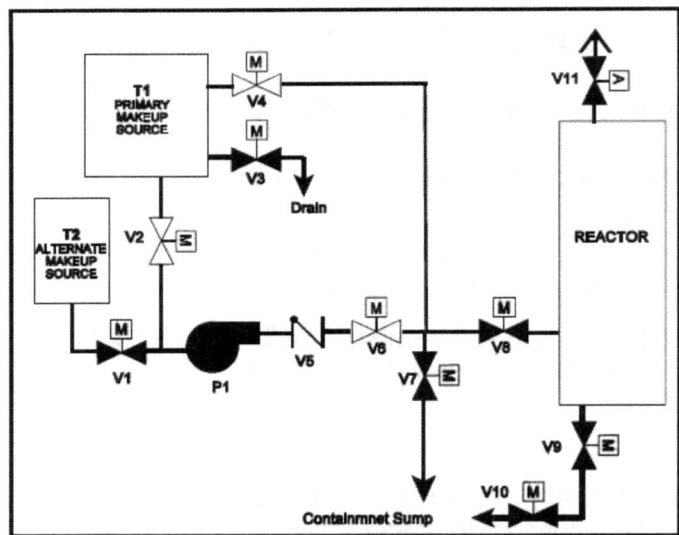

Figure A-1 Simplified Shutdown System Flowpath

Associated Circuits of Concern

The achievement of safe-shutdown is dependent on ensuring the active control of some components and preventing the maloperation of other components. A post-fire safe-shutdown analysis should be a bounding analysis that identifies the range of possible fire impacts within each fire area (vulnerabilities) and ensures that appropriate measures are in place to prevent this damage from affecting the ability to safely shutdown the plant. Therefore, it is not sufficient to only consider the effects of fire damage to cables of equipment needed to ensure operation of credited shutdown *systems* (required circuits). The scope of a successful shutdown strategy will consider the effects of fire damage to nonessential equipment and systems of which inadvertent or spurious actuation could impact the shutdown capability (associated nonsafety circuits).

The principal staff guidance related to the potential impact of fire-induced circuit failures in "nonessential" or "associated" circuits is contained in GL 81-12, dated February 20, 1981, and its subsequent clarification, dated March 22, 1982. As described in these documents, there are three specific configurations of associated circuits of concern to post-fire safe-shutdown:

• Nonessential circuits that share a **common power supply** (e.g., SWGR, MCC, Fuse Panel) with circuits of equipment required to achieve and maintain safe-shutdown; or,

• Nonessential circuits that share a **common enclosure**, (e.g., raceway, conduit, junction box, etc.) with cables of equipment required to achieve and maintain safe-shutdown

• Circuits and cables that have a connection to equipment of which **spurious operation** would adversely affect the shutdown capability.

With few exceptions, most licensees successfully resolve common power supply and common enclosure associated circuit concerns on a plant-wide basis through the performance of generic evaluations of electrical protective devices (e.g., fuse/breaker coordination studies). When resolved in this manner, the types of circuits/cables of concern to post-fire safe-shutdown are then reduced to two specific classifications:

(1) *Required Cables*: Circuits/cables of equipment needed to ensure the proper operation or functioning of shutdown systems defined/designated in the SSA.

(2) *Spurious Nonsafety Cables*: Circuits/cables of systems and equipment that are not needed to ensure the operation of shutdown systems credited in the SSA, but of which inadvertent (spurious) actuation or maloperation could impact the shutdown capability.

In its clarification of GL 81-12, the staff defined the scope of the spurious operation associated circuit concern as those circuits/cables that could impact the safe-shutdown capability if they are damaged by fire. As shown in Figure A-2 (Reference GL 81-12 Clarification, Enclosure 2), a fundamental presumption of the GL is that circuits/cables of equipment that could prevent operation or cause the maloperation of redundant shutdown systems (i.e., required circuits) are provided with fire protection features sufficient to meet Section III.G.2 of Appendix R, and therefore, would remain *free of fire damage*. As shown in Figure A-2, however, even when redundant trains of "required" cables meet III.G.2 criteria, fire damage to circuits/cables of "nonessential" systems and equipment (i.e., not needed to ensure operation of the defined/credited shutdown systems) may significantly impact the shutdown capability.

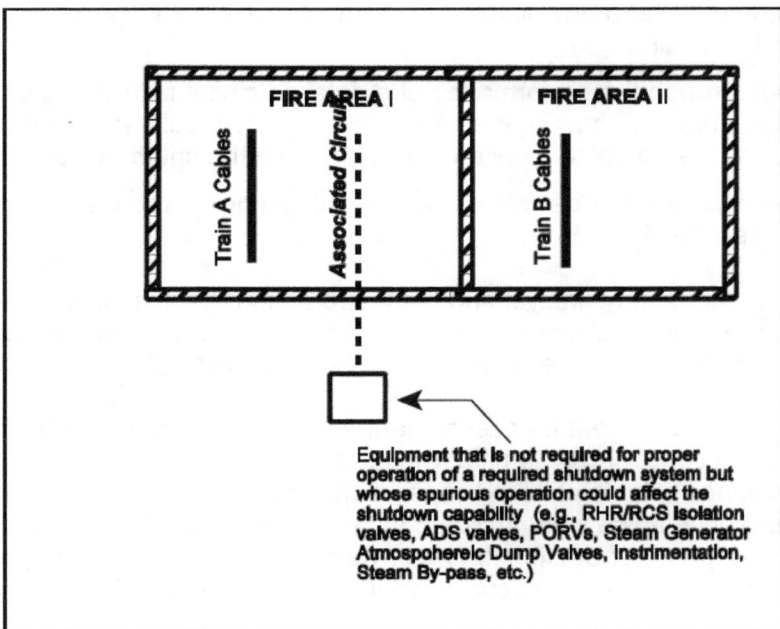

Figure A-2 Spurious Operation Associated Circuits of Concern

As described in this document, a common approach for ensuring that the SSA sufficiently bounds the range of circuit failures of concern to post-fire safe-shutdown starts by defining *shutdown success paths* (redundant and alternative), where each path is comprised of a set of systems (i.e., credited shutdown systems) capable of accomplishing each of the required shutdown functions (e.g., reactivity control, DHR). With the shutdown paths and systems defined, equipment needed to ensure the proper operation of the credited shutdown systems (required components) *and* nonessential/nonsafety equipment or systems of which spurious actuation could impact the shutdown capability are then identified and documented on a SSEL. As a result of this process, the SSEL will include all components (essential and nonessential) that could impact the shutdown capability if they are damaged by fire, and will not be limited to only those components needed to ensure the operation of the defined shutdown systems. From this comprehensive listing of equipment, the SSEL can then serve as a starting point for identifying circuits and cables of concern to post-fire safe-shutdown in each fire area.

A.2 Resolving Identified Vulnerabilities

When circuits/cables of concern to post-fire safe-shutdown are found to be located in a specific fire area under evaluation, the analyst has several options for ensuring that an appropriate level of fire safety is achieved, as illustrated by the following examples:

(1) Assuming that fire damage to affected circuits/cables will cause connected equipment to fail in an undesired manner and providing fire protection features sufficient to satisfy Section III.G.2 of Appendix R (while this approach requires no additional analysis it may not be cost-effective),

(2) Revising the shutdown strategy developed for the specific fire area under evaluation (e.g., use of other equipment)

(3) Demonstrating, through the performance of a detailed circuit failure mode and effects analysis (circuit analysis), that the credible range of circuit faults (as described in Chapter 6) to all exposed circuits/cables of concern will not impact the shutdown capability

(4) Requesting an exemption or deviation from specific technical requirements of regulatory requirements (see Section 4.5)

The challenge to the fire safety analyst and plant operating organization is to determine the best solution possible based on its ability to provide cost-effective protection against the threat of fire in a manner that is consistent with regulatory criteria and the plant's fire protection licensing basis.

During the initial stages of a fire area assessment, it is not uncommon to identify a large number of cable/circuit "interactions" or "cable hits." Since each "interaction" or "hit" represents a potential noncompliance with established separation/protection requirements, all interactions must be resolved. This may be accomplished by either the installation of additional fire protection features (e.g., meet Section III.G.2 of Appendix R), or through a rigorous analysis of the effect of fire damage to each circuit/cable involved in the identified interactions (circuit analysis).

Since it is typically not desirable to perform unnecessary plant modifications, most plants elect to perform a comprehensive analysis of each interaction. Since such an analysis can also be a time-consuming, resource-intensive process (particularly if excessive engineering effort is expended in the evaluation of circuits/cables that would not impact safe-shutdown if they are damaged by fire), it is desirable to limit its scope to only those circuits/cables that could actually impact the shutdown capability if they are damaged by fire. In cases where the SSEL is sufficiently comprehensive to bound the range of circuit failures of concern to post-fire safe-shutdown, licensee's have shown that the number of circuits/cables requiring a detailed review can be significantly reduced by considering the function, normal operating mode/status, and desired operating mode/status of components related to the identified cable/circuit interactions. The application and benefits of this screening technique are illustrated in the following example.

As discussed in Chapter 6, not all cable/circuit failures identified as "potential interactions" will impact the ability of connected equipment to function as needed for post-fire safe-shutdown. For example, since MOVs fail to the "as-is" position upon a loss of motive power, a loss of power to "normally closed" MOVs V-3, V-7, V-9 and V-10 (shown in Figure A-1), will not impact the shutdown capability. A loss of motive power to these valves will only cause them to remain "closed" which is their desired position for post-fire safe-shutdown. Additionally, if spurious actuation (opening) would not result in a LOCA, (i.e., the valves do not comprise a high/low pressure interface boundary) the power cables may be screened from further consideration for spurious actuation concerns. For the example shown, this would include power cables for valves V-3 and V-7. Since valves V-9 and V-10 comprise a high/low pressure interface, their power cables can not be screened at this point in the evaluation.

A.2.1   Use of Operator Manual Actions

Section III.G.2 of Appendix R requires that circuits that could prevent the operation or cause maloperation of redundant trains of safe-shutdown equipment have one of the specified fire protection features. Operator manual actions to respond to maloperations are not listed as an acceptable method for satisfying this requirement. However, the NRC has previously accepted plant-specific operator manual actions in formal exemption/deviation requests and in SERs. Rulemaking is currently in progress to codify the use of acceptable manual operator actions as discussed below.

Based on inspection results and industry comments the NRC determined that licensees have, without request for exemption/deviation from the code, implemented operator manual actions where the specified requirements of Section III.G.2 cannot be met. The staff concluded that rulemaking would be required to allow licensees committed to Appendix R to substitute operator manual actions in lieu of Section III.G2 compliance. For an interim period, while rulemaking is in progress, the staff determined that acceptance criteria can be developed which would facilitate evaluations of certain operator manual actions. Authority to approve a licensee methodology that does not meet NRC regulations is not delegated to the inspectors. However, inspectors will ensure that plant-specific operator manual actions meet the following guidelines:[34]

---

[34]   NRC Inspection Procedure 71111.05, March 6, 2003.

- **Diagnostic Instrumentation**

  Adequate diagnostic instrumentation, unaffected by the postulated fire, is provided for the operator to detect the specific spurious operation or maloperation that occurred. Additional instrumentation beyond that identified in IN 84-09 may be needed to properly assess a spurious operation. Annunciators, indicating lights, pressure gages, and flow indicators are typical examples. Sufficient instrumentation should also be available to verify that the operator manual action accomplished the intended objective.

- **Environmental Considerations**

  The environmental conditions the operator may encounter while accessing and performing the operator manual action have been fully considered. Radiation levels should not exceed normal 10 CFR Part 20 limits. Emergency lighting should be provided as required in Appendix R, Section III.J or by the licensee's approved FPP. Temperature and humidity conditions should be reviewed to ensure that temperature and humidity do not affect the capability to perform the operator manual action. Fire effects should be reviewed to ensure that smoke and toxic gases from the fire do not affect the capability to perform the operator manual action.

- **Staffing**

  Adequate qualified personnel are on shift and available perform the required operator manual actions and to safely operate the reactor.

- **Communications**

  If operator manual action coordination with other plant operations is required, then communications capability must be protected from effects of a postulated fire.

- **Special Tools**

  If special tools are required they are dedicated for use and readily available from an accessible nearby location.

- **Training**

  Operators are trained on the operator manual actions and the procedure is adequate and current.

- **Accessability**

  Operator is capable of reaching the required location without personal hazard. If a ladder or other special access equipment is needed, it should be readily available.

- **Procedures**

  Procedural guidance has been developed to implement the operator manual actions. Operators should not rely on having time to study normal plant procedures to find a method of operating plant equipment that is seldom used.

- **Verification and Validation**

  All operator manual actions have been verified and validated (V&V) by plant walkdowns using the current procedure. The licensee has adequately evaluated the capability to perform the operator manual action in the time available before the plant will be placed in an unrecoverable condition.

A.3 Plant-Specific Examples of Successful Implementation

The following six examples show how cable/circuit vulnerabilities have been successfully identified and resolved by licensees. The examples are based on actual problems that were identified by licensees during recent re-evaluations for Appendix R compliance. In addition to illustrating the potential impact that fire-induced circuit failures may have on the ability to achieve and maintain safe-shutdown conditions, the examples also illustrate the extent and depth of the analysis.

Case 1   Potential for Secondary Fire Initiation

Problem   During a reevaluation of its Appendix R program in 1997, a licensee of a PWR discovered that fault currents generated as a result of fire damage to power cables could be larger than the interrupting capability of the connected SWGR. If the associated SWGR is located in a different fire area, then this overcurrent condition could lead to another, secondary fire. This condition is unacceptable because the SSA assumes the occurrence of a single fire. The capability of the plant to achieve and maintain safe-shutdown for fires in multiple fire areas had not been demonstrated.

Resolution   In order for the failure scenario described above to occur two conditions must exist: (1) the fault current must exceed the interrupting capability (rating) of the SWGR and (2) the fire must occur in a fire zone other than where the SWGR is located. Since cable impedance (which is generally proportional to cable length) will reduce the magnitude of fault current, the licensee performed an evaluation to determine the minimum distance away from the SWGR that a fault must occur for the cable's impedance to reduce the magnitude of fault current to a value within the rating of the SWGR. In addition, the routing of each cable was reviewed to determine whether the cable's route took it through different fire areas than that in which the SWGR was located. As a result of this review, the licensee identified six fire zones where the initiating fire had a potential to cause a secondary fire at the associated SWGR. As an immediate corrective action the licensee implemented compensatory measures to establish a roving fire watch in each of the six identified fire zones. As a permanent corrective action, the licensee implemented design changes to ensure that the subject SWGRs are capable of interrupting fault currents that may be generated during a fire.

Case 2   Inadequate Coordination Could Disable Essential Instrumentation

Problem   In 1997, during a review of electrical cable routing, a PWR licensee discovered that a 125 VDC power cable was exposed to the effects of fire damage. Fire-induced faults (short to ground) in this cable, coupled with a lack of circuit breaker coordination on the 125 VDC system, could result in a loss of power to instrumentation that is essential for achieving and maintaining post-fire safe-shutdown. The licensee determined that this condition was caused by an inadequate review of a plant modification for Appendix R concerns. (See Chapter 7.) The modification routed a new "associated circuit" cable without verifying the adequacy of circuit breaker coordination.

Resolution   Compliance with Section III.G.2 of Appendix R was achieved by implementing a plant modification to enclose the power cable in a 1-hour rated fire wrap.

Case 3   **"Hot Short" Could Result in a Loss of the Service Water System**

Problem   In year 2000 the licensee of a BWR discovered that a fire-induced circuit fault resulting from fire in the CSR could lead to a loss of all service water cooling to essential shutdown systems. Although three sources of water to the service water pump seals are normally available, all three sources could be lost as a result of fire damage in the cable spreading room. The specific vulnerability involved a multi-conductor cable that carries 24 VDC start control circuits for the pump that is credited in the licensee's analysis for providing cooling water to the gland seals of the service water pumps. A conductor-to-conductor short, either between individual conductors of the multi-conductor cable, or between conductors of the multi-conductor cable and conductors of two other cables located inside the same conduit, could cause the 24 VDC start control circuits to be energized by 120 VAC power. This condition could disable the automatic starting and running of the pump relied on to provide cooling water to the service water pump gland seals. The service water pumps are required to operate during and after a fire to supply cooling water to essential shutdown equipment. The loss of the service water system would prevent the plant from achieving and maintaining safe-shutdown conditions.

Resolution   The licensee has developed modifications to eliminate this vulnerability. In the interim, the licensee posted a continuous fire watch in the cable spreading room.

Case 4   **Multiple Circuit Faults Could Cause a Loss of all Makeup/Charging Capability**

Problem   During a re-evaluation of its Appendix R analysis a licensee of a PWR discovered that a fire could result in damage to any of the operating charging pumps. The charging system provides makeup water to the RCS, reprocesses water letdown from the RCS, and provides seal water injection to the reactor coolant pump seals. During normal plant operations two pumps are running and the third pump is secured in standby. At least one pump must be available to support safe-shutdown. A temporary loss of charging is acceptable as long as one pump can be restored within 30 minutes with full pump capacity. However, if the running pump(s) is the only credited pump available (i.e., other pumps are unavailable because of fire-induced failures), its failure/loss as a result of fire would lead to a total loss of all charging capability.

The normal suction supply to the operating charging pumps is from the volume control tank (VCT). During its re-evaluation the licensee discovered that multiple circuit faults could cause a loss of all charging capability. Specifically, a hot short on the control cable of an MOVs located in the VCT supply line could cause the valve to shut. Although an alternative source of water is available from the refueling water storage tank (RWST), the same fire could also damage cables for the charging water supply valve and prevent that valve from opening. The spurious actuation (close) of the VCT isolation valve and a failure of the RWST valve to open would result in a loss of suction and subsequent pump damage.

Resolution   The licensee identified the specific fire zones where this scenario may occur and installed modifications to correct cable routing and separation deficiencies.

Case 5 **Potential Loss of All Vital Buses As a Result of Multiple Faults In Ungrounded DC Control Circuits**

Problem   The alternative shutdown strategy developed by a licensee of a PWR relied on operator manual actions to isolate 125 VDC control power to breakers of 4 kV SWGR. This was accomplished by opening the feed breaker to the bus. With the control power isolated, the licensee had assumed that the 4kV breakers could then be manually operated as needed. During a recent, 1999 reassessment of its safe-shutdown analysis, however, the licensee discovered that in the event of fire in certain alternative shutdown areas, cables associated with the 125 VDC control circuits could experience fire damage resulting in an external hot short on the positive side of the open/close coils. If this fault were to occur in combination with multiple grounds on the negative legs of the 125 VDC circuit, the closing or trip coils would become energized. Fire-induced shorting/grounding of 4 kV circuit breaker 125 VDC control circuits could result in inadvertent opening or closing of these breakers, or inability to locally position these breakers manually. This scenario could lead to a loss of all three vital buses.

Since the 125 VDC system was ungrounded, the licensee had assumed that a review of these circuits for spurious actuation was not required. At the time of its original analysis, operator manual actions to remove 125 VDC control power from the breakers was considered adequate to isolate the 4 kV breakers from the alternative shutdown areas and allow manual manipulation of the breakers. During its re-evaluation, however, the licensee recognized that this assumption was not consistent with staff guidance described in Question 5.3.1 of GL 86-10, which requires an analysis of sufficient depth to determine the adverse impacts of hot shorts, shorts to ground, or open circuits on safe shutdown related control circuits and their associated logic.

Resolution   The licensee intends to implement corrective actions necessary to resolve compliance with Appendix R as part of its corrective action program.

Case 6 **Spurious Opening of Multiple Safety Relief Valves**

Problem   During a reevaluation for compliance with Appendix R to 10 CFR Part 50 the licensee of a BWR determined that a control room or relay room fire could cause multiple SRVs to spuriously open resulting in rapid depressurization and inventory loss. The cables associated with the SRVs share a common cable tray, and single hot short will result in the spurious opening of each SRV. Given the potential for fire-induced failures high-volume makeup systems capable of mitigating this event [CR, RHR, low-pressure coolant injection (LPCI) and HPCI] may not be immediately available. The consequence of multiple SRV failures without the availability of a high-volume injection system could lead to core uncovery.

There are 11 DC-operated SRV, of which seven in the ADS are automatically controlled by relay logic circuits. The remaining four SRVs are manually controlled. For each valve, one of the two solenoids is operable from the control room. The other solenoid is operated from the local SRV control panel located in the reactor building. The solenoids are powered from redundant DC power sources. In the event of fire requiring control room evacuation, all eleven SRVs can be operated manually at the Local SRV control panel. However, since there was no provision for isolating the SRV solenoids from the control room, a control room or reactor building fire could induce a hot short and spuriously open these valves irrespective of the position of control switches located in the control room.

Resolution   To ensure SRV operation in the event of a control room fire the licensee implemented plant modification to install a dedicated isolation switch for each of the eleven SRVs in a new auxiliary shutdown panel located outside the control room. In addition, the licensee modified the circuitry of the seven ADS valves and the four manual SRVs to provide additional isolation capability in the event of a reactor building or control room fire.

APPENDIX B.
SPECIFIC CIRCUIT ANALYSIS ISSUES

APPENDIX B.
SPECIFIC CIRCUIT ANALYSIS ISSUES

This appendix discussed certain circuit analysis issues that specific have been the subject of much confusion and debate and include: multiple spurious actuations, fire damage to nonessential systems, and multiple circuit faults. The discussion is provided in terms of "real world"examples of technical issues that were identified during the review of safe-shutdown analyses developed by various licensees.

B.1   Multiple Spurious Actuations

In Question 5.3.10 of GL 86-10, the staff provides a response to a question posed by industry regarding the type of plant transients that should be considered in the *design of the alternative or dedicated shutdown systems*. In its response the staff states, in part: *"the safe shutdown capability should not be adversely affected by any one spurious actuation or signal resulting from a fire in any plant area."*

The intent of the guidance contained in the staff's response is to ensure that the *design* of the alternative or dedicated shutdown capability is sufficiently robust to be capable of m tigating the occurrence of one worst-case spurious actuation prior to isolation of potentially affected circuits from the fire-affected area. In certain instances, however, the staff's response has been mis-interpreted to mean that only a single spurious actuation need be considered for any fire area, without any further consideration of the number, type, function, or specific location of potentially affected circuits and cables. This misunderstanding appears to have been further complicated by the fact that this approach (i.e., assumption of a single spurious actuation per fire event) has been accepted in several NRC safety evaluations of plant-specific post-fire safe-shutdown methodologies. While the fire protection licensing basis for these facilities would only require consideration of a single spurious actuation, it should be noted that certain licensees recognize that the application of this assumption could result in a shutdown strategy that is inconsistent with the fundamental objective of ensuring that one train of systems needed to achieve and maintain hot-shutdown conditions remains free of fire damage. For example, although the "single spurious actuation per fire event" assumption was accepted by the staff in a safety evaluation of a BWR, an NRC inspection of this facility did not identify any cases where the potential for fire to cause multiple spurious actuations had not been sufficiently evaluated. Specific cases of how this "single spurious actuation per fire event" assumption can impact the shutdown capability are illustrated by the following examples:

- At one PWR cooling water flow to the EDG may be provided by one of two parallel flowpaths. Since a "normally open" MOV is located in each flowpath, at least one of these valves must remain open to ensure an adequate supply of cooling water is supplied to the EDG. Based on its interpretation of Question 5.3.10 of GL 86-10, however, the licensee had not considered the potential for both valves to spuriously change position as a result of fire damage. In lieu of identifying the routing of cabling associated with both valves by fire area and evaluating for the potential effects of fire damage to these circuits/cables within each fire area, the licensee had dispositioned this potential vulnerability on the assumption (per its interpretation of GL 86-10 Question 5.3.10) that only one spurious actuation would occur per fire event. As a result of its interpretation, the potential for fire to cause both valves to inadvertently change position as a result of fire damage was not considered in the analysis.

- As described in Chapter 6, the SSEL identifies equipment that is needed to ensure the successful accomplishment of essential shutdown functions. Based on its assumption that only one spurious actuation would occur per fire event, the shutdown methodology developed by a licensee of a 4-loop Westinghouse PWR relied on operator intervention to mitigate this "one" actuation should it occur. Since no action is taken before fire damage occurs, the successful implementation of this approach is largely predicated on the operators' ability to detect the spurious actuation and perform manual actions in a timely manner to defeat its effect on safe-shutdown capability. Based on this approach, the SSEL did not include any automatically actuated flow-path valves MOVs or air-operated valves (AOVs)] that were in their desired position for post-fire safe-shutdown during normal plant operations (e.g., a normally open MOV in the flowpath of a required shutdown system). Since the SSEL serves as a starting point for identifying circuits and cables that could impact the shutdown capability if they are damaged by fire, the routing of cables associated with these components was not considered. As a result, the potential for fire to cause more than one automatically actuated valve to spuriously change position in an undesired manner for post-fire safe-shutdown had not been evaluated for each fire area.

- A review of the SSA submitted by the licensee of a BWR identified examples where redundant components may be subject to spurious actuations (i.e., undesirable change of position or operating state) as a result of a single hot short on each of their respective control circuits. Although the control circuits of the redundant MOVs were subject to damage by a single fire, in its evaluation of this issue, the licensee stated: *"For both valves to open simultaneously, a hot short on each valve is required. NRC GL 86-10 does not require the assumption of multiple hot shorts for non-high/low-pressure interfaces. Therefore, one of these two valves is assumed to remain closed."* In subsequent meetings and correspondence, the staff informed the licensee of its concern that the application of this assumption may result in an inability to adequately demonstrate compliance with Sections III.G.2 and III.L of Appendix R to 10 CFR Part 50. In a subsequent response, the licensee submitted revised criteria it had developed and employed for the analysis of potential spurious operations. Under its revised methodology, all circuits which could cause undesirable spurious operations were identified and evaluated for potential fire damage. With the exception of components which comprise a high/low pressure interface boundary the licensee's evaluation considered any and all spurious operations that may occur as a result of a single fire, on a one-at-a-time basis (i.e., sequential, nonconcurrent). That is, for each fire area all potential spurious operations that may occur as a result of a postulated fire were identified, and corrective actions were implemented as needed on a one-at-a-time basis. Fire-initiated faults were assumed to exist until action was taken to negate their effects. The fire was not postulated to eventually clear the faults. For redundant components which form a high/low pressure interface boundary, the evaluation considered the potential for concurrent, simultaneous, spurious operations. When cables or equipment of which spurious operation could affect safe-shutdown were identified, they were included as required cables in the licensee's Appendix R separation analysis.

- The licensee of a BWR used the single spurious actuation per fire event assumption as a basis for not providing fire protection features for redundant trains of shutdown equipment. In this case, although redundant suction valves of the RCIC system were identified as being required to achieve and maintain hot shutdown conditions and their cables were located in close proximity [<4.56 m (<15 ft)], the licensee did not consider the separation requirements of Section III.G.2 to be applicable on the basis that both valves must fail (spuriously actuate to the closed position) in order to cause a total loss of makeup capability.

Section III.G of Appendix R to 10 CFR Part 50 requires, in part, that circuits and cables that could prevent operation or cause maloperation of SSCs important to safe-shutdown be provided with a level of fire protection necessary to ensure that such circuits will remain free of fire damage. Consistent with the deterministic approach described in Chapter 6, circuits and cables which lack a suitable level of fire protection (as delineated in Section III.G.2 of Appendix R) must be assumed damaged by their exposure to fire and this damage should be expected to cause one or a combination of circuit faults to occur between conductors of each cable or circuit that may be affected by the fire. Accordingly, if, because of a lack of fire protection features, there is a potential for multiple cables or circuits to be faulted, it follows that faults between the conductors of the affected cables or circuits may lead to the occurrence of one or more (i.e., multiple) spurious actuations. In a letter to the NEI dated March 11, 1997, the staff reiterated the deterministic approach where the number of spurious signals or changes in operational configuration that may be expected to occur as a result of fire damage to unprotected cables or circuits cannot be predicted.

As described in Chapter 6 and Appendix A to this document, licensees have historically identified equipment (safety-related and nonsafety-related) of which spurious operation could impact the safe shutdown capability described in the plant-specific SSA. If it can be demonstrated that the occurrence of all credible circuit failure modes (hot shorts, open circuits and shorts to ground), will not cause the connected equipment to spuriously actuate or malfunction in a manner that would adversely impact the post-fire safe-shutdown capability, no further analysis is necessary and the component may be screened from further evaluation. For example, a review of plant P&IDs may indicate that the spurious actuation (opening) of two, series connected, MOVs has the potential to impact the shutdown capability by creating an undesired diversion (i.e., loss) of process coolant flow. If it can be shown that this failure mode (both valves open) would not impact the shutdown capability (e.g., if the amount of flow lost was small compared to the makeup capability of the system) the components (MOVs) can be screened from further consideration. However, if this initial evaluation determines that spurious actuation of the components (opening of both MOVs) could impact the shutdown capability (flow loss in excess of makeup capability), a detailed circuit analysis that considers the impact fire damage to connected circuits and cables is necessary.

As discussed in Chapter 6, with the exception of components that comprise a high/low pressure interface boundary, the evaluation should consider any and all spurious operations that may occur as a result of a single fire, on a one-at-a-time basis. That is, for each fire area, all potential spurious operations that may impact the shutdown capability should be identified. While it is not assumed that all such spurious actuations will occur instantaneously at the onset of fire, the analyst must consider the possibility for each spurious actuation to occur in a sequential manner, as the fire progresses, on a one-at-a-time basis. Since it is not assumed that the fire will clear the fault(s) that caused the undesired actuation (Reference GL 86-10, response to Question 5.3.2), the potential for sequentially occurring failures to result in the concurrent failure of two or more components (such as the MOVs described above) must be considered. Accordingly, if control cables of two components (e.g., normally-closed MOVs) are subject to damage, the potential for both valves to spuriously actuate (open) as a result of fire damage cannot be ignored. Since the control cable of neither valve is ensured to remain free of fire damage, it is considered credible that both valves could spuriously open sequentially during a fire event. It is expected that such conditions would be identified where they may exist and appropriate preventive or mitigating actions implemented.

Although they do not satisfy the certain technical requirements of Appendix R, the use of operator manual actions to mitigate this event may provide an acceptable resolution (see Appendix A). For example, the licensee's evaluation of a control room fire at one BWR found circuits of three valves to be susceptible to fire damage. Since the spurious opening of all three valves would result in a drain down of the suppression pool, the potential for all three valves to spuriously actuate could not be ignored. To mitigate this event, the licensee implemented procedural changes which require one of the valves to be ensured closed by operator manual actions.

B.2 Fire Damage to Nonessential Systems

Fire damage to systems that are not needed to perform essential shutdown functions (i.e., nonessential or nonsafety systems) can have a significant impact on shutdown capability, as illustrated by the following examples:

- Inadvertent initiation of the HPCI system: The analysis performed by one BWR revealed that inadvertent initiation of the HPCI system and concurrent loss of the 137.16-cm (54-in.) high-water trip for HPCI as a result of a control room fire could, in a short time period (approximately 3 minutes), cause a vessel overfill condition to the point where HPCI would be disabled and the main steam lines would be filled with high pressure water.

- Inadvertent feedwater initiation: Certain BWRs employ steam-driven feedwater pumps in their design. Since these pumps are not electrically powered they will continue to provide flow during feedwater system coast down as long as sufficient steam is available. The concern with this configuration is that a fire-induced spurious signal on the feedwater pump control circuit (typically located in the control room) could cause a false demand for the steam-driven pumps to inject coolant at maximum capacity. If this were to occur, operators would have a very short time frame to implement mitigating actions, such as closing the MSIVs, closing of the feedwater discharge valves, and tripping the feedwater turbine from outside the MCR.

- The normal charging line to the RCS was not credited for post-fire safe-shutdown by the licensee of a PWR. This flowpath, which branches off the credited RCP seal injection flowpath, includes four normally open valves before entering the regenerative heat exchanger. The pressurizer auxiliary spray valve (PASV), which is located downstream of the regenerative heat exchanger, is a normally closed MOV. Since the normal charging flowpath was not credited for safe-shutdown, none of the valves in its flowpath were included in the SSEL. As a result, none of the cables associated with these valves were fully evaluated for the effects of fire damage. While not needed to perform an essential shutdown function, the spurious opening of PASV as a result of fire-induced faults in its control circuitry could have a significant impact on the shutdown capability by causing a collapse of the steam bubble in the pressurizer and rapid depressurization of the RCS.

- The shutdown strategies developed by most PWRs do not credit the use of pressurizer heaters. While not needed for safe-shutdown, fire damage that causes the heaters to inadvertently actuate (load) at a time when power is being supplied from the onsite source of electrical power (e.g., EDG) could significantly impact safe-shutdown capability if the EDG was not capable of supplying this additional load (EDG overload).

As discussed in Chapter 6 and Appendix A, the achievement of safe-shutdown is dependent on ensuring the active control of some components and preventing the maloperation of other components. A post-fire safe-shutdown analysis should be a bounding analysis that identifies the range of possible fire impacts within each fire area (vulnerabilities) and ensures that appropriate measures are in place to prevent this damage from affecting the ability to safely shutdown the plant. Therefore, it is not sufficient to only consider the effects of fire damage to cables of equipment needed to ensure operation of credited shutdown systems. The scope of successful shutdown strategies also includes consideration of the effects of fire damage to nonessential equipment and systems of which inadvertent or spurious actuation could impact the shutdown capability.

B.3   Multiple Circuit Faults

In GL 81-12 and GL-86-10, the NRC established that either physical protection from fire (per Section III.G.2 of Appendix R), or detailed electrical circuit analyses may be used to demonstrate that fire will not cause equipment to mal-operate in a manner that could adversely affect the post-fire safe-shutdown capability of the plant. While either approach is acceptable, the use of analytical techniques places greater importance on the assumptions, criteria, and review methodology which form the basis of the analysis. Also in GL 86-10, the NRC staff defined the circuit failures to be considered. Specifically, in Question 5.3.1 the staff provided the following guidance:

> Sections III.G.2 and III.L.7 of Appendix R define the circuit failure modes as hot shorts, open circuits, and shorts to ground. For consideration of spurious actuations, all possible functional failure states must be evaluated, that is, the component could be energized or de-energized by one or more of the above failure modes (emphasis added). Therefore, valves could fail open or closed; pumps could fail running or not running; electrical distribution breakers could fail open or closed...

In accordance with this guidance, when performing a circuit failure analysis, one or more circuit failure modes (e.g., multiple hot shorts, a hot short combined with a ground or open circuit etc.) must be considered. When considering the effects of fire damage to a multi-conductor cable, the potential for fire to cause multiple hot shorts between individual conductors must be considered. The failure to fully evaluate the potential for fire to cause more than a single fault in each circuit/cable under consideration may have potentially significant consequences on the plant's shutdown capability.

For example, the circuit analysis performed by a licensee of a BWR was found to arbitrarily limit the number of failure modes to one hot short, or one short to ground, or one open circuit on an individual device or component basis. As a result of this approach, the potential for fire to cause electrical contact between individual conductors of two twisted-pairs of conductors located within a single multiconductor cable was not considered credible by the licensee. In this case, an instrument cable contained two pairs of twisted conductors. If fire were to cause the individual conductors of the twisted pairs to short together (i.e., a short between conductors of twisted pair No. 1 and a short between conductors of twisted pair No. 2) two false high RCS pressure signals would be generated. The two high pressure signals would cause all 16 SRVs to fully open to rapidly de-pressurize the reactor. In addition, the fault current associated with these two circuit failures would not be large enough to open the protective fuse. Fire test data provided by the cable vendor showed that the wires could short in about 3 minutes when exposed to a test fire.

This page intentionally left blank.

NRC FORM 335
(2-89)
NRCM 1102,
3201, 3202

U.S. NUCLEAR REGULATORY COMMISSION

BIBLIOGRAPHIC DATA SHEET

(See instructions on the reverse)

1. REPORT NUMBER
(Assigned by NRC, Add Vol., Supp., Rev., and Addendum Numbers, if any.)
NUREG-1778

2. TITLE AND SUBTITLE

Knowledge Base for Post-Fire Safe-Shutdown Analysis

Draft Report for Comment

3.	DATE REPORT PUBLISHED	
	MONTH	YEAR
	January	2004
4. FIN OR GRANT NUMBER		

5. AUTHOR(S)

M.H. Salley

6. TYPE OF REPORT
Technical

7. PERIOD COVERED *(Inclusive Dates)*

8. PERFORMING ORGANIZATION - NAME AND ADDRESS *(If NRC, provide Division, Office or Region, U.S. Nuclear Regulatory Commission, and mailing address; if contractor, provide name and mailing address.)*

Division of Systems Safety and Analysis
Office of Nuclear Reactor Regulation
U.S. Nuclear Regulatory Commission
Washington, DC 20555-0001

9. SPONSORING ORGANIZATION - NAME AND ADDRESS *(If NRC, type "Same as above"; if contractor, provide NRC Division, Office or Region, U.S. Nuclear Regulatory Commission, and mailing address.)*

Same as above

10. SUPPLEMENTARY NOTES

S.D. Weerakkody, NRC Project Manager

11. ABSTRACT *(200 words or less)*

Every operating nuclear power plant is required to have a program that demonstrates the capability to safely shut down and maintain the reactor in the event of a fire. The U.S. Nuclear Regulatory Commission (NRC) initially issued its requirements in the Fire Protection Rule set forth in Title 10, Section 50.48, of the Code of Federal Regulations (10 CFR 50.48) and Appendix R to 10 CFR Part 50. The NRC has since issued numerous related generic communications over the past 20 years. The purpose of this document is to facilitate understanding of this technically challenging process and the regulatory framework upon which it is based by compiling all essential information in a single source. This document also lays the groundwork for future risk-informed activities in the post-fire safe-shutdown area.

12. KEY WORDS/DESCRIPTORS *(List words or phrases that will assist researchers in locating the report.)*

Circuit analysis
Post-fire safe-shutdown
Appendix R
Risk-informed approach

13. AVAILABILITY STATEMENT
unlimited
14. SECURITY CLASSIFICATION
(This Page)
unclassified
(This Report)
unclassified
15. NUMBER OF PAGES
16. PRICE

Printed
on recycled
paper

Federal Recycling Program

www.ingramcontent.com/pod-product-compliance
Lightning Source LLC
Chambersburg PA
CBHW080242180526
45167CB00006B/2383

Das Bild

Theorie – Geschichte – Praxis

Akademie Studienbücher

Kulturwissenschaften

Matthias Bruhn

Das Bild

Theorie – Geschichte – Praxis

Akademie Verlag

Der Autor:
Dr. Matthias Bruhn, Jg. 1966, Wissenschaftlicher Mitarbeiter des Hermann von
Helmholtz-Zentrums für Kulturtechnik an der Humboldt-Universität zu Berlin

Bibliografische Information der Deutschen Nationalbibliothek
Die Deutsche Nationalbibliothek verzeichnet diese Publikation in der Deutschen
Nationalbibliografie; detaillierte bibliografische Daten sind im Internet über
http://dnb.d-nb.de abrufbar.

ISBN 978-3-05-004367-8

© Akademie Verlag GmbH, Berlin 2009

www.akademie-studienbuch.de
www.akademie-verlag.de

Einband- und Innenlayout: milchhof : atelier, Hans Baltzer Berlin
Einbandgestaltung: Kerstin Protz, Berlin, unter Verwendung des Gemäldes
    Die Erfindung der Malerei (1830) von Karl Friedrich Schinkel. Von der Heydt-
    Museum, Wuppertal.
Satz: Druckhaus „Thomas Müntzer" GmbH, Bad Langensalza
Druck und Bindung: CS-Druck Cornelsen Stürtz GmbH, Berlin

Printed in Germany

Das Bild
Theorie – Geschichte – Praxis

1 Das Bild vom Bild: Sehweisen, Redeweisen

Abbildung 1: Nicolas Poussin: *Armida entführt den eingeschläferten Rinaldo*, Öl auf Leinwand, Aufnahme bei Tageslicht (um 1637)

Abbildung 2: Nicolas Poussin: *Armida entführt den eingeschläferten Rinaldo*, Aufnahme mit Röntgenstrahlung (2003)

In dem französischen Gemälde des 17. Jahrhunderts wird eine literarische Begebenheit geschildert: Zu sehen ist eine figurenreiche Szene, mit intensiver Farbigkeit und großer Plastizität komponiert, in deren Gestalten und Handlungen ein belesener Betrachter einen Moment aus dem berühmten Epos „La Gerusalemme liberata" („Das befreite Jerusalem", entstanden 1559–75) von Torquato Tasso wiedererkennen kann. In diesem Werk wird berichtet, wie die Zauberin Armida den Kreuzfahrer Rinaldo einschläferte, um ihn auf ihre Insel zu entführen wie einst Circe die Gefährten des Odysseus – ein Mythos, der auch literarisch und musikalisch häufig verarbeitet worden ist.
Dasselbe Gemälde, eine großformatige Leinwand (120 × 150 cm), könnte auch auf ganz andere Weise wahrgenommen werden, z. B. in einer Restaurierungswerkstatt. Die Röntgenaufnahme legt im Untergrund der Malerei äußerlich unsichtbare Kompositionen frei und lässt auf diese Weise ältere Strukturen und spätere Ergänzungen erkennen. So zeigt sich z. B., dass hinter der mittleren Figur der Armida auf dem Weg im Hintergrund eine ausschreitende Person vorgesehen war. Überdeutlich treten auch der Rahmen und Strukturen der Leinwand hervor.

Jedes Bild kann auf unterschiedliche Weise und im Hinblick auf verschiedene Sinnschichten und Funktionen betrachtet und verwendet werden. Aus diesem einfachen Umstand folgen bereits erhebliche Schwierigkeiten, die deutsche Bezeichnung „Bild" auch nur annähernd zu definieren. Als umso wichtiger und hilfreicher erweisen sich begriffliche Klärungen. Während die Bedeutung bildlicher (oder visueller) Kommunikation für Kunst, Unterhaltung oder Werbung stets unstrittig war, ist in den letzten beiden Jahrzehnten auch in anderen, z. B. naturwissenschaftlichen Bereichen, die Aufmerksamkeit für den instruktiven und kommunikativen Wert visueller Repräsentation stark gewachsen. Als „Bildwissenschaft" zeichnet sich dabei in diversen Disziplinen eine Diskussion um das eigenständige Vermögen von Bildern ab, Wissen zu erzeugen und zu reflektieren und das Verstehen zu fördern. In dieser Diskussion wird wiederum deutlich, wie vielfältig die Auffassungen, Funktionen und Anwendungsbereiche des Bildlichen sind.

1.1 Betrachtungsweisen von Bildern
1.2 Begrifflichkeiten
1.3 Bild vs. Text und Sprache
1.4 Bildwissenschaft: Disziplinäre Grenzen

1.1 Betrachtungsweisen von Bildern

Menschen können auf ein einzelnes Ereignis oder auf einen einzelnen Gegenstand bekanntlich sehr unterschiedlich reagieren. Wenn eine interessierte und vorgebildete Besucherin im Kunstmuseum an ein Gemälde wie Nicolas Poussins *Armida entführt den eingeschläferten Rinaldo* (→ ABBILDUNG 1) herantritt, so wird sie darin wahrscheinlich eine historische oder mythologische Szene geschildert sehen, deren einzelne Figuren sie zu identifizieren versucht, indem sie Details wie Kleidungstücke, Attribute oder Gesten studiert und ausdeutet. Vielleicht wird sie dabei erkennen, dass hier ein bestimmter literarischer Stoff verarbeitet wurde, der die Figuren miteinander verbindet. Womöglich interessiert sie sich aber auch mehr für die Handschrift des Malers, dessen Stilmittel und Darstellungstechniken sie schätzt, oder für die intensive und kostbare Farbigkeit, die das Werk ausstrahlt.

Ein anderer Besucher wird sich vielleicht im Vorbeigehen, weil er die Darstellung reizvoll findet, die Beschilderung ansehen, um sich über das Motiv und die Entstehungszeit zu informieren. Eine Schulklasse läuft unterdessen im Hintergrund achtlos an der Tafel vorbei, unterwegs in die Sammlung zeitgenössischer Kunst. Ein Werbegrafiker auf Durchreise hat vor dem Gemälde eine spontane Eingebung, während eine Videokünstlerin bei einem Interviewtermin im Museum auf das Bild zeigt und ironisch anmerkt, sie sei froh, wenn ihre Arbeit in 400 Jahren ebenfalls so präsentiert würde.

Es gibt aber auch die Restauratorin, die eine Stunde später vorbeischaut, eine in Teilen beschädigte oder übermalte Leinwand vor sich findet und die den Erhaltungszustand des Gemäldes zum Anlass nimmt, eine technische Untersuchung daran vorzunehmen. Mit radiologischen Aufnahmen oder chemischen Analysen wird sie die Leinwand begutachten und zur eigenen Überraschung feststellen, dass sich unter der äußerlich sichtbaren Ölfarbe eine abweichende Komposition verbirgt (→ ABBILDUNG 2).

Die Besucher und die Restauratorin sehen denselben Gegenstand, ein „Bild", präziser „Gemälde". Im einen Fall ist es eine bestimmte gestalterische Seite, im anderen Fall die Verbindung von Bildträger und Malpigmenten, die bei der Betrachtung im Mittelpunkt steht. Wenn in der Restaurierungswerkstatt Farbproben entnommen werden, werden noch dazu völlig andere Schichten des Gemäldes freigelegt: als Querschnitt, im Streiflicht, als mikroskopisches Detail.

Diese Mehrdimensionalität gilt nicht nur für Bildformen, die sich unter späteren Übermalungen oder hinter rätselhaften Motiven verstecken, sondern für jedes einzelne Bildmotiv, seine mediale und funktionale Einbettung und seine betrachtungsabhängige Interpretation. So kann das simple Bildmotiv eines Adlers sowohl als Repräsentation einer Tierart als auch als ein nationales Wappentier verstanden werden, es kann mit verschiedenen Techniken (Zeichnung, Fotografie) aufgenommen sein und unterschiedliche Verwendungen erfahren. Als Bild in einem Lehrbuch erfüllt es eine andere Funktion und erfährt eine andere Stilisierung als z. B. bei einem Tattoo, das in der Haut seines Trägers eingespritzt ist, um als Schmuck und persönliches Symbol zu dienen.

1.2 Begrifflichkeiten

Uneindeutigkeit des Begriffes

In Anbetracht solcher vielfältigen Schichtungen und Möglichkeiten der Betrachtung und Nutzung kann der Begriff „Bild" kaum in einem absoluten, sprach- und kulturübergreifenden Sinne für einen einheitlichen Gegenstand stehen. Was als ein Bild gilt und wahrgenommen wird, hat gesellschaftliche wie individuelle Hintergründe, hängt von intellektuellen und psychologischen Faktoren, von langfristigen Sprachtraditionen und langlebigen Konnotationen ab. Bilder lösen Leidenschaften aus, aber in Abhängigkeit von Vorkenntnissen, Interessen und Aufmerksamkeiten durchaus verschiedene. Ein Porträt hat als privates Erinnerungsstück im Portemonnaie andere Adressaten und Bedeutungen denn als Wandplakat. In der Werbung oder Kunst erfüllen fotografische Aufnahmen andere Funktionen als in der Kriminalistik. Derart weit auseinander liegende Felder können sich jedoch in anderer Hinsicht auch sehr nahe kommen, etwa in allgemeinen ‚visuellen' Fragen, die mit dem Begriff „Bild" nur unzureichend beschrieben wären. Der Begriff, sein Inhalt und auch seine Aktualität sind historisch und kulturell bedingt. Diese Bedingungen sind daher viel eher der geeignete Gegenstand der Untersuchung.

Übergänge zu anderen Medien

Es ist dabei zu berücksichtigen, dass viele der nachfolgend beschriebenen Probleme ihre Entsprechung durchaus in anderen Medien haben und dass das Bild zudem selten allein in Erscheinung tritt, sondern in wechselnden Kombinationen mit Musik und Text oder als Bestandteil größerer Ensembles und Strukturen (z. B. eines Fotoarchivs; → KAPITEL 10). Infografik oder Computertomografie können beispielhaft für eine weitergehende Ausdehnung des Gegenstandsfeldes „Bild" auf alles stehen, was mit Augen gesehen werden kann – und darüber hinaus. Schon die „bildende" Kunst ist nie auf das Bild allein reduziert gewesen, sondern hinterlässt Formen, Artefakte, Werke in besonderen Räumen oder biografischen Zusammenhängen (→ KAPITEL 6).

Aufschlussreich sind in diesem Zusammenhang die historischen Transformationen des Begriffes. „Bild" konnte über Jahrhunderte für das plastische Kultobjekt stehen, um in der Neuzeit zu einem Synonym für Tafelgemälde und Wandmalereien zu werden. In diesen kommen andere ästhetische Mittel zum Einsatz als in der dreidimensionalen Skulptur, auch mit Auswirkungen auf die Theoriebildung: Die berühmte Definition des Künstlers und Kunsttheoretikers Leon Battista Alberti (1404–72), wonach das Bild als Blickrahmen eine Perspektive fixiert (*De Pictura;* vgl. Alberti 1435, § 19), schließt nun die Skulptur aus, der eine eigene Darlegung gewidmet wird (*De statua;* vgl. Alberti 1440). Seit dem 20. Jahrhundert dürften die meisten Menschen mit „Bild" eher Fotografien oder andere technische Reproduktionen assoziieren (→ KAPITEL 2.1): „Haben Sie schon die Bilder aus New York gesehen?" kann von einem Zeitungsfoto bis zu einer TV-Übertragung fast alles einschließen, was einen sichtbaren Gegenstand hat. Was genau aber meinen und sehen die Sprecher, wenn sie fragen, ob jemand anders „es" auch gesehen hat?

Begriffsverschiebungen

Wenn dagegen absolut festgelegt werden sollte, ob eine Formel, ein plastisches Objekt oder ein Straßenbild als „Bild" zu gelten hätten, so würde sich entweder nur ein sehr kleiner gemeinsamer Nenner finden lassen (etwa, dass das Bild eines Dinges von diesem Ding verschieden ist), oder die Definition wäre so offen, dass sie mit traditionellen Vorstellungen brechen und zu anderen Bestimmungen aufbrechen müsste:

> „Sofern Bilder [pictures] definiert werden sollten anhand von durchschnittlichen Beispielen, so wären dies eher Piktogramme als Gemälde. Ein Bild [image], zufällig ausgewählt, ist sehr viel wahrscheinlicher ein Ideogramm, eine Steinritzung oder eine Börsenkurve denn ein Gemälde von Degas oder Rembrandt, so wie ein Tier viel wahrscheinlicher als Bakterium oder Käfer in Erscheinung tritt denn als Löwe oder Mensch." (Elkins 1999, S. 4, Übersetzung des Autors)

Die Mehrheit der Bilder: Grafiken

Die Etymologie kann hier nur bedingt weiterhelfen. Im Germanischen wird *bilaþja* für Bild genommen, mit einer Wurzel in *bil-* (für Zwischenraum) oder *bila* (unterscheiden; diese Unterschieds- oder Ähnlichkeitsbeziehung ist noch präsent in Wörtern wie „Zwilling", germanisch *billingr;* vgl. Köbler 1980). Das *Deutsche Wörterbuch* der Gebrüder Grimm kennt allein zehn verschiedene, deutende Her-

Etymologien

leitungen (vgl. Grimm 1854, Sp. 8–16), darunter diejenigen, wonach das Bild für Verlebendigung, für die „Abspiegelung" in der Natur stünde, oder für die Formung durch Künstlerhand. Das deutsche „Bild" kann sowohl für *species* (lateinisch für: Scheinbild, temporäres Bild, Larve) als auch für *imago* (lateinisch für: dauerhaftes Bild) stehen.

Antike — Ein Großteil lexikalischer Bestimmungen setzt in der griechischen Antike mit Platon an und dessen Behauptung in der *Politeia*, dass das menschengemachte Bild gegenüber der Idee immer nur nach- und zweitrangig sein könne (Platon 2004, 596e–598d). Als solches „Abbild" wird der Begriff auf das griechische Wort *eikon* oder das lateinische *simulacrum* zurückgeführt. Dieses beschreibt auch das Erinnerungsbild, z. B. das Bildnis der Verstorbenen wie in der *Naturalis historia* von Plinius d. Ä. (Plinius 1997, S. 15–19). Darin folgt es dem griechischen Wort *eidolon* (in der Homerischen *Ilias* das Seelenbild der Toten im Hades), das dann in der weiteren Folge, bedingt durch die nachantiken bildtheologischen Debatten, zum „Idol" im Sinne des verführerischen Götzenbildes wurde (→ KAPITEL 2, 9).

Bilder als Gestaltung — Das Vermögen von versierten Gestaltern, die sichtbare Welt mit Linien, Punkten und Farbverläufen wiederzugeben, Unsichtbares sichtbar zu machen, mit Formen eine veränderte und sublimierte Welt zu schaffen, wurde allerdings zu allen Zeiten auch bewundert. „Wer die Malerei nicht schätzt, verschmäht die Wahrheit", behauptete daher treffend der griechische Philosoph Philostratos (um 200 n. Chr.) in seiner Sammlung von Bildbeschreibungen, die er unter dem Titel *Eikones* (Die Bilder) verfasst hat (Philostratos 1968, S. 85).

Bedeutung der Anschaulichkeit — Auf ihn und seine Zeitgenossen bezogen sich in der Folge Theorien, die in der Gestaltung von Bildern den intellektuellen Wert wie ihre Faszinationskraft schätzten. Als Mittel der ästhetischen Kommunikation waren sie stets und in den meisten Gesellschaften präsent, im öffentlichen Bauwesen, als Kirchendekoration oder politische Werbung (→ KAPITEL 7). Die Beschäftigung mit Bildern als Bildungsgut gewann mit dem Aufstieg der Pädagogik (Johann Amos Comenius, 1592–1670) an gesellschaftlicher Anerkennung (→ KAPITEL 10.3). Dies gilt auch in der weiteren Geschichte der Kunstbetrachtung und Ästhetik, etwa im 18. Jahrhundert, wo Begriffe wie Sehen, Betrachtung, Gestalt gegenüber dem Bild an Bedeutung gewannen (→ KAPITEL 4.2). Die Einrichtung und schrittweise Öffnung von Museen seit dem 17. Jahrhundert, die Begründung von Studienfächern wie Archäologie und Kunstgeschichte und die Einführung des schulischen und akademischen Zeichenunterrichts förderten eine breitere gesellschaftliche Aufmerksamkeit für das Visuelle, welches sich wiederum nach 1800 in der Gründung von „Bilderbögen" (vgl. Brakensiek/Rockel 1993) und illustrierten Zeitschriften für gebildete Schichten bemerkbar machte, noch ehe diese fotografisch ausgestattet waren (→ KAPITEL 2.2). Gegenüber dem Logozentrismus, also dem Vorrang des Wortes, wurden im 19. Jahrhundert in Kirchen oder Bildungsvereinen die Universalität des Bildes und seine Anschaulichkeit betont, die noch heute als Argument für Lehrfilme oder Powerpoint-Präsentationen vorgebracht werden.

Indem die Jahrhunderte während theologische Diskussion den Kultwert und die Möglichkeit von Offenbarung und Gottesschau (griechisch *theoria*) im Bild infrage gestellt hatte, wurde womöglich die Auffassung weiter befestigt, wonach Bilder durch ihre Materialität vom abstrakten Zeichen unterschieden sind (→ KAPITEL 3.2). Nicht nur Bilder werden angeschaut, Bilder beschreiben jedoch das Geformte, wie es auch im deutschen Wort „Bildung" oder im englischen *building* nachklingt. Besonders das Englische hat diese Lesart durch Unterscheidung von *picture* und *image* mitvollzogen, indem es *picture* als einzelne Darstellung begreift, während unter *image* allgemeine oder kollektive Phänomene wie z. B. das „Erscheinungsbild" in der Öffentlichkeit fallen. Im Deutschen ist die Verbindung von Material und Motiv unverändert präsent. Ein Bild ist auch im digitalen Zeitalter und trotz der „Fernsehbilder" weiterhin für viele, was auf das Papier gebracht, aus dem Netz geladen, ausgedruckt, von der Wand genommen, in das Buch befördert werden kann.

Sobald der Begriff „Bild" hingegen auf andere Bereiche, etwa den des Mentalen, Imaginativen oder Metaphorischen ausgedehnt wird, müssten Erinnerungs- und Gedächtnisbilder ebenso hinzugezählt werden wie flüchtige Träume, Spiegelungen und Halluzinationen. In einer solchen übertragenen Verwendung von „Bild" bleibt offen, ob das „Bild von einer Stadt" eine Skyline, eine ortstypische Straßenszene oder eine Assoziation bezeichnet.

Selbst innerhalb eines einzelnen Gebiets wie der Psychologie wird der Begriff „Bild" auf unterschiedliche Weise verwendet: In den Bildern der Traumwelt spiegeln sich Bedürfnisse oder Ängste des Selbst (vgl. Freud 1899). Bilder jeglicher Art können nach Jacques Lacan auch zum Brennpunkt eines „begehrenden Blickes" werden, der etwas verlangt, weil er es nicht bekommt (Lacan 1996; vgl. Blümle/von der Heiden 2005). Die psychologische Sehweise berührt sich zudem an vielen Stellen mit anderen Disziplinen, etwa der Philosophie,

Materialität des Bildes

Mentale Bilder

Ethnologie oder Kunstgeschichte (→ KAPITEL 3–7): Als gestaltete Kunstform speichern und eröffnen Bilder gestische Energien, Bildnisse dienen der Stellvertretung, Identitätsbildung und Biografisierung. Bilder werden so auch zum Gegenstand von Ethnologie und Anthropologie (vgl. Belting 2004; Kaschuba 2008). Bilder liefern für die Phänomenologie die sinnlichen Impulse, die Schwelle und Schnittstelle des Denkens sind (vgl. Waldenfels 2001).

Unmöglichkeit einer Definition

Wenn das Wort „Bild" trotz seiner verschiedenen Gegenstände und Diskurse von der Mehrzahl der (deutsch sprechenden) Menschen verstanden wird, so lässt es sich gleichwohl nicht in einer eindeutigen, abstrakten und wissenschaftlich verbindlichen Weise definieren. Entsprechende Versuche führen unvermeidlich zu theoretischen Einengungen:

> „Neben der Vagheit kann an der Frage ‚Was ist ein Bild?' die essentialistische Suggestion irreführend sein, es müsse ein Wesen des Bildes und daher auch nur eine Hauptfrage in Bezug auf den Bildbegriff geben." (Scholz 2004, S. 15)

Ausdehnung des Objektbereiches

In den Vorworten von Studien, die sich mit dem Bild in einem weiteren historischen Sinne befassen, werden dagegen die Aufzählungen der Medien und Formen, die zum Gegenstandsbereich gehören und ihn fassbar machen sollen, immer länger. Jeder dieser Texte deutet an, dass das „Bild" nur ein Stichwort sein kann, welches seine Transformationen und Unschärfen hat und sich in der Kunst, in der wissenschaftlichen Repräsentation, in den Massenmedien weiter entfaltet. Sie lassen zwar erkennen, dass es eine wiederkehrende Rede vom Bild gibt, aber auch wie sehr diese Rede durch zeitgenössische soziale, technische oder ökonomische Blickweisen geprägt ist.

1.3 Bild vs. Text und Sprache

Abgrenzungen

Obwohl das „Sprachbild" eine geläufige Wendung ist, erfolgt die Charakterisierung des Bildlichen auch über die Abgrenzung von Sprache oder durch den Vergleich mit Text. Dies wurde in der Poetik des Horaz, die den Wettstreit der Kunstgattungen als Motor ihrer Ausdifferenzierung begreift, in die berühmte Formel „Ut pictura poiesis" (wie die Malerei, so die Dichtung; *Ars poetica*, 361–65) gebracht. Es ist aber auch medienhistorisch bedingt, dass Sprache durch Laut und Schrift andere Formen der Aufzeichnung und Archivierung erfahren hat als Bilder und selbst im Digitalen noch durch Dateiformate und Ausgabekanäle von diesen unterschieden wird.

Bilder und Symbole

Bilder gestatten außerdem nach allgemeiner Ansicht eine symbolische Darstellung von Objekten auch ohne verbale Übersetzung, erschließen sich dem Verstand oder der Emotion durch die hohe Geschwindigkeit der optischen Aufnahme und entwickeln eine spezifische Dynamik des Erkennens und Erinnerns, sagen mehr als tausend Worte. Visuelle Informationen können eine Schnelligkeit und Informationsdichte beanspruchen, die durch die Rede (als zeitliche Abfolge von Wörtern und als Übersetzung in abstrakte Inhalte) nicht eingeholt wird, wie schon der britische Naturforscher Robert Hooke (1635–1703) in seinem berühmten illustrierten Traktat *Micrographia* von 1694 wissen lässt:

> „Denn auch wenn eine Beschreibung in Worten uns eine unvollkommene Vorstellung und Idee des Dinges geben kann, das so beschrieben wurde, so kann uns doch keine Beschreibung durch Wörter eine so vollständige Darstellung der wahren Form dieses Dinges geben, wie es eine Skizze oder Abzeichnung desselben auf Papier vermag." (Hooke 1726, S. 293)

Diese Leistung ist nicht so sehr bild-, als vielmehr seh-spezifisch, weil sie für jede Form des Sehens zutrifft. Sie trifft gleichwohl in besonderem Maße auf Bilder zu, wenn mit diesen eine bestimmte Form in einer vergleichbar visuellen Weise wieder- und weitergegeben werden kann.

Sofern Bilder und Texte von vornherein als geschlossene grafische Einheiten voneinander unterschieden werden (etwa als Zeitungsfoto mit Zusatztext oder als Comic mit Sprechblasen), lässt sich aber über deren Wechselbeziehung, über Sinnveränderungen und Sinnzuwächse sprechen. Auf Urkunden, an Denkmälern, in illustrierten Zeitschriften weisen Beschriftungen in höchst fantasiereicher Form den Weg durch das visuelle Etwas, und seit dem Altertum erscheinen Bilder – als Vasenschmuck, als Buchmalerei, auf Medaillen, im Flugblatt – in Verbindung mit Schriftzeichen aller Art (→ KAPITEL 7.3, 8, 10.2). Die schriftlichen Begleiter dienen der Erläuterung, verändern die Gesamtwirkung eines Motivs, überliefern Bildbedeutungen. Im Archiv werden Abbildungen vornehmlich über Suchbegriffe wiedergefunden. Wenn der heute meistzitierte Theoretiker der technischen Reproduktion, Walter Benjamin, im Jahre 1931 feststellte, die Welt der Gegenwart sei eine „beschriftete", weil zu jedem Pressefoto schon eine Bezeichnung hinzukomme, die ihm einen Sinn gebe (vgl. Benjamin 1931, S. 372), so ist dies insofern treffend, als die Beschriftung durch das Illustrationswesen von einer Gewohnheit zu einer Notwendigkeit geworden ist.

Beschriftungen

Übergänge

Die Berührungen von Schrift und Bild sind zahlreich und fließend, und es bestehen Beziehungen zwischen visuellen und verbalen Phänomenen auf allen Ebenen. Text und Schrift haben eine Reihe von bildlichen Eigenschaften, Sprache insgesamt hat eine komplexe Audiovisualität (vgl. Wenzel 1995). Die fließenden Grenzen von Bild und Nicht-Bild sind nicht nur in Typografie und Kalligrafie, im Comic oder in der Werbung überdeutlich, sondern auch an den Übergängen von Bild, Schrift und Zahl. Ziffern, Formeln, Tabellen oder Aufzeichnungen wie EEG und EKG sind Mittel der grafischen Notation, die nur unzureichend mit dem Begriff „Schriftkultur" beschrieben wäre, weil es sich zugleich um eine Zahlen- und Zeichenkultur handelt, eine Kultur der symbolischen Repräsentation. Unter dem Begriff der „Schriftbildlichkeit" wird daher zum Ausdruck gebracht, dass Visualität neben mathematischer und begrifflicher Erkenntnis eine wesentliche Form der Erschließung ist (vgl. Krämer / Bredekamp 2003).

Bilder als „Sprache"

Hinzu kommt, dass Bilder aufgrund ihrer kommunikativen Eigenschaften auch selbst als „Sprache" angesehen werden (→ KAPITEL 3, 10) und wie der Text einen informativen oder argumentativen, memorativen oder emotionalen Charakter haben können, auch wenn sie andere Sinne ansprechen und anders aus- und eingeübt werden. Beide Ausdrucksformen dienen Funktionen, die nicht auf ein bestimmtes Trägermedium begrenzt sind, und beide können ihre Ausdrucksleistung erbringen oder diese verfehlen. Das meiste, was visuell erfassbar, aber nicht verbal ist, fällt unter eine sehr offene Kategorie von Bildlichkeit. Seit einiger Zeit wird daher anstelle von Bild und Text auch alternativ zwischen „Sichtbarkeit" und „Sagbarkeit" unterschieden (vgl. Voßkamp / Weingart 2005).

Ikonische Wende?

Die seit nunmehr über ein Jahrzehnt andauernde Diskussion um einen möglichen *pictorial turn* (vgl. Mitchell 1992a) oder *iconic turn* (vgl. Boehm 1994; Bredekamp 2005, S. 15f.) stellt dagegen die bestehende Auffassung vom Bild als naivem Ersatzmittel zu Schrift oder Zahl infrage. Sie beschreibt – mit unterschiedlichen Akzenten – einen grundsätzlichen Bruch mit der logozentrischen, d. h. sprachzentrierten Definition von Wissen, reagiert auf die angeblichen Anschauungskrisen in den Naturwissenschaften, aber auch auf Veränderungen in den Künsten und in der Technik. Ihr Ausgangspunkt ist nicht der quantitative Aufschwung visueller Medien im Allgemeinen, deren Dauerpräsenz in der Gesellschaft seit über einhundert Jahren unübersehbar wäre. Das Bild findet vielmehr als kulturelle Leistung seine

Ikonische Differenz

Bestimmung in einer „ikonischen Differenz" zur Umwelt (vgl. Boehm 1994, S. 11–38), indem es eine Unterscheidungs- oder Distanzie-

18

rungsleistung vom Betrachter einfordert und sich von einer bloßen „Sichtbarkeit" abhebt. Auf den Unterschied zu anderen sinnlichen Kanälen reagiert Bildlichkeit mit rhetorischen und ästhetischen Mitteln, die den visuellen Sinn kultivieren (vgl. Naumann / Pankow 2004; Mersmann / Schulz 2006 u. a.); sie ist dem Text weder nachrangig noch nachgeordnet. Der *turn* beschreibt ferner das Politikum, dass die heutige Wissensgesellschaft, die sich unverändert skeptisch gegenüber dem Bild als „Abbildung" äußert, ihr Wissen gerade durch Bilder erzeugt und kommuniziert. Dieser Emanzipation des Bildlichen als eigenständiger Denkform ist in den letzten Jahren eine Fülle von Studien und Sammelbänden gewidmet worden, die den wissensgenerierenden Eigenwert des Bildlichen betonen (vgl. z. B. Kemp 2003; Lefèvre 2003; Breidbach 2005) und die bereits als Vorergebnisse einer neuen Bildwissenschaft verstanden werden könnten.

Bildliches Wissen

1.4 Bildwissenschaft: Disziplinäre Grenzen

Bildmedien und -typen haben durch den Aufstieg neuer Massenmedien im 19. Jahrhundert und durch den Siegeszug elektronischer Medien reale Veränderungen der Nutzung und der Distinktion in Kunst, Werbung oder privatem Gebrauch erfahren und diversifizieren sich weiter. Auch scheint die Zahl der wissenschaftlichen Bereiche und Disziplinen, die mit Bildern operieren oder sich mit ihnen pflichtmäßig befassen, zu wachsen. Durch die Zunahme von Datenmengen und -speichern ergeben sich daraus grundsätzlich neue Fragen der Prozessierung, Strukturierung und Archivierung von Informationen (→ KAPITEL 3.4), so wie auch deren soziale Funktionen sich verändern.

Veränderungen der Nutzung

Auch aus der Dauerpräsenz des Visuellen hat sich die Notwendigkeit einer wissenschaftlich-theoretischen Aufarbeitung ergeben, die sich im deutschen Sprachraum seit Ende der 1990er-Jahre als Bildwissenschaft begrifflich formiert (vgl. Sachs-Hombach 2005). Gegenstand und Zielsetzung der Diskussion sind jedoch trotz namhafter Protagonisten und profilierter Beiträge weiterhin unbestimmt. Zum einen ist dies dadurch bedingt, dass eine Bildwissenschaft dem Titel nach sämtliche Bildformen in allen Bereichen einschließlich der damit berührten ästhetischen und technischen Aspekte einbeziehen müsste – ein Vorhaben, das praktisch wie theoretisch uneinlösbar wäre. Problematisch ist auch, dass die Grenzerweiterung selber die engere Bestimmung visueller Medien nicht mehr zulässt (vgl. Mitchell 2005a).

Gegenstand der Bildwissenschaft

Visuelle Kompetenz

Zum anderen lässt die Bezeichnung Bildwissenschaft offen, ob damit eher das wissenschaftliche oder praktische Studium von Bildern gemeint ist oder vielmehr die kompetente Nutzung von Bildmedien, wie sie auch unter den Begriffen *Visual Literacy* (vgl. Messaris 1994; Kress 2006; Elkins 2008) oder „Visuelle Kompetenz" (vgl. Doelker 2002; Huber u. a. 2002) diskutiert wird. Der Gegenstandsbereich schließt damit so weit auseinanderliegende Aspekte wie die Computervisualistik, Fragen der Kognitionsforschung oder der philosophischen Ästhetik oder die technischen Probleme der Bildbearbeitung oder -archivierung ein.

Historische Bildfragen

Die in der Debatte um Sinn und Zweck einer Bildwissenschaft zum Ausdruck kommende gesellschaftliche Aufmerksamkeit für Bilder und Bildtechnologien beruht wiederum auf älteren kulturhistorischen Voraussetzungen. Was heute als Bildwissenschaft eingeführt werden soll, ist genealogisch gesehen das Produkt einer jahrhundertealten Diskussion um die politische oder kulturelle „Macht" des Bildes in der Gesellschaft, die ihre Wurzeln im religiösen Bilderstreit, im künstlerischen Wettstreit der Gattungen oder im frühneuzeitlichen Aufstieg von Wissenschaften hat, die gezielt auf das Bild als Argument und Wissensquelle setzten, etwa bei Galileo Galilei (1564–1642) oder dem erwähnten Robert Hooke.

Eben diese Macht, die immer auch auf die spontane und affektive Wirkung anspielt, wirkt andererseits individuell verschieden und wandelt sich historisch. Die heutige Allgegenwart von Bildern sollte nicht vergessen lassen, dass auch die wissenschaftliche oder gewerbliche Nutzung von Bildern einerseits schon vor Jahrhunderten einsetzte und sich andererseits auch gegen Anfechtungen behaupten musste. Erst innerhalb bestimmter Kulturen der Visualität können auch bestimmte Bilder gegenüber anderen eine größere öffentliche Aufmerksamkeit auf sich ziehen oder als kompakte Vermittler komplexer Sachverhalte oder als diagnostisches Instrument akzeptiert und verstanden werden (vgl. Schuck-Wersig 1993). Hierdurch wie durch die Möglichkeiten technischer Reproduktion von Abbildungen wurde der Bildbegriff insgesamt allmählich erweitert und sein Einzugsgebiet auf alltägliche Bereiche ausgedehnt.

Gegen die Bilderfeindlichkeit

Eine dezidiert kunstwissenschaftliche Richtung arbeitet vor diesem Horizont ausdrücklich für eine historische Einbettung der bildwissenschaftlichen Fragestellungen. Sie argumentiert zugleich gegen einen unverändert deutlichen „Anikonismus" in den Wissenschaften, indem sie die konstruktive Seite des Bildlichen und sein konfliktreiches Vermögen zur Wissensstiftung herausstellt (vgl. Bredekamp 1997).

Sie insistiert darauf, dass aus der Geschichte der Bilder sowohl Erkenntnismomente wie Fragen folgen und dass sich Bildlichkeit mit Sprachlichkeit die Eigenschaft teilt, immer wieder neu gelernt werden zu müssen. Damit wird eine ältere Forderung nach Aufhebung normativer Ästhetiken und der Trennung von *high & low* (→ KAPITEL 3.3) ebenso weitergeführt wie die pädagogische Würdigung des Bildes für ein mehrkanaliges Lernen (vgl. Vester 1975). Eine anthropologische Ausrichtung der Bildwissenschaft sieht Bildlichkeit wieder als Grundkondition menschlicher Existenz (vgl. Belting 2006, S. 11).

Weitgehend unberücksichtigt in den bildwissenschaftlichen Debatten blieben bisher Überschuss- und Sättigungsphänomene, Medienfacettierungen und die unzähligen individuellen Nutzungen, die zur industriellen Versorgung der Konsumgesellschaft mit Bildern gehören, ebenso wie mikro- und makroökonomische oder juristische Fragen der gewerblichen Bestimmung von Bildlichkeit. Diese sind mit dem Schlagwort der Bilderflut vollkommen falsch beschrieben. Während in einem globalisierten Güter- und Informationsaustausch immer unübersichtlicher wird, welchen Stellenwert eine Darstellungsweise in einem einzelnen Sprach- oder Kulturraum hat, existieren unter Agenturen oder Verlagen grenzüberschreitende Bewirtschaftungsformen, die ihre eigene Definition des Bildlichen darin gefunden haben, dass sie für jede Bildverwendung einen Tarif ansetzen können (→ KAPITEL 8). Diese ökonomischen Bestimmungen sind eine direkte Weiterführung von Wertzuweisungen und gestalterischen Ansprüchen, die sich aus der Geschichte der Bildmedien selber ergeben haben. Was in die Kategorie „Bild" fällt, bestimmt sich auch aus erfolgreichen Technologien, aus der Begegnung von Produzenten, Märkten und Konsumenten, aus Rechtskonflikten. Eine zeitgemäße Bildwissenschaft muss derartige Aspekte grundlegend einbeziehen.

Bilder als Wirtschaftsgut

Je nach Berufsfeld kann das Bild in den Naturwissenschaften, in Werbung und in Politik, in der Kunst, in der Software-Entwicklung, im Webdesign oder in der Spiele-Industrie damit etwas höchst divergierendes beschreiben, und selbst innerhalb dieser Felder wandelt sich der Begriff im Laufe der Zeit. Durch das TV-Bild, das auf ein Massenpublikum trifft, die Handy-Fotografie, bei der Bilder mit Kommunikationsnetzwerken kurzgeschlossen werden, oder durch das Internet, das sich zu einem Verkaufsprospekt gewandelt hat, bilden sich unerwartete Stile, Märkte und Nischen aus, die wiederum ihre Spezialisten und Anwendungen haben. Durch deren Praxis wird auch das in Bildern transportierte Wissen erschlossen oder ihre Qualität definiert.

Professionelle Bestimmungen

Der Begriff „Bild" ist hilfreich, um Unterscheidungen zu treffen zwischen einer einzelnen Darstellung und allgemeinen Begriffen wie Sichtbarkeit, Visualität, Wahrnehmung. Unter Verweis auf ihn lassen sich so unterschiedliche Aspekte und Themen wie die Wirkung des Films, der Informationscharakter des Nachrichtenbildes, die etappenreiche Geschichte der Fotografie, die Nutzung von Handydisplays oder die Bedeutung von Papierformaten und Drucktechniken ansprechen und differenzieren. Die Verwendung des Wortes muss dann wiederum berücksichtigen, dass je nach Sachgebiet und Fragestellung ein unterschiedlicher Kanon von Objekten beschrieben wird, der zur Analyse eine entsprechende Detailkenntnis verlangt.

Die Vielfalt von bildlichen Kunst- und Kulturtechniken könnte im Gegenzug terminologisch durchaus genauer bestimmt werden, auch ohne dass sie auf einen einzigen Generalbegriff reduziert werden muss. Die Mittel zur sprachlichen Differenzierung stehen bereit. Eine Handzeichnung ist kein Scan, ein Nacktfoto nicht automatisch Pornografie, ein Monitor ist kein Fernseher. Einführungen in die Fotografiegeschichte haben schon vor einem Jahrhundert Dutzende von Bildtechniken und -funktionen unterschieden (wie das Luftbild, das Messbild, die Radiografie), um auf solche Eigenarten und die Konstruktivität technisch erzeugter Aufnahmen aufmerksam zu machen.

Die Vielzahl von methodischen und empirischen Aspekten, die mit solchen fachspezifischen Betrachtungen einhergehen, legen daher nahe, auch eine Bildwissenschaft als Vielfalt von Disziplinen zu begreifen, die je nach Sachgebiet ihre eigenen und operablen Bilddefinitionen einbringen und setzen müssen. Diese sind oftmals mehr konnotativ als streng unterscheidbar: So haben Visualisierung oder Sichtbarmachung im naturwissenschaftlichen Bereich eigene Bedeutungen (→ KAPITEL 14), Bilder gibt es in der Philosophie wie in der Physik, und sie sind auch innerhalb dieser Fächer wiederum vielgestaltig einsetzbar. Auf dem Weg zu einer gemeinsamen Bildwissenschaft sollte daher nicht vergessen werden, auf welcher breiten Basis die Forschung in Technik-, Wissenschafts- und Kulturgeschichte, Journalistik, Neurologie und anderen Fächern bereits vorangeschritten ist (vgl. die Übersicht bei Elkins 2007 sowie die folgenden Kapitel) und dass Bilder stets eine Fülle von Ausprägungen und Bedeutungsebenen besitzen. Daher sollten bildliche Fragen nicht mehr als nötig eingeengt werden auf theoretische Modelle. Sie sollten der Sensibilisierung für ein Problemfeld dienen, das sich stets durch eine Pluralität von Perspektiven ausgezeichnet hat.

Fragen und Anregungen

- Nennen Sie Gründe, warum Bilder zu allen Zeiten eine wichtige Rolle spielten, aber erst im späten 20. Jahrhundert der Bedarf nach einer Bildwissenschaft artikuliert wurde.

- Was könnte eine Bildwissenschaft von etablierten Studienfächern wie Visuelle Kommunikation oder von alternativen Konzepten wie Bildkompetenz unterscheiden?

- Auch Spiegelungen werden im Deutschen zumeist als „Spiegelbilder" bezeichnet, obwohl ihr Inhalt flüchtig ist und nur dem jeweiligen Gegenüber sichtbar. Woher rührt die Bezeichnung?

- Der deutsche Begriff „Bild" ist dem englischen „image" nicht gleichzusetzen. Warum?

- Kann eine wissenschaftliche Forschung mit qualitativen und quantitativen Methoden überhaupt noch die Einsatzgebiete und Medien überblicken, die heute mit visuellen Mitteln operieren? Wie könnte eine mögliche Eingrenzung der bildwissenschaftlichen Fragestellungen aussehen?

- Inwiefern könnten sich Bedeutung und Verwendung der Begriffe „Bild" und „Bildlichkeit" unterscheiden?

Lektüreempfehlungen

- **Gottfried Boehm (Hg.): Was ist ein Bild?,** München 1994, 4. Auflage 2006. *Das am meisten zitierte Werk der jüngeren Diskussionen zur Frage nach dem Eigenwert des Bildlichen, mit einer Reihe historischer Beiträge vornehmlich philosophischer Ausrichtung.*

- **James Elkins: The Domain of Images,** Ithaca/New York 1999. *Studie zum Wechselverhältnis von Kunst und Wissenschaft und zum Einzugsbereich einer bildwissenschaftlichen Forschung aus US-amerikanischer Sicht, mit Schwerpunkt auf technischen und diagrammatischen Darstellungen.*

- **Vilém Flusser: Ins Universum der technischen Bilder,** Göttingen 1995, 6. Auflage 2000. *Einflussreicher Essay zur Theorie und Geschichte der modernen Bildkultur am Übergang zu digitalen Medien.*

- **Ralf Konersmann (Hg.): Kritik des Sehens,** Leipzig 1999. *Grundlegende Zusammenstellung theoretischer Texte zur Emanzipation des Sehens, der Visuellen Kultur und ihrer Bildmedien.*

- **Klaus Sachs-Hombach: Bildwissenschaft,** Frankfurt a. M. 2005. *Einer von mehreren Sammelbänden des Autors zur interdisziplinären Diskussion um Aufgabenfelder der Bildwissenschaft, der zwar im Ergebnis offen bleibt, jedoch mit den fachspezifischen Diskursen bekannt macht.*

2 Träger und Rahmen: Geschichte des Bildes und seiner Medien

Abbildung 3: Konrad von Soest: *Wildunger Altar mit spätgotischem Altarkreuz* in der Stadtpfarrkirche Bad Wildungen (1403)

Abbildung 4: Kunstdrucke im Rahmen, Angebote aus einem Berliner Möbelhaus-Werbeprospekt (Ausschnitt) (2006)

Der Altar des Konrad von Soest für die evangelische Stadtkirche von Bad Wildungen, entstanden 1403, besteht aus einer Mitteltafel mit zwei quadratischen Seitenflügeln, die ursprünglich nur zu Sonn- und Feiertagen geöffnet wurden, um dann ein elegantes Gesamtbild aus bemalten Tafeln und Rahmen zu präsentieren. Der Altar ist in einem hellen Kirchenchor aufgestellt und dort in angenehmer Blickhöhe unterhalb eines spätgotischen Altarkreuzes zu bewundern. Im Unterschied zu zahlreichen Kirchenausstattungen, die im Laufe der Kunstgeschichte aus konfessionellen oder anderen Gründen zerstört oder bewegt worden sind, ist der Altar bis heute an seinem originalen Aufstellungsort erhalten.
Eine überaus mobile Variante gerahmter Kunst präsentiert dagegen der Verkaufsprospekt eines Möbelhauses, welcher Wandbilder als Kunstdrucke für die Wohnraum- und Bürodekoration anbietet. Die Drucke imitieren den Stil und die Themen schlichter Ölmalereien, welche in den 1950er-Jahren populär gewesen sind und sich vornehmlich der Kunst des 19. Jahrhunderts (impressionistische Landschaften, Idyllen) bedienten. Wegen der kleinformatigen Wiedergabe im Prospekt gibt ein begleitender Werbetext Auskunft, dass es sich hierbei um „hochwertige Reproduktionen" handle, die auf Keilrahmen aufgespannt seien.

Die Wahl von Motiven und die gestalterischen Besonderheiten einer Darstellung stehen immer in Relation zu äußeren Bedingungen, zum Beispiel den flächenmäßigen Begrenzungen und Rahmen eines Bildfeldes, der Beschaffenheit von Trägermedien oder dem Aufstellungsort einer Tafel. Daher hat auch der historische Wandel der Medien Auswirkungen auf die Gestaltung von Bildern gehabt. Aus der Mobilisierung und Vervielfältigung von Bildern, etwa durch leichte und flexible Träger wie Leinwand und Papier, ergaben sich neue Formate, wurden Themen und Darstellungsstile über weite Räume transportierbar und damit auch konventionalisiert. Dies führte insgesamt zu einer veränderten Auffassung des Bildes und seiner Möglichkeiten. Gerade weil sie mobil und reproduzierbar geworden sind, wurden Bilder immer mehr als immaterielle, beliebig verfügbare Gegebenheiten angesehen, auch wenn ihre Form und Entwicklung nachhaltig von Materialien, von Darstellungstechniken oder den Orten der Präsentation geprägt worden ist.

2.1 **Vom Bildträger zum Bildmedium**
2.2 **Formate und Standards**
2.3 **Neue Techniken, neue Fokussierungen**
2.4 **Entgrenzungen: Das Bild als Thema des Bildes**

2.1 Vom Bildträger zum Bildmedium

Nur einer Minderheit von historischen Bildwerken, die auf einem Untergrund aufgemalt, in Holz geschnitzt oder in Bronze gegossen worden sind, war es vergönnt, an ihrem Herstellungs- oder ersten Aufstellungsort zu bleiben. Der Begriff „Bildträger" nimmt dadurch **Bildträger** in gewisser Weise eine ambivalente Bedeutung an. Einerseits meint er den Untergrund einer Darstellung, andererseits deutet er an, dass das Bild als Objekt im Laufe der Geschichte der Bildkünste selber mobil geworden ist, indem es transportabel, reproduzierbar und zu einem Objekt des Handels, des Geschenk- und Sammelwesens oder auch des Diebstahls wurde. Vor diesem Hintergrund ist Konrad von Soests Altar für die Stadtkirche von Bad Wildungen von 1403 nicht nur wegen seiner historisch hochrangigen Malerei und der eleganten Dekoration bedeutsam, sondern auch aufgrund seiner unverändert erfahrbaren originalen Aufstellung.

Vielleicht ist es dem Umstand, dass diese originale Situation erhalten blieb, zu verdanken, dass bei fotografischen Aufnahmen des Altars auch der Rahmen und die architektonische Umgebung mit eingefangen **Rahmungen** werden – dies geschieht ansonsten in der Literatur (etwa in Reiseführern, Katalogen, Fachbüchern und anderen illustrierten Werken) nämlich immer seltener. Meist werden natürliche Lichtsituationen vor Ort ebenso ausgeblendet wie Sockelzonen, Einrahmungen und Verzierungen oder andere Teile, die als ein Beiwerk verstanden werden, welches den Blick auf das vermeintlich Wesentliche (etwa die figürliche Komposition der Bildtafeln) ablenkt und daher in gedruckten Wiedergaben abgeschnitten und durch den weißen Papierhintergrund ersetzt wird.

Dies ist umso nachteiliger, wenn der Rahmen in einem solchen erweiterten Sinne nicht nur ästhetischer Bestandteil des Werkes ist, sondern auch einiges über spätere Veränderungen und Rekontextualisierungen verraten könnte; seine Auslassung auf Abbildungen in Büchern und Zeitschriften unterschlägt dann, dass es sich um Reproduktionen von Objekten handelt, die eine spezifische und für die Argumentation wichtige Materialität, Dreidimensionalität oder Ortsbin- **Materialität** dung besitzen – gleichgültig ob als weltberühmtes Wandfresko oder als Dekoration einer Automotorhaube. Ob der Altar einer Kirche als Andachtsobjekt, Kunstwerk oder bemaltes Holz angesehen wird, hängt nicht nur vom Betrachter, sondern auch vom Ort und seiner Geschichte ab. Der Gegenstand selber kann außerdem, als Holzstück, als malerische Technik, als Heiligenlegende, ein Element ganz unterschiedlicher Chronologien sein.

Bildorte   Auch der Buchdeckel, das Zierbild auf dem Pokal oder die moderne Verpackungsillustration (→ KAPITEL 10.2) sind konkrete Bildorte, in denen grafische Motive als Zusätze eine enge Bindung mit Objekten eingehen und diesen vorgeblendet sind. Vergleichbar dem figurengeschmückten Tympanon (Bogenfeld) über einem Kirchenportal, das dem Besucher beim Eintreten ein Bildprogramm auf den Weg gibt, werden damit Imaginationsräume eröffnet, die dem Gegenstand, Raum oder Thema eine zusätzliche Bedeutung oder Ausstrahlung geben.

Wo die Ursprünge dessen, was heute Bild genannt wird, liegen (etwa in der vorzeitlichen Höhlenmalerei, im asiatischen Schattenspiel, im göttlich inspirierten Heiligenbild oder, biografisch gesehen, in der Kinderzeichnung), ist eine Frage der Definition. Jedoch kann bis in die ältesten Kulturen zurückverfolgt werden, wie figurale Schmuckformen, häusliche Wandbilder, astronomische Darstellungen oder politische Dekorationen unterschiedlichster Art durch ihren Bezug zur Umgebung, etwa durch die Ausfüllung bestimmter Flächen oder durch ihre räumlich-plastische Form, immer auch jenen Maßstab erhalten haben, der ihre Wahrnehmung beeinflusst hat und der für das später verbreitete, insbesondere europäische Verständnis von Bild, Bildlichkeit oder Bildhaftigkeit bestimmend wurde. Schon in der frühindustriellen Produktion von Bildern (im serienmäßig hergestellten Andachtsbild des Mittelalters, im Flugblatt, später in der Malerei oder Fotografie) war eines unstrittig: Bildlichkeit entsteht aus einer Struktur, die in Bezug auf einen äußeren Rahmen, einen Fokus oder einen erkennbaren Umriss zu denken ist, welcher die Innen- oder Außengrenzen einer Form festlegt:

> „Was uns als Bild begegnet, beruht auf einem einzigen Grundkontrast, dem zwischen einer überschaubaren Gesamtfläche und allem, was sie an Binnenereignissen einschließt. [...] Bilder – wie immer sie sich ausprägen mögen – sind keine Sammelplätze beliebiger Details, sondern Sinneinheiten." (Boehm 1994, S. 29f.)

Bildflächen,   Bildflächen können sich bieten auf überschreibbaren Wachstafeln,
Bildgrenzen   großflächigen Wandfeldern, auf zerbrechlichen Gebrauchsgütern – meist gibt es dabei einen Bereich, der für die bildliche Darstellung eigens reserviert ist und der eine fokussierende Rahmung hat, auf welchen hin bestimmte Figurationen, Kompositionen oder Perspektiven bezogen sind und sinnvoll werden. Die in einem eigens gestalteten Rahmen ausgedrückte Beziehung von Bild und Umgebung, Bild und Raum, von Bildebenen oder von nebeneinander liegenden Szenen durch formale Abgrenzung lebt bis heute, im Wechselrahmen wie im Comic, fort.

Zusätzliche Bedeutung für diese Verselbstständigung des erzählerischen Bildfeldes hatte auch das halbplastische Relief, das als Schmuck von Grabsteinen, Schreinen und Kanzeln, als Elfenbeinschnitzerei oder Bauschmuck seit der Antike geschätzt wurde.

Relief

Eine zentrale mediengeschichtliche Rolle spielte nicht nur in Europa die Entwicklung mobiler Bildträger. Bilder waren dadurch auch Formen des transkulturellen Austausches. Asiatische Dekorformen und Stile gelangten als Buchmalereien im frühen Mittelalter nach Westeuropa, ebenso osteuropäische Ikonen. Andere kostbare und mobile Bildträger, die mit steigendem Wohlstand seit dem Spätmittelalter aufkamen, waren bemalte Möbel (insbesondere Truhen), Geschirr (wie Teller und Krüge) und Teppiche, wobei besonders die letzteren die funktionale und politisch überaus nützliche Verbindung von transportablem Schmuck, repräsentativer Bildform und Raumgestaltung beibehielten.

Mobilisierung

Seit dem Altertum sind bereits Stoffe als Maluntergründe bekannt gewesen. Sie wurden auch im Mittelalter noch für Fahnen oder als Ersatz für Webteppiche verwendet. Als Untergrund für figurative Malereien dienten dagegen vor allem bespannte Holzplatten, die aber klima- und transportempfindlich waren. Um 1500 begann man daher Holz durch Leinwand zu ersetzen, vor allem in Italien (bei Andrea Mantegna oder Sandro Botticelli), doch hat sie sich erst nach über hundert Jahren auf dem Kontinent durchgesetzt. Sie war leichter, flexibler und in größerem Umfang herstellbar. Dabei diente das elastischere Öl – ebenfalls schon in der Antike bekannt – dazu, das bisher bevorzugte Eigelb als Bindemittel abzulösen.

Die flexible Leinwand

Bedingt durch den wachsenden Reichtum vor allem bürgerlich-kaufmännischer Schichten erhöhte sich der grenzüberschreitende Transport jener Kunstwerke quer durch Europa stetig. Wertvolle Altäre oder Teppiche gelangten nach 1400 von Brügge und Antwerpen bis nach Florenz, und zwar bereits auf Bestellung. Die Krynitz'sche Enzyklopädie von 1775 definiert Bilder daher unter anderem als „Dinge, die meistens in Klöstern von Nonnen gemacht werden, und womit vornehmlich die auf den Märkten in ganz Europa herumziehenden Augspurger, Nürnberger und Tyroler handeln." (Krynitz 1775, Bd. 5, S. 292) Die europäische Kultur der Spätrenaissance führte seit der Expansion Portugals und Spaniens einen Handel globalen Ausmaßes, der in seiner Ausdehnung die Grenzen der antiken Weltreiche übertraf und Kontinente und Kulturräume miteinander verband.

Europäischer Handel

Globaler Handel mit Bildern

Der Handel mit Kunstwerken berührte sich auch mit der Tradition des Geschenkaustauschs zwischen den Höfen (Diplomatie, Mitgift, Information, Bildnisverkehr) und der adligen und bürgerlichen Repräsentationskultur, welche ihre Aufwendungen für die Selbstdarstellung von der Frühen Neuzeit in Italien (vgl. Goldthwaite 1993) bis zur Revolutionszeit des 18. Jahrhunderts immer weiter steigerte, die einen Höhepunkt des ostentativen Konsums aus Geltungssucht darstellt (vgl. Veblen 1899).

Konsum

Mit der fortschreitenden Kunstproduktion, besonders in den Niederlanden des 17. Jahrhunderts, entstanden außerdem zunehmend Kunstwerke jenseits von kirchlichen und höfischen Aufträgen für einen anonymen Markt (vgl. Montias 2006 u. a.). Sie gehören zu Gegenständen der heutigen Kultursoziologie (→ KAPITEL 6.3), weil dabei Themen und Motive zirkulierten, für die ein allgemeiner Bedarf vorausgesetzt wurde und die entsprechende Repräsentationsbedürfnisse erkennen lassen. Es ist daher beispielsweise vermutet worden, dass die Steigerung des bürgerlichen Privatbesitzes eine bevorzugte Produktion von Bildnismalereien, von Scherenschnitten oder von Porträtfotografien zur Folge gehabt habe (vgl. Lippincott 1995; Freund 1968, S. 13) und mythologische oder historische Szenen der privaten Besitzfunktion des Tafelbildes immer weniger entsprochen hätten. Motivwelten wie Stillleben, Landschaften, die Tiermalerei u. a. erfuhren verstärkt Nachfrage. Bildkunst wurde Teil einer Geschichte der Kapitalmehrung und Kapitalvernichtung, und mit der Geschichte des Konsums veränderten sich auch die Berufsfelder ihrer Hersteller und Verkäufer (→ KAPITEL 6.3).

Einschneidend für die Entwicklung der europäischen Bildgeschichte waren neben dieser politischen und gesellschaftlichen Bedeutung vor allem die religiösen Ansprüche an das Bildwerk. Der Bilderdienst, aus der antiken Welt an spätere Gesellschaften übertragen und in den Religionen unterschiedlich stark reflektiert, wurde besonders im Christentum zu einem Gegenstand heftiger Auseinandersetzungen. Der christliche Bilderstreit hat seit den Anfängen der Heiligenverehrung in der Spätantike die Frage der Darstellung des Göttlichen und seiner grundsätzlichen Legitimität umkreist (vgl. Nordhofen 2001). Nach dem Bildverbot durch Leo III. von Byzanz 726 und dem Konzil von Hierera, die beide gegen den schwunghaften Handel mit Devotionalien gerichtet waren, erlaubte das zweite Konzil von Nizäa 787 wieder die Verehrung durch Bilder (Ikonodulie), im Unterschied zur simplen Anbetung von Bildern (Idolatrie). In der westlichen Kirche wurde unter dem Pontifikat Papst Hadrians I. († 795) die Bildverehrung wieder zugelassen.

Bildverehrung, Bildanbetung

Über das Mittelalter hinaus blieb die Bildfrage präsent in der theologischen Lehre (so bei Albertus Magnus, um 1200–80, und bei Thomas von Aquin, um 1225–74), welche mit Aristoteles (Aristoteles 1991, VII, S. 27f. [1032a]) und im Unterschied zu Platon im menschengemachten Bildwerk nicht mehr nur den schwachen Abglanz der göttlichen Idee gesehen hat. In der Zeit der Reformation, in der ersten Hälfte des 16. Jahrhunderts, konfessionalisierte sich dieser Streit um den Anbetungs- und Vertretungscharakter von Bildern (vgl. Bredekamp 1975; Warnke 2002). Die verehrungswürdigen „Idole" (von griechisch *eidolon*), die hierdurch und in den nachfolgenden Kriegen in erheblichem Umfang zerstört wurden, waren insbesondere Heiligenbildnisse. Andererseits sorgten die konfessionellen Streitigkeiten für einen Zuwachs an gedruckter Bildpropaganda in Form von Flugschriften, vergleichbar den späteren sozialen und politischen Revolutionen des 18. oder 19. Jahrhunderts. Seit der Zeit des Barock nahm der Aufwand für die Gestaltung öffentlicher Kirchenräume sowohl in plastischer wie malerischer Form wieder beständig zu (→ KAPITEL 6). Er wurde in der katholischen Kirche durch das Konzil von Trient reguliert, um die übermäßige und falsch verstandene Bildverehrung wieder einzudämmen. **Bildersturm**

Parallel zur Frage des Bildes im Kirchenraum etablierte sich im Italien des 15. Jahrhunderts eine neue, weltliche Kunsttheorie und Ästhetik, deren Bestimmung von Kunstgattungen (Malerei, Skulptur, Architektur) zugleich die allmähliche Transformation des Bildbegriffs vom plastischen Objekt zur Tafelmalerei beschreibt. Durch ihren ausdrücklichen Bezug zu den antiken Quellen wurden ältere Vorstellungen wie von Plinius d. Ä. in seiner *Naturalis historia* wiederbelebt (vgl. Plinius 1997, S. 15–19), wonach das Bild die Natur oder eine Person möglichst genau abzubilden habe oder dazu diene, berühmte Menschen unvergesslich werden zu lassen. **Wandlung des Bildbegriffs**

Nach Ansicht des Philosophen Georg Wilhelm Friedrich Hegel (1770–1831), dessen ästhetische Theorie in die Anfänge der heutigen Kunstgeschichtsschreibung fällt, ist die Entwicklung der europäischen Kunst maßgeblich geprägt worden durch den Wechsel des Bildverständnisses vom religiösen Andachtsbild hin zu einem Anschauungsobjekt, welches als künstlerisches und zugleich bewegliches Ding wahrgenommen wird: **Andacht vs. Kunst**

„[…] mögen wir die griechischen Götterbilder noch so vortrefflich finden, und Gott Vater, Christus, Maria noch so würdig und vollendet dargestellt sehen: es hilft nichts, unser Knie beugen wir doch nicht mehr." (Hegel 1832, Bd. 1, S. 151)

31

Da der moderne Mensch (d. h. in der Zeit um 1800) Religion mit Vernunft begreife, verstehe er auch religiöse Bilder nun als ästhetische Objekte, als Verwirklichung einer Geistesgeschichte, welche ihrer Schönheit oder Bedeutung wegen geschätzt werden. Bilder werden so zum Objekt einer Kunstgeschichte. Mit dem sogenannten Reichsdeputationshauptschluss von 1803 und dem Ende des Heiligen Römischen Reiches 1806 (gut 15 Jahre vor Hegels Vorlesungen zur Ästhetik) wurden zahllose Klöster und Kirchen aufgelöst, religiöse Werke entfernt und zerstört.

2.2 Formate und Standards

Eine Geschichte des Bildes im strengen Sinne ließe sich nicht darstellen, weil sie sich aus mehreren Strängen, aus der Entwicklung unterschiedlichster Gattungen zusammensetzt, die jeweils ihre eigene Chronologie haben.

Die illusionistischen Wanddekorationen der Antike, die noch heute *in situ* in den ruinierten Villen von Pompei und Herculaneum studiert werden können, sind beispielsweise im Mittelalter unbekannt gewesen. Ihre Verbindung von Wandflächen und gemalten Ansichten, die das Bild als eine bestimmte Fläche definierte, war aber gleichwohl präsent in den Bildprogrammen italienischer Kirchen, deren architektonisch gegliederte Teile (Wandfelder, Bögen, Portale, Gewölbe) zu Bildfeldern wurden, die mit Motiven ausgeschmückt und dann auch erzählerisch aneinandergereiht waren, sodass sich aus dem Gesamtraum größere Zusammenhänge (zum Beispiel Heiligenlegenden) ergaben.

Bauglieder als Bildfelder

Auch aus der Zusammenfügung von Andachtsbildern, etwa den gemalten frühchristlichen Ikonen oder geschnitzten Elfenbeintäfelchen, entwickelten sich mehrteilige Altäre mit einheitlich abgegrenzten Bildfeldern, Sockeln und Rahmenwerken, wie im Falle des Wildunger Altars. Vor allem in der sogenannten Predella, dem Untersatz von Altären, lässt sich beobachten, wie sich die Darstellung von Historien, d. h. von Begebenheiten oder Erzählungen, aus dem zumeist breiten, querliegenden Format der Tafel heraus entwickelte, darin vergleichbar den schon erwähnten Holztruhen, die ebenfalls mit Bilderzählungen dekoriert waren. In der nordeuropäischen Altarmalerei entwickelte sich die Bildtafel so zu einen eigenen Bildraum, der neben den verbreiteten Themen – den Viten von Heiligen oder Szenen des Jüngsten Gerichts – Platz für die Darstellung von Nebenereignissen oder auch für fantastische Landschaftswiedergaben bot.

Erzählerische Bildräume

Die Gliederung von Wandbildern und Polyptichen (d. h. mehrteiligen Bildtafeln) lässt einerseits erkennen, dass die Begrenzung der Bildfläche als gegeben vorausgesetzt wurde, andererseits wurde sie auch zum kompositorischen Problem. Oftmals stoßen sich die dargestellten Gestalten regelrecht ihre Köpfe an den Rahmen aus Holz, die der Bildtafel vorgeblendet sind. Wenn sich in der mittelalterlichen Malerei die Erzählungen auch über die Grenzen von Bildtafeln hinweg- und fortsetzen, so wurde aber dadurch bereits auch mit der Begrenzung des Bildfeldes und den beiden Realitätsebenen von Rahmen und Bildfeld gearbeitet (vgl. Kemp 1996; Puttfarken 2000). Die Bildtafel wurde zugleich immer mehr als festes Format anerkannt, das seine Mitte, seine Ränder und seine Grundstrukturen hat, welche mit perspektivischen und kompositorischen Mitteln zu füllen sind. Besonders der Italiener Giotto di Bondone (1266–1337) steht für diese Etablierung des Kompositionsgedankens, die Entwicklung des klar gegliederten, erzählerischen Gemäldes und eine ‚innere Monumentalität‘ der Figuren, bei der eine Gestalt aufgrund ihres Größenverhältnisses, ihrer klaren Konturen oder durch perspektivische Mittel als solitär erscheint und die später besonders von Raffael (1483–1520) perfektioniert wurde.

Vor allem das Tafelbild, das zunächst die Holztafel, dann jede Form der beweglichen Malerei beschrieb, entwickelte sich zu einer separaten Form, die ihre eigenen, dafür geeigneten Aufstellungsorte und Präsentationsweisen erhielt, etwa in der Bildergalerie. Die neuen Maltechniken und die Preise für die Herstellung erschlossen der Malkunst neue, auch anonyme Abnehmerschichten. Dies hatte wiederum erhebliche Auswirkungen auf die Verbreitung und Konventionalisierung von Kunststilen, Formaten und Motiven und auch auf die Vorstellung davon, was in europäischen Diskursen unter einem Bild (französisch *tableau*, englisch *image*), einem Kunstwerk oder unter dem „Schönen" verstanden wurde. Künstler entwickelten bestimmte inhaltliche oder formatmäßige Spezialisierungen, einige arbeiteten durchweg in einem festen Leinwandformat und wählten selber bereits die Rahmung aus.

Mit der Illustrierung von Pergamentrollen und -büchern wurde schon in der Spätantike eine eigene Bildform geschaffen, die zunächst in Miniaturen und Handzeichnungen, in der Neuzeit in gedruckten Illustrationen Verbreitung fand. Hier wie auch bei Flugblättern, Andachtsbildchen, Briefmarken, Spielkarten oder Geldscheinen wurde dabei der Papierschnitt maßgeblich, etwa die schrittweise Halbierung von rechtwinkligen Bögen in einheitlichem Format, wie sie bis heute

Verselbstständigung des Bildfeldes

Das Tafelbild als Standard

Papierformate

erhalten ist. Sie war vergleichbar den später aufkommenden festen Plattenformaten der Fotografie und anderen technisch bedingten Maßeinheiten.

Schon im 17. Jahrhundert war das Schema der wissenschaftlichen Illustration auf einer separaten Seite, mit durchnummerierten Detailansichten eines Objektes und umgeben von einem Rahmen, selbstverständlich. Aus der Verbindung von Trägerformaten und Illustrationen ergaben sich außerdem Darstellungsgenres wie die Tabelle (als mathematisiertes Informationstableau mit Titelangaben, Spalten und Zeilen).

Expansion des Illustrationswesens

Mit der Auflagensteigerung von Druckerzeugnissen, der Erweiterung der Druckverfahren (Lithografie, Stahlstich, Autotypie) und dem Aufschwung des Illustriertenwesens im 19. Jahrhundert, später mit der Einführung der fotografischen Reproduktion in Zeitschriften und Zeitungen, traten dann neuerliche Rezeptions- und Nutzungsveränderungen von Abbildungen ein; die Lehrbuchillustration und der Bildatlas, die Wand- und Schautafel wurden populär (→ KAPITEL 10.3), die neuen ‚Massenbilder‘ wurden zum alltäglichen Konsumprodukt (→ KAPITEL 3).

Popularisierung von Kunst

Auch die Reproduktion von berühmten Werken der bildenden Kunst wurde zum Ende des 19. Jahrhunderts durch das Verlagswesen und die Einrichtung reprografischer Spezialbetriebe erheblich verstärkt (vgl. Krause 2005), wodurch sich die Begriffe Kunst und Bild wieder sehr eng verbanden. In Kunstdrucken wurden Meisterwerke der großen Kunstgalerien für einen breiten Markt reproduzierbar und traten an die Stelle von Kupferstichen. Zur Jahrhundertwende wurde mit der Herstellung von Ölgemälden auf Leinwand im Umdruckverfahren experimentiert (vgl. Pietsch 2006).

Bildtechniken

Daneben hat sich die Wahrnehmung von Bildformen mit der Entwicklung der Technik kontinuierlich verändert: Der Bedeutungszuwachs der illustrierten Presse hat reprografische Fertigkeiten bis ins Äußerste gesteigert und auch bildjournalistische Innovationen vorangetrieben; die Massenreproduktion von Motiven und Produkten hat Sehkonventionen verändert und verfestigt. Transportable Medien wie die Postkarte verbreiteten sich um 1900 aufgrund ihrer Formateinheitlichkeit in rasanter Geschwindigkeit und in globalem Maßstab. Die Reproduktion von Bildern wurde auch im Privatleben eine Selbstverständlichkeit, die inzwischen multimedial erweitert ist (Handyfotografie, Fotoblogs). Mit dem Live-Fernsehen verband sich ein neues Versprechen auf Öffentlichkeit und Partizipation.

In einem solchen globalen Markt macht sich andererseits auch deutlich bemerkbar, dass Medien als Produkte bestimmter Gesellschaften konditioniert sind:

„Man kann sich vorstellen, daß, wären Fernsehen und Video nicht in den westlichen Industrienationen, sondern beispielsweise ausgehend vom asiatischen Kulturraum als neuer Standard eingeführt worden, die Ausrichtung des Zeilenaufbaus anders verlaufen wäre." (Spielmann 2005, S. 80)

Der Großteil dieser Bildformen hat die enge Kopplung von Formaten und Motiven (in der deutlichen Mehrzahl rechtwinklig, Porträts hochformatig, Historien- und Legendendarstellungen querformatig) weiter verstärkt. Dies macht sich bemerkbar in Formaten wie der „Bilddatei" oder der rechtwinkligen Gestalt von PC-Monitor oder Kinoleinwand, für die nicht unbedingt konstruktive Gründe verantwortlich sind (vgl. Manovitch 2001, S. 95ff.). Die rechtwinklige Form von Bildern hat beispielsweise gegenüber dem Rundbild, welches als Münze oder Tondo stets parallel dazu existent war, ein deutliches Übergewicht erhalten. Das Radar (als Vorläufer des PC-Displays) oder das Oszilloskop sind als Anzeigegeräte ursprünglich Rundformen, und auch beim Einsatz von Linsen in der *camera obscura,* in Mikroskopie oder Teleskopie entstehen auf optischem Wege eigentlich runde Ansichten, die nur durch den Zuschnitt von Bildträgern (Metall, Glas, Papier) oder in der Reproduktion kantig wiedergegeben werden.

Erfolg des rechten Winkels

2.3 Neue Techniken, neue Fokussierungen

Normierungen und Standardisierungen wirken als intellektuelle Formatierung, als ‚Schere im Kopf', noch in jedem Foto-Schnappschuss nach, und zwar immer dann, wenn intuitiv der vermeintlich richtige oder entscheidende Ausschnitt gesucht wird, der einer bestimmten Schönheitsnorm entspricht oder einen bestimmten Bildwitz erzeugt, der einer eingeübten Bildtradition genügt oder diese sichtbar durchbricht.

Der Begriff „Rahmung" kann daher auch in einem viel allgemeineren Sinne die Formen und Mechanismen der Aufmerksamkeitssteuerung beschreiben, die durch derartige Fokussierungen erfolgt. Was als Rahmen, Format oder Träger bezeichnet wird, bezeichnet nicht allein eine dingliche und sichtbare Einfassung oder Begrenzung, etwa ein hölzernes Rahmenwerk oder die Hoch- oder Querformate einer Wandmalerei. Der Rahmen kann ebenso in einem inhaltlichen Bezug des Objektes zur Sammlung, in einem größeren „Kontext" bestehen und durch Beschilderungen, Lichtführungen und anderes un-

Ort und Kontexte als Rahmen

terstützt werden. Ein Sockel, eine Nische, eine Ädikula (lateinisch für: Häuschen) oder eine ganze Platz- oder Gartenanlage kann den Rahmen einer Skulptur abgeben.

Größe und Distanz

Die Einrahmung kann räumliche Perspektiven, Ausschnitte oder Betrachterdistanzen einschließen, etwa als *close up* in der Fotografie, als Totale oder Halbtotale im Film: Hier wird z. B. auch durch den Zoom entschieden, ob eine Darstellung ein Porträt, eine Menge von Gesichtern oder eine Straßenszene einfängt und wiedergibt. Ein Passepartout, das eine Fotografie umgibt, wird ebenso Teil der Wahrnehmung wie die Wahl des Zeitpunkts und Ausschnitts. Auch Größenverhältnisse machen sich bemerkbar. Eine Skulptur oder Plastik, auf doppelte Größe skaliert, würde selbst bei unveränderter Kontur eine veränderte Perspektive, Beleuchtung und Oberflächenbeschaffenheit zur Folge haben (→ KAPITEL 4.1). Eine kleine Gemäldetafel, als Plakat abgedruckt, lässt detaillierte Pinselführungen in überdimensionaler Form erkennen, und viele *special effects* im Kino verraten sich dadurch, dass kleine Modelle (etwa ein Schiffchen, das bei Nahaufnahme von zentimeterhohen Wellen bewegt wird) als solche erkannt werden, weil Zeit-, Maß- oder Lichtverhältnisse nicht passen.

Auch zusätzliche Informationen sind „rahmende" Bestimmungen, die dem Angezeigten zugleich ihre spezifische Bedeutung geben und die Aufmerksamkeit steuern sollen, wo die Darstellung alleine eine andere Wahrnehmung erführe. Dies gilt für den Maßstab, der die absolute Größe eines Gegenstandes im Bild angeben soll. Er kann in

Maßstäbe, Skalen und Legenden

Form von Vergleichsfiguren im Bild (Personen vor einem Gebäude), von Skalen am Bildrand oder durch Legenden angegeben werden, welche erklären sollen, dass ein scheinbarer Steinklumpen in Wirklichkeit aus Sandkörnern besteht oder dass ein im Satellitenfoto zu sehendes Hausdach, in einer Größe von 10 cm in einer Zeitschrift abgedruckt, in Wirklichkeit eine Länge von 10 m besitzt und aus einer Höhe von 250 km aufgenommen wurde.

In traditionellen Begrifflichkeiten des Bildes nur schwer zu fassen ist ein im computergestützten Grafikdesign üblich gewordenes Mittel, das ursprünglich der fotografischen Mehrfachbelichtung entstammt

Überblendungen

und bei dem Elemente in halbtransparenten Ebenen übereinander geblendet werden, z. B. indem Schrift über blasse Hintergrundabbildungen gelegt wird. Auch hier bleiben komplexe Einheiten erhalten, die als Schrift und Bild nach wie vor unterschieden und verstanden werden, sofern sich bestimmte Konfigurationen zu Sinneinheiten zusammenfügen: Gleichwohl ist ihre Definition nicht über einen Rahmen im herkömmlich materiellen Sinne möglich, darin vergleichbar den *frames* pro Sekunde, die im Filmbild die Verbindung von Kamera und Filmverlauf beschreiben (→ KAPITEL 13).

Schließlich wird überall da, wo Sucher, Linsen, Zirkel oder Schablonen zum Einsatz kommen, das Bild schon in einem konkreten, formatierten Sinne mitgedacht: Jegliches Instrument, das ein visuelles Ereignis fokussiert oder projiziert, ist seinerseits als Rahmen zu begreifen und Teil des Bildes, das es miterzeugt. Ein Beispiel dafür ist die medizinische Diagnostik, bei der in der Sonografie hochfrequente Schallwellen ausgesendet werden, deren radiale Ausstrahlung in der Anzeige erkennbar bleibt. Ein komplexeres Beispiel sind die logistischen Anforderungen des globalen Bildmarkts als eines ‚Apparats‘, der nach bestimmten Bildstilen oder und Dateiformaten verlangt und damit im weitesten Sinne als Rahmenbedingung in das Erscheinungsbild der Werbung oder der Massenmedien allenthalben eingeschrieben ist (→ KAPITEL 8).

Apparate als Teil des Bildes

2.4 Entgrenzungen: Das Bild als Thema des Bildes

Doch auch Entgrenzungen und Sprengungen bisheriger Rahmen sind in der darstellenden und bildenden Kunst ein wiederkehrendes Motiv. Mit der zunehmenden Emanzipation des Tafelbildes seit dem Spätmittelalter wurde dessen „ästhetische Grenze“ (so der Titel der Studie von Ernst Michalski 1932; vgl. Krüger 2001) nun selber zum Thema der Malerei. Diese behauptet den Realitätsstatus des Dargestellten und stellt ihn mit malerischen Mitteln zugleich infrage.

Die ästhetische Grenze

So zeigt schon das Altarbild des Konrad von Soest auf seiner voll entfalteten, sogenannten Sonntagsseite, wie sie in → ABBILDUNG 3 zu sehen ist, Szenen aus der Leidensgeschichte Jesu, geschildert in kraftvollen Farben, während auf den Außenseiten der Flügel, die werktags den Altar verschließen, Darstellungen von Heiligen in *Grisaille*-Malerei (d. h. graue, wie aus Stein wirkende Figuren) aufgetragen sind. Mit dieser Trennung von Werktags- und Sonntagsseite werden zwei unterschiedliche Realitätsebenen eingeführt, wie es für viele Altäre der Zeit üblich war. Durch den malerischen Kunstgriff einer scheinbaren Steinfigur wird die Grenze zwischen Bildraum und Körper, zwischen Bild und Umwelt überblendet und zugleich aufrechterhalten: Das Bild thematisiert sich selber, es wird „selbstbewusst“ (vgl. Stoichita 1998, S. 22, 109) und erweist gerade in dieser Künstlichkeit seine Kunst. Die Wahrnehmung von Malerei als Malerei, von Bildern als Bildern, und auch die daraus folgenden zerstörerischen, doktrinä-

Spiel mit Realitätsebenen

ren und paradoxalen Momente sind der Geschichte der Kunst einge-schrieben (vgl. Marin 1977).

Ephemere Kunstformen

Dieses Spiel von Grenze und Entgrenzung wurde durch die Be-liebtheit anderer, vor allem ephemerer (flüchtiger) Kunstformen des Festwesens, z. B. Umzüge, Feuerwerke, Wasserspiele, Gartenarrange-ments und *tableaux vivants* (französisch für: lebende Bilder, d. h. von Personen nachgestellte Motive der Kunstgeschichte) stets begleitet, auch durch die Weiterentwicklung der darstellenden Künste in Thea-ter, Tanz und Oper. In der malerischen Ausgestaltung von Architek-tur durch umfassende Freskenprogramme des 17. Jahrhunderts erga-ben sich aus der Verbindung von Architektur und Malerei, in Verbindung mit illusionistischen und szenografischen Effekten, neue

Neudefinition des Betrachters

Übergänge zwischen Bild und Raum, die den Standpunkt und die Rolle des Betrachters neu definierten und die nach 1800 im Panora-ma (Rundbild) und Diorama (dem durchleuchteten Wandbild) wie-der aufgegriffen wurden (→ KAPITEL 12, 13). Sie machten auch deutlich, wie sehr die Wahrnehmung von Bildern an Atmosphären, Synästhe-sien, an musikalische Begleitung und anderes geknüpft ist.

In der Kunst der Moderne wurde das selbstreflexive Spiel mit der Ein- und Ausgrenzung des Betrachters zu einem Hauptmoment der Darstellung selbst. Aus der intellektuellen Beschäftigung mit der Be-grenzung des Sichtbaren, dem bedeutungsvollen Einsatz von Blick-punkten und Perspektiven, hat das Bild neue Möglichkeiten und Ef-fekte der Komposition, Fokussierung und Aufmerksamkeitssteuerung gewonnen, die am Ende zur Dekomposition führen konnten, aber auch zu einer modifizierten Wiederkehr traditioneller Motive und Formate. Der insbesondere für die Moderne vielfach diagnostizierte

Ausstieg aus dem Bild?

„Ausstieg aus dem Bild" (Ausst.-Kat. Karlsruhe 2005) meint daher vor allem den Ausstieg aus dem einheitlichen und verbindlichen Ta-felbild des Museumsraums des 19. Jahrhunderts und ist zugleich Ausgangspunkt für eine Suche nach neuen Feldern und Räumen von Sichtbarkeit (→ KAPITEL 6).

Das Bild als Bild

Insbesondere wurde damit die Frage, was das Bild überhaupt zum Bild macht, wieder aufgeworfen. Bilder eröffnen Räume und Hand-lungsfelder, indem sie Leitmotive und Ikonen liefern, Reize und Auf-merksamkeiten erzeugen, Erinnerungen speichern. Daher ist der „Ausstieg aus dem Bild" schon rein technisch gesehen kein völliger Abschied von Bildlichkeit und Sichtbarkeit. Selbst ein Kinofilm mit *split screen* hat eine Leinwand oder einen größeren Screen zur Vo-raussetzung, und auch eine abstrakte Malerei ist als Malerei zu er-kennen, die einen materiellen wie intellektuellen Rahmen hat.

Die Zahl solcher bildbestimmender Parameter, die besser als Formatierung oder Rahmenbedingung bezeichnet werden, scheint angesichts der möglichen vielfältigen Überwindungen und Konterkarierungen endlos (vgl. Goodman 1984, S. 131), zumal wenn sämtliche performativen und akustischen Aspekte, Architekturen und Perspektiven hinzugezählt werden. Andererseits erfährt das Bild noch keine völlige Entgrenzung, wenn sich eingeübte Grenzen lediglich verschieben oder ausdehnen. Im Bereich des Bildlichen kollidieren immer wieder künstlerische Strategien der Rahmensprengung, technische und soziale Rahmenbedingungen, industrielle Formatierungen und Zwänge oder mediale Umbrüche. Der Rahmen markiert dabei die sich stetig verschiebende Differenz von grenzloser Sichtbarkeit und bewusst eingeschlossener Bildlichkeit.

Entgrenzung des Bildes?

Fragen und Anregungen

- Überlegen Sie, ob sich Bilder im Laufe der Geschichte vollständig von ihren materiellen Trägern abgelöst haben.

- In welcher Weise könnten sich im Bilderstreit der Reformation religiöse und ökonomische Ansprüche an das Bild begegnet sein?

- Nennen Sie Beispiele für nichtmaterielle Begrenzungen von Bildern.

- Wieso könnte die Mobilisierung von Bildformen auf deren Inhalte und Themen Einfluss gehabt haben?

- Was meint der Begriff „Historienmalerei"?

Lektüreempfehlungen

- **Hans Belting / Christiane Kruse: Die Erfindung des Gemäldes. Das erste Jahrhundert der niederländischen Malerei,** München 1994. *Kunsthistorische Darstellung, die einführend auch die Genese der Bildform behandelt und weiterführende Literatur zusammenträgt.*

- **Manfred Faßler (Hg.): Geschichte der Medien,** München 1998. *Eine kompakte Zusammenstellung von Aufsätzen zu grundlegenden Medienformen.*

- Michel Frizot: Neue Geschichte der Fotografie, Köln 1998. *Voluminöses und reich bebildertes Einführungswerk in deutscher Übersetzung, das zentrale Aspekte der Fotogeschichte, deren Entstehung, Verwendung und soziale Gebrauchsweisen, in Einzelbeiträgen untersucht.*

- Thomas W. Gaehtgens / Uwe Fleckner (Hg.): Historienmalerei, Darmstadt 2003. *Die Historienmalerei galt über Jahrhunderte als vornehmste Gattung. Der Band versammelt antike und frühmoderne Texte zum Thema und bietet so zugleich einen Kanon kunst- und bildtheoretischer Autoren.*

- Walter Koschatzky: Die Kunst der Graphik. Technik, Geschichte, Meisterwerke. 14. Auflage, München 2003. *1974 erstmals erschienene, als Taschenbuch mehrfach aufgelegte Einführung in Kunstform und Technik der Druckgrafik, mit Beispielen sowie technischen und begrifflichen Erläuterungen zum Nachschlagen.*

- Kaspar Maase: Grenzenloses Vergnügen. Der Aufstieg der Massenkultur 1850–1970, Frankfurt a. M. 1997. *Die Industrialisierung hat eigene Bildformate wie Postkarte, Stereoskop und Kino hervorgebracht (vgl. Buddemeier 1970) – Maase geht diesem Prozess im Bereich der Bildmedien und in den Gattungen nach.*

3 Geschichte, Sprache, Diskurs

Abbildung 5: Beschädigte Filmkopie mit überlagerten Szenenbildern aus Asta Nielsens Spielfilm *Afgrunden* (1910)

Abbildung 6: Beispielseite aus der Online-Datenbank einer Bildagentur zum Suchwort „Krieg": Atombombentest Nevada 1957 (7.9.2007)

Das Filmstill stammt aus dem dänischen Spielfilm „Afgrunden" (Der Abgrund) von 1910, in der Regie von Urban Gad mit Asta Nielsen in der Hauptrolle. Die Schauspielerin tanzt in der hier reproduzierten, seinerzeit sehr gewagten Szene mit einem heimlichen Verehrer. Die im Dänischen Filminstitut erhaltene Kopie des Films wurde im Laufe der Jahrhunderte durch chemische Prozesse stark beschädigt. Am unteren rechten Rand zeigt sie in einigen Sequenzen schattenhafte Figuren, die das Ergebnis von Nitratzerfall im Trägermaterial sind, welcher dazu geführt hat, dass die Filmspule verklebt ist und Bilder aus unterschiedlichen Abschnitten ineinander kopiert wurden.
Darunter findet sich das Ergebnis einer Recherche im Webarchiv einer internationalen Fotoagentur, die große Bestände von Standardmotiven (zu Themen wie Reise, Beruf, Alltag, Natur u. a.) anbietet. Zum Schlagwort „Krieg" wird dem Kunden im sogenannten Kreativ-Bereich der Datenbank eine Auswahl von Illustrationen zum Lizenzerwerb angeboten, die nach Ansicht der Agentur für symbolische Aussagen geeignet sind. Motive der jüngeren Geschichte wie die berühmte Aufnahme der Testzündung einer Atombombe in der Wüste Nevadas 1957 können ebenfalls darunter fallen.

Eine historische Betrachtung von Bildern ist nicht allein auf das Problem beschränkt, ältere Bildmotive und die in ihnen dargestellten Ereignisse oder Personen zu dechiffrieren, um sie den schriftlichen Quellen an die Seite zu stellen. Sie führt auch zu der grundsätzlicheren Frage, welche Bildformen und -gattungen überhaupt in den Untersuchungsbereich historischer Forschung fallen. Fächer wie die Geschichtswissenschaft oder Ethnologie, sofern sie an der gesamten Vielfalt von Bildquellen interessiert sind, müssen zum Beispiel auf die kontinuierliche Vermehrung von gedruckten, filmischen und elektronischen Erzeugnissen mit quantitativen und qualitativen Methoden reagieren. Darin nähern sie sich wiederum Ansätzen der Archäologie und Kunstgeschichte an, die auch jenseits von Quellen- und Zuschreibungsfragen die eigenständige Entwicklung von Traditionen, Darstellungsstilen und Bildfunktionen beschreiben.

3.1 **Bildkunde: Sichtbarkeit der Geschichte**
3.2 **Sichtung, Überlieferung, Bewahrung**
3.3 **Massenbild, Populärkultur, Kunst**
3.4 **Lesbare Strukturen: Codes, Archive, Diskurse**

3.1 Bildkunde: Sichtbarkeit der Geschichte

Geschichte erscheint im Bild, Bilder sind Medien der Geschichtsschreibung. Als Vergangenheitsvision, Epochenbild, nostalgisches Erinnerungsstück und Souvenir sind sie zugleich ein Motor für die individuelle historiografische Beschäftigung. Die Geschichte der Bilder ist dabei nicht auf einzelne Dokumente der Ereignisgeschichte oder auf prominente Personen der Zeitgeschichte beschränkt, sondern deckt den gesamten Bereich visueller Produktion ab. Wer Interesse an historischen Bilddarstellungen hat, muss sich methodisch auf eine große Vielfalt an Medien, an Motivwelten und gesellschaftlichen Bereichen einstellen.

Geschichte im Bild

Der historische Blick auf Bilder verbindet zudem Fächer, die grundsätzlich verschiedene Erkenntnisinteressen verfolgen können. So sollte eine Fotografiegeschichte nicht nur die Evolution einer Technik untersuchen, sondern auch die von ihr beförderten Bildsprachen oder ihr zugrundeliegenden sozioökonomischen Bedingungen; Archäologie oder Kunstgeschichte sind nicht auf eine Auswahl von Werken der Hochkunst beschränkt, sondern analysieren und vergleichen visuelle Leistungen unterschiedlichster Art und Qualität unter dem Gesichtspunkt ihrer ästhetischen oder konstruktiven Eigenart, um bildgeschichtliche Zusammenhänge oder vergessene Bildfunktionen zu rekonstruieren. In der Ethnologie oder Soziologie kommen historische Ansätze zum Tragen, sobald das Bild als Zugang zu früheren Gesellschaftsformen gewählt wird.

Interdisziplinärer Charakter

Gemeinsam ist all diesen Blickrichtungen, dass sie durch die bildhistorische Perspektive ihren eigenen Ausgangspunkt jeweils um Fragen erweitern, die sich mit benachbarten Disziplinen berühren. Denn jedes Bildmotiv ist in Traditionen verankert, hat spezifische Hersteller und Medien und ist als Auswahl aus größeren Erinnerungskatalogen visueller Kommunikation zu begreifen. Entsprechend komplex und interdisziplinär gestaltet sich seine historische Rekonstruktion und Auslegung. Dies wird besonders deutlich, wenn ein Motiv aus größerem zeitlichen Abstand oder ohne Vorkenntnisse nicht mehr verständlich ist. Die Bedeutung von Heiligenattributen, von Wappen und Abzeichen, von historischen Gerätschaften gerät in Vergessenheit, Gesten wirken unverständlich, Karikaturen überzeichnen Themen und Gestalten, die auf temporäre Situationen und Anschauungen Bezug genommen haben.

Dechiffrierungsprobleme

Dies führte zur Entwicklung von wissenschaftlichen Methoden, um Bilder auf Grundlage formaler und stilistischer Analysen mit ge-

Wissenschaftliche Bildgeschichte

sicherten Zuschreibungen zu vergleichen, um hierdurch (wie etwa in der Archäologie) die Zuordnung von Bildnisbüsten zu bekannten Persönlichkeiten oder die nähere Bestimmung von uneindeutigen Bildthemen zu gestatten. Die Institutionalisierung der Kunstgeschichte als akademischem Studienfach in Deutschland zu Beginn des 19. Jahrhunderts verdankt sich der Zielsetzung, durch das Studium von Schriftquellen wie durch Sammlung und Vergleich die Orts- oder Zeitspezifik von Kunstwerken zu fassen. Damit konnte einerseits das Programm der philosophischen Ästhetik konkretisiert werden (vgl. Dilly 1979; Beyrodt 1991), wonach sich in den Künsten umfassendere historische Prozesse abzeichnen; andererseits wurde so im Besonderen und Individuellen nach Merkmalen und Bildleistungen gesucht, welche diese Prozesse (und ihre Betrachung) motivieren. Hierdurch wurden zugleich neue Gattungen und Themen erschlossen, z. B. die Elfenbeinschnitzereien des Mittelalters oder Flugblätter der Renaissance, die zuvor als außerkünstlerisch, als nicht bewahrenswert oder als Gebiet einer speziellen Mittelalter-Archäologie gegolten hatten.

Diese Ansätze antworteten auch auf den Ausbau von Museen und die erleichterte Reproduzierbarkeit von Kunstwerken. Schon die archäologische Stilkunde, ein Produkt vor allem des 18. Jahrhunderts (Comte de Caylus, Johann Joachim Winckelmann, Christian Gottlob Heyne), ist in der Rückschau als Reaktion auf die systematische Reproduktion von Bildwerken und die darin zum Ausdruck kommende Industrialisierung des historischen Blicks anzusehen. Durch bildliche Wiedergabe in Form von Zeichnungen, Abgüssen, Kupferstichen (und später auch Fotografien) konnten Apparate und Klassifikationssysteme erstellt werden, mit denen sich Bilder nicht nur nach Datum oder Urheber, sondern ebenso nach inhaltlichen Stichworten deuten und ordnen ließen.

Auch von Seiten der Geschichtswissenschaft wurde nach einer solchen ikonografischen (motivkundlichen) Lehre verlangt, um unabhängig von Wappen, Siegeln oder Namenszusätzen bildliche Themen und Zeichen verbindlich beschreiben zu können. 1928 wurde dazu auf dem Internationalen Historikertag in Oslo die Internationale Ikonographische Kommission ins Leben gerufen, die sich zum Ziel setzte, alle erdenklichen und verfügbaren Bilder entsprechend aufzunehmen und zu klassifizieren (vgl. Kämpfer 1997, S. 8).

Die Identifikation von Abbildungen unterschiedlicher Provenienz ist seither einer der Schwerpunkte der historischen Bildkunde geblieben, welche als Zweigdisziplin der Geschichtswissenschaft untersucht, wie eine Darstellung (ob in Form eines gezeichneten Porträts

Kunst als Geschichte *(margin note)*

Reproduzierbarkeit als Voraussetzung *(margin note)*

Ikonografie *(margin note)*

Historische Bildkunde *(margin note)*

oder einer technisch erzeugten Dokumentarfotografie) als Quelle dienen kann (vgl. Tolkemitt/Wohlfeil 1991; Jaritz 1996; Burke 2003). Hierfür ist unter anderem zu klären, in welcher Form und Medialität ein Bild vorliegt und wie zuverlässig Beschriftungen, Herkunfts-, Inhalts- oder Datumsangaben sind. Darüber hinaus ist auch zu fragen, wo es entstanden, in welchem Zustand es überliefert ist und inwieweit es damit Vergangenes repräsentieren kann.

3.2 Sichtung, Überlieferung, Bewahrung

Um die konkrete Bedeutung dieser Frage vor Augen zu führen, ist hier eine filmische Szene wiedergegeben (→ ABBILDUNG 5), die wegen zeitbedingter Zerfallsprozesse nicht nur an Klarheit verloren hat, sondern in die sich aufgrund chemischer Vorgänge auch neue Bilder eingeschrieben haben. Die Verklebung von auf der Filmspule übereinanderliegender Bilder durch Nitratzerfall führt im Film *Afgrunden* zu Figuren, die im Verlauf des Films schemen- oder geisterhaft auftauchen und wieder verschwinden.

Zerfallsprozesse

Hieraus ergeben sich unmittelbare Überlieferungs- und Konservierungsprobleme. Obwohl der erfolgreiche dänische Film in zahllosen Kopien zirkuliert wurde, sind heute nur noch wenige beschädigte Reste verfügbar. Es ist also sogar möglich, dass massenhaft vorhandene Bildformen spurlos verschwinden können, gerade weil sie für endlos verfügbar und beliebig kopierbar gehalten werden. Auch die Fotografie, die als eine so naturgetreue wie leicht kopierbare Form der Darstellung begrüßt wurde (vgl. Kemp/Amelunxen 2006, S. 101), hat sich im Verlauf der Zeit als ein weiteres verfallsanfälliges Medium erwiesen (→ KAPITEL 9). Von ganzen Bildwelten der Vergangenheit sind aufgrund solcher Verfalls- oder Sedimentierungsprozesse, auch trotz und wegen möglicher Massenproduktion, nur noch Spuren erhalten. Langzeiterprobte und unkomplizierte Sicherungsmedien sind die Ausnahme und, wie die Mikroverfilmung, nur für einige grafische Formen einigermaßen hinreichend.

Unsichtbare Bildbestände

Demgegenüber gibt es keine wissenschaftlich einheitlichen Bewertungsregeln für die Erhaltung und Erhaltungsnotwendigkeit von Bildern, da nicht die gesamte Bildgeschichte in allen Medien konserviert werden kann und mit jedem neuen Medium (der elektronischen Videoaufzeichnung, der Computertechnik) wiederum neue Konservierungsfragen einhergehen. Zahllose Pressebilder oder Bilder der Tech-

Konservierungsprobleme

nik, über Jahrzehnte intellektuell wie konservatorisch vernachlässigt, sind z. B. in weiten Teilen nur aufgrund individueller Initiativen erhalten. Auch wenn Kunstwerke oder Kulturgüter, einer gesellschaftlichen Vereinbarung zufolge, um jeden Preis gepflegt werden sollen, so bleibt angesichts begrenzter finanzieller Möglichkeiten ihrer Aufbewahrung stets die erinnerungspolitische Frage bestehen, welche davon vorrangig und welche überhaupt dazu zählen. Kunsthistorische Kanonisierungen, Notfallpläne von Museumsdirektionen, Welterbe- oder Denkmalschutzlisten spiegeln einen temporären Konsens wider.

Wandel von Wertvorstellungen

Gleiches galt in früheren Jahrhunderten für das Sammlerwesen, ohne das die heutigen Museumsbestände aus Antike und Mittelalter nicht existieren würden, das aber oft im Windschatten politischer und kultureller Aufmerksamkeiten stand oder gerade aus der Achtlosigkeit gegenüber dem Altertum hervorging. Der Dichter Clemens von Brentano beklagt 1810 in einem Brief an den Maler Philipp Otto Runge:

> „Hier, wo zu gleicher Zeit ein tüchtiger und redlicher Philolog und Philosoph seinen Schülern und Freunden die Ästhetik und Kunstgeschichte und das Lob der alten Meister nach den neuesten Ansichten fortwährend vorträgt, kaufte ich am Tag nach meiner Ankunft einen ganzen alten Altar mit vielen sehr schönen Bildern um zwei Gulden, den die Bürger hinauswerfen ließen, um sich einen elenden architektonischen Altar, den sie aus einer zerstörten Abtei gekauft, hinsetzen zu lassen [...]." (Brief vom 21. Januar; Runge 1982, S. 316f.).

Antiquarische oder kunstinteressierte Sammlungsbemühungen und individuelle stilgeschichtliche Forschungen haben durch den detailgenauen Vergleich und die Unterscheidung von Objekten jenes nachhaltige gesellschaftliche Interesse am Bildlichen befördert, das heute als selbstverständlich gilt. Praktische Grenzen einer massenhaften Bildsammlung wurden aber schon im 19. Jahrhundert, etwa im Zuge von großen Editionsprojekten oder beim Aufbau von Kunstkabinetten, sichtbar, und auch der Neubewertung einzelner Medien (des Films, des Fernsehens) stehen immer wieder beträchtliche Probleme der Archivierung gegenüber. Das Bewusstsein für die Erhaltung von Bildern unterschiedlichster Zeichensprache, Machart und Technik setzt voraus, dass ihr kultureller und intellektueller Wert überhaupt ernstgenommen wird. Der historische Blick auf Bilder stellt den geschichtlichen Filterprozessen eine begründete Auswahl entgegen, die durch ihre selbstkritische Analyse ihrerseits darüber mitbestimmt, was als intellektuell tatsächlich verfügbar gelten kann.

Sammlungs- bemühungen: Bilder als Kulturerbe

3.3 Massenbild, Populärkultur, Kunst

Die Dauerhaftigkeit der Massenreproduktion von Bildern stellt inzwischen eine eigene Entwicklungsstufe und Qualität von Bildlichkeit dar. Eine Bedingung für die systematische Analyse visueller Konsumformen war daher auch die Aufhebung von Grenzen der Hoch- und Populärkultur im Verlauf des 19. Jahrhunderts. Dadurch wurden historische Bildmedien als Ausdrucksformen aufgewertet und die Einzugsgebiete der Wissenschaften wesentlich erweitert, die Sammlung und Auslegung von Bildern wurde ausgebaut zu einer an die Naturwissenschaften angelehnten Beschreibung von Stilentwicklungen: In der „Kunstwissenschaft" nach 1900 sollten evolutionäre Gesetze auch jenseits der Kunst und über räumliche Grenzen hinweg aufgespürt werden, die einer streng vergleichenden Prüfung standhalten könnten; die bildenden Künste wurden dahingehend erweitert verstanden, dass sich in ihnen das „Sehen" einer Gesellschaft abzeichne (→ KAPITEL 4.2).

Kunstwissenschaft

In der sozialpsychologischen Theorie der Zeit um 1900 wurde das „Kollektive" (Durkheim 1895) zu einem leitenden Begriff, um gesellschaftlich verankerte, auch verborgene Ideen und Prozesse zu beschreiben. Die Forschungen Aby M. Warburgs (1866–1929), aufbauend auf den religionsphilosophischen Lehren Hermann Useners, synthetisierten diesen Blickwinkel mit einer kunsthistorischen Auffassung von Formprinzipien und mit textbasierten Untersuchungen zu literarischen, politischen und wissenschaftsgeschichtlichen Motiven. Warburgs Methode unterschied sich ausdrücklich von der Kulturgeschichte des 19. Jahrhunderts, die das Bildliche zwar für die Historiografie nutzte, aber nicht thematisierte. Dafür war sie intensiv mit dem Phänomen der kollektiven Funktion und Weitergabe von Formen befasst, die eine Bildform historisch begründe und energetisch auflade, und zwar bis hinein in die eigene massenmediale Welt. Hieraus entwickelte Warburg ein wegweisendes Programm, das insbesondere dem „Nachleben der Antike" in der Kultur der Gegenwart gewidmet war und das im Vordergrund der von ihm aufgebauten Kulturwissenschaftlichen Bibliothek in Hamburg (auf die das Warburg Institute in London zurückgeht) stand.

Das Kollektive

Für seine Methode wird allgemein der Begriff der Ikonologie angewendet, auch wenn er nicht durch Warburg eindeutig definiert wurde. Das Wort ist der Epoche des Manierismus um 1600 entlehnt, die mit einer *iconologia* buchförmige Verzeichnisse von Bildthemen, gedacht zur geistreichen Ver- und Entschlüsselung von Kunstwerken,

Ikonologie

47

meinte. Der Aufstieg der Bildkunst jener Zeit zu einem europäischen Phänomen gestattete es, Kunst als internationale Bildersprache zu verstehen und zu transportieren (→ KAPITEL 2.1). Aus demselben Grunde wurde die ikonologische Lesart aber auch als wortfixiert kritisiert (vgl. Pächt 1977, S. 235–238), weil sie individuelle Ausdrucksweisen zugunsten von abstrakten Botschaften ignoriere. Erst Warburgs Nachfolger haben Ikonografie (als Bestimmung bildlicher Sachthemen) deutlicher von der Ikonologie (als Bestimmung bildlicher Gebrauchsformen) geschieden (vgl. Panofsky 1939, S. 42).

Die Frage, wie Bilder unabhängig von der Zuordnung zu Hochkunst oder Kitsch (als *high* und *low art*) in einem breiteren gesellschaftlichen Sinne zu fassen wären, bestimmt auch die Forschungsrichtung zur Visuellen Kommunikation. Heute in weiten Teilen mit der gestalterischen Ausbildung (Werbung, Grafik- und Produktdesign) befasst, entsprang sie einer Forderung der Kulturtheorie besonders der 1970er-Jahre nach einer Kritik der funktionalen und ökonomischen Gesetze und Strategien der Bildproduktion (→ KAPITEL 10.2). Ziel war die Analyse der populären Bildsprache, welche die gegenwärtige Gesellschaft in ihrer Breite bestimmt. Mit ihrer Fokussierung auf das kommerziell und politisch werbende Medium und in Reaktion auf Entwicklungen in Kunst und Technik wandte sie sich ebenfalls gegen einen engeren Bildbegriff und richtete sich auf das Visuelle, das jegliche sichtbare Erscheinung einschließt, welche zur Selbstrepräsentation einer Gesellschaft gehört. Wie schon die Kunst- und Kulturwissenschaft nach 1900, erschloss sie sich damit erneut die als *low culture* betrachteten Medien, etwa die Reklame, den Comic oder die als banal oder flüchtig angesehenen elektronischen Medien wie Fernsehen, Video und Computerspiel. Im Gegenzug war sie mit dem Problem der „Masse" beschäftigt, die in doppelter Hinsicht negativ konnotiert ist.

Einerseits speist sich aus dem Begriff der Masse die fortgesetzte Klage von der Bilderflut, der Banalisierung, Verkitschung und Abwertung, die mit der technischen Reproduktion von Bildern einhergeht. Die Kritik der Fotografie und des Illustriertenwesens der 1920er-Jahre erkannte in der Flüchtigkeit der Bilder ein Mittel gezielter ideologischer Blendwirkung. Wie Walter Benjamin in seinem Aufsatz *Das Kunstwerk im Zeitalter seiner technischen Reproduzierbarkeit* (1936) ausführt, verlagerte sich durch Illustrierte und Kino die Wirkung des Kunstwerkes von dessen Originalität hin zu dessen breitenwirksamer Leuchtkraft (→ KAPITEL 9.3). Diese Gedanken blieben in der Kritischen Theorie der Nachkriegszeit, gerichtet gegen die Ent-

Visuelle Kommunikation

Topos der Bilderflut

48

fremdung der Kulturindustrie, dauerhaft präsent. Aus den Möglich-keiten der Reproduktion folgt wiederum nur eine neue Sehsucht oder „Ikonomanie" (Bilderwut; Anders 1956, S. 56–59).

Mit dem Begriff der Bilderflut ist auch die Verführungskraft der Bilder als solcher angesprochen, die das Publikum angeblich nur als Schein, Schimäre, Schatten, Abglanz oder Ersatz erreichen und die einen voyeuristischen Blick bedienen, den sie zugleich unbefriedigt lassen. Deutlich beeinflusst sind hiervon auch Aussagen zur angeb-lichen Bilderfeindlichkeit einzelner Kulturen, Religionen oder Gesell-schaften (vgl. Freedberg 1989, S. 54). Dass sich der Topos der „Flut" nicht auf das (ebenfalls wiederholbare und in hoher Auflage ge-druckte) Wort, sondern bevorzugt und wiederkehrend auf Bilder richtet, hängt offenbar mit dem Platon'schen Urteil von der Zweit-rangigkeit des Bildes (→ KAPITEL 1.2) zusammen. Daran lässt sich auch die Langlebigkeit eines Topos ablesen, der schon vor mehr als ein-hundert Jahren verbreitet und reaktionär aufgeladen war (vgl. Maase 2001, S. 9ff.).

Eine empirisch ausgerichtete „Massenbilderforschung" (Brückner 1973; vgl. Bringéus 1982) versucht dagegen positiv und auf der Grundlage einer breiten Materialbasis, die Ordnungsstrukturen ge-sellschaftlich verankerter Bilder zu entschlüsseln. Denn Massenme-dien überhöhen durch Wiederholungseffekte, plakative Sprache, durch Druck- oder Sendeformate ihre jeweiligen Inhalte. Dies trägt dazu bei, dass die visuelle Kommunikation der heutigen Konsumwelt eine bestimmte Struktur erhält, und zwar im Hinblick auf den sicht-baren Bildermarkt wie im Hinblick auf dessen verborgene Regeln. Nur eine größere Masse von Bildern kann vorherrschende Codes oder Standards freilegen, die sich als erfolgreiche und vielgefragte Muster in ihr reproduzieren und die sich auch im Aufbau von Sammlungen und Archiven widerspiegeln. Aus der Häufung von Bildmotiven ergeben sich so für die Forschung auch Plausibilitäten und Vergleichsmöglichkeiten (→ KAPITEL 10.3); durch Untersuchung von Postkartensammlungen, Unterhaltungsfilmen, Zeitschriften oder Sammelbildern können ethnografische oder sozialhistorische Einbli-cke in die frühere Industriegesellschaft gewonnen werden (vgl. Mayor 1971; Pieske 1989). Wegen der breiten Ausgangsbasis werden in diesem Kontext auch von der Wissenschaft eher vernachlässigte Bereiche und Gattungen einbezogen.

Die quantitative Auswertung ist aber durch qualitative Bewertun-gen aufzufangen: Ein 30-prozentiger Anteil an Bäumen auf archivier-ten Landschaftsfotos besagt nicht, dass diese der Anlass für die Auf-

Massenbilder: Quan-titative Verfahren

Grenzen quantita-tiver Forschung

nahme waren. Auch das Bildnis eines Herrschers auf einem Siegel ist nicht nur Motiv oder Erkennungsmarke, sondern ein in sehr bestimmter Weise gestaltetes Bild von Herrschaft. Eine Fotografie von 1850 ist trotz gleicher Basistechnologie kein Foto von 1900. Dies führt zu einem Bedarf nach präziseren zeitlichen Unterscheidungen, wie sie die „historische Bildforschung" (Jäger 2000) einfordert, die ausdrücklich die Entwicklungen der Technik und der Gesellschaft mit dem Wandel von Bildordnungen verbinden will.

Problem
der Eingrenzung
Eine visuelle Analyse, die sich ihrem Gegenstand im Stile ethnologischer Feldforschung nähert, stößt daher immer wieder auch auf die (kunst)historiografische Frage, wie Bilder der Massenkultur qualitativ (z. B. in geschmacklicher, sozialer, kultureller Hinsicht) auszuwählen sind und welche Formen überhaupt begründeter Gegenstand einer quantitativen Zählung sein sollten. Die historische Bildbetrachtung weitet sich über diese Fragestellungen zu einer eigenen Form der Kulturgeschichte aus und trifft in methodischer Hinsicht wieder auf die Kunstgeschichte (vgl. Oexle 1997; Haskell 1995, S. 220ff.). Ebenso wie sich der geschichtliche Blick nicht mehr allein aus ereignisgeschichtlichen Schriftquellen speist, lässt er sich von den Ansätzen einer Kunst- oder Kulturgeschichte nicht mehr allein dadurch abgrenzen, dass er etwa einen anderen Gegenstandsbereich oder andere Motive als diese untersucht. Stattdessen wird der Übergang zu anderen bildhistorisch interessierten Fächern fließend, etwa zur Wissenschafts- und Technikgeschichte (vgl. Dommann 2004), zur Soziologie (vgl. Chaplin 1995; Volk 1996), Anthropologie (vgl. Edwards 1997) oder Erziehungswissenschaft (vgl. Ehrenspeck/Schäffer 2003). In der Folge wurde eine Erweiterung der historischen Bildkunde, in Anlehnung an die Erforschung visueller Kultur jenseits fester Bild- oder Kunstbegriffe, auch im Deutschen als *visual history* diskutiert (vgl. Mietzner 2005; Paul 2006).

3.4 Lesbare Strukturen: Codes, Archive, Diskurse

Besonders in den nachfrageorientierten Sortierungsregeln von gewerblich agierenden Fotoagenturen und ihren Bilddatenbanken treten begriffliche Ordnungen zutage, die die Ausfilterung von Bildformeln sowohl dokumentieren als auch ihrerseits weiter begünstigen. So wächst die Bedeutung des Schlagwortes, und zwar auch im elektronischen Raum, wenn die meisten Bestände oder Portale über Suchbegriffe abgefragt werden. Aus der Strukturierung und Gliederung

Archivstrukturen

von Themenbereichen wie Sport, Politik oder Beruf (Berufsleben → Berufstätige Frau → berufstätige Frau mit Kind), die sich seit gut 100 Jahren im Pressebereich etabliert haben, resultiert eine weitere Filterwirkung; ebenso aus der Unterstützung bestimmter Rechercheformen (etwa die Suche nach Ereignissen und Personen, nach Urhebern und Rechteinhabern usw.).

→ ABBILDUNG 6 zeigt das Ergebnis einer Recherche im Webarchiv einer Fotoagentur, die sich auf Standardmotive zu Themen wie Reise, Beruf, Alltag, Natur spezialisiert hat und zu Suchbegriffen sogenannte Symbolbilder anbietet, für deren Verwendung der Kunde eine Lizenz erwerben kann und die dem Bestreben nach piktogramm-artiger Einfachheit nahekommt, indem für gesuchte Motive wie „Bauernhaus", „Familie am Strand" oder „Snowboarding" eine Mischung aus vorhersehbaren Standards und ungewöhnlichen Aufnahmen bereitsteht. Zum Schlagwort „Krieg" wird eine Auswahl von Illustrationen angeboten, die sowohl Motive der Zeitgeschichte zeigen als auch abstrakte und stellvertretende Motive (Aufnahmen von Waffen, zerstörten Gebäuden). Die berühmt gewordenen Aufnahmen von Atombombentests in der US-amerikanischen Wüste sind inzwischen soweit historisch gefiltert, dass sie nach Ansicht der Agentur als zeitlose Illustrationen für das Thema Krieg oder für Moderne fungieren könnten und entsprechend verschlagwortet werden, mit dem Ergebnis, dass das Bild einer Explosion, die weder einen Krieg noch dessen Opfer zeigt, in Zeitungen zur Verbildlichung des Themas Krieg verwendet wird.

Symbolbilder

Eine marxistische Ideologiekritik sieht hierin gewerbliche Interessen und Technologien (als Produktionsverhältnisse und -apparate) in die Bildwelt eingeschrieben. Aus Sicht einer linguistisch geprägten Semiotik (Zeichentheorie), die seit den 1950er-Jahren auch auf die Bildformen der Massenmedien angewendet wurde, zeigt sich im Massenbild eine ‚Sprache', die ihre allgemein gültigen Zeichen findet (vgl. Eco 1987). Hierauf basiert z. T. auch die Analyse von Diskursen, welche aus der begrifflichen Gliederung von Archiven und Sammlungen die Denkweisen einer Gesellschaft herausliest (wie formuliert vom französischen Philosophen Michel Foucault; Foucault 1971). Auf Bilder angewandt, gehen mit den Ordnungsstrukturen auch konkrete Bildtechnologien und Motivpräferenzen einher (vgl. Tagg 1988; Sekula 1989). Dies wurde besonders in den kriminalistischen, anthropologischen und psychiatrischen Aufnahmekampagnen des 19. Jahrhunderts deutlich (→ KAPITEL 10.4).

Ideologie, Diskurs

Gerade die Fotografie ist, als technisch erzeugtes Bild, ein solcher komplexer Spiegel von Zuständen, Vorstellungen und sozialen Produktionsmöglichkeiten und daher bevorzugter Gegenstand einer kritisch-diskursiven Analyse geworden: Ihr Zeige- oder Verweischarakter (Indexikalität), ihr scheinbar unmittelbarer Dingbezug dient als Beweismittel und wird zugleich täglich durch Manipulation oder veränderte Beschriftungen infrage gestellt. Dies hat andererseits nie zu einer nachlassenden Produktion geführt, zumal weil Fotografie als populäres Medium andere gesellschaftliche Verwendungen und Kodierungen erfahren konnte als etwa die Gemäldekunst, wie Pierre Bourdieu in seinem vielzitierten Essay *Un art moyen* (Eine illegitime Kunst, 1985) festgehalten hat.

Durch die Analyse der Diskurse und Technologien, die über ein Bild und seine Auswahl bestimmen, erhielten in den 1980er-Jahren Begriffe wie Aufmerksamkeit, Wahrnehmung und Sichtbarkeit („visuality"; Foster 1988) eine neue Aktualität. Sichtbarkeit bezeichnet hier die Gesamtheit von Phänomenen, die sich aus den verborgenen Strukturen einer Gesellschaft ergeben. Im englischsprachigen Raum haben sich daraus die *Visual Culture Studies* als neues Lehr- und Studienfach entwickelt, welche das gesamte Feld sichtbarer Formen und Medien einbeziehen, um die Sehkultur einer Zeit und die in Sehweisen ausgedrückten Vorstellungen und Erwartungen zu beschreiben; das „Sehen" und der (klassen-, kultur-, oder geschlechterspezifische) „Blick" werden als Regime oder Ordnungen begriffen, welche die Welt und ihre historische Überlieferung nicht nur wiedergeben, sondern selbst dominieren. Der Begriff „Bild" erweist sich als Produkt einer normativen Ästhetik des 19. Jahrhunderts, die es demnach zu überwinden gelte.

Im Zuge dieser breitangelegten Kritik visueller Kultur im „Zeitalter der Physiologie" des 19. Jahrhunderts (Crary 1996, S. 16) wurde auch den Instrumenten und Apparaturen des Sehens und Aufzeichnens eine größere Aufmerksamkeit zuteil. So steht beispielsweise die *camera obscura* nach Ansicht des Kunsthistorikers Jonathan Crary für das starre Sehmodell einer Epoche, in welcher der Mensch sich als kameragleicher Betrachter einer unveränderlichen Natur wähnte. Infrage gestellt durch die Physiologie und Psychologie nach 1800, begann dieses Sehen an Selbstverständlichkeit zu verlieren. Die Erforschung des Augensinnes führte schließlich zu einer Idee von instrumenteller und instrumentalisierbarer Sichtbarkeit, in der die Kenntnis von Sehvorgängen zugleich Herrschaft über den Menschen bedeutet, der vom Seher zu einem Bediener seiner eigenen Beobachtungsapparate, ob als Auge oder als Fotokamera, wird.

Infolge allgemeiner logistischer Umbrüche (weltweite Vernetzung, Digitalisierung von Information, Aufstieg von Suchmaschinen u. a.)

verändern sich auch die kollektiven Bildspeicher und deren Probleme. Daher sind nicht nur die ökonomischen Bedingungen und Marktmechanismen zu studieren, sondern auch jene reproduktionsbedingten Effekte und Strukturen, die sich aus der fortlaufenden Technisierung des Bildes und der mit ihr verbundenen medienspezifischen Ausdifferenzierung von Sparten und Zielgruppen, Bildfunktionen und Sendeformaten ergeben (→ KAPITEL 8.2). Wie beim gedruckten Wort oder in der Musik, deren Aufzeichnung sich in vielfältiger Weise auf ihre Wahrnehmung oder Darbietung auswirkt, führt die bildliche Reproduktion zu quantitätsbedingten Phänomenen der Rückkopplung und Konvergenz (Bevorzugung einzelner Anbieter, Motive oder Stile), die durch die Struktur des Bildmarktes und seine Arbeitsabläufe befördert werden: Der erfolgreiche Verkauf eines Bildes führt z. B. dazu, dass nach diesem Bild häufiger gefragt wird, weil es für einen Verkaufserfolg steht oder dem Redakteur erinnerlich bleibt.

Konvergenz

Trotz großer Unterschiede in Semantik und Motivwelt ähnelt die moderne Bildversorgung darin der barocken Emblematik, dass zu jedem Thema ein konventionalisiertes Motiv aufgerufen werden kann: Milch steht für Gesundheit, ein lachendes Kind für Lebensfreude. Im anonymisierten, gewerblichen Illustrationswesen muss nach der Bildordnung ein alter Mensch erkennbar alt, ein Franzose erkennbar französisch sein. Durch die Dauerpräsenz von Bildern filtern sich wiederum auch beim Betrachter bestimmte Motivtypen oder Einzelmotive aus, die später als symbolische Bilder des Jahrhunderts, Bilder der Zeitgeschichte usw. verstanden werden. Wenn diese Betrachter nun ihrerseits Bildverwender sind, z. B. im Schulbuchverlag, werden wiederum einzelne Motive bevorzugt (vgl. Kaufmann 2000).

Emblematik

Eines dieser Motive, das in Schulbüchern Verwendung gefunden hat, ist der Eingangsturm des Konzentrationslagers Auschwitz (hier im Titel eines *Spiegel*-Covers (→ ABBILDUNG 7) mit dem gezeigten Motiv des Atompilzes drastisch überblendet). Als symbolische Darstellung wirft es die Frage nach der Möglichkeit der Abbildung des Holocaust auf. Schon die SS hat ihre Lager fotografiert, aus Furcht vor Vergeltung aber die Bilder zum Kriegsende vernichtet. Die alliierten Befreier haben 1945 die Spuren des Grauens in Aufnahmen von ausgemergelten Gefangenen und Leichenbergen festgehalten, in der Hoffnung auf eine schockierende Wirkung und um Beweise für die Massenvernichtung erbringen zu können. Der französische Regisseur Claude Lanzmann hat in seinem Filmwerk *Shoah* von 1985 dagegen die grundsätzliche Unmöglichkeit zum Ausdruck gebracht, das Grauen

Darstellung des Grauens

*Abbildung 7: Spiegel-*Titelblatt: Bilder des 20. Jahrhunderts (November 1998)

vor Augen zu führen. Seiner Auffassung steht inzwischen die Sorge gegenüber, dass die Totalität faschistischer Vernichtung durch Bilderlosigkeit aufgehoben oder gar bestritten werden könnte (vgl. Hoffmann 2004; Didi-Huberman 2007). Die Betrachtung von Schrecken und Leid sind zwar Motor ältester Kunstformen (und bergen zugleich das Potenzial eines besonderen Voyeurismus, vgl. Hartwig 1986; Sontag 2003), die historiografische Frage nach der Mitteilungskraft von Bildern oder nach der Möglichkeit und Angemessenheit bildlicher Darstellung berührt damit aber auch tiefste ethische Dimensionen, wenn Gewalt und Tod durch sie sowohl dokumentiert wie auch präsent gehalten werden.

Ikonen der Vernichtung Indem aus der symbolisch verwendeten Fotografie historische „Ikonen der Vernichtung" (so der Titel einer Studie von Brink 1998) werden, noch dazu in Konkurrenz zu weiteren Bildern des 20. Jahrhunderts, besteht dabei durchaus die Gefahr, dass sie als bloße Zeichen für Vergangenes weitgehend relativiert sind. Die historische Forschung muss diese spröden ökonomischen Transformationsvorgänge in Betracht ziehen, die über die Ikonisierung einzelner Bilder – auch diejenigen des Holocaust – mitentscheiden.

Fragen und Anregungen

• Was meint eine „qualitative Analyse" von Bildern?

• Überlegen Sie, ob eine historische Fotografie als Quelle beweiskräftiger oder zuverlässiger ist als eine gezeichnete oder gemalte Szene.

• Schon ein großer Teil frühneuzeitlicher Altertumskundler war zugleich Sammler von Münzen, Bildnissen und anderen bildhaften Objekten. Woher könnte Ihrer Ansicht nach der häufig geäußerte Eindruck rühren, dass Historiografie dennoch ein ‚bildfernes' Fach sei?

• Worin unterscheidet sich der Begriff der „Sichtbarkeit" gegenüber Bezeichnungen wie „Bild" oder „Bildlichkeit", und welche Vor- und Nachteile bringt er mit sich?

• Welches könnten die wichtigsten Folgen einer digitalen Ordnung von Archiven und Dokumentationen für die Arbeit der Geschichtswissenschaften sein?

Lektüreempfehlungen

• Frank Büttner / Andrea Gottdang: Einführung in die Ikonographie. Wege zur Deutung von Bildinhalten, München 2006. *Methodische Einführung, die zugleich mit den Motivwelten der kunst- und bildhistorischen Forschung vertraut macht.*

• Francis Haskell: Die Geschichte und ihre Bilder. Die Kunst und die Deutung der Vergangenheit, München 1995. *Die profunde und umfangreich illustrierte Studie des britischen Historikers verfolgt die Anfänge der historischen Bildkunde in der Zeit der ersten Bildreproduktion und das Entstehen der frühen Kunst- und Altertumsforschung.*

- Martina Heßler: **Bilder zwischen Kunst und Wissenschaft. Neue Herausforderung für die Forschung,** in: Geschichte und Gesellschaft 3, 2005, S. 266–292. *Resümierender Aufsatz zur Frage, welche Bilder als Quellen historischer Forschung infrage kommen, wenn diese als interdisziplinäres Problemfeld begriffen werden.*

- Gerhard Paul (Hg.): **Visual History. Ein Studienbuch,** Göttingen 2006. *In diesem geschichtswissenschaftlich orientierten Sammelband werden exemplarisch die aktuellen Ansätze zur historischen Analyse von Bildern verschiedener Epochen, Herkunft und Medialität behandelt.*

- Herta Wolf (Hg.): **Fotokritik am Ende des fotografischen Zeitalters. Bd. 1: Paradigma Fotografie,** Frankfurt a. M. 2002; **Bd. 2: Diskurse der Fotografie,** Frankfurt a. M. 2003. *Kompakte Zusammenstellung jüngerer fototheoretischer Schriften, darunter in diesem Zusammenhang z. B. die diskursanalytischen Beiträge von Sekula, Tagg und Phillips zur Bedeutung der industrialisierten Bildtechnik der Fotografie für die gesellschaftliche Wahrnehmung.*

4 Formen, Wahrnehmungen

Abbildung 8: Georg Kolbe: *Großer Wächter* (auch: *Junger Wächter*), Standbild für die Flak-Abteilung Lüdenscheid (1936/37)

Abbildung 9: Henry Moore: *Large Two Forms*, Bonn (Aufstellung 1979)

*Die Bronzeplastik „Großer Wächter" von Georg Kolbe zeigt eine
nackte, männliche Figur, die sich halb im Gehen auf das rechte Knie
stützt und mit ernster Miene in die Ferne blickt, dabei in der linken
Hand einen Bogen trägt und die rechte Hand zur Faust ballt. Kräftige
Muskeln zeichnen sich ab, die der Figur Anspannung verleihen. Durch
das Entstehungsdatum (Deutschland 1936) und den mythisierenden
Titel lässt sich die Figur aus heutiger Perspektive als Verklärung der
NS-Ideologie lesen. Diese Deutung wird befördert durch ein martia-
lisches Körperideal der Zeit, das auch hier verwirklicht scheint.*

*Vierzig Jahre später wurde auf einer Freifläche vor dem Bonner
Kanzleramt, das seinerzeit wie ein Hochsicherheitstrakt abgeschirmt
war, eine zweiteilige Bronzeplastik „Large Two Forms" des britischen
Bildhauers Henry Moore aufgestellt, die aus zwei etwa vier Meter hohen, unregelmäßigen und blasenartigen Bügeln besteht. Die ungegen-
ständliche, weiche und je nach Standpunkt anders wirkende Form
steht im Kontrast zu der kantigen Wirkung des Gebäudekomplexes
und zur Freifläche. In ihrer fließenden Erscheinung bleibt sie scheinbar unbestimmt.*

Die formale Analyse und Beschreibung von Artefakten ist der primäre und schwierigste Schritt der Annäherung an sichtbare Phänomene,
zumal wenn sie bereits als Träger von Bedeutungen und Anschauungen begriffen werden. Dies gilt für das Tafelbild wie für die dreidimensionale Gestalt. Am Problem der Formbestimmung und -be-
schreibung lässt sich zeigen, dass Bilder stets ihre zeitlichen und
räumlichen Beziehungen haben; veränderliche Aufstellungsorte und
Milieus, historische und biografische Zusammenhänge sind ebenso
ausschlaggebend für die Deutung des Gesehenen wie eingeübte Dar-
stellungs- und Zeichentraditionen. Welche konkrete oder abstrakte
Form jedoch welche politisch-gesellschaftlichen oder individuellen
Anschauungen transportiert, hängt zugleich von grundlegenderen Er-
kennungsvorgängen ab, welche die Wahrnehmung von Figurationen
und Oberflächen betreffen. Diese sind ebenso zu reflektieren und zu
üben wie die Verbalisierungsleistungen, welche die große Form wie das
kleine Detail in Worte fassen müssen.

4.1 Probleme der Formbestimmung
4.2 Erkennung und Erkenntnis
4.3 Gestalt, Gestaltung
4.4 Beschreibung

4.1 Probleme der Formbestimmung

Georg Kolbes *Großer Wächter* von 1936 ist Ausdruck und Sinnbild einer zeittypischen Vorstellung von Männlichkeit, Härte und Entschlossenheit. So zumindest lässt sich die bronzene Plastik deuten, die das Bild eines Mannes zeigt, dessen nackter und breitschultriger Körper, in Verbindung mit einer unruhigen und schimmernden Oberflächenbearbeitung, eine athletische Erscheinung bietet. Die Figur ist, halb im Gehen, auf das rechte Knie gesunken oder im Begriff, sich aufzurichten. Die rechte Hand ist zur Faust geballt, die linke umschließt einen Bogen. Der kantige Kopf sitzt auf einem kompakten Hals, tiefe Augenbrauen betonen die strenge Miene. Unter der Körperhülle zeichnen sich Muskeln ab, die der Figur zugleich Vitalität und Anspannung verleihen.

Durch Entstehungsdatum und Titel wird dem heutigen Betrachter leicht suggeriert, dass die Figur des deutschen Bildhauers die Verherrlichung einer politischen Ideologie, zumindest eines damit verbundenen Körperideals widerspiegelt. Damit wird der Hersteller eines Standbildes, das – gegenständlich gesehen – nichts weiter als einen nackten Mann zeigt, in eine Mitverantwortung genommen. Dies wird noch bestärkt durch den Umstand, dass Kolbe zusammen mit Arno Breker und anderen Künstlern, die dem Regime des Dritten Reiches in prominenter Stellung dienstbar waren, seine Arbeiten öffentlich ausgestellt und offizielle Aufträge angenommen hat. Ist demnach auch diese Figur Kolbes, eines der bekanntesten Bildhauer des frühen 20. Jahrhunderts, eine vom Faschismus kontaminierte Figur?

Einbindung in NS-Propaganda

In Bildern der Arbeit oder des Sports, wie sie seit der Jahrhundertwende Gewerkschaftsbauten, Badeanstalten oder öffentliche Plätze in ganz Europa zierten, finden sich durchweg vergleichbare, strenge und pathetische Formen, in denen körperliche Tüchtigkeit oder Reinlichkeit als Ideal der gesellschaftlichen Erneuerung propagiert worden sind. In Materialwahl und -behandlung ist Kolbe nicht grundverschieden von Käthe Kollwitz, Gustav Seitz oder Ernst Barlach. Die leicht überdehnten Proportionen und die unruhige Oberfläche seiner Figur ist außerdem deutlich weniger auf jene fotografisch massenreproduzierbare Klassik bedacht, wie sie die makellosen, kraftstrotzenden Figuren Brekers oder Josef Thoraks zelebrierten. So gesehen, scheint der erste Eindruck der Gestalt des *Großen Wächters* vor allem vom historischen Kontext bestimmt zu sein.

Die Figur im historischen Kontext

Während die meisten ‚Dekorateure‘ des Nationalsozialismus heute schon aus qualitativen Gründen keinen Eingang in den kunsthistori-

schen Kanon fänden, haben einzelne Vertreter, auch gegen den Vorwurf ihrer Verstrickung in das politische System, ihre Anhänger. Im
Falle umstrittener Künstler wie dem Hitler-Porträtisten Breker wird
der Disput dann zumeist in eine langfristige historische Perspektive
verschoben (gemäß der Idee, dass auch die großen Künstler der Renaissance Despoten zu Diensten gestanden hätten). Die formale oder
ästhetische Qualität des Kunstwerks wird so von politischen Motiva-
tionen abgekoppelt. Auf diese Position beruft sich auch, wer die Idee
einer gesellschaftlich sanktionierten „künstlerischen Freiheit" vertritt,
wonach die Kunst nur dann bildend oder kritisch in die Gesellschaft
zurückwirken könne, wenn sie ungegängelt bleibe – also auch im
Hinblick auf politisch unliebsame Äußerungen. Die zeitliche Verhaf-
tung hat gleichwohl dazu geführt, dass ein prominenter Künstler wie
Kolbe erst nach jahrzehntelanger Absenz wieder wissenschaftliche
Aufmerksamkeit erfahren hat (in Ursel Bergers erstem Verzeichnis
seiner Werke; vgl. Berger 1990).

Paradoxie der künstlerischen Freiheit

Ein Vergleichsbeispiel: Eine Büste Mussolinis von Hans Wimmer
(heute in der Städtischen Galerie München) zeigt den italienischen
Diktator mit in sich gekehrter Miene. Die Oberfläche wirkt geradezu
zärtlich gestaltet. Von ihr heißt es in einer späteren Ausstellung:
„man kann sich wundern, dass es 1942 möglich war, das Porträt auf
der Münchner Kunstausstellung öffentlich zu zeigen und dass es im
Palazzo Venezia in Rom aufgestellt wurde". Die Frage richtet sich
fast automatisch auf den Konflikt von Kunstfreiheit und Kunstauf-
trag und wendet diesen um: „Gerade weil es die Figur des Scheiterns
beschreibt, ist es ein großes Kunstwerk" (Ausst.-Kat. Berlin 1983,
S. 176). Möglich wäre es aber ebenso, von einer Strategie der Verhüllung zu sprechen, die den Diktator bewegte, sich als geistvollen Me-
lancholiker zu inszenieren. In jedem Falle handelt es sich hierbei um
Bewertungen, die von der bereits interpretierten Form auf deren
Konflikt mit der Umwelt schließen.

Strategien der Verhüllung

Unabhängig davon wird sich eine Interpretation von Formen stets
auf Zusatzinformationen stützen. Hierzu gehören biografische Daten,
welche die Einheit von Leben und Werk oder deren Brüche darstellen. Im Falle Kolbes wäre dies besonders der Umstand, dass der
Künstler den Hauptteil seines Werkes weit vor der Machtergreifung
der Nationalsozialisten erarbeitet und seinen Stil auch während ihrer
Herrschaft nicht grundlegend verändert hat. Kolbe hat modern ge-
wohnt, sich mit avantgardistischen Kunstformen identifiziert. Seine
Bildnisse zeigen Persönlichkeiten, die dem Naziregime kritisch gegenüber standen.

Biografie und Werk-zusammenhang

Mit dem Standbild Kolbes wird, trotz des pathetischen Titels, keine Partei beworben, Gewalt nicht verherrlicht. Hätte es einen anderen Namen, etwa aus der griechischen Mythologie (Orion, Aktäon), könnte der Bildsinn sofort eine neue Wendung nehmen. Die Figur erregt außerdem an ihrem Aufstellungsort in einem Park von Lüdenscheid heute kaum noch Aufsehen. Zeitbedingt war Kolbes Arbeit mit dem System dennoch kompatibel. Er erhielt den Auftrag aus dem Reichsluftfahrtministerium und konnte die Arbeit auch in der Frühjahrsausstellung der Preußischen Akademie der Künste, deren Mitglied er war, unter dem Titel *Junger Wächter* mit großem öffentlichen Erfolg präsentieren.

Wechselnde Titel

Der Männerkörper ist sichtbarer Ausdruck einer Vorstellung von viriler Kraft und gespannter Sprungbereitschaft. In seiner Nacktheit könnte er auf die Urform des antiken olympischen Wettkampfs anspielen. Der Geist der Nachkriegs- und Rezessionszeit der 1920er-Jahre, wonach das Leben ein elementarer Antagonismus aus Leben und Tod sei, den nur der Stärkere überlebe und der selbst die sportive Betätigung zur sublimierten Übung eines Daseinskampfes mache, muss gar nicht im Titel ausgesprochen oder durch den originalen Aufstellungsort – eine Militärkaserne – deutlich werden.

Bilder des Antagonismus

Vergleichbar der heutigen Mode- oder Nahrungsmittelwerbung und ihren Normen der Schlankheit und Jugend oder vergleichbar der Gewaltdarstellung im Film und Computerspiel könnte die Form einer solchen Plastik durchaus als beschönigend angesehen werden; je nach Blickwinkel kann die leicht verzerrte Kontur und Oberfläche aber ebenso eine Spannung von Kraft und Verletzlichkeit zum Ausdruck bringen, die den gigantischen Leibern anderer zeitgenössischer Künstler völlig abgeht. Es bleibt damit die Frage offen, ab welcher graduellen Veränderung aus dem Bild eines menschlichen Körpers die Verherrlichung einer Körper- oder gar Rassenideologie wird. Es wird außerdem deutlich, dass der Streit um die ethische Dimension der Form nicht auf den engen Begriff eines zweidimensionalen Bildes begrenzt bleibt.

Ethische Dimension von Form

Dies wird durch ein anderes Kontrastbeispiel deutlich. Im Jahre 1979 wurde vor dem Bonner Kanzleramt die Bronzeplastik des britischen Bildhauers Henry Moore (1898–1986) unter dem deskriptiven Titel *Large Two Forms* aufgestellt (→ ABBILDUNG 9). Aufgrund ihrer Lage im Garten des Kanzleramtes war sie zwar allein dem privilegierten Besucher unmittelbar zugänglich, wurde aber durch Fernsehübertragungen zu einem Wahrzeichen des Gebäudes, welches vom damaligen Bundeskanzler Helmut Schmidt als „Girozentrale" abgetan

Das Kunstwerk als Wahrzeichen

worden war und optisch aufgewertet werden sollte. Mit dem Umzug der Bundesregierung nach Berlin verblieb Moores Plastik am Ort und wurde der Stadt Bonn 1999 unter Bundeskanzler Gerhard Schröder als Geschenk überlassen.

Geliefert wurden zwei etwa vier Meter hohe, unregelmäßige Bügel, die eng zusammenstehen und ineinander zu fließen scheinen, sodass sich erst durch eine Umrundung die fließende zweiteilige Struktur erschließt. Eine schimmernde Oberfläche lässt die perfekte Materialbehandlung erkennen, die mit zunehmender Patinierung an Reiz gewinnt und gleichzeitig durch ihre wechselnden Ansichten absolute Größenverhältnisse aufhebt. Die Figur Moores ist also keineswegs formlos, auch wenn sie ungegenständlich ist – sie gestaltet Raum und Material.

Abstraktion: Ungegenständlichkeit als Form

Wegen ihrer abstrakten Gestalt könnte sie ohne Weiteres ein biochemisches Labor schmücken, wo sie als Symbol für Zellgebilde oder organische Prozesse verstanden würde. Diese alternative Deutung wäre insofern nicht abwegig, als die Figur auf einer früheren Miniaturfassung von 1966 beruht und schon 1967, weit vor der Aufstellung in Bonn, in Bronze gegossen wurde (vgl. Wenk 1997). Moores Arbeit zeigt zudem eine Stufe in der Stilentwicklung des Künstlers, und sie ist Teil einer internationalen Richtung der plastischen Kunst. Erst durch den bundesdeutschen Auftrag wurden die Bügel daher mit einer Bedeutung versehen, die es gestattete, sie als politisches Symbol (z. B. der deutschen Teilung) zu lesen. Die scheinbare Beliebigkeit war jedoch kein Mangel, sondern war selber Ausdruck einer pluralistischen Demokratie, deren Regierungen Kunst nicht mehr als Hoheitssymbol begreifen, sondern nach internationalen Marktregeln bestellen.

Bedeutungswandel

4.2 Erkennung und Erkenntnis

Aufwertung durch Formung

Kunst leistet die konkrete Aufwertung eines Materials durch Formung und geht damit perfektionierend über die bloße Nachahmung von Natur hinaus, wie es laut Leon Battista Alberti (1404–72) und seinem Buch *Über die Malerei* schon die Meister der Antike bewiesen haben:

"Elfenbein, Gemmen und all jene teuren Dinge werden noch kostbarer durch die Hand des Malers. Gold selbst, wenn es mit Malerei verziert ist, erhält einen noch höheren Wert als zuvor; sogar Blei, das geringste unter den Metallen, wenn es von einem Phidias oder Praxiteles in ein Bild gearbeitet wurde, kann wertvoller sein als rohes Silber." (Alberti 1435, Buch 2, 11r–v)

Hier äußert sich ein konkreter Formbegriff, der die Veränderung eines Stoffes von Menschenhand, ein sichtbares Artefakt beschreibt. Daneben steht Form aber auch für musikalische Verläufe, narrative Strukturen und ist nicht notwendig visuell. Sobald etwas als etwas wahrgenommen wird, geschieht dies innerhalb bestimmter Formen. Der Begriff ist in dieser erweiterten Lesart so offen wie die Fächer und Gebiete, in denen er Verwendung findet.

Geordnete Wahrnehmung

In einem absoluten Sinne gibt es Form ebenso wenig wie Formlosigkeit. Ob sie eine Einbildung des menschlichen Geistes oder Realität ist, muss jedoch immer wieder neu begriffen werden. Form war schon seit den Anfängen ein Kernproblem der Philosophie, etwa in der Erkenntnislehre oder in der ästhetischen Theorie; seit Aristoteles (384–322 v. Chr.) wurde die begriffliche Unterscheidung von *morphe* und *hyle* (griechisch für: Form und Stoff) vorgenommen, wie sie noch heute verbreitet ist. Demokrit (um 460–370 v. Chr.) vermutete, dass das Auge unsichtbare Abdrücke oder Ausstrahlungen aller Dinge aufnehme und zu Bildern zusammensetze. Hierdurch sollte erklärlich werden, wie eine Form im Auge sichtbar sein könne, ohne als Ding selber darin zu sein.

Entdeckung der Form

Im Mittelalter ist Form als Ausdruck der göttlichen Schöpfung und als *vis formativa* (lateinisch für: formgebende Kraft) in der Vielfalt der Tiere, Pflanzen und Mineralien verstanden worden. Die Materie ordnet sich nach Regeln, die in Fauna und Flora, in der Anordnung von Kristallen, im Blatt- oder Skelettbau erkannt und miteinander verglichen werden können. Abweichungen von der regelhaften Welt wurden daher später als Anzeichen der verborgenen Regeln gedeutet und gesammelt. Aus der Idee einer „Lesbarkeit der Welt" (vgl. Blumenberg 1999; → KAPITEL 10.4) ging im 16. Jahrhundert auch die Signaturenlehre der Mediziner Paracelsus (1493–1541) und Giambattista della Porta (1535–1615) hervor, denen zufolge die Natur himmlische Kräfte spiegele und z. B. Pflanzen diejenigen Organe oder Symptome heilen, denen sie durch ihr Äußeres gleichen.

Natürliche Eigenschaften

In der Naturphilosophie wurde schon im 17. Jahrhundert Kritik an der ‚Lektüre‘ einer Natur laut, welche nicht dechiffrierbar im Sinne schriftlicher Zeichen sei. Dies bedeutete nicht, dass das Studium der Natur mit der Frühen Neuzeit von einer naiv-visuellen Weltschau zu einer nichtvisuellen oder mathematisierenden Forschung übergegangen wäre. Die formkundliche Naturbetrachtung wurde gleichermaßen durch Studien von Künstlern wie durch das Sammlungswesen

Naturstudium der Frühen Neuzeit

weitergeführt, wie sie auch die Systematisierung von Muschelformen von Nehemiah Grew (→ ABBILDUNG 10) bezweckte. Als die technische und wissenschaftliche Entwicklung zu Grews Zeit erkennen ließ, dass zu den Formen des Makrokosmos noch andere, nämlich mikroskopische Gegenformen existieren, wurde der Formbegriff grundlegend herausgefordert; doch auch in der Welt des Mikroskops wurden Formen des Makrokosmos wiedererkannt.

Morphologische Tradition

Die visuelle Erfassung der Natur kulminierte im System des Schweden Carl von Linné (1707–78), dessen Beschreibung und Klassifikation der Tier- und Pflanzenarten noch heute gebräuchlich ist (→ KAPITEL 11.2). Die Auseinandersetzung mit den fließenden Übergängen innerhalb der Tier- und Pflanzenwelt und die Frage der zeitlichen Entwicklung von Körpern führte jedoch zur Frage, inwieweit sich hinter der laufenden Verwandlung des Einzelwesens oder der Wesen ineinander (von Petrus Camper 1778 als „Metamorphose" beschrieben) ein ideales Wesen verberge. Die Suche nach der inneren Verwandtschaft von Gestalten (etwa Tierarten) wurde zum Kernthema der Naturspekulation des 18. Jahrhunderts. Johann Wolfgang Goethe kritisierte die starre Bedeutung des Begriffs „Gestalt" (→ KAPITEL 9, 10.3). Eine neue Wissenschaft mit Namen „Morphologie" müsse helfen, den dynamischen Wandel der Form zu beschreiben (vgl. Goethe 1807, Bd. 6, S. 358). Die Frage, was Form sei, wenn diese sich organisch wandelt, wurde mit dem Fortgang der naturwissenschaftlichen Forschung immer wieder aktuell (vgl. Thompson 1917; Whyte 1951).

Bildhafte Natur

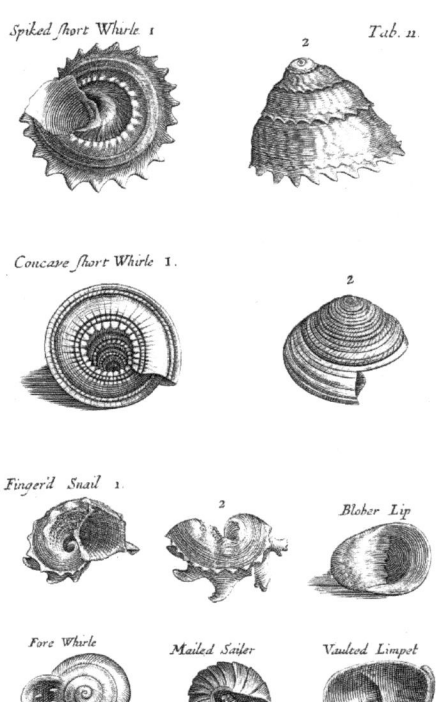

Natur erweist sich in dieser intellektuellen Geschichte, die besonders im deutschen Sprachraum zahlreiche Konjunkturen und Wandlungen erfuhr, als zutiefst bildhaft. Der visuelle Zugang zu ihrer Mannigfaltigkeit bleibt eine Voraussetzung für

Abbildung 10: Nehemiah Grew: *Musæum Regalis Societatis* (1686), Tafel 11: Muschelformen

den Vergleich z. B. von Gesteinsformationen oder von Krankheitssymptomen. Die wissenschaftliche Beschreibung von Form hat sich daher bis in das 20. Jahrhundert hinein, selbst wo sie aus physikalischen Kräften oder aus chemischen Prozessen heraus die Gestalt oder den Organisationsgrad von Lebewesen erklärt sollte, immer wieder auf das Beispiel Goethes und seiner Nachfolger bezogen.

4.3 Gestalt, Gestaltung

Nach Aristoteles (→ KAPITEL 2.1) kann Kunst eigenständige Bilder erzeugen, die den geschaffenen Formen der Natur (lateinisch *natura naturata*) nicht nachstehen; dies wurde in der Kunsttheorie besonders seit dem 16. Jahrhundert als Legitimation für die Schöpfungskraft der Kunst und den Status des Künstlers als ein ‚zweiter Gott' angesehen, der wie die tätige Natur (lateinisch *natura naturans*) Dinge hervorbringt, die aber von ihr verschieden sind (→ KAPITEL 5.2, → ASB KELLER). Die Schöpfung wird Ausdruck einer Gestaltung, deren veränderlicher oder gar spielerischer Charakter sich in der Kunst wiederholt. In deren Raum wird das Geheimnis der Natur, als Zufall oder Fantasie, nachahmbar. Leonardo da Vinci umschreibt dieses Vermögen in seinem Malerei-Traktat mit der Entstehung des Bildes aus einem Farbschwamm, der an einer Wand sinnvolle Muster erzeugt; damit war das Problem benannt, dass das Sehen von Mustern das Hineinsehen, die Ein-Bildung verlange. Zudem war darin bereits die enge Beziehung von Kunst und Psychologie ausgedrückt.

Kunst als zweite Natur

Augen- und nervenkundliche Forschungen, wie sie im Zeitalter des Barock von René Descartes (1596–1650) betrieben wurden, bestärkten die Vermutung, dass die Wahrnehmung von Mustern nicht nur unter passiver Beteiligung des Auges, sondern auch des Gehirns stattfindet. Dem vermittelnden Auge kommt nach Ansicht der Naturkundler eine eigene Intelligenz als Organ zu; es ist nicht auf die Funktion einer Kamera reduziert:

Kooperationen von Optik, Auge und Gehirn

> „Man kann sich gar nicht vorstellen, wie und auf was für eine Art die außer uns befindlichen Dinge gesehen werden, wenn man von dem Werkzeuge des Sehens, von dem Auge, keinen deutlichen Begriff hat." (van Musschenbroek 1741, S. 548)

Die Zeit nach 1800 führte zu einem rasanten Aufstieg dieser sehphysiologischen Forschung, einsetzend mit Johannes Müllers *Zur vergleichenden Physiologie des Gesichtssinns* (1826). Aus der Wirkung optischer Experimente (→ KAPITEL 13.3) wurde geschlossen, dass das

Physiologie

Auge konstitutiv für das Sehen von Form sei und als Sehsystem auf äußere Reize (auch Druck o. Ä.) stets mit visuellen Eindrücken reagiere. Müllers Forschungsfeld erlangte höchste gesellschaftliche Anerkennung durch den Physiker Hermann von Helmholtz (1821–94), der in Experimenten nachwies, dass Sehen aus einer Vielzahl diffuser Eindrücke und schneller Augenbewegungen stabile Sinneinheiten herstelle und diese unbewusst ergänze. Im Laufe des 19. Jahrhunderts haben neue Druckverfahren gezielt auf diese natürliche ‚Unschärfe‘ des Auges gesetzt, wie z. B. das Rasterprinzip der Autotypie um 1880, das den Abdruck von Fotografien erleichterte. Auch der Mehrfarbendruck, der Film oder die Bildschirmzeile beruhen auf Kooperationen von Lichtimpuls, Auge und Gehirn, die gemeinsam erzeugen, was als sichtbar bezeichnet wird.

Was durch einen Betrachter gesehen wird, ist außerdem durch die Kombination von intellektuellen Faktoren unterschiedlicher Komplexität (Gedächtnis, Wünsche, Traditionen) vorstrukturiert. Hierauf richtete sich besonders die Psychologie, die im späteren 19. Jahrhundert zu einer Basiswissenschaft aufstieg und dabei mit den Fragen der Ästhetik im Austausch stand, zumal sie die historische und zeitgenössische Kunstproduktion in neuer Weise zu deuten erlaubte.

Psychologie und Kunstwissenschaft

So wurde auf der einen Seite erörtert, inwieweit sich Formen auf den Betrachter im Sinne eines inneren Nachvollzugs unmittelbar auswirken (darin heutigen Studien zu Spiegelneuronen vergleichbar) und auf diese Weise die „Einfühlung" in das gegebene Bild ermöglichen (so beschrieben von Friedrich Theodor Vischer 1857). In eine andere Richtung zielte eine Kunsttheorie, die dem Konzept der abbildenden Nachahmung die „reine Sichtbarkeit" des Kunstwerkes (vgl. Fiedler 1887) gegenüberstellte. Im Kunstwerk entfalte sich eine Geistigkeit, indem das Bild zugleich Mittel der Reflexion über seine Gestaltung sei. Dies war ein Grundstein für die spätere gezielte Absage an nachahmende Funktionen von Bildern, bis hin zur völligen Abstraktion in den westlichen Avantgarden nach 1900, wie sie auch bei Moore wirksam ist.

Bilder des Geistes

Die philosophische Ästhetik des 18. Jahrhunderts markiert die Entdeckung der Zeitlichkeit der Form. Der „Stil", in der Altertumswissenschaft des 18. Jahrhunderts von Johann Joachim Winckelmann entwickelt (→ KAPITEL 3.1, → ASB D'APRILE/SIEBERS), erwies sich dabei als ein Kernbegriff für das Prinzip der sichtbaren Formung. Unter dem Einfluss der industriellen Produktion haben wichtige Strömungen des 19. Jahrhunderts damit die geschmacksbildende Bedeutung von Formgebung für den Alltag und die wirtschaftlich relevante Ge-

Kunst und Industrie

staltung von Produkten unterstrichen. Weltausstellungen und neue Sammlungstypen wie das Kunstgewerbemuseum (Victoria & Albert Museum, London 1852) sollten die Wertschätzung für Gebrauchsgüter steigern. Dies erhöhte auch das Bewusstsein für die Zeitabhängigkeit der Form, die Pluralität historischer Stile und die Bedeutung handwerklich-traditioneller Herstellungstechniken und Materialien, die im Zuge gesellschaftlicher Erneuerungsbewegungen (*arts and crafts*, Jugendstil) gefördert wurden.

Schon der Florentiner Künstler und Kunstbiograf Giorgio Vasari (1511–74) hatte auf den Unterschied von Handschrift (italienisch *maniera*), Epoche und Entstehungsort hingewiesen. Durch Theorien des Milieus (Kulturraum, Kunstlandschaft), der Zeit (Geist, Epoche, Mode) oder des Einflusses (Strömungen, Generationen) wurde im 19. Jahrhundert nach weiteren Bestimmungen gesucht, die für das konkrete Aussehen einer Gestaltung verantwortlich sein könnten. In der historischen Entwicklung von Kunst wurde ein Beleg für den Anteil eines gestalteten und gestaltenden Sehens gefunden, das die Wahrnehmung auch jenseits von Kunst mitbestimmt.

Gestaltetes Sehen

Alois Riegl (1858–1905), ein Vertreter der Kunstwissenschaft Wiener Prägung, suchte einen Stilbegriff, der die psychologischen Faktoren der Gestaltung mit einer historischen Perspektive verband. Stil stehe demzufolge für das „Kunstwollen" (vgl. Riegl 1901) des Menschen, das selbst in einem manufakturmäßig erzeugten Alltagsprodukt jenen ästhetischen Überschuss verursache, welcher sich in der Rückschau bemerkbar mache. Damit ging er über die materialistische Definition des Architekten Gottfried Semper (1860) hinaus, wonach Stil aus der Übertragung von Formen und Zwecken auf veränderte Materialien resultieren würde.

Kunstwollen

Die psychologisierende Kunstbetrachtung führte zu übergreifenden Studien wie Wilhelm Worringers kunsthistorischer Arbeit *Abstraktion und Einfühlung* (1908) und zur Begründung der kulturwissenschaftlichen Methode Aby Warburgs (1866–1929), der nach den zentralen gestischen Impulsen in der Geschichte der Künste gefahndet hat und sie als „Pathosformeln" bezeichnete (Warburg 1905). Einzelne Formen könnten nach Warburg soweit Eigenleben entwickeln, dass sie sich unabhängig von Bildbedeutungen, Motiven und Medien übertrügen und in jedem Bild ihre eigene Energie entfalteten.

Pathosformeln

Wie Menschen ihre Wahrnehmung konkret artikulieren, war auch Gegenstand der Forschungen Heinrich Wölfflins (1864–1945), der auf der Grundlage der bildenden Kunst und Architektur eine „Geschichte des Sehens" (Wölfflin 1915, S. 11f.) entworfen hat. In Be-

Kategorien des Sehens

griffspaaren wie „malerisch/linear" oder „hell/dunkel" sollten jene Kategorien benannt werden, zwischen denen sich Sichtbarkeit unweigerlich bewegen müsse. Wölfflins Auswahl blieb von der europäischen Hochkunst bestimmt, war aber von simplen Stilunterscheidungen z. B. nach Epochen (gotisch, barock) bereits weit entfernt.

Gestaltforschung

Diese neueren kunst- und kulturwissenschaftlichen Forschungen standen in enger Beziehung zur Experimentalpsychologie, die sich mit Formerkennung, Assoziationsvermögen oder Gedächtnisleistungen befasste und in großem Umfang mit visuellen Versuchen operierte. Von erheblicher Ausstrahlung waren die Grazer und Berliner Schulen der Gestaltpsychologie, die unter Berufung auf den Philosophen Christian von Ehrenfels und dessen Buch *Über Gestaltqualitäten* (1890) die Erkennung von Formen als „Seelenbilder" deuteten, bei denen mentale Prozesse und sinnliche Reize untrennbar ineinandergreifen.

Das Leben der Form

Die Verbindung von Psychologie und Kulturwissenschaft im deutschen Sprachraum wurde mit dem Nationalsozialismus unterbrochen und in der Emigration, vor allem von Rudolf Arnheim (*Gestalt and Art*, 1943; *Art and Visual Perception*, 1954) und Ernst H. Gombrich (*Kunst und Illusion*, 1960) weitergeführt. Vergleichbare Traditionen hatten sich daneben in England und den USA (vgl. Fry 1920; Buswell 1935) und in der französischsprachigen Theorie entwickelt, für die *Das Leben der Form* (1939) des Kunsthistorikers Henri Focillon oder das Werk des Phänomenologen Maurice Merleau-Ponty (1908–61) angeführt werden könnten. Auch in der Werbelehre blieb die Verbindung von Gestaltung und Wahrnehmung eng verknüpft (→ KAPITEL 10.2).

Kognition

Nach 1945 wurde in den westlichen Staaten und unter dem Eindruck der aufkommenden Elektronisierung eine exakte Ästhetik eingefordert, die Form als ‚Information' berechenbar macht (vgl. Bense 1969). Dieser Ansatz lebt z. T. in den kognitionspsychologischen Forschungen fort, die seit den 1990er-Jahren erneut zu einer Welle von Publikationen geführt haben (vgl. Hubel 1990; Hoffman 2000), in denen Form das dynamische Produkt von Sinnesinformationen ist. Doch auch diese neuronalen Modelle werfen die Frage auf, inwiefern Sehen durch kulturelle Faktoren strukturiert ist. Kategorien wie Form, Farbe, Raum oder Bewegung erweisen sich als Aspekte, die von Traditionen geprägt sind – und die auch in diesem Buch nur beispielhaft als Zugang zu Problemfeldern behandelt werden.

Ästhetische Erziehung

Als Mittel zur Sozialreform zielte die ästhetische Erziehung besonders auf diesen kulturellen Wert visueller Form. Die Kunstpädagogik etablierte sich als Sehschulung gegen eine reizgesteuerte und standardisierte Umwelt. Das Weimarer Bauhaus von Gropius schloss in der Namensgebung bewusst an die mittelalterliche Bauhütte an, um in Verbindung mit gesellschaftlichen Erneuerungsbewegungen ein neues Gemeinschaftswerk zu kennzeichnen (vgl. Kerbs 2001, S. 380f.).

Die Idee einer zeitgemäßen Gestaltung fand ihren Widerhall im Werk des US-Architekten Louis H. Sullivan (1856–1924), der mit dem Ausdruck *form follows function* eine naturnahe Funktionalität beschrieben hat. Formale Reduktion und Zweckmäßigkeit sind zwar kein Spezifikum der Moderne, in Verbindung mit Industrialisierung und moderner Gestaltung erwuchs hieraus aber die Doktrin, wonach die nützliche Form schmucklos zu sein habe.

Form follows function

Gerade dadurch aber ließen auch der Gebäude-Kubus der 1920er-Jahre oder das Industrieprodukt ihre zeitliche Bindung erkennen. Der Formbegriff erwies sich als normativ, seit er die Zweckhaftigkeit mit dem Anspruch verband, den Gegenstand überformen zu können, also auch unabhängig von Zwecken. Seit neue Materialien wie Stahl und Kunststoff der formgebenden Fantasie keine Grenzen mehr zu setzen schienen, wurde die Gesellschaft insgesamt zu einer *tabula rasa*, jener leeren Malfläche, die von der Imagination aufgefüllt werden kann (vgl. Wagner 2004). Formung erwies sich als Kampfbegriff, seit Kunst und Politik selbst beanspruchten, Gestalterinnen sozialer Massen zu sein. Sie wurde zum Schlagwort avantgardistischer Kunstrichtungen wie dem Futurismus, die durchaus die gewaltsame Veränderung suchten, auch in Rückübertragung von Gesetzen der Natur, welche selbst eine stetige Veränderin und Zerstörerin sei.

Gestaltung des Sozialen

Georg Kolbes plastische Arbeit ist eingebettet in solche parallelen, auf Form und Formung bezogenen avantgardistischen Diskurse, so wie sie in die Bildkultur einer Zeit gehört, die zwischen Pomp und radikaler Modernität, einer explosiven Mischung von Massenbewegung und Führeridee, nationalem Pathos und sozialistischen Gesellschaftsvorstellungen oszillierte und in Technik, Wirtschaft, Unterhaltungskultur oder Städtebau rasanten Veränderungen unterworfen war.

Bildkultur

4.4 Beschreibung

Das Wissen um die Kontexte eines bestimmten Kunstwerkes verringert nicht die Schwierigkeiten, ein Objekt nach formalen Kriterien zu erfassen. Eine erste Hürde liegt schon in dem Umstand, dass die Betrachtung visueller Formen in der Umgangssprache mit dem Makel

Geringschätzung der visuellen Form?

der Äußerlichkeit („rein formal") behaftet ist, so als scheine sie nur unter Ausklammerung sozialer oder technischer Faktoren oder auf der Grundlage bestimmter Normen und Geschmäcker gewürdigt werden zu können. Durchaus in dieser Denktradition wurde z. B. der „Formalismus" in der stalinistischen Kunstpolitik als Schlagwort gegen die Abstraktion ins Feld geführt. Abstrakte Kunst, welche Äußerlichkeit selbst zum Thema macht, erregt nach wie vor die öffentliche Meinung und stößt auf Ablehnung bei denjenigen, denen eine figurative Wiedergabe oder Nachahmung als Hauptzweck von Bildern gilt.

Erweiterungen der Form

Längere Zeit als überspezialisierte ‚Kennerschaft' abgetan, gehört die Unterscheidung und Beschreibung von Formen aber zum Geschäft eines jeden, der eine Filmkritik formuliert, eine Person identifiziert oder als Richter festlegen soll, ob ein bestimmtes Foto ein anderes plagiiere. Die Analyse visueller Form der Kunst bezieht dabei idealerweise gesellschaftliche und mediale Faktoren in ihre Darstellung ein. So soll ausgeschlossen werden, dass aus der Betrachtung einer Form bereits auf deren Stellenwert oder Intention geschlossen wird und dabei wechselnde Situationen, Publikumsreaktionen oder Rekontextualisierungen von Objekten ausgeblendet bleiben. Anschläge werden auf Denkmäler nicht allein wegen ihrer Gestaltung verübt, sondern auch, weil sie von Räumen Besitz ergreifen, eine Person vertreten oder an bestimmte Ereignisse erinnern – doch dies ist ein Teil ihrer Form.

Formwahrnehmung verändert sich unter dem Einfluss von Techniken der Aufzeichnung und Reproduktion, durch kulturelle Bedingungen, durch Bildung und Erziehung, sie ist ein Produkt individueller Aufmerksamkeiten oder Gedächtnisleistungen. Die reflektierte Wahrnehmung von Form setzt daher eine Beschreibung voraus, die das Detail würdigt, das Gedächtnis übt und so weit als möglich die äußeren Faktoren bedenkt, welche die eigene Sicht bestimmen. Da kaum etwas so anspruchsvoll ist wie die Übersetzung von visuellen Formen in sprachliche Form und dies stets von neuem zu lernen ist,

Ekphrasis

galt die Bildbeschreibung, als *ekphrasis*, schon in der Antike als eigenständige Kunst, wurde in der Literatur der Frühen Neuzeit wie in der Salonkultur des 18. Jahrhunderts geschätzt und parallel zur bildenden Kunst stetig weiterentwickelt, z. B. als Beschreibung von Impressionen oder Gemütszuständen bei der Betrachtung von Bildern.

Jede Autopsie muss durch die Definition und Vereinheitlichung von Begriffen konventionalisiert werden; sie folgt daher gleichermaßen aus der vergleichenden Anschauung wie aus den Kommunikationen darüber (vgl. Inhetven / Kötter 1996; → KAPITEL 10). Sobald Bewertungen eingefordert werden, ob eine Darstellung „diffamierend" wirke oder eine Aktienkurve eine „Tendenz" erkennen lasse, stellt sich die Frage nach der Form als Grundlage einer Kommunikation über sichtbare Sachverhalte. Es ist keineswegs paradox, dass diese ihrerseits der präzisen Begriffe bedarf.

Fragen und Anregungen

- In welchen Hinsichten wird durch den Formbegriff der Bildbegriff erweitert?

- Welche unterschiedlichen Bedeutungen hat der Begriff des Stils im Laufe seiner theoretischen Diskussion angenommen?

- Aus welchen verschiedenen Gründen wurde Form besonders im Zeitalter der Industrialisierung zu einem besonderen Thema?

- In welchen technisch-naturwissenschaftlichen Bereichen könnte der Formbegriff heute einen besonders hohen Stellenwert haben?

Lektüreempfehlungen

- **Karl R. Gegenfurtner: Gehirn und Wahrnehmung**, Frankfurt a. M. 2003. *Die Zahl der Publikationen zu visueller Wahrnehmung, Neurologie des Sehens oder Neuroästhetik wächst derzeit rasant. Gegenfurtner bietet hier eine recht aktuelle Übersicht über die – durchaus verschiedenen – Bereiche der Forschung zur visuellen Informationsverarbeitung im Gehirn.*

- **Ernst H. Gombrich: Kunst und Illusion. Zur Psychologie der bildlichen Darstellung**, Stuttgart / Zürich 1978. *Gombrich, der Verfasser der meistverkauften Einführung in die Kunstgeschichte, hat hier eine materialreiche Studie zum Wahrnehmungswandel vorgelegt, der sich aus der Interaktion von Kognition und Kultur bei der Produktion von Bildern ergibt.*

- **Hans Holländer (Hg.): Erkenntnis, Erfindung, Konstruktion. Studien zur Bildgeschichte von Naturwissenschaft und Technik vom 16. bis zum 19. Jahrhundert**, Berlin 2000. *Sammelband mit kompakten Aufsätzen Holländers, die als Einführungen in das frühneu-*

zeitliche Naturstudium und die künstlerische Theorie und in die Entstehungsgeschichte technischer Bilder dienen können.

- **Udo Kultermann: Geschichte der Kunstgeschichte. Der Weg einer Wissenschaft,** München 1996. *Auch wenn der Titel eine reine Disziplingeschichte erwarten lässt, hat der Autor in dieser mehrfach erweiterten Chronik vor allem die Entdeckung der Form, Struktur und Komposition und die zahlreichen Bezüge von Kunst, Ästhetik und Psychologie in gut lesbarer Form verfolgt.*

5 Farbe und Linie

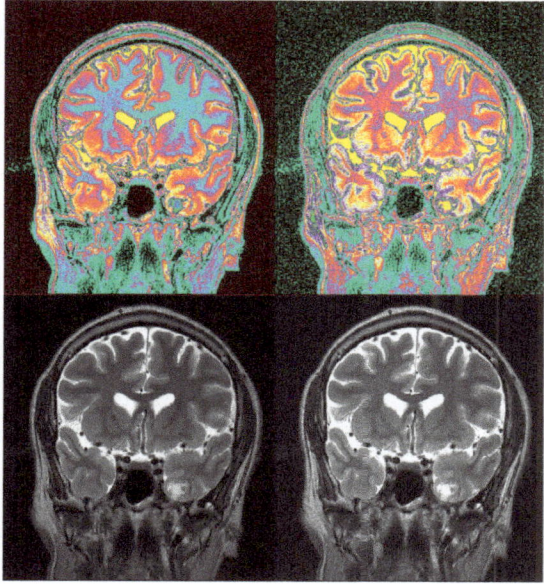

Abbildung 11: Federico Barocci: Hals- und Schulterpartie eines Mannes, der einen Stab hält (vor 1590), Rötel und farbige Stifte auf blauem Papier

Abbildung 12: MRT mit koronalen Längsschnitten durch den Kopf eines Tumorpatienten, farbige und schwarz-weiße Darstellung (2006 / 2002)

In der Vorzeichnung für ein Gemälde, das Federico Barocci Ende des 16. Jahrhunderts angefertigt hat, kommt eine frühneuzeitliche Technik des Pigmentauftrags zum Einsatz, bei der mit einigen bunten Stiftstrichen eine effektvolle Überhöhung von Details erreicht wurde. Durch die partielle Kolorierung der Haut scheint sich der gezeichnete Körper greifbar und plastisch vom Untergrund abzuheben: Farbe wird zum materiellen Ausdruck von Lebendigkeit, obwohl die Zeichentechnik selbst nur auf wenigen, routiniert aufgetragenen Linien beruht.

In den Aufnahmen einer Schädeluntersuchung per Magnetresonanz-Tomografie (MRT) werden krankhafte Veränderungen des Hirngewebes in einer Gegenüberstellung zweier Aufnahmen kontrastiert, um die unterschiedliche Erkennbarkeit farbkodierter und schwarzweißer Visualisierungen zu demonstrieren. So zeigt sich eine minimale Veränderung im unteren rechten Bereich in der schwarzweißen Darstellung besonders deutlich. Der Vergleich wird dadurch erleichtert, dass der Betrachter intuitiv und aufgrund entsprechender Gewohnheiten aus dem identischen Umriss der Schnittbilder auf eine mögliche Verwandtschaft zwischen beiden Aufnahmen schließt.

Farbigkeit wird, auch wenn sie sich synästhetisch auf andere Phänomene wie die ‚Klangfarbe‘ übertragen lässt, als eine der zentralen Kategorien bildlicher Darstellung angesehen. Umso bemerkenswerter ist, dass viele Bildformen auch ohne Buntwerte auskommen, etwa ein großer Teil von Pressebildern oder Grafiken, aber auch jede Bleistiftskizze, die einer ‚farbigen‘ Umsetzung (d. h. einer Ausführung mit bestimmten Pigmentstoffen) oft vorausgeht. Auf dieser über die Jahrhunderte eingeübten technischen und sprachlichen Unterscheidung von bunten und schwarzweißen Darstellungen beruhen auch Bildkonzepte, die zwischen einer farbigen Oberfläche und einer zeichnerisch-linearen Struktur trennen, welche das Wesentliche übermittelt und eine besondere Intelligenz von Auge und Hand erfordert. Damit drücken sie, wie bei der Differenzierung von Form und Inhalt, ein vergleichbar komplexes Wechselverhältnis aus. Aus der Beziehung von Zeichnung und Farbgebung in Kunst und visueller Kommunikation, Technik und Naturwissenschaft lässt sich ablesen, wie tief deren Verständnis kulturell-historisch verankert ist.

5.1 **Bunt und Schwarzweiß**
5.2 **Zwischen Malerei und Zeichnung**
5.3 **Paletten, Farbstoffe**
5.4 **Farbsemantiken**

5.1 Bunt und Schwarzweiß

Farbe meint im alltäglichen Sprachgebrauch etwas, das korrekterweise begrifflich als Buntheit, als Pigment (Farbstoff), als Graustufe oder als Wellenlänge bezeichnet werden müsste. Genaugenommen ist alles Farbe, was innerhalb der sichtbaren Wellenlängenbereiche von 400 bis 700 nm vom menschlichen Auge wahrgenommen werden kann: Eine weiße Wand wird mit Wandfarbe bemalt, und auch eine grau bemalte Figur hat bestimmte Töne und Schattierungen, ganz unabhängig davon, ob und wie das Auge, seine Netzhaut oder das Gehirn an der Wahrnehmung derselben beteiligt sind. **Was meint Farbe?**

Wie schon das Beispiel des Wildunger Altars (→ ABBILDUNG 3) im zweiten Kapitel belegt, haben sich im Laufe der Geschichte dennoch begriffliche wie praktische Unterscheidungen zwischen farbigen (bunten) und grauen (oder monochromen) Darstellungen entwickelt, welche auch die künstlerisch-ästhetischen Diskussionen, insbesondere zum Unterschied von Zeichnung und Malerei, beherrschten und die durch das Aufkommen von Techniken wie Fotografie und fotomechanische Reproduktion noch einmal verstärkt wurden.

Vor allem der Buchdruck und der Holzschnitt haben den Druck mit schwarzer Farbe auf weißem Papier zur Regel gemacht. Schon früh wurden Drucke jedoch auch koloriert oder in verschiedenen Farben mit mehreren Druckstöcken hergestellt. Seit dem 18. Jahrhundert wurde verstärkt erkundet, wie durch die Überlagerung von Farbschichten eine bunte Grafik produziert werden könnte (vgl. Graselli/Philipps 2003). Einen ersten Höhepunkt dieser Versuche und Überlegungen stellte Godefroy Engelmanns Chromolithografie von 1837 dar, die auf dem lithografischen Flachdruckverfahren Alois Senefelders und anderer (um 1800 einsetzbar) beruhte. Das Bild wurde dazu in Farbwerte zerlegt und stufenweise übereinander gedruckt. Generell erlaubte die Lithografie durch Auftrag einer flachen präparierten Platte die zeichnerische Erstellung der Druckvorlage (im Unterschied zum Kupferstich) und einen Abdruck in höherer Auflage. Parallel dazu haben sich die Techniken der Schattierung und Schraffur im 19. Jahrhundert bis ins Äußerste verfeinert, um den Eindruck von Buntheit auch im Einfarb-Druck zu simulieren, vor allem in der sogenannten Reproduktionsgrafik zur Wiedergabe von Kunstwerken (vgl. Bann 2001). **Einfarbiger Druck**

Optische Experimente zur Brechung des Lichts mithilfe von Prismen durch Isaac Newton (1643–1727) und Studien zur Physiologie des Auges (→ KAPITEL 4.3) führten zu einer Relativierung bisheriger

Vorstellungen von der Natur der Farbe. Newton ging noch davon aus, dass das regenbogenfarbige Spektrum sich im Rotbereich zu einem Kreis schließen würde. Daher blieb die Diskussion des Farbspektrums von zahlreichen Fehllektüren geprägt (vgl. Hentschel 2002). Die Farbenlehre Johann Wolfgang Goethes (1749–1832) und das Modell der Farbkugel von Philipp Otto Runge (1770–1810) lassen zudem erkennen, dass in der Theorie der Farbe stets Erfahrungen mit dem materiellen Gestaltungsmittel enthalten waren.

Andererseits gewann die Farbe zur Zeit der europäischen Entdeckungsreisen und der sich auf Weltmaßstab ausdehnenden Naturstudien um 1800 (James Cook, Alexander von Humboldt) immense Bedeutung für die exakte Bestimmung von Naturalien. Die Färbung von Zeichnungen und Drucken spielte schon vorher eine erhebliche Rolle für Botaniker und Zoologen, deren Illustrationen (besonders prachtvoll bei Vögeln und Schmetterlingen) manuell koloriert wurden (vgl. Nickelsen 2006). Farbe erwies sich jedoch als uneindeutig und instabil, was prinzipiell schon seit der Antike bekannt und später sogar eine Voraussetzung für die Entwicklung der Fotografie war. Da die Färbung von Pflanzen- und Tierpräparaten ebenso wie ihre bildliche Darstellung natürlicherweise ausbleicht und die Wahrnehmung von Farben sich noch dazu, z. B. je nach Medium oder Lichtverhältnissen, gravierend ändern kann, mussten für wissenschaftliche Zwecke Techniken zur Speicherung von Farbinformationen entwickelt werden, meist in Form von Skalen und Diagrammen, deren Farbwerte mit Benennungen versehen oder zahlenmäßig kodiert wurden, wie in → ABBILDUNG 13 aus einem botanischen Handbuch von 1815, das die Abmischung von leuchtenden Farben mit Grautönungen in minutiös handkolorierten Kreismodellen vorstellt (vgl. Freedberg 1994; Altmann 2006).

Als Beispiel für die komplexe Geschichte der Farbe kann die Verbreitung fotografischer Techniken im 19. Jahrhundert angesehen werden. Historische Fotografien sind als Schwarzweißbilder im allgemeinen Gedächtnis verankert, charakterisiert durch grobe Kontraste und geringe Binnendifferenzierung. Zwar wird auch in den Anfängen der Fotografie vielfach von Farbe(n) gesprochen, weil die ersten Versuche auf den Verfärbungen von Chemikalien wie Silbernitrat beruhten und gelbe oder blaue Bilder lieferten. Farbe meinte hier aber nicht Buntheit, sondern die Färbung der Substanz oder des Bildträgers selber.

Seinerzeit war das Verfahren der Daguerrotypie unerreicht in der Abstufung von hellen und dunklen Werten und wurde darin den gra-

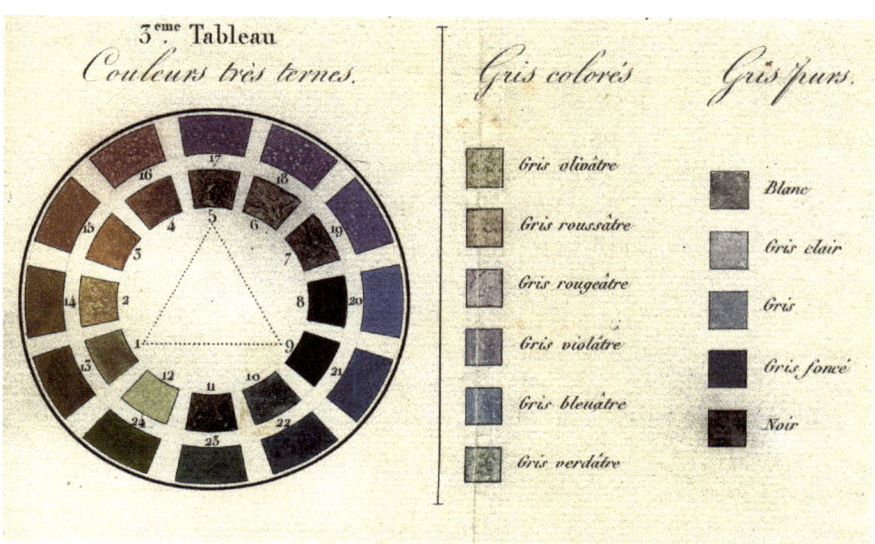

Abbildung 13: Charles François Brisseau-Mirbel: Kupferstich mit handkoloriertem Farbsche-ma aus seinem botanischen Lehrwerk *Elémens de Physiologie Végétale* (1815)

fischen Künsten schon als ebenbürtig, ja überlegen angesehen. Doch wurde auch in der Anfangsphase der Fotografie das Fehlen natürli-cher Buntheit bemerkt und durch Kolorierungen behoben, z. B. bei Johann Baptist Isenring (1796–1860). James Clerk Maxwell wandte 1861 die Hypothese von Thomas Young (1773–1829) und Hermann von Helmholtz (→ KAPITEL 4.3), wonach das Augensehen auf Rot, Grün und Blau beruhe, auf die Fotografie an und stellte mithilfe von gefilterten Aufnahmen und deren Lichtprojektion Farbfotografien vor. In fotografischen Betrieben und Druckereien wurde mit Licht-quellen und der unterschiedlichen Empfindlichkeit von Chemikalien und Messgeräten für einzelne Spektralbereiche experimentiert, was auch zu neuen naturwissenschaftlich-technischen Analyseverfahren führte.

Um 1900 war Farbfotografie somit durchaus geläufig (Ausst.-Kat. Köln 1981, S. 71), sie war jedoch nicht massenhaft im Druck repro-duzierbar. Der Abdruck von Fotografie im Buch wurde zwar in der zweiten Hälfte des 19. Jahrhunderts in industriellem Maßstab mög-lich, blieb aber monochrom. Besonders im Bereich des Kunstdrucks, in staatlichen oder verlagseigenen Versuchs- und Lehranstalten, wur-de daher an Verfahren zur Wiedergabe von Farben und an Farb-kodierungen gearbeitet. In einigen Fällen wurden transparente Folien mit Beschriftungen in das Buch eingefügt, die auf künstlerische Pig-

Frühe Farbfotografie

Probleme der Farbreproduktion

mente und Erdarten (wie Ocker und Umbra) bezug nahmen. Bis heute üblich ist der Einsatz von Farbskalen in der fotografischen Aufnahme oder an Druckrändern zum Abgleich des Reproduktionsergebnisses.

Mit Kenntnis optischer Gesetze, wonach die Farbigkeit von Gegenständen durch Auszug oder Nichtreflexion einzelner Spektralbereiche entsteht (subtraktiv) und daher im Farbdruck auf Papier anders gemischt werden muss als z. B. bei einem Projektor mit mehreren Filtern oder Röhren (additiv), entstanden unterschiedliche **Farbräume** und konkurrierende Modelle von Farbmischungen und -räumen (wie dem CMYK-Farbraum, einem Standard im heutigen Druckwesen), die weiterhin Gegenstand der Forschung und der Softwareentwicklung sind. Neue Materialien, Mess- und Kalibrierungssysteme sollen die Schwankungen von Druckfarben, Farbfilmen oder Scannern ausbalancieren, um dem Anspruch auf eine korrekte Wiedergabe von Farben zu genügen.

Farbe bleibt variabel Farbtreue ist aber überhaupt nur möglich, sofern Farbe allein über Wellenlängen definiert ist und damit nicht über die individuelle Wahrnehmung. Auch wenn es nur eine Skala von Wellenlängen gibt, so gibt es nicht nur ein einziges mögliches System zur Farbgenerierung am Bildschirm oder im Druck, schon weil es nicht nur eine Art gibt, Farbe zu definieren oder zu sehen.

5.2 Zwischen Malerei und Zeichnung

In der scholastischen Philosophie des Mittelalters war die Frage nach dem Charakter und Entstehungsort von Farbe wesentlich für die ästhetische Definition von Dinglichkeit und Wahrnehmung überhaupt. Sie umkreiste, ungetrübt durch physikalische Präzision, die Frage, ob **Farbe als Zutat?** Farbe Teil eines Stoffes und seiner Form sei oder nicht vielmehr ein Attribut, d. h. eine äußere Zutat zu einem ideal gedachten Körper (vgl. Crone 1999). Dies ließ sich leicht annehmen, solange die Argumentation auf Erfahrungen mit Figuren aus Stein oder Holz oder auf der Bemalung von Wänden beruhte, die mit Farbe zum Leuchten gebracht werden; gerade das Mittelalter kannte nicht nur eine umfassende Farbdiskussion, sondern auch einen umfassenden Einsatz farbiger Medien wie Buntglas oder Textilfarben.

In dieser Tradition von Form und Farbe hat der Griffel des Zeichners und Grafikers in der Geschichte der Kunsttheorie vor allem für **Doktrin der Zeichnung** eine abstrahierende Leistung gestanden. Die Linie wurde begriffen als jene konstruktive, tragende Seite der Sichtbarkeit, in der sich gleichsam das Skelett der Natur zeigt. Die Zeichnung zeigt Formen und Verläufe, die in der Natur nicht dinglich existieren und doch visualisierbar sind.

In der Kunsttheorie des 16. Jahrhunderts erhielt das menschliche Vermögen, die Welt in linearen Formen zu erfassen, durch den Begriff des *disegno* eine langfristig wirksame Aufladung (vgl. Kemp 1974), da sie das Typische, das Ideale, den Kern oder das Konzept einer Sache freilegen könne (vgl. Panofsky 1924). In jeder Handzeichnung oder Skizze äußert sich das Vermögen der Kunst, mit der eigenen Handbewegung und damit im Einklang mit der Natur ihrem Wesen näherzukommen. Hierauf gründete sich eine weitreichende Kultur der bildnerischen Erfassung der Welt, die auf die Analogie von innerer und äußerer Bildung abzielt. Die Zeichnung wurde daher nicht nur als Mittel der Abstraktion im theoretischen Diskurs hochgeschätzt, sondern auch im allgemeinen Zeichenunterricht gefördert (vgl. Kemp 1979; Bermingham 2000).

Disegno

Obwohl die *camera obscura* als Lochkamera eigentlich farbige Ansichten der Welt liefert, wenn auch mit Kontrastverlusten, so erklären lehrbuchmäßige Darstellungen der Frühen Neuzeit den Gebrauch solcher Apparate stets, indem sie einen Zeichner beschreiben, der vor dem Gerät platziert ist. Bekannt ist zwar, dass schon Maler des Barock die Technik selbstverständlich nutzten, so Jan Vermeer, jedoch ist ungeklärt, inwieweit sie diese nur als Anregung oder zum Festhalten linearer Strukturen verwendeten, oder mit großen und entsprechend lichtstarken Kamerakonstruktionen direkt auf eine Leinwand projizierten, um darauf auch mit Farbpigmenten zu malen (vgl. Steadman 2002).

Camera obscura

Die Idee zeichnerischer Klarheit blieb demgegenüber, auch jenseits der Kunst wie z. B. in der anatomischen Darstellung, über lange Zeit einflussreich und ist unverändert präsent, etwa in der Karikatur, in der technischen Zeichnung oder der Computergrafik; obwohl die Fotografie prinzipiell verfügbar ist, wird die Zeichnung weiterhin zur Reinzeichnung verwendet, um verallgemeinerbare Merkmale herauszuarbeiten. Dies wird noch begünstigt durch den Umstand, dass Grauwerte im menschlichen Auge feiner differenziert werden und auch einer natürlichen Sichtbarkeit, vergleichbar der Wahrnehmung im Halbdunkel, entsprechen.

Zeichnung als Herausarbeitung

In dieser ästhetischen Tradition wurden der Kupferstich, der Schattenriss und Scherenschnitt, die von Farbresten befreite Skulptur oder ihre bereinigte Kopie im 18. Jahrhundert zu den Trägern eines klassizistischen Schönheitskanons. So führte im 19. Jahrhundert die archäologische Feststellung, dass die Bildwerke der Antike durchaus in Farbe gefasst waren und ihre gipsweiße Reproduktion eine spätere Bereinigung darstelle, zu eine langanhaltenden Debatte um die Polychromie (Buntheit) der Antike (vgl. Brinkmann 2007).

Die Auffassung von Linie und Farbe als einem Gegensatzpaar kann demnach leicht in die Irre führen: Der legendäre Streit um künstlerische Meisterschaft zwischen den antiken griechischen Malern Protogenes und Apelles, bei dem es darum ging, wer die feinste aller Linien malen könne, zeigt zugleich, dass auch mit einem Pinsel **Übergänge von Linie und Fläche** eine Linie hergestellt werden kann und der Übergang vom breiten Auftrag zum feinen Strich also kein absoluter ist – genauso wie eine Folge von Bildschirmpixeln sowohl eine Linie als auch eine Fläche erzeugen kann.

Farbe als Zeichnung Auch im Falle Federico Baroccis (→ ABBILDUNG 11) begegnen sich Techniken, die aufgrund unterschiedlicher Materialien (Tusche, Ölfarbe, Kreiden, Stifte) gewöhnlich als grafische und malerische Techniken unterschieden werden, die aber nur zusammen ihr Resultat ergeben. Zur Vorbereitung eines größeren Gemäldes, das die *Beschneidung Christi* zum Thema hat und das sich heute in Paris befindet, hat der Künstler die Schulter eines Mannes mit einem Stab gezeichnet und anschließend in einer für ihn sehr typischen Weise mit wenigen Farbstrichen betont. Gerade durch den Kontrast von rötlicher Hautfarbe zu einem blaugrauen, monotonen Untergrund gewinnt die Schulterpartie an einigen Stellen an Plastizität, wirkt sie besonders verlebendigt.

5.3 Paletten, Farbstoffe

Am Beispiel der Zeichnung Baroccis lässt sich leicht ersehen, dass diese Lebendigkeit sich nicht der vermeintlichen Naturnähe verdankt, sondern dem bildinternen Kontrast von Farbstrichen und grauem Hintergrund. Sie zeigt auch, dass die Abgrenzung von Linie und Farbe eine rein theoretische ist. Jede Bleistiftzeichnung ist einfarbige Wirklichkeit, jede Malerei künstliche und überhöhende Farbgebung **Farbgebung als höchste Kunst** – ihr Pigment strahlt jedoch umso mehr, wenn sie als Gegensatz zum monochromen Grund gesehen wird. Durch den praktischen Umgang mit Paletten und Farbaufträgen, mit Bildträgern und Lichteffekten wird die Farbe als eine Wirklichkeitsform greifbar, die ihre eigenen semantischen Werte besitzt und darin auch die Natur übersteigt. Farbe ist Material und wesentlicher Teil der Form, Farbgebung eine Kunst für sich.

Von Marcello Malpighi (1628–94), einem der Begründer der modernen Embryologie, stammen Handzeichnungen, die seinen gedruckten Werken zugrundelagen. Sie sind mit Bleistift, Kreide und Rötel gearbeitet, als stammten sie aus der Hand des Malers Barocci. Die künstlerisch ausgereifte und einstudierte Technik des Zeichners wird dadurch als ein wesentliches Mittel der Erkenntnisgewinnung aufgewertet (vgl. Bredekamp 2007), das indirekt auch das Wissen über die Gestalt der Natur unter dem Mikroskop beeinflusste: Zeichnung und Farbe werden die eigentlichen Träger von Wissen.

Farbe als
Erkenntnismittel

Zum Ausdruck kommt dies auch in dem Umstand, dass Farbe als aufwendig zu produzierender Farbstoff, als Pigment einen zählbaren materiellen Wert besitzt: Gold oder Lapislazuli stehen für kostbarste Farbstoffe, die aufgrund ihrer seltenen Verwendung auch symbolische Aufladung erfahren haben und nur für bestimmte Zeichen verwendet wurden, z. B. Blau für das Gewand Mariens oder für die himmlische Sphäre. Blaue Pflanzenfarbstoffe (Waid-Blau, Indigo-Blau) waren kompliziert zu gewinnen und sehr teuer. Indem sie ein Textilstück auch real durchwirken kann, gehört Farbe zur „Wirklichkeit" der Bilder (vgl. Baxandall 1999, S. 108), die lange Zeit auch die Ikonografie von Bildfiguren bestimmt hat. Erhalten geblieben ist diese Bewertung im Druckwesen („Jetzt mit 120 Abbildungen in Farbe"), wo der Buntdruck mit mehreren Farben teurer ist als der Einfarbendruck; farbige Aufnahmen haben auch eine höhere Informationsdichte als Graustufen und erfordern in digitaler Form mehr Speichervolumen. Hier werden jahrhundertelange technische Erfahrungen mit Pigmenten, Oberflächen, Textilien, Fixierungen weitergegeben, die einen substanziellen Teil der künstlerischen Bildgebung ausmachen und nicht nur eine äußere Zutat sind.

Materieller Wert
von Farbstoffen

Der Bedarf nach Farbe, von der Textilfärbung über den Buntdruck bis zum Farbfernsehen, hat stetig neue Medien erfasst. Aus der industriellen Umsetzung von Farbabbildungen und ihren technischen Begrenzungen wurde abgeleitet, dass sich Kunst und Gestaltung in ihrer expressiven Farbwahl bereits „den rohen Erfordernissen des Farbdrucks angepaßt" hätten (Wind 1960, S. 78), um durch Massenreproduktion größere Bekanntheit zu erzielen. Das Illustrationswesen, die Produktwerbung und besonders das *comic feature* arbeiten in der Tat mit vereinfachter Palette, was durch die *pop art* mit ihren grobgerasterten Farbgebungen auch eigens thematisiert wurde – besonders prominent bei Roy Lichtenstein (1923–97), der

Herstellungsbedingte
Vereinfachungen

Abbildung 14: Roy Lichtenstein: *Frau mit Hut* (1963)

Winds Vorwurf dahingehend umdrehte, dass er Motive im Stile der Kubisten produziert hat, an deren Adresse Winds Kritik gerichtet war (→ ABBILDUNG 14).

Bestimmte Gattungen der Kunst und des Kunstgewerbes haben stets ein reduziertes Farbschema verwendet: So wurde für die Produktion von Teppichen nach Vorlagekartons verlangt, die sich auch in textiler Form umsetzen lassen, ebenso wie der Herstellungsprozess von Freskomalereien die Abgrenzung von Farbbereichen begünstigt hat, die als sogenannte Tagewerke bruchlos abgeschlossen werden können.

Durch ihre Materialität, die fortschreitende Verfügbarkeit von Ölfarben seit dem 17. Jahrhundert oder durch das Bedürfnis, eingeübte Motive zu überhöhen, konnte im 18. Jahrhundert der Farbstoff selber als Malweise zum Mittel des Ausdrucks werden, ehe die Bildkünstler in expressiver oder sogar gewalttätiger Form im 19. und 20. Jahrhundert mit Farbtuben, Eimern oder Sprühtechniken zu experimentieren begannen, bis hin zur vollständigen Absage an mimetische Strukturen wie im Tachismus, dem Informel und den *drip paintings* bei Jackson Pollock; dem Pigmentträger kam aber auch ohne diese aggressive Anwendung in der Malerei stets der Status des formbaren Materials zu (vgl. Krüger 2007).

Bedeutung des Farbstoffs

Sobald sich die technischen Verhältnisse verändern, kann die Situation überraschende Wendungen nehmen, wie es im Bereich der Fotografie der Fall ist, bei der die Nachfrage nach Farbfotografien die Schwarzweiß-Fotografie in preislicher Hinsicht weitgehend verdrängt hat. Gerade die Schwarzweißfotografie hat dessen ungeachtet ihre eigene Ästhetik entwickelt, die unverändert die Diskussion unter Fotografen und Grafikern bestimmt, weil die Darstellung als Helldunkel andere Kontraste erzeugt und die Beschreibung der Umwelt in eigener Weise konstruktiv verändert; eine Diskussion, in der die Spannung um *disegno* und *colore* fortdauert.

Ästhetik des Schwarzweißfotos

Die Entwicklung neuartiger, industrieller Pigmente, vor allem in Forschung, Medizin und Industrie von Bedeutung, ist unterdessen weiter fortgeschritten, bis hin zur Erzeugung fluoreszierender oder radioaktiver Markierungen, denen noch weitere Techniken der Einfärbung (etwa von Zellgewebe, → KAPITEL 10.4) entsprechen und auf Erfahrungen mit ihnen beruhen, selbst wo sie für das menschliche Auge unter Tageslicht unsichtbar sind und sich von einem landläufigen Farbbegriff vollständig verabschiedet haben.

Farbe als Markierung

In der Druckvorstufe oder Fotografie wird dagegen bemängelt, dass der mögliche Farbreichtum elektronischer Medien nicht zu einer gesteigerten Kompetenz in der Farbpraxis geführt habe, sondern dass die erschwinglichen, aber oft unzulänglichen Darstellungs- und Druckmöglichkeiten von Druckern, Bildschirmmedien oder Scannern dieses historisch gewachsene Verständnis für die Komplexität von Farben, weitgehend haben vergessen lassen. Allerdings kann selbst eine nach den Regeln der heutigen Druckkunst präzise kalibrierte und ausgesteuerte Wiedergabe von Farben nicht das Problem beheben, dass z. B. ein Gemälde je nach Lichteinwirkung, Zustand, Ta-

Alltäglichkeit technischer Farbgebung

geszeit oder der individuellen Einstellung der Zuschauer immer verschieden aussehen wird. In dieser Frage der Zuverlässigkeit chromatischer Information berührt sich die Diskussion auch mit den Erfahrungen in Medizin oder Naturwissenschaft.

5.4 Farbsemantiken

Farben dienen dem Schmuck, der Werbung und Warnung oder der Orientierung. Sie erzeugen Stimmungen, haben Signal- und Blicksteuerungswirkung, tragen oder unterstreichen Botschaften. Als Wappen, Abzeichen oder Fahne geben sie Dynastien und Parteien zu erkennen, sollen sie Heere oder Nationalitäten voneinander sichtbar unterscheiden. Stadtpläne, Wegeschilder, Infografiken mit statistischen Inhalten arbeiten mit farblichen Mitteln. Einer der gewollten Effekte ist die schnelle Orientierung, beispielsweise über mögliche Häufungen, Tendenzen, Symptome, Regelmäßigkeiten oder Abweichungen und die daraus zu ziehenden Schlüsse (→ KAPITEL 11).

Besonders Kartierungen zeigen an, inwieweit Bilder durch Farbe funktionalisiert sind und Farbsemantiken eine zusätzliche Sinndimension einführen. Bestimmte Angaben von Einwohnerzahlen, Wahlergebnissen oder Häufigkeiten von Kriminalitätsfällen können sich durch eine Übertragung in Farbwerte auch psychologisch zu Intensitäten verdichten, die in Verbindung mit deutlichen Farbabstufungen Gefahrenzonen signalisieren. Wahlkreise, die auf einer Landkarte nebeneinander in gleicher Farbe repräsentiert sind, suggerieren, dass das Wahlergebnis in der gesamten dargestellten Region gleich hoch ausgefallen ist. Farbkodierungen für Klimazonen oder Höhenangaben beruhen auf Konventionen, die oft an natürliche Gegebenheiten angelehnt sind, ihnen aber nicht zwingend entsprechen (z. B. Rot für Hitze).

Farben oder Farbkombinationen haben im Laufe der Geschichte nicht immer dieselbe Bedeutung, sondern entfalten eine eigene Kulturgeschichte (vgl. Gage 2007), so wie auch ikonografische Regeln zu Schematisierungen der Farbwahl führen. Dies wurde insbesondere in universellen Zeichensystemen und im globalen Marketing als Problem erfasst (vgl. Birren 1945). Hier beruht die Verwendung einzelner Farben einerseits auf allgemeinen Vermutungen, wonach diese eine höhere, evolutionär bedingte Signalwirkung haben (etwa im Falle der Farbe Rot als Abwehrreaktion auf den Anblick des Blutfarbstoffs). In Farbgebräuchen machen sich jedoch auch kulturelle Gren-

84

zen bemerkbar (z. B. im Unterschied von Trauerfarben). Internationale Markenprodukte müssen je nach Land eine andere Tönung haben, um lokal akzeptabel zu sein. Farbe besitzt daher nicht nur statische Werte, sondern kann ihre materiellen wie semantischen Eigenschaften auch wechseln.

In Verbindung mit bildgebenden Verfahren hat Farbe zusätzliches diagnostisches und erkenntnisförderndes Potenzial gewonnen, offenbart aber auch hier wahrnehmungsbedingte Grenzen. → ABBILDUNG 12 zeigt Aufnahmen einer Schädeluntersuchung per Magnetresonanz-Tomografie (MRT), bei der Veränderungen des Hirngewebes einander gegenübergestellt werden, um die Eigenarten farbkodierter und schwarzweißer Visualisierungen zu demonstrieren. In allen Aufnahmen handelt es sich um rechnerisch zusammengesetzte Darstellungen eines Kopfschnittes. Eine minimale Veränderung im unteren rechten Bereich zeigt sich aber vor allem in der schwarzweißen Darstellung.

Grenzen der Farbwahrnehmung

Eine kritische Analyse solcher Verfahren muss daher abwägen, wann eine Verwendung von Buntwerten hilfreich ist und ob sie Messergebnisse sinnvoll verdichtet (vgl. Hennig 2006), zumal wenn Farbbedeutungen je nach Fachgebiet und Zweck (Infografiken, Klimadarstellungen) spezialisiert sind. In jedem Falle ist die in der Bildgebung verwendete Farbe ausdrücklich eine Falschfarbe zur künstlichen Überhöhung und Erzeugung von Information. Auch in den Naturwissenschaften ist bis heute keine einheitliche Semantik formuliert worden, die verbindliche Aussagen über die angemessene Verwendung von Farben wie Rot oder Blau treffen könnte. Im Begriff der Falschfarbe kommt aber bereits zum Ausdruck, dass jede Kolorierung als eigenwertige, gestaltete Produktion von Wirklichkeit begriffen werden kann und muss.

Falschfarben

Fragen und Anregungen

- Was meinen die Begriffe „additiv" und „subtraktiv" in Bezug auf Farbsysteme?

- Aus welchen Gründen wird auch eine schwarzweiße Darstellung als naturnah akzeptiert?

- In welchen – künstlerischen wie wissenschaftlichen – Bereichen spielt Zeichnung heute nach wie vor eine maßgebliche Rolle?

- Kann sich die Interpretation oder Signalwirkung einzelner Farben grundlegend verschieben, und wenn ja, durch welche Faktoren?

Lektüreempfehlungen

- Stephen Bann: Parallel Lines. Printmakers, Painters and Photographers in Nineteenth-Century France, New Haven 2001. *Eine informative wie anspruchsvolle Studie zur jüngeren Geschichte der Reproduktionsgrafik, welche u. a. die Versuche zur Simulation von Farbwerten und Schattierungen durch Stichtechniken darstellt.*

- John Gage: Kulturgeschichte der Farbe. Die Sprache der Farben, Leipzig 2007. *Das Standardwerk zur gemeinsamen Geschichte von Farben und Farbtheorien zwischen Wissenschaft und Kunst.*

- Wolfgang Kemp: Disegno. Beiträge zu einer Geschichte des Begriffs zwischen 1547 und 1607, in: Marburger Jahrbuch für Kunstwissenschaft 19, 1974, S. 219–240. *Nach wie vor der am meisten herangezogene und grundlegende Aufsatz zum Aufstieg der Zeichnungs-Doktrin in den Künsten.*

- Matthias Krüger: Das Relief der Farbe. Pastose Malerei in der französischen Kunstkritik 1850–1890, München / Berlin 2007. *Aktuelle Arbeit zur Materialität der Farbe als Thema der Kunst.*

- Pam Roberts: 100 Jahre Farbfotografie, Berlin 2007. *Empfehlenswerte Einführung in die Farbfotografie mit Informationen zur Theorie und Geschichte der Farbe.*

6 Systeme

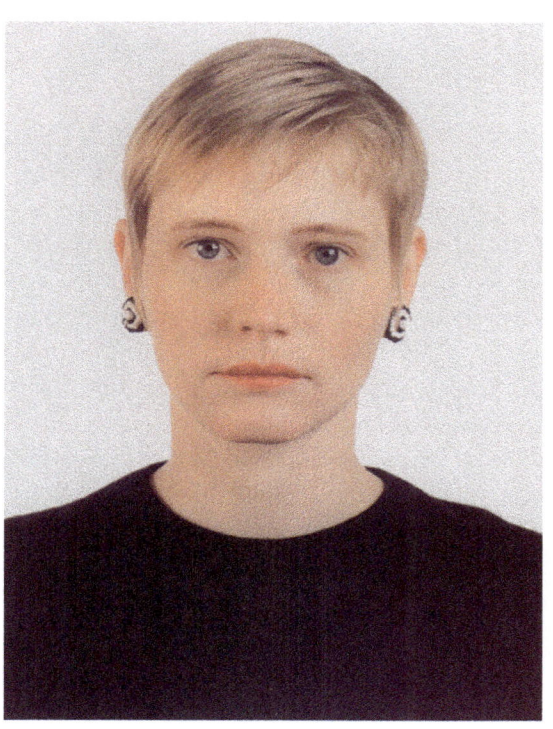

Abbildung 15: Thomas Ruff: *POR 090, Porträt*
(A. Knobloch) (1990)

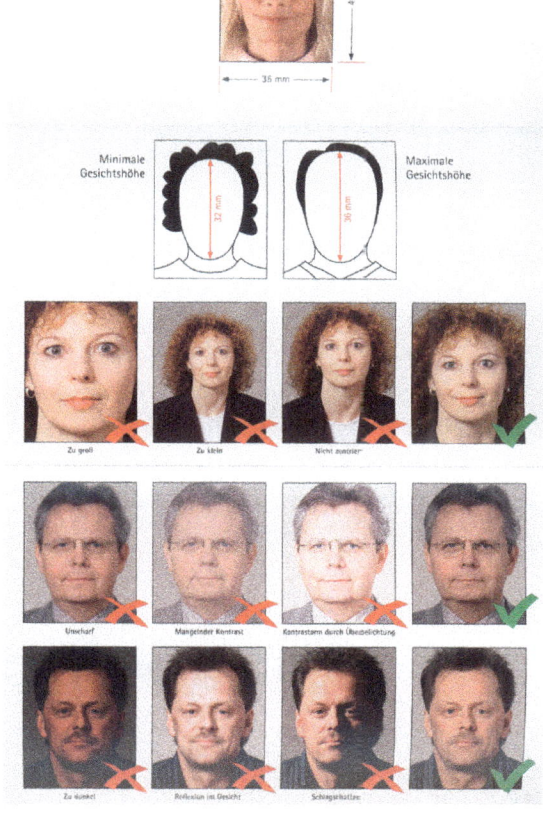

Abbildung 16: Auszug aus der Normtafel der Bundesdruckerei zur Herstellung von biometrischen Passbild-
aufnahmen (2005)

Der deutsche Fotokünstler Thomas Ruff hat in größeren Werkzyklen Menschen unterschiedlichen Alters und Aussehens abgelichtet und die Aufnahmen in übergroßen Formaten im Kunstraum präsentiert. Das hier gewählte Beispiel zeigt das Gesicht einer jungen Frau in Nahsicht (im Original eine Fotografie von über zwei Metern Höhe), die mit größtmöglicher Präzision von vorn porträtiert wurde.
Die Fototafel der Bundesdruckerei dient seit 2005 als Muster für sogenannte biometrische Passbilder in der Bundesrepublik Deutschland. Im Hinblick auf die frontale Aufnahme oder die gleichmäßige Beleuchtung werden schnell Parallelen zur Porträtform bei Ruff deutlich. Aufgabe der Tafel ist jedoch die Information darüber, nach welchem Musterbild Passfotos zu erstellen sind. Grund hierfür ist, dass die Pässe identifizierende Gesichtsinformationen über den Inhaber des Ausweises in doppelter Form aufnehmen sollen, nämlich als Porträtfoto, aber auch als sogenannte biometrische Information auf einem eingearbeiteten elektronischen Chip.

Bilder, die sich hinsichtlich Motiv oder Komposition deutlich ähneln oder zumindest zu ähneln scheinen, können je nach Funktion, Hersteller, Bezeichnung oder Entstehungszusammenhang grundverschieden sein; gesellschaftliche Kontexte und Systeme wie der Kunstmarkt, die Politik oder die Wissenschaft bestimmen mit darüber, wie ein bestimmtes Bild zu verstehen ist oder welche Rolle ihrem Hersteller (etwa dem Künstler) innerhalb der gesellschaftlichen Aufgabenteilung zukommt. Historische, soziale oder funktionale Bedingungen, die über die Intention eines Bildes Auskunft geben, sind entsprechend in eine erweiterte Bildanalyse einzubeziehen. Andererseits behalten Bilder auch in ihren jeweiligen Verwendungszusammenhängen und als instruktives Mittel ihre besonderen Eigenarten, sind auf weitergehende Traditionen bezogen und können dadurch Systemgrenzen überschreiten als auch befestigen.

6.1 Spezialisierung von Bildfunktionen

Eine Fotografie von 1900 sieht gewöhnlich nicht so aus wie eine Fotografie von 1950, selbst wenn sie denselben Gegenstand zeigt. Diese simple Beobachtung der historischen Stilkunde (→ KAPITEL 3, 4) hat eine Reihe von durchaus benennbaren Gründen. Die Aufnahmetechnik hat sich weiter entwickelt, das Trägermaterial ist gealtert, die Sehweisen und Interpretationen haben sich verschoben, es haben sich unmerkliche, langfristig wirksame Stilmerkmale ausgebildet, die zu Darstellungskonventionen oder zu Gegenbewegungen führen können. Andererseits kann es jedoch ebenso gut vorkommen, dass eine Fotografie genauso aussieht (oder auszusehen scheint) wie eine andere, obwohl sie etwas völlig anderes zeigt oder funktional und intentional grundverschieden ist.

Aussehen und Bedeutung

Ein solcher Eindruck kann besonders dann entstehen, wenn Porträts von Personen miteinander verglichen werden, die *enface,* also frontal fotografiert wurden (vgl. auch ein weiteres Beispiel in → KAPITEL 9.1). An diesem Bildtyp oder Schema lässt sich daher jener unsichtbare kontextuelle Unterschied darstellen, der aus zwei scheinbar gleichen Aufnahmen zwei unterschiedliche Bilder macht und der daher nach einer Differenzierung der entsprechenden Betrachtungs- und Deutungsebenen verlangt.

Enface-Bildnisse

Der deutsche Fotokünstler Thomas Ruff porträtiert in größeren Werkzyklen Menschen unterschiedlichen Alters und Aussehens und präsentiert Abzüge seiner Aufnahmen in großen Formaten. Seine Arbeiten sind in internationalen Ausstellungen und Sammlungen prominent vertreten, Ruff gehört zu den führenden Kunstfotografen deutscher Herkunft und genießt besonderes Ansehen und Wertschätzung in der Kritik und auf dem Kunstmarkt. In seinen Arbeiten zeichnet er sich trotz der simpel wirkenden Machart durch eine besondere Qualität der Aufnahme, vor allem die durchdringend scharfe Nähe zum Porträtierten und die Neutralisierung von Hintergründen durch leuchtend weiße Flächen aus. Außerdem folgt er konsequent seinem Aufnahmekonzept und nimmt so gezielt auf den Eindruck scheinbarer Wiederholung oder Gleichförmigkeit Bezug.

Das Beispiel Thomas Ruff

Das Foto mit dem abstrakten Titel *POR 090* ist einer Serie von Porträts entnommen, die mit einer Großbildkamera aufgenommen und als sogenannter Cibachrome-Print (mit überaus stabilen Farben, die bereits in der Filmschicht vorliegen und nicht erst durch die Entwicklung entstehen) auf knapp über zwei Meter Höhe vergrößert wurden. Es zeigt ein Gesicht in überdimensionaler Form und in einer

Nahsicht, die mit größtmöglicher geometrischer Exaktheit von vorn aufgenommen wurde. In der Struktur ähnelt es nicht zufällig dokumentarischen Aufnahmen, wie sie aus kriminalistischen Fahndungskarteien seit dem 19. Jahrhundert bekannt sind. Diese Ähnlichkeit wird dadurch noch verstärkt, dass die Reproduktion im vorliegenden Abdruck wiederum im verkleinerten Maßstab erfolgt, die nicht dem Ausstellungskontext entspricht: Wird die Aufnahme Ruffs noch dazu, wie hier geschehen, mit Fotografien kontrastiert, die als Muster für ein sogenanntes biometrisches Passbild dienen sollen, lassen sich leicht Parallelen herstellen.

Ende 2005 wurden für die Bundesrepublik Deutschland sogenannte ePässe vorgeschrieben, in denen identifizierende Gesichtsinformationen über den Inhaber des Ausweises in doppelter Form enthalten sind: als biometrische, maschinenlesbare Datei auf einem eingearbeiteten Chip sowie als frontales Porträt, das einem strengen Schema der Internationalen Luftfahrtbehörde genügen soll. Hierfür gibt es einen Musterkatalog mit Vorschriften zur Kopfposition und Kleidung der Porträtierten oder zur Farbqualität von Aufnahmen. Die Identifikation einer Person anhand von Gesichtsproportionen kann nur festgestellt werden unter der Bedingung, dass das Gesicht unverhüllt zu sehen ist und bestimmten Normen der Regelmäßigkeit und Symmetrie nicht zuwiderläuft. Grobe Abweichungen von der Norm und damit einher gehende Probleme beim Grenzübertritt werden mit einer schriftlichen Ausschlussklausel, der „Lichtbildbelehrung", belegt, die dem Passinhaber die Verantwortung dafür überlässt (vgl. Meyer 2006, S. 65).

Das Passfoto, welches die Überwachung und Identifikation einer Person z. B. beim Grenzübertritt gestattet, ist zugleich ein kriminalisierendes Foto, wenn es der vorsorglichen oder rastermäßigen Datensammlung oder der anthropologischen Forschung auf Grundlage von Reihenaufnahmen dient, wie es besonders die polizeilichen Täterkarteien des 19. und frühen 20. Jahrhunderts demonstriert haben (→ KAPITEL 11), oder wenn es als Steckbrief genutzt wird, wie es schon seit dem Mittelalter geschieht (vgl. Groebner 2004). Der unmittelbare Vergleich von Kunst- und Passbild bestätigt auch hier einige grundsätzliche kompositorische Ähnlichkeiten, selbst wo Unterschiede im Hinblick auf das Modell oder auf Details bestehen: Form und Technik sind verwandt, beide Produkte zeigen Porträts in Frontalsicht. Die Passbilder verzichten lediglich auf die Wiedergabe der Schulterpartie.

90

6.2 Kontexte des Bildes

Gleichzeitig sind wichtige funktionale Unterschiede in den sichtbaren Strukturen nicht enthalten: Einerseits der Ausstellungskontext, der Künstlername, das Fachpublikum, andererseits die polizeilichen Aufgaben oder die aus dem Bild ausgelesenen elektronischen Daten zur Grenzüberwachung oder Strafverfolgung. Die einzelne Aufnahme bei Thomas Ruff genießt im Galerieraum und darüber hinaus eine andere Wertschätzung und Wahrnehmung als eine Ausweisfotografie, die in den meisten Gesellschaften zur Grundausstattung des bürgerlichen Lebens gehört. Wenn solche äußeren Bedingungen notwendig sind, um die Bedeutung eines Bildes zu verstehen, die diesem in seiner ausschnitthaft reproduzierten Form nicht anzusehen sind, so könnte daraus folgen, dass die Analyse und Interpretation einer Bildform der Betrachtung solcher Bedingungen nachgeordnet ist (→ KAPITEL 4). Gewöhnlich wird dann vom Kontext gesprochen, dem ein Bild entstamme oder in dem es gesehen werden müsse, darin vergleichbar dem wörtlichen Zitat, das nicht aus dem Zusammenhang gerissen werden dürfe.

Andererseits hängt das, was als Kontext eines Bildes gelten kann, zunächst davon ab, um welches Bild es überhaupt geht. Auch führen spätere Veränderungen eines Kontextes zur veränderten Wahrnehmung dieses Bildes, sie erzeugen selber ein anderes oder gänzlich neues Bild. In einer historisch erweiterten Betrachtungsperspektive ist der Kontext eines einzelnen Bildes diesem durchaus sichtbar eingeschrieben.

Auch der Begriff des Kontextes selbst bedarf weiterer Präzisierungen. Zu ihm gehören sowohl individuelle oder gesellschaftlich eingeübte Verwendungsformen als auch technische Konventionen oder die Gepflogenheiten, Werturteile oder Routinen einzelner Berufsgruppen oder Disziplinen: Die durchschnittliche Passbildaufnahme wird in der Regel in einem Fotostudio auf der Grundlage praktischer Erfahrungen hergestellt, in denen sich die ästhetischen Ansprüche von Berufsfotografen und ihren Kunden an ein gelungenes Bild mit den polizeilichen Vorgaben arrangieren müssen. Die Arbeit des Künstlerfotografen kann dagegen einer persönlichen Regel folgen, die sich erst in jenem größeren Werkzusammenhang abzeichnet, nach der die künstlerische Praxis üblicherweise strebt. Im Falle des Ausstellungskünstlers Ruff, der seine Aufnahmen nicht in beliebiger Form zirkuliert, sondern sie auswählt, vergrößert und im Galerieraum inszeniert, kann auch der sogenannte Kunstbetrieb als ein solcher Kontext

Unsichtbare Kontexte

Bild und Kontext gehören zusammen

Werkzusammenhang

91

bezeichnet werden. Ausgeschlossen ist dadurch nicht, dass Ruffs Arbeiten erkennungsdienstliche Bildformate übernehmen; die plakatgroße Ausstellung im Kunstraum kann den Normierungscharakter von Passbildern sogar noch betonen. Dadurch, dass seine Arbeiten in formaler und technischer Hinsicht vergleichbar, aber in systematischer Hinsicht verschieden sind, beziehen diese möglicherweise noch zusätzliche semantische Energie.

Ob Ruffs Arbeit als Teil einer Serie durch andere Porträts variiert werden sollte, ist ihr nicht anzusehen. Doch auch als Einzelstück kann sie Traditionen und Techniken wiedergeben. Als individuelles Porträt ist sie Ausdruck einer uralten Bildgattung, und angesichts der Perfektion moderner digitaler Überarbeitungs- und Rendering-Techniken stellt sie auch die Frage nach der Identität einer Person, solange nicht bestimmt werden kann, ob die Arbeit überhaupt eine reale Person wiedergibt oder nicht vielmehr eine virtuelle Person, deren Äußeres am Computer erzeugt oder aus anderen Gesichtern generiert wurde. Erst im Kunstraum wird jedoch der überdimensionale Print als Einzelstück mit einer besonderen Sichtbarkeit versehen und diesen Fragen die Aufmerksamkeit eines ausgewählten Publikums zuteil.

Der Kontext bleibt unbestimmbar Die Frage nach dem Kontext oder Ort stellt sich schon deshalb, weil Bilder durch ihre Betrachter wahrgenommen, gedeutet und erklärt werden. Dies gilt sowohl im Hinblick auf die intellektuelle Botschaft eines Kunstwerkes als auch im Hinblick auf den Informationswert einer medizinischen Aufnahme. Im Unterschied zu konkreten örtlichen und zeitlichen Situationen und Rahmenbedingungen (etwa die Auftraggeberschaft oder der Aufstellungsort eines Denkmals, → KAPITEL 4.1), ist der weitere gesellschaftliche oder politische Kontext eine diffuse Größe – er beschreibt eine Fokussierung des Blicks, die die Deutung unweigerlich mitbestimmt, aber selber unsichtbar bleibt. Er verhält sich damit in gewisser Weise selber wie das Bild, das dem Betrachter entweder einen Gegenstand zeigt oder für ihn ein Gegenstand ist, aber nicht beides im selben Moment. Von Boris Groys wurde zu bedenken gegeben, dass Kontexte der Kunst immer nur eine abstrakte Größe bleiben, weil jeder Kontext eines Werkes seinerseits in einem größeren Kontext gesehen werden kann (vgl. Groys 1994). Damit ist wiederum das historiografische Problem angesprochen, dass eine Betrachtung von Gegenständen in ihrem historischen Kontext das Problem nicht umgehen kann, eine Geschichte zu schreiben, die stets von einzelnen Aspekten ausgehen und andere ausblenden muss.

Kontextualisierung als Idealziel Die Forderung nach Kontextualisierung, also nach Einbeziehung eines Kontextes oder Ambientes, dem ein konkretes Bild entstammt, bleibt damit vor allem eine programmatische und ideale Forderung nach Erweiterung der Blickperspektive, nach einer pluralistischen Geschichte (vgl. Kemp 1991), als Aufnahme von weiteren technischen, wissenschaftlichen oder kulturellen Informationen, welche die eigene Sicht ergänzen und relativieren – auch wenn das ‚ganze Bild' immer undarstellbar bleiben muss.

6.3 Bildkunst als System

Aus einer materialistischen Sicht auf die Produktionsverhältnisse erweist sich die Kunstproduktion als Betrieb, welcher der Herstellung einer Ware dient, wenn auch einer besonderen. In der soziologischen Totale betrachtet, fungiert Kunst als Symbol eines finanziellen oder intellektuellen Aufwands, der bestimmte Erwartungen und Ansprüche transportiert und zum Spiegel gesellschaftlicher Verhältnisse wird (vgl. Hauser 1951). Eigenverantwortliches künstlerisch-industrielles Unternehmertum, wie es bei Rembrandt, Antoine Watteau oder William Hogarth zu beobachten ist, reagiert auf diese gesellschaftliche Verfassung. Seit dem Verlust an Aufgaben, den die Kunst um 1800 im Bereich kirchlicher und höfischer Ausstattungsaufgaben erfahren hat, und seit dem Aufstieg von Salons und öffentlichen Museen als Ort der Präsentation von Kunst sind die bildenden Künste in zunehmendem Maße auf eine anonyme Gesellschaft und ihren Markt bezogen (vgl. Crow 2000; → KAPITEL 2.2). Die Selbstdarstellung des modernen Künstlers (vgl. Krieger 2007) muss ihrerseits helfen, die fortlaufenden Veränderungen der Kunst zu überbrücken.

Kunst als Spiegel von Verhältnissen

In einer solchen Entwicklungslinie wurde die Bildkunst des 20. Jahrhunderts auch als *art world* (vgl. Becker 1982) begriffen, als ein in sich geschlossenes System mit exklusiven Codes und Wertmaßstäben, das einem elitären Zirkel von Teilhabern – dem *white cube* des Galerie- und Ausstellungswesens – vorbehalten ist (vgl. Lingner 1994; Danto 1996). Diese Entwicklung der Kunst steht jedoch nicht im Widerspruch zu ihrem modernen Anspruch auf Kritik oder gesellschaftliche Veränderung. Vielmehr wird sie selber zum Motor der intellektuellen Reflexion, da seither auch die Selbst- oder Innenbetrachtung des Kunstsystems ein Thema der eigenen Arbeit abgibt. Kunst ist dahingehend gesellschaftliche Avantgarde, als sie ihre eigenen Funktionen als Bildgeber radikal infrage stellt, Sehkonventionen durchbricht. Sie verdichtet sich dadurch zu einem System, dass sie alltägliche Gegenstände durch unerwartete und provokative Ausstel-

Kunst als Ausstellung

lung in einem Museum, durch verfremdende Montage, durch irritierende Beschriftungen ihrerseits zum Zerrspiegel einer zuschauenden Gesellschaft macht.

System Der Begriff „System" bezeichnet die Ordnung und sinnvolle Gliederung der Teile in Bezug auf ein Ganzes und darüber hinaus auch das (gedachte oder bildliche) Schema der Wissensorganisation. Eng verbunden ist der Begriff mit der Klassifikationsarbeit Carl von Linnés von 1735 (→ KAPITEL 4.2). Durch den Kosmologen und Mathematiker Johann Heinrich Lambert wurde er um 1764 zu einer allgemeinen Theorie der Wissenschaften ausgebaut. In industriellen Bereichen meint er heute die Abstimmung von Produkten aufeinander, z. B. als Systemmöbel, Systemgastronomie. Unter physikalischen oder kybernetischen Gesichtspunkten bezeichnet das System einen Raum mit den sich darin selbst regulierenden Kräften, in der Biologie die Selbstorganisation von Lebensformen.

Bilder der Gesellschaft In der Folge ließ sich der Systembegriff auch auf Zivilisierungsprozesse, auf politische und wirtschaftliche Ordnungen übertragen. Innerhalb der arbeitsteiligen Gesellschaft kommt es zu Schichtungen, Spezialisierungen und Gruppenzugehörigkeiten, die sich auch zu eigenen Sphären (wie Politik, Kunst, Wirtschaft) mit spezifischen Praktiken, Codes, Werten und Betrachtungsstandpunkten ausweiten, die aber – im Unterschied zu einer ständisch hierarchisierten Gesellschaft – auch die Relativität und damit Beweglichkeit der einzelnen Position einschließt: Der Geschäftsführer ist selber auch Supermarktkunde. Die Vorstellungen der Gesellschaft von sich selber sind nach Niklas Luhmann daher das Ergebnis einer Kommunikation, die darüber bestimmt, welche Position einer Person oder Sache in einem bestimmten Raum zukommt und die nur zulässt, was der Systemerhaltung dient (vgl. Luhmann 2002). Übertragen auf das symbolreiche Feld der Kunstproduktion, wäre es eine systematische Frage, ob ein Nacktporträt als Pornografie oder Kunst zu werten ist oder ob der Ankauf eines Bildes ein kulturelles Ereignis oder einen Geschäftsvorgang darstellt.

Alles ist Kunst Während besonders der ideologie- und kapitalismuskritische Diskurs der Nachkriegszeit noch geleitet war von der Idee, dass sich Kunst durch ihre Komplexität und Autonomie gegen die Vereinnahmung durch Kommerz oder Politik behaupten müsse, wurden derartige Oppositionen durch Kunstformen der 1960er-Jahre wie die Pop Art oder Aktionskunst, die auf Massenmedien, auf soziale Interaktion oder Intervention setzten, wieder unterlaufen. *Fluxus*, Konzept- und Projektkunst, die im öffentlichen oder digitalen Raum agieren und in der gesellschaftlichen Kommunikation aufgehen, spielen mit wechselnden Orten und Wirklichkeitsebenen. Aufzeichnungen und Spuren aller Art, bis hin zur Person des Künstlers selbst, können zum Werk oder Ausstellungsstück werden.

Unterdessen wird die Gesellschaft, welche sich nach offizieller Diktion für die Förderung von Kunst und Kultur interessieren soll, je nach Kontext abwechselnd als Kritiker, Steuerzahler, als Künstler, als Museumsbesucher oder als Sammler angesprochen. Bisherige Unterscheidungen von Akteuren und Zuschauern, von Ursachen und Wirkungen weichen auf. Die berufliche Kategorie des Künstlers ist mit seiner Rolle als Bildgeber nicht mehr deckungsgleich, weil Kunst nicht mehr nur eine Menge von „Bildern" produziert, sondern als Ideengeber prinzipiell auch unbildlich möglich ist. Globalere Wertmaßstäbe der gesellschaftlichen Ökonomie wie Aufmerksamkeit und Nachhaltigkeit erfahren neue Bedeutung auch im Kunstsektor, der verstärkt ökonomisch begriffen wird (vgl. Franck 1998).

Bildlose Kunst

6.4 Übergänge: Kunst und Wissenschaft

In der Folge wurde in den letzten Jahren die Überzeugung vernehmlicher, dass die jüngeren Ausdrucksformen der bildenden Kunst eine intensivierte Begegnung mit (oder sogar Parallelität zu) anderen Bereichen der modernen Gesellschaft erkennen lassen, insbesondere mit den Naturwissenschaften, zumal Kunst wie auch Wissenschaft experimentell arbeiten, Repräsentationsprobleme zu bewältigen haben und Strategien und Verfahren jenseits der herkömmlichen Sichtbarkeit entwickeln.

Kunst begegnet Wissenschaft

Kunst und Wissenschaft haben demnach nicht nur traditionell gemeinsame Felder (wie z. B. die Anatomie), sondern arbeiten an einem gemeinsamen Erkenntnisziel, wenngleich mit unterschiedlichen Mitteln. Hier werden einerseits Traditionslinien zur älteren Geschichte der Bildkünste sichtbar, wie das Beispiel des Universaltalents Leonardo da Vinci prominent verdeutlicht (vgl. Kemp 2005). Untersucht wird andererseits auch, wie sich Wissenschaft ästhetisch vermittelt (vgl. Wechsler 1988). Es ergeben sich Interdependenzen aus der Arbeit an und mit Bildern, die ihre besonderen Experten und Expertisen haben (vgl. Baigrie 1996). Als Ausdruck eines sozialen Prozesses ist auch Wissenschaft von historischen Wandlungen und Krisen betroffen, kann weder theoretisch noch praktisch so streng verfahren, wie sie hofft oder vorgibt (vgl. Feyerabend 1980), und ist darin dem

Gemeinsame Erkenntnisziele

Ungewissheitszustand der Künste durchaus vergleichbar. Während angewandte Bereiche wie Werbung, Politik oder Technik ihre eigenen Strategien der Visualisierung entwickeln, definiert sich Kunst ihrerseits als Grundlagenfrage und Denkraum über künstlerische Medien und ihre gesellschaftliche Funktion.

Dies wurde seit den späten 1980er-Jahren z. B. im Rahmen maßstabsetzender Ausstellungen und Festivals wie *Ars Electronica* in Linz und *Interface* in Hamburg, *Sieben Hügel* in Berlin oder durch die Gründung von Einrichtungen wie dem *Zentrum für Kunst und Medientechnologie* in Karlsruhe thematisiert. Der jüngste, weltweite Boom der *Science Center* und Erlebniswelten sowie die Renaissance der Weltausstellungen, deren Rhetorik an diese intellektuelle Bewegung anschließt, zeigen ebenfalls, welche Bedeutung dem Visuellen und Sinnlichen in einer industrialisierten *technoscience* wie in der theoretisch hochgerüsteten Gegenwartskunst nach wie vor für die Vermittlung zukommt. Die bisherigen Grenzen von Disziplinen oder von Laborwelt und Gesellschaft scheinen in dieser neuen Bilderwelt zu zerfließen.

Dinge als Grenzüberschreiter

Die Grenzüberschreitung wird noch verstärkt durch das aktuelle wissenschaftshistorische Interesse an Bildern und materiellen Objekten, etwa an Laborzeichnungen oder naturkundlichen Präparaten. In diesen äußern sich nicht nur Formen und Fortschritte des Wissens, sondern sie markieren eine Neubewertung des Konkreten, das seine eigene Geschichte, Haptik und Widerspenstigkeit mitbringt und das auch Labor- und Fachgrenzen überschreiten kann, von unterschiedlichen Spezialisten verschieden angesehen wird und Wissen auch in einem nichtschriftlichen Sinne und über methodische Umbrüche hinweg speichert. Aus der ethnologisch geprägten Richtung der jüngeren Wissenschaftsforschung betrachtet (vgl. Latour/Woolgar 1979; Lynch 1985 u. a.), nähern sich der kulturwissenschaftliche und der technische Begriff des „Artefakts" (als menschengemachtes Werk einerseits, als störende Selbsteinschreibung andererseits) wieder einander an. Hieraus ergeben sich auch für die Kunst neuerliche Anregungen, die sich z. B in wissenschaftsbezogenen Themen oder in der Arbeit mit naturwissenschaftlichen Geräten und Materialien äußert.

Unveränderte Grenzen

Bilder, als Mitteilung wie als Objekt, sind jedoch nach Ansicht der Wissenschaftssoziologie auch für das Markieren von Grenzen verantwortlich (vgl. Knorr-Cetina 1999). Wenn in gegenwärtigen Debatten das intensive Wechselverhältnis von Kunst (als *techne*) und Wissenschaft (als Experiment) hervorgehoben wird, so bleibt ein systematischer oder perspektivischer Unterschied weiterhin eingeschrieben, da es einen Unterschied macht, ob zwei Felder einen gemeinsamen Kern besitzen oder ob sich zwei begrifflich und beruflich streng geschiedene Felder in praktischer Hinsicht annähern. Ein Mikroskop hat im Kunstraum einen anderen Status als in einer Technikausstellung. Auch stellen sich nicht, wie es Installationen der sogenannten Medienkunst nahelegen, mit jedem neuen Medium automatisch künstlerische Revolutionen ein (vgl. Reck 2002).

Als radikale Auseinandersetzung mit den Symbolen und Ideen der Gesellschaft und mithin als soziale Praxis wäre Kunst schon deshalb nicht von anderen Feldern geschieden, wenn sie, nach eigenem Verständnis, aktiv an politischen und gesellschaftlichen Entwicklungen mitwirkt. Sie würde dementsprechend auch nicht das ethische Gegenstück der Technomoderne repräsentieren. Künstler, die sich neuen Formen ästhetischer Produktion im Umfeld von Labor-Objekten verschreiben, etwa durch die Einfärbung eines Hasen mit dem *Green-Fluorescent-Protein* (→ ABBILDUNG 17), den der Künstler Eduardo Kac mit seinem *GFP Bunny* aus dem Jahre 2000 spektakulär vorgeführt hat (vgl. Reichle 2005), müssen sich mit dieser *transgenic art* oder Cyberforschung in anderer Weise gegenüber den Ausprägungen industrialisierter Wissenschaft positionieren, wenn sie von dieser noch unterschieden werden wollen.

Die moderne ästhetische Praxis steht vor dem Paradox, Ausdrucksformen jenseits eingeübter Motive zu suchen und trotzdem als Kunst, als Intervention oder Kritik wahrgenommen zu werden. Auch eine Kunst, die im Austausch mit Wissenschaften entsteht oder gar als Wissenschaft gedacht wird, muss sich durch soziale Auszeichnungen wie Signaturen, Ausstellungsorte und ausgewählte Betrachtergruppen weiterhin von anderen Systemen unterscheiden; sie umgeht überdies nicht den Konflikt von Autonomie und ökonomischer Abhängigkeit, der seit Jahrhunderten konstitutiv ist für ihr Selbstverständnis. Kunst beschreibt somit eine eigene Gesellschaftsform, die die Frage zu klären

Kunst als soziale Praxis

Soziale Auszeichnungen

Abbildung 17: Eduardo Kac: *GFP-Bunny* (2000)

97

hat, wem sie nutzt und wer sie fördert (vgl. Claus 1970; Grasskamp 1981). Zugleich schildert sie eine immer weiter getriebene Reflexion darüber, wie sich aus dem Produktions- und Konsumptionsdrang eines historisch gewachsenen Systems heraus Gestaltungswünsche und Kreativität, Kunstmarktinteressen und Sammlerleidenschaften in allen Varianten begegnen.

Die Kunst der Forschung

Andererseits erfordert der künstlerische Bedarf nach Reflexion und selbstverantworteter Gestaltung jenen vorwettbewerblichen Freiraum, den auch das Forschungslabor in Anspruch nimmt, sodass sich bildende Kunst zunehmend als Forschung – nicht notgedrungen als Wissenschaft – im eigentlichen Sinne des Wortes versteht. Von kategorischen Unterscheidungen unbenommen, gibt es weiterhin fließende Übergänge zwischen den Bereichen, bedingt durch die zahllosen Akteure, Positionen und Rezeptionsweisen, so wie auch in der Praxis der Übergang von geschlossenen Kreisen namens „Kunst" oder „Wissenschaft" in die Öffentlichkeit ein kontinuierlicher bleibt.

Fragen und Anregungen

- Inwiefern unterscheidet sich Thomas Ruffs Arbeit von einer biometrischen Fotografie?

- In welchen Bereichen waren Bildmedien schon in Mittelalter und früheren Epochen in technischen und naturwissenschaftlichen Kontexten in Gebrauch?

- Hat sich die Rolle bildender Künstler als Gestalter grundlegend verändert, und wenn ja, wann oder wodurch?

- Naturwissenschaftliche Darstellungen, etwa mikroskopische oder teleskopische Aufnahmen, werden oft auch als „ästhetisch" bezeichnet, in Ausstellungen und / oder kostspieligen Katalogen vorgeführt: Ist diese Bewertung Ihrer Ansicht nach berechtigt?

Lektüreempfehlungen

- **Martin Hellmold u. a. (Hg.): Was ist ein Künstler? Das Subjekt der modernen Kunst,** München 2003. *Die Beiträge des Sammelbandes gehen der Frage nach, inwiefern traditionelle Vorstellungen kreati-*

ver Subjektivität an den Problemen moderner Kunstproduktion vorbeigehen.

- Caroline Jones / Peter Galison (Hg.): **Picturing Science. Producing Art,** New York / London 1998. *Der Reader verbindet durchaus heterogene, soziologische, wissenschafts- und kunsthistorische Beiträge, welche die möglichen Gemeinsamkeiten der Herangehensweise (z. B. bei Ginzburg, Freedberg oder Latour) sondieren.*

- Wolfgang Ruppert: **Der moderne Künstler. Zur Sozial- und Kulturgeschichte der kreativen Individualität in der kulturellen Moderne im 19. und frühen 20. Jahrhundert,** Frankfurt a. M. 1998. *Diese Studie zur Entwicklung des Künstlerberufes macht zugleich mit wichtigen Strömungen der jüngeren Kunstgeschichte und den grundlegenden Fragen der kultursoziologischen Forschung bekannt.*

- Anja Zimmermann (Hg.): **Sichtbarkeit und Medium – Austausch, Verknüpfung und Differenz naturwissenschaftlicher und ästhetischer Bildstrategien,** Hamburg 2005. *Aus der Fülle rezenter Publikationen zum Verhältnis von Kunst und Wissenschaft sticht dieser Band durch seine Bereitschaft hervor, auch die beträchtlichen systematischen Unterschiede wissenschaftlicher und künstlerischer Praktiken zu erörtern.*

7 Repräsentationsfunktionen, Kommuni-kationsräume

Abbildung 18: Andrea Mantegna: *Camera degli Sposi* in Mantua, Palazzo Ducale, Westwand (1465–75)

Abbildung 19: Antonio Lafreri: *Rom, Benediktion auf dem Petersplatz,* Kupferstich aus dem Buch *Speculum Romanae Magnificentiae* (ca. 1574)

Der Mantuaner Hofkünstler Andrea Mantegna (1431–1506) schuf mit seinem Ehesaal, der „Camera degli Sposi" (auch „Camera picta" genannt), sein wohl bekanntestes und prächtigstes Werk. In dieser Arbeit, die den gesamten architektonischen Raum einbezieht, werden Szenen aus der Familiengeschichte der Mantuaner Herrscherdynastie der Gonzaga vorgeführt, die einerseits der Selbstdarstellung der Auftraggeber dienen sollten, andererseits nur für ein höfisches Publikum gedacht und zugänglich waren.

Der historische Kupferstich zeigt den überfüllten Petersplatz in Rom, bei dem die Massen den Segen des Papstes empfangen. Der römische Bildhauer und Architekt Gianlorenzo Bernini wurde zwei Jahrhunderte später und nach dem Neubau Sankt Peters damit beauftragt, zur Aufnahme der wachsenden Besucherströme einen Platz zu schaffen, der sowohl der Öffnung des Kirchenraums nach außen dienen als auch die ankommenden Besucher umfangen sollte. Im Zusammenhang mit späteren urbanistischen Eingriffen zur Schaffung von Schneisen wurde das gesamte Stadtbild in diese Gestaltung einbezogen, um auf den Kirchenbau zu fokussieren und ihn zu einem bildhaften Gegenstück der Stadt werden zu lassen.

Die europäische Geschichte der politischen und kirchlichen Selbstdarstellung ist gekennzeichnet durch eine stetige Ausdehnung von Räumen und Zielgruppen auf das Bauwesen und die Künste, auf die Presse, ja selbst auf ganze Städte und Landschaften. Künstler und Gestalter unterschiedlichster Spezialisierung werden einbezogen, um bildhafte Mitteilungen zu schaffen, in denen sich Auftraggeber und Betrachter, Regierende und Regierte begegnen, und um neue politische Ansprüche in eine verständliche Bildsprache zu übersetzen. Mit der quantitativen wie qualitativen Ausdehnung visueller Kommunikation auf die unterschiedlichsten Kunstformen und gesellschaftlichen Bereiche werden zugleich wachsende Betrachterkreise eingebunden, entwickeln sich neue Orte und Medien der Kommunikation, die in der Folge auch die bildrhetorischen Mittel der Ansprache, das Bild der Öffentlichkeit und den Begriff des Politischen und der Repräsentation selbst allmählich verändern.

7.1 Von der Dekoration zum Bildprogramm

7.2 Öffentlichkeit und Rhetorik

7.3 Hoheitszeichen

7.4 Repräsentation

7.1 Von der Dekoration zum Bildprogramm

Der sogenannte Ehesaal (*Camera degli Sposi* oder auch *Camera picta* genannt) des Künstlers Andrea Mantegna war ursprünglich nicht wie heute allgemein zugänglich, sondern allein der Familie der Auftraggeber und ihren offiziellen und diplomatischen Gästen, also dem Hof, vorbehalten. Eine größere Öffentlichkeit hatte keinen Zutritt.

In der ersten Abbildung ist die Wandseite zu sehen, welche die Begegnung des Fürsten Ludovico Gonzaga mit seinen Söhnen Federico und Francesco als Ganzfigurenbildnisse vor einer prächtigen Landschaftsszene zeigt. Die Wandfläche ist Teil eines größeren Bildprogramms, das einen würdevollen Hintergrund für feierliche Begegnungen bieten sollte, indem es Begebenheiten der Familienchronik zu einem historischen Bericht verknüpft und so den dynastischen Anspruch der Auftraggeber unterstreicht. Aus dem Zusammenspiel von Bildfeldern und Standpunkten ergeben sich für den Betrachter wechselnde Bezüge und Leserichtungen. Doch auch die einzelnen Felder stellen anspruchsvolle Kompositionen dar, die den Übergang vom dekorativ geschmückten Wandfeld zur Historienmalerei markieren und eine einzelne Herrschaftsgruppe in die Sphäre des Weltgeschichtlichen rücken.

Die Wandgestaltung steht auch für die familienpolitische Einbindung der Hochkunst und den gestiegenen inszenatorischen Aufwand der Fürstenhöfe. Durch ihre Auftraggeber werden die Darstellungen zu Projektionsflächen für gesellschaftliche Ideen und Vorstellungen glücklicher Herrschaft oder kluger Vermählungsstrategien. Daneben werden auch die großen Wandbilder und Ausstattungen städtischer Gebäude (z. B. die großen Ratssäle italienischer Stadtstaaten oder nordeuropäischer Hansestädte) zur Schnittfläche von Erwartungen einzelner Schichten, Berufsstände und Zünfte, die sich jeweils in historischen und allegorischen Programmen zu verwirklichen suchen (vgl. Belting/Blume 1989).

Die schönen Künste waren den staatlichen Lenkern in praktischer Hinsicht nützlich. Sie gaben den Regierenden das Gesicht, welches sich auch als Münze verbreiten oder als Denkmal errichten ließ. Künstler richteten die großen Zeremonien und Turniere aus oder entwarfen Festarchitekturen. Im Laufe der Jahrhunderte gewannen einzelne Gattungen besonders an Wert, in denen sich praktischer Gebrauch und Repräsentationsbedarf begegneten, z. B. in der Tapisserie, in Rüstungen, Pokalen, Möbeln, und besonders in der Baukunst.

Vom Wandbild zur Historienmalerei

Bildprogramme

Bedeutung der Künstler als Gestalter

Funktionen der Kunst

Die aufwendigen Dekorationsprogramme von Höfen und Kirchen stehen damit zugleich für den Aufstieg ihrer Gestalter. Um die entsprechenden künstlerischen Leistungen vollbringen zu können, bedurfte es kluger und technisch versierter Regisseure. Die Künste waren erforderlich, um politische Siege, Krönungen oder Begegnungen zu verewigen, um Friedensbotschaften auszusenden, Hochzeiten anzubahnen, ihre Auftraggeber als klassische Tugendhelden (als Herkules, als Heiligen, als Philosophen) einzukleiden, Gegner einzuschüchtern oder die lokale Bevölkerung gewogen zu stimmen. Hieraus folgte wiederum im Verlauf des 16. bis 19. Jahrhunderts der steigende Bedarf an kundigen und geschickten Künstlern. Diese mussten auch da, wo Themen vorgegeben und vertraglich geregelt waren, über motivisches Wissen verfügen, den Formenschatz der Kunst beherrschen und diplomatisch geschickt sein. Zuweilen waren gelehrte Ratgeber erforderlich, um eine Idee von Herrschaft in raffinierte oder unverfängliche Allegorien zu übersetzen.

Emanzipation des Künstlerstandes

Die Gründung von Akademien und die Vergabe von Aufträgen durch aufstrebende Höfe führten dazu, dass das Monopol der Handwerkszünfte gebrochen wurde, welches den Künstlern vormals Beschränkungen auferlegt hatte. Die bei Hofe beschäftigten Künstler wurden teilweise mit dem Adel gleichgestellt oder als Gesandte, Militärexperten und Gutachter engagiert. Da diese Hofkünstler in einem gesellschaftlich sanktionierten Freiraum operierten, sind sie als Vorläufer der modernen Kunst gedeutet worden (vgl. Warnke 1996, 2005). Ihr Emanzipationsprozess ist jedoch nicht nur auf den höfischen Raum begrenzt geblieben. Auch die Städte und Kirchen waren als Auftraggeber von beträchtlicher Bedeutung für den Unterhalt des Künstlerstandes, zumal wenn sie – wie der Vatikan und seine Kardinäle – zugleich als weltliche und geistliche Macht auftraten. In jedem Fall war die Bedeutung der Bildkunst für die politisch-gesellschaftliche Kommunikation unübersehbar, während sie sich durch diese auch selbst stets neu definiert hat (→ KAPITEL 6.3).

Die Kulturgeschichte zeigt außerdem, dass politische Kommunikation mehr ist als eine Summe von Bildern mit einer werbenden Funktion und parteilich-bekenntnishaften Aussage: Der scheinbar eindeutige Begriff „politischer Kunst" beschreibt nicht so sehr eine bestimmte Form engagierter oder instrumentalisierter Kunst, sondern auch die Entstehung kommunaler oder höfischer Räume und machtorientierter Wirklichkeitsbereiche, in denen sich soziale Ordnungen und Erwartungen ausgedrückt haben. Bildkunst war für die Befestigung wie Überwindung von Gesellschaftsformen gleichermaßen wesentlich, die Errungenschaften der bildenden Kunst seit der Frühen Neuzeit sind als Produkte von Auseinandersetzungen um politische Orte, um Hoheiten und Zuständigkeiten zu lesen. Diese werden hier anhand einiger Etappen der europäischen Kunstgeschichte behandelt, auch wenn sie sich seit Menschengedenken ebenso in anderen Kulturen und Epochen vollzogen haben.

7.2 Öffentlichkeit und Rhetorik

Dies gilt auch für das Entstehen einer Öffentlichkeit und die Wandlung ihres Begriffs. Größere Bevölkerungsgruppen, etwa die Kirchengemeinde oder die Einwohnerschaft einer Stadt, waren stets Adressat kirchlicher, fürstlicher oder kommunaler Kommunikation. Durch das Bauwesen wurden jedoch auch neue Räume, etwa Marktplätze, Ansprache- und Versammlungsorte, geschaffen und damit neue Interaktionen möglich, welche die Grundlagen schufen für die moderne Selbstwahrnehmung der Gesellschaft als ‚Öffentlichkeit‘. Hierin drückt sich ein Strukturwandel der Gesellschaft aus, wie er im 19. Jahrhundert in vergleichbarer Weise durch das Aufkommen von Printmedien und deren Leserschaften markiert worden ist (vgl. Habermas 1962).

Die großen Bildprogramme der frühneuzeitlichen europäischen Kunst wie die Sixtinische Kapelle oder die *Camera picta*, deren malerischer Aufwand die vorangehenden Leistungen der Kunst in den Schatten gestellt hat, waren dagegen nicht öffentlich im heutigen Sinne, sondern an einen Hof gerichtet, der die ständische Gesellschaft als Ganzes vertrat. Andererseits wurden Kirchendekorationen, Grab- und Denkmäler, architektonische und städtebauliche Maßnahmen mehr und mehr in die Selbstdarstellung von Fürstengeschlechtern und Kommunen einbezogen, ebenso das Fest- und Theaterwesen. Neue Medien wie das Flugblatt führten zu einer grenzüberschreitenden Ausweitung visueller Regime.

Im Zeitalter des Absolutismus zeigte sich Herrschaft als privilegierter Blick von oben, als eine machtvolle, gestaltende Planung (→ KAPITEL 12.3), welche Parkanlagen oder Wasserwege in Blickachsen weiterführt und Stadt- und Landräume überformt hat (vgl. Jöchner 2003). Im 19. und 20. Jahrhundert führte dies wiederum, auch unter veränderten sozialpolitischen Verhältnissen, zum Abriss und Neuaufbau ganzer Stadtviertel und zur Anlage großer Boulevards und Sichtachsen. → ABBILDUNG 20 zeigt als Beispiel die zum Petersplatz führende

Entstehung der Öffentlichkeit

Privilegierte Blicke

Abbildung 20: Blick auf Sankt Peter und die Kolonnaden des Petersplatzes in Rom (1997)

Straßenachse, die im 20. Jahrhundert in den Stadtkörper geschlagen wurde, um den Platz zum Zielpunkt einer politischen Prozession werden zu lassen.

Visuelle Rhetorik   Durch ihren Bedeutungszuwachs in der Kommunikation der Frühen Neuzeit gewann die seit altersher als politisches Instrument geschätzte Kunst der Rhetorik auch für das Visuelle an Gewicht (vgl. Brassat 2005). Als Mittel der Erregung oder Besänftigung wurde die Bildkunst im Zuge der Konfessionalisierung im 16. und 17. Jahrhundert zur Belehrung, Erziehung oder Indoktrination breiterer Bevölkerungsschichten eingesetzt, in der Hoffnung, dass Bilder als Gegenstand intimer Andacht oder vor größerem Publikum emotionale, anleitende oder abwehrende Kräfte entfalten, die sich unmittelbar auf den Betrachter übertragen. Gesucht wurde auch die affektive therapeutische Wirkung, die aus der klassischen Dramentheorie bekannt war und die durch die Wahl geeigneter Motive, durch erzählerische oder dramatische Kompositionen oder durch schiere Quantität den Betrachter überwältigen oder erschüttern sollte. In diese Rhetorik waren nicht nur das gemalte oder gedruckte Bild, sondern auch die Gesten von Predigern oder die Wirkung von Zeremonien und Musik einbezogen. Unterdessen wurde die bildkritische Befürchtung aufgegeben, wonach das Bildnis der Heiligen als falsches Idol angebetet werden (→ KAPITEL 2.2) oder auch dem Armutsgebot der Orden widersprechen könnte, um der Wirkung von Bildern bei der psychophysischen Glaubensintensivierung Platz zu machen.

Insbesondere die kirchlichen Programme bezeichneten frühe Formen der wohldosierten ‚Propaganda' für die Bevölkerung. Dieser Begriff wurde in der Ordenspolitik des 17. Jahrhunderts geprägt und schloss einfache Mittel zur Ansprache breiter Bevölkerungsschichten (wie die illustrierten ‚Armenbibeln' und Katechismen), aber auch komplexe Programmbilder ein, die zu anspruchsvollen Ausdeutungen einluden. Die gewaltige Architektur der Kirchen und ihre Ausstattung sollte die Gesamtheit der Gesellschaft auf visuellem Wege umfangen.

,Propaganda'

Auch der Kirchenraum war in diesem Geflecht ein gesellschaftspolitischer Ort und Teil von innerstädtischen oder landesweiten Rivalitäten um Einfluss und Mehrheiten. Durch den Bau gewaltiger Dome und Klöster entstanden Versammlungsflächen, die für viele gesellschaftliche Gruppen lange Zeit die einzigen öffentlich zugänglichen Plätze blieben. Wände und Decken von Kirchengebäuden wurden zu Demonstrationen kirchlicher Pracht oder von speziellen programmatischen Aussagen.

Kirchen als öffentliche Stätten

Zugleich wurde der Raum selber, etwa als bauliches Ensemble (→ KAPITEL 12), wieder zum Bild gemacht und so auch in der Bildkunst als Vedute (Stadtansicht) wiedergegeben. Die Bernini'sche Platzanlage aus Kolonnadenumgang, Heiligenfiguren und Obelisk gibt den Rahmen ab für die Schauseite des Doms, dessen Fassade (vgl. italienisch *facciata,* englisch *face*) ebenfalls Bildcharakter hat. Als Versammlungsort heute von Millionen Menschen aufgesucht und durch Fernsehübertragung einer der bekanntesten Stätten weltweit, beruht sein Erfolg auch auf der schrittweisen Einbettung in ein städtisches Ganzes, das im frühen 20. Jahrhundert durch brutale Umbaumaßnahmen herbeigeführt wurde. Er ist damit das Ergebnis einer Geschichte der Ansprache durch visuelle Mittel, die stets darauf abzielte, langlebige und historisch abgesicherte Bilder der Größe und Macht zu erzeugen.

Vom Raum zum Bild

7.3 Hoheitszeichen

Kirchenräume wurden seit dem ausgehenden Mittelalter auch als Ort der Selbstdarstellung reicher Stifter genutzt und durch Umbaumaßnahmen aufgewertet, mit Altarbildern und Grabmälern ausgestattet und durch Wappen in Anspruch genommen, die sich von den Bild-zeichen der Gemeinde, etwa Marienbildnissen und Votivtafeln, deutlich abhoben.

Aber auch andere öffentliche Räume wurden zum Austragungsort von Reklamierungen und Konkurrenzen aller Art: Wappen und Insignien zierten Gebäude, Denkmäler sollten die Präsenz eines Staatswesens nach innen und außen anzeigen, Grenzsteine demarkierten Straßenverläufe und Landschaften. Stadtplätze strotzen noch heute von den Abzeichen vergangener Herrscherdynastien, die um den öffentlichen Raum gestritten haben, welcher zum Territorium eines frühmodernen *branding* oder *sponsoring* gemacht wurde.

In der römischen Baukunst und Stadtmöblierung haben sich darin mehrere päpstliche Familien hervorgetan, besonders die Familie der Chigi. Papst Alexander VII. Chigi (1599–1667) dürfte zu jenen Nachfolgern Petri gehören, welche die meisten Erkennungsmarken auf Gebäuden der Ewigen Stadt hinterlassen haben, um für ihr dynastisches Nachleben zu sorgen, so auch auf dem Tor an der Grenze der nördlichen Altstadt bei der Kirche S. Maria del Popolo, nach der später die alte Porta Flaminia in Porta del Popolo umbenannt wurde (→ ABBILDUNG 21).

Wappen und Abzeichen

Papst Alexander VII. als Stadtfürst

Abbildung 21: Herrscherwappen der Chigi-Monti auf der Porta del Popolo in Rom (2003)

Das Tor ist wie eine Tafel in den Verlauf der aurelianischen Mauer integriert und imitiert dabei einen antiken römischen Triumphbogen mit drei Durchgängen und einem oberen Abschluss, der nach Art von Schrifttafeln an den Auftraggeber der ersten Neugestaltung, Papst Pius IV. Medici (1499–1565), erinnert. Darüber erhebt sich ein krönender Mauerteil mit den *monti* (italienisch für: Hügel) und dem Stern, dem Wappen der Chigi. Alexander hatte das Tor durch Bernini erneuern lassen, um den festlichen Rahmen für einen triumphalen Einzug der schwedischen Königin Christina zu schaffen, die nach ihrem Übertritt zum Katholizismus 1655 von Norden her nach Rom gelangte. Derartige politische Rituale haben so jahrhundertelang Spuren hinterlassen.

Mit den heraldischen Zeichen wurde einem über tausend Jahre alten kommunalen Bau der Name einer bestimmten Dynastie umso nachdrücklicher eingeprägt, als sich hier jemand über den Namen eines Vorgängers setzte – beides Päpste, deren Amt durch Schlüssel und Tiara abgebildet wurde, und beides Persönlichkeiten, die sich an einem der prominentesten Orte der Stadt als Kunstförderer in deren Gedächtnis einschrieben. **Erinnerungsformen**

Die Bedeutung und das Alter der Anlage, der Ruhm ihres zweiten Architekten und die Bekanntheit des Ortes konnten sicherstellen, dass eine solche Strategie der Aneignung langfristig bemerkbar blieb. Nachhaltig wirksam war auch der Kunstgriff, das heraldische Zeichen selber zum Gestaltungsmittel zu machen, ohne welches das Gebäude gar nicht mehr auszukommen scheint. Als Anzeiger in den Innenraum des Platzes teilt das Wappen dem Zuschauer nun einen mäzenatischen wie einen konservatorischer Kunstanspruch mit. Hoheit über das Gebäude meint sowohl Schutz wie Ordnungsmacht. **Strategien der Aneignung**

Dadurch wird verdeckt, dass in dieser Beschriftung des Gebäudes ein Streit um kulturelle Signaturen ausgetragen wird, der – würde er sich heute wiederholen – einen Aufruhr größten Aufmaßes verursachen müsste. Ist es vorstellbar, dass ein Kreditinstitut oder Bekleidungshersteller sein Firmenlogo nicht nur temporär auf einem jahrhundertealten städtischen Gebäude anbringt, sondern bei dieser Gelegenheit auch das Gebäude auf Dauer nach Prinzipien der *corporate identity* umbaut? Tatsächlich ist das Wappen der Chigi Teil der Architekturform geworden und nicht nur ein verzichtbarer Zusatz. **Kunst als Markenzeichen**

Es wird in der Kulturgeschichte als Indiz der Modernisierung angenommen, dass einzelne Individuen für sich herausgenommen haben, eine Sache mit ihrem Namen zu kennzeichnen. In Abgrenzung zum vermeintlich namenlosen Mittelalter schält sich das Subjekt als

Künstler, als Bürger, als politisch-militärischer Anführer aus einer anonymen Gemeinschaft und meldet nach dem Vorbild der Antike einen Anspruch auf Überdauerung an. In der Tat hat sich im Laufe der Kunstgeschichte neben dem Auftraggeber und seinen heraldischen Devisen und Wappen auch das Selbstbildnis und vor allem **Künstlerische Signaturen** die Signatur des Gestalters etabliert, die sich dem Auftraggeber oder Dargestellten und seinem Namen an die Seite stellt, ja diesen überschreibt, bis hin zu Fällen, in denen ein Künstler bis heute bekannt ist, sein Modell aber nicht mehr identifiziert werden kann (wie z. B. bei vielen Gemälden van Dycks oder Rembrandts). Auf Medaillen oder Denkmälern stehen Signaturen in einem wechselseitigen Spannungsverhältnis, ebenso bei politischen Wahlplakaten des frühen 20. Jahrhunderts, in denen namhafte Künstler zugleich eine Partei unterstützten und für die eigene Gestaltung warben.

Anonyme Kunst Dem gegenüber sind die Autoren in weiten Teilen der historischen Bildproduktion jedoch auch namenlos geblieben (→ KAPITEL 8.3): Viele Kupferstecher und Münzschneider, Maler und Zeichner sind biografisch kaum greifbar und allein über ihre Monogramme bekannt. Obwohl also zu erwarten wäre, dass herausragende Herrscher sich den namhaftesten Künstlern, vornehmlich ihren Hofangestellten und Leibmalern, anvertraut hätten, sind ganze Herrscherikonografien überkommen, ohne dass deren Künstler bekannt wäre.

Im konkreten Fall der Porta del Popolo ließ ein Kirchenfürst seinen Namen nachträglich auf ein Werk schreiben, weil er den Umbau beauftragt hatte. Diese bei kostspieligen Bauvorhaben geläufige Praxis konkurriert nachdrücklich mit jener Praxis der Beschriftung, wie sie dann in der Malerei üblich geworden ist, bei der vom Künstler oder seiner Werkstatt die Unterschrift oder ein Firmenzeichen (wie im Falle der Malerfamilie Cranach) als Zertifikat geradezu verlangt wurden, um die Qualität oder Eigenhändigkeit einer Arbeit zu bestätigen.

Konflikt von Auftrag und Gestaltung Daraus ließe sich folgern, dass die Anbringung eines Wappens oder einer Künstlersignatur lediglich einen Akt der reviermäßigen Inbesitznahme darstellt. Doch beanspruchen Bildwerke und Gebäude, Platzanlagen und Kirchenräume etwas, das sie zugleich stiften. Eben weil sie umstritten sind, entfalten Kunstwerke in gleichem Maße gesellschaftliche Spreng- wie Tragkraft. Dieser Konflikt von Auftragskunst und Freiheit, von Käufer und Hersteller, Investor und Idee, der das System der Künste unverändert kennzeichnet (→ KAPITEL 4.1, 6), ist nicht nur virulent im Bereich der politischen Werbung, sondern ebenso im Sponsoring, im privaten Sammler- und Mäzenatentum und in jeder Initiative, bei der sich private und öffentliche Ansprüche begegnen. Bilder können werbend in dieses Austauschverhältnis eintreten und selber dafür stehen.

7.4 Repräsentation

Verschiedene historische Disziplinen wie die Rechts- oder Kulturgeschichte haben dieses Problemverhältnis analysiert, indem sie in den Bildformen des Öffentlichen den Zusammenhang von Bild und Brauchtum erkannten und die Funktion von Effigien (Stellvertreterbildnissen, vgl. Brückner 1966), von Denkmälern, Flugblättern, Zeremonien und anderen breitenwirksamen Bildformen als Ausdruck einer Gesellschaftsverfassung deuten. Jedes staatliche Denkmal, jedes repräsentative Gemälde ist nur deshalb existent, weil ihm offenbar eingeräumt wird, jemanden oder etwas zu vertreten. Da es Wahlen beeinflussen kann, wird es zum Bestandteil einer modernen Politikführung, deren Wirkung sich empirisch analysieren lässt (vgl. Hofmann 1999).

Dies wird auch durch den Begriff „Repräsentation" (als Aufführung, Darbietung, Vorstellung) ausgedrückt, einen so zentralen wie vieldeutigen Begriff zur Beschreibung von stellvertretenden Erscheinungsformen. In der philosophischen Ästhetik und Semiotik meint *repraesentatio* schon seit der Antike eine Wiedergabe mit bildlichen, gestischen oder verbalen Mitteln, die einem Gegenstand soweit nahekommt, dass damit eine Stellvertretung möglich wird. In diesem Sinne wird Repräsentation auch in der mittelalterlichen Theorie bei Thomas von Aquin als Signifikation verstanden (vgl. Zimmermann 1971; Hofmann 1974). Eine solche Definition kann jegliche Form von Darbietung einschließen, die eine Beziehung zu einer anderen Sache erkennen lässt. Bezogen auf das Rechtswesen meint Repräsentation dann auch die Vertretung von Bürgern oder Handwerkern in Kammern und Räten oder eine angemessene Vertretung vor Gericht. In der Kunsttheorie wurde die Idee der Stellvertretung (zum Beispiel von abwesenden Personen) oftmals verknüpft mit der Idee der Ähnlichkeit oder täuschenden bildlichen Nachahmung von Gegenständen. Mit der Frage, wie der menschliche Geist sich sichtbare Dinge oder unsichtbare Gedanken innerlich vorstellt, wurde (innere) Repräsentation auch zu einem Schlüsselbegriff der Psychologie (→ KAPITEL 4.2).

Besonders im Englischen oder Französischen meint das jeweilige Wort für Repräsentation auch die Aufführung (oder das Sich-Auffüh-

Bilder als Stellvertreter

Repräsentation

Darstellende Kunst

ren einer Person) im Sinne des Theaters, in welchem Akteure Scheinhandlungen ausführen. Diese Ambivalenz ist im Deutschen in den Übersetzungen „Darstellung" oder „Vorstellung" enthalten. Möglicherweise durch die Konjunktur des Theaters und der lebenden Bilder (*tableaux vivants*) seit dem Barock haben sich die Begriffe „Darstellung" und „Bild" zunehmend verschliffen. Die Historienmalerei des französischen *Ancien Régime*, z. B. bei Jean-Honoré Fragonard (1732–1806), konkurriert absichtsvoll mit der Leibhaftigkeit des Theatralischen und will dessen Darbietung auf die Leinwand bringen.

Im theoretischen Kontext hat sich dagegen gezeigt, dass Repräsentation ein eher unspezifischer Terminus ist, da alles auch alles repräsentieren kann (etwa der Körper den Gen-Code, das Polizeifoto einen Tatort usw.), dies aber nichts über die Qualität der Repräsentation aussagt. Der Begriff ist daher auch Gegenstand naturwissenschaftlicher Debatten zum Status von Abbildungen und Bildgebungsverfahren geworden. Repräsentation / Darstellung könnte hier suggerieren, dass Befunde oder Daten am Bildschirm in unveränderter Form „wiedergegeben" sind, oder im Gegenteil implizieren, dass diese lediglich scheinhaft „aufgeführt" werden (→ KAPITEL 14).

Technische Repräsentation

Im deutlichen Unterschied dazu dürfte die politische Repräsentation mehrheitlich als bewusste Selbstdarstellung von Herrschaft verstanden werden, welche im Auftrage von Regierungen, Unternehmen oder Individuen und durch die Instrumentalisierung der Künste steuernd in die Gesellschaft hineingetragen wird (→ KAPITEL 8.3). Repräsentation ist demnach nicht der äußere Anschein, dem ein Bild als bloßes Vermittlungswerkzeug dient, sondern eine Wirklichkeitsform. Der theatralisch-performative Aspekt ist in der Idee eines ‚Staatstheaters' enthalten, das als Zeremoniell, Protokoll oder Liturgie das politische Handeln reguliert (vgl. Meyer 1998). Repräsentation meint ferner den symbolischen Aufwand (bei Architektur, Kleidung, Festwesen), der auf einen gesellschaftlichen Anspruch hindeutet. Hierzu gehört auch der Einsatz der Künste im Sinne von Prachtentfaltung, der heute ebenfalls im Begriff des „Repräsentativen" – im Sinne von aufwendig, geräumig, für öffentliche Auftritte angemessen – und im Begriff „Statussymbol" enthalten ist.

Politische Repräsentation

Der negative Beiklang, der im Politisch-Repräsentativen vor allem das Maskenhafte sieht, darf nicht vergessen machen, dass Öffentlichkeit stets ein bestimmtes Auftreten nach sich zieht, das in eine Erwartungshaltung gegenüber dem Auftretenden umschlägt, der im Namen vieler agiert. Ritual und Schauspiel sind aus ethnologischer Sicht eine

Repräsentation als Vermittler

Alltagspraxis – politische Kommunikation ist daher nicht reduzierbar auf ein einfaches Verhältnis von Sendern (Herrschenden) und Empfängern (Beherrschten), zwischen denen Bilder ausgetauscht würden, um die Menschen zu informieren oder zu indoktrinieren, sondern bleibt auf gemeinsame „regulative Fiktionen" (Koschorke 2007, S. 77) angewiesen, die sich symbolisch konkretisieren.

Auf der Ebene der Hochkunst, besonders im Historiengemälde oder Staatsporträt der europäischen Kunst, haben sich daher Zeichen und Ideen zu feingliedrigen und rezeptionsbewussten Ikonografien verdichtet, um Anschauungen zu vermitteln und zu gestalten. Weil sie temporären Zwecken dienten und dienen, ist von Seiten der politischen Theorie vorgebracht worden, dass eine politische Kunst (z. B. eine tagesaktuelle Karikatur) geringe Haltbarkeit besitze, da sie ohne Zeitbezug nicht mehr verständlich sei (vgl. von Beyme 1998, S. 30). Aus dieser These müsste folgen, dass zahllose Kunstwerke von Michelangelo Buonarotti über Honoré Daumier bis John Heartfield unlesbar wären, sofern sie eine kurzlebige Botschaft transportierten, während ein veritables Kunsthandwerk dauerhaft gültig bleibe.

Kurzlebigkeit politischer Kunst?

Bilder besitzen jedoch stets Funktionen, die sie kulturell und historisch in einer Gesellschaft verankern. Als Objekte gesellschaftlichen Disputs sind sämtliche Bilder so politisch, wie sie als Tauschobjekt ökonomisch sind (→ KAPITEL 8). Aus der historischen Perspektive ist unübersehbar, dass sich in Grabmälern, Ahnentafeln und Kunstprogrammen weit mehr mitteilt als das Prunkbedürfnis eines Auftraggebers gegenüber den Zeitgenossen und darin Ideen von Herrschaft und Gesellschaft zum Ausdruck kommen, dass es also bei visuellen Strategien nicht allein um die Wahl bestimmter Zeichen geht, die einen Glaubenssatz übermitteln, sondern Politik insgesamt visuell wird. Auch das Bild ist als symbolische, visuelle Kommunikation um sämtliche Formen erweitert zu denken, in denen sich eine Gesellschaft entwirft.

Alle Bilder können politisch sein

Fragen und Anregungen

- Inwiefern könnten auch religiöse Malereien und Bildthemen als politisch bezeichnet werden?
- Warum ist der Repräsentation seit dem Mittelalter eine zunehmend politische Funktion zugekommen?

- Inwieweit ist es einem politischen System möglich oder unmöglich, das Erscheinungsbild eines Kunstwerkes vollständig zu kontrollieren?

- Wie erklärt sich Ihrer Ansicht nach das gewandelte Verständnis von Repräsentation im Sinne einer Darbietung hin zu einer Repräsentation von politischer oder ökonomischer Macht?

Lektüreempfehlungen

- Sabine R. Arnold / Christian Fuhrmeister / Dietmar Schiller (Hg.): **Politische Inszenierung im 20. Jahrhundert. Zur Sinnlichkeit der Macht**, Wien u. a. 1998. *Die Autoren analysieren in Fallstudien das Feld der Repräsentation als selbstbestätigende Inszenierung, bei der die Aufführung zur Verwirklichung von Macht wird.*

- Peter Burke: **Ludwig XIV. Die Inszenierung des Sonnenkönigs**, Berlin 1993. *In einer exemplarischen Studie hat Burke die gezielte Etablierung des französischen Absolutismus mithilfe von Kunst und Architektur studiert.*

- Bram Kempers: **Kunst, Macht und Mäzenatentum. Der Beruf des Künstlers in der italienischen Renaissance**, München 1989. *Eine Einführung in das System der neuen Höfe und Dynastien der italienischen Renaissance, ihren Repräsentationsbedarf und den damit verbundenen Aufstieg des Malerberufes.*

- Roman Schneider / Wilfried Wang (Hg.): **Macht und Monument. Moderne Architektur in Deutschland 1900 bis 2000**, Band 3, Ostfildern-Ruit 1998. *Der 3. Teil dieser Trilogie befasst sich mit den Repräsentationsfunktionen von Architektur seit dem Kaiserreich und diskutiert leitende visuelle Prinzipien.*

- Paul Zanker: **Augustus und die Macht der Bilder**, München 1987. *Die exemplarische archäologische Studie weist den gezielten Einsatz von Bildern zur Propagierung eines neuen augusteischen Zeitalters und zur visuellen Steuerung des gewaltigen römischen Territoriums nach.*

8 Märkte

Abbildung 22: Tagesschau-Ausschnitt vom 13. 5. 1991, 20 Uhr

Abbildung 23: Der französische Präsidentschaftskandidat Nicolas Sarkozy zu Pferde (2007)

Eine Nachrichtensprecherin in adretter Kleidung erscheint vor einem eingeblendeten Hintergrund, in welchem eine leicht verwackelte Farbaufnahme, versehen mit einer Orts- und Datumsangabe und einer dazugehörigen Schlagzeile, auf ein aktuelles Geschehen hinweisen soll (hier eine Begegnung des damaligen Bundeskanzlers Helmut Kohl mit Bürgern in Halle, die durch Demonstranten gestört wurde). Die Sprecherin begrüßt das Publikum und verliest die erste Nachricht, ehe ein Videobericht das Gesagte weiter vertiefen wird.

Das zweite Beispiel zeigt Pressevertreter bei dem Versuch, den französischen Präsidentschaftskandidaten Nicolas Sarkozy von einem Traktor aus, der für diesen Zweck eigens bereitgestellt wurde, beim Reiten zu filmen. Die marktmäßige Nachfrage nach derartigen Motiven oder ihr journalistischer Wert müssen so groß sein, dass selbst absurde protokollarische Vorgaben keine Proteste mehr auslösen. Die Profi-Fotografen, die als „rat pack" zur Entourage jedes Prominenten gehören, sind vielmehr in so großer Zahl auf der Ladefläche versammelt und in ihr in Tagesgeschäft vertieft, dass sie selber von Kollegen abgelichtet und zu einem ironischen Motiv der Bildberichterstattung werden.

Innerhalb der letzten hundert Jahre hat sich eine globale Maschinerie der Bildberichterstattung etabliert, die selbsttätig jeden Winkel der Erde auszuspähen scheint. Nicht nur Bereiche wie Werbung und Unterhaltung oder das Verlagswesen, sondern auch die öffentlichen Nachrichtenagenturen geraten dadurch in den Sog ökonomischer Kräfte, die sich aus dem Wert bildlicher Information, als Symbol gesellschaftlicher Erwartungen, herleiten. Professionelle Bewertungskriterien wie Auflagenhöhe, Geschwindigkeit oder Zuschauerzahl, die dem Betrachter verborgen bleiben, gewinnen an Bedeutung. Unklarer wird auch, wer innerhalb eines solchen Systems noch steuernd auftreten oder als Autor einer Bildnachricht bestimmt werden kann. Angesichts der rasanten, weltweiten Bildberichterstattung unterliegt die Presse selbst ohne Zensur und protokollarische Reglements den logistischen Strukturen der Massenmedien, die für die Auswahl einzelner Motive oder Standpunkte und die Entwicklung bestimmter Stile verantwortlich sind.

8.1 **Sendeformate**

8.2 **Bilder als Content: Material, Tarif, Lizenz**

8.3 **Professionalität und Autorschaft**

8.4 **Ausschuss, Bearbeitung, Zensur**

8.1 Sendeformate

Nachrichtensendungen des Fernsehens sind über Jahrzehnte hinweg hinsichtlich Ablauf und Ästhetik unverändert geblieben und haben sich erkennbar vom übrigen Programm als selbstständiges Format abgekoppelt, um ihren quasi-amtlichen Charakter zu unterstreichen. Bis heute äußern sich im Nachrichtenwesen und seinen Bildversorgungsstrukturen die historisch gewachsenen Bedingungen des Pressewesens und seiner journalistischen, verlegerischen und politischen Erwartungen. Im hier gezeigten Beispiel von 1991 ist eine Sprecherin vor einem eingeblendeten Studiohintergrund aus Foto, Schriftelementen und Hintergrundfarbe erschienen, die den ersten Bildbericht, einen kommentierten Videofilm, ankündigt.

Die Nachricht als Rundfunkformat

Seither haben sich im Zuge fortschreitender Digitalisierung und Programmspezialisierung gestalterische Einzelheiten grundlegend verändert: Die Studiowände wurden durch virtuelle Landschaften ersetzt, die Sprecher reden im Stehen statt im Sitzen oder treten paarweise auf, wobei sie sich auch miteinander unterhalten, Grafik- und Schriftelemente werden animiert. Im Ganzen aber hat sich das Erscheinungsbild von Nachrichtensendungen als Mischung aus Ansagen, Bildhintergründen und Filmblöcken über lange Zeit und sogar unabhängig von den politischen Systemen erhalten.

Der vorgelesene Text ist wesentlicher Bestandteil einer TV-Nachricht, obwohl er für viele Zuschauer weniger relevant ist als das Bildmaterial. Dass die Ansprache durch eine Person im Zeitalter von Liveberichterstattung und Ton-Bild-Kopplung eigentlich entbehrlich wäre, die Nachrichtensprecherin jedoch als Bezugsperson (*talking head*) dient, ist bekannt: In der Fernsehanalyse wird aber untersucht, ob das Auftreten oder die Kleidung von Moderatoren und Sprechern ablenkend oder einnehmend wirken oder welche Folgen es hat, wenn zu einer Nachricht bestimmte Hintergrundfotos gezeigt werden.

Transformation von Mediengewohnheiten

Dass der grundsätzliche Aufbau und das Erscheinungsbild der Sendung, das Schema von Moderator und Hintergrundbild beibehalten werden, folgt aus der Gliederung von Funk und Fernsehen durch Sendeprogramme. Diese formatieren den Ablauf der Beiträge zu wiedererkennbaren Einheiten, die als redaktioneller Nachrichtenteil oder gewerblicher Werbeblock getrennt sind, selbst wenn sie zunehmend ineinanderfließen oder sogar identisch sind (weil z. B. auch eine Nachricht werbend sein kann).

Grundsätzlich kann daher eine Nachrichtensendung um 20 Uhr gesendet und unverändert um 23 Uhr als Aufzeichnung abgespielt werden, ohne dass dies zu Irritationen führt. Da der Zeitpunkt einer Nachrichtensendung oder Unterhaltungsserie jedoch auch den Alltag von Millionen Zuschauern strukturiert und damit Aufmerksamkeiten kanalisiert, ist die Zeitfrage im Hinblick auf den Informationswert keine Nebensächlichkeit. Nicht nur mit zunehmender Alterung ändert sich daher der Nachrichtenwert eines Beitrages. Eine erfolgsorientierte Programmpolitik konfektioniert die Zusammenstellung der Nachrichtenbeiträge auf den zeit- und zielgruppengerechten Konsum.

Vor allem aber sind die in unmittelbarer Nähe der Nachricht verkauften Sendezeiten für Werbespots um ein Vielfaches teurer, weil sie
in der *prime time* mit der größten Zuschaltquote gebracht werden. In der Folge ist das, was als Bild gleichermaßen in Information, Werbung und Unterhaltung in Erscheinung tritt, ein intellektuelles Gut, dessen Wert innerhalb einer größeren Medienökonomie festgelegt wird. So erhöht sich der Preis einer Abbildung in der Regel, sobald sie vergrößert, farbig oder mehrfach gezeigt wird; und in Verbindung mit einem hohen Nachrichtenwert oder hoher ästhetischer Qualität gelingt auch der „Sprung auf das Titelblatt" oder *scoop,* der einer Darstellung und ihrem Fotografen ein Vielfaches an Beachtung verschaffen kann. Im Gegenzug wird die Werbeanzeige, die auf einer Umschlagseite eines Magazins erscheint, zu einem höheren Preis gehandelt als die Anzeige auf der Innenseite

Die Mehrzahl von Menschen hat sich an eine derartige marktmäßige Organisation kommerzieller Bildnutzung offenkundig gewöhnt, wenn sie abendliche Dauerwerbesendungen oder Pay-TV ebenso akzeptiert wie den Eintrittspreis für ein Kino oder ein öffentliches Museum. Zudem ist das zugrundeliegende Prinzip nicht völlig
neu. Verknüpfungen von Material und Wert (→ KAPITEL 4), Symbolkraft und Aufmerksamkeitsökonomie (→ KAPITEL 6, 7) sind in der gesamten Kunst- und Kulturgeschichte zu beobachten.

8.2 Bilder als Content: Material, Tarif, Lizenz

Nachhaltig verändert wurde der Bildhandel in der zweiten Hälfte des 19. Jahrhunderts durch die Ausstattung von Zeitschriften und Zeitungen mit Illustrationen (→ KAPITEL 2, 3) und besonders nach 1900 durch die Möglichkeit des Abdrucks von Fotografien. Sogenannte Presseillustrationsbetriebe schalteten sich zu dieser Zeit ein, um im
Auftrag der Redaktionen und nach dem Vorbild der ersten Nachrichtenagenturen (AP, Reuters; vgl. Bruhn 2007) die Bildbeschaffung zu

zentralisieren und den Kontakt zwischen Anbietern und Verlagen zu organisieren, wobei bereits mit den Möglichkeiten der interkontinentalen Bildtelegrafie experimentiert wurde. Eine zweite große Globalisierungswelle erfuhr die Bildbranche durch das Distributionswesen des Kinos und der Wochenschauen Mitte der 1920er-Jahre. Der grenzüberschreitende Handel mit Bildern, als Fotografien oder als Film, nahm ökonomisch wie rechtlich andere Dimensionen an.

Die Kopplung von Information, Unterhaltung und Werbung wurde jedoch schon vorher, zur selben Zeit wie die Erfindung der Fotografie, durch Zeitungen neuen Typs eingeführt (*penny press*), die ihre billigen Massenauflagen durch Einwerbung von Inseraten finanzierten. Der Zeitungskonsum wurde dadurch über die Auflage gemessen – ein Geschäftsprinzip, das der Presse nach wie vor zugrunde liegt. Auch wurde dem industriellen Kernfaktor Zeit bei der Beschaffung und Verbreitung von Nachrichten oder durch Faktoren wie die Laufzeit von Anzeigen höhere Priorität zugewiesen.

Auflage und Laufzeit

Durch das System der illustrierten Presse und der Produktwerbung wurde Bildlichkeit in weiten Teilen neu definiert – der Begriff Bild wurde synonym mit reproduzierbaren Gattungen wie Grafik, Fotografie oder Film, deren Wirkung und Anziehungskraft nun über Masse definiert werden konnte. Walter Benjamin hat diesen Prozess in seinem erst spät rezipierten Aufsatz zum *Kunstwerk im Zeitalter seiner technischen Reproduzierbarkeit* (Benjamin 1936) aus seiner Erfahrung mit dem Massenmedium des Kinos gelesen und darin, wenn auch verkürzend, die Geschichte der technischen Reproduktion und des dialektischen Verhältnisses von Original und Kopie verhandelt (vgl. Locher 2008, S. 49f.; → KAPITEL 9.3).

Für den Handel mit visuellen Produkten hatte dies zur Folge, dass nichtbildliche Faktoren preisrelevanter wurden: Bilder erfahren Kursschwankungen, ohne dass dies durch die sichtbare Machart oder Qualität begründet sein müsste. In die Preisbestimmungen fließen neben Kriterien wie dem Produktions-, Beschaffungs- oder Neuigkeitswert einer Nachricht oder dem Beliebtheitsgrad eines Motivs auch abstrakte Kriterien der Konsumforschung ein, die dem Paradigma folgen, dass quantifizierbare Daten (wie Ausschnitt, Reichweite, Farbigkeit, Textposition u. Ä.) über den Wert einer Information mitentscheiden. Mit zunehmender Facettierung und Virtualisierung der Medien werden technische Kriterien wie Auflösung, Übertragungsformate, Downloadfähigkeit, Speicherplatz ebenfalls relevant.

Aufmerksamkeitsmessung

Durch das digitale Zeitalter wurde wiederum eine neue Form des *visual content* (vgl. Frosh 2003) definiert, der das Bild in rechtlicher, technischer und logistischer Hinsicht kompatibel macht mit anderen Produkten der Medienindustrie wie der Musik und dem Text, mit denen es sich multimedial immer weiter verschränkt. Lichtbildwerke unterschiedlichster Art sind inzwischen so weit mit anderen Produkten korreliert, dass sie faktisch nur noch in Verbindung mit diesen wahrgenommen werden. Die visuellen Artefakte, die in Presse, Werbung und Unterhaltung anzutreffen sind, sind seither Teilbausteine einer marktmäßig organisierten Versorgung, die je nach Einsatzort in Werbeprospekten, Zeitschriften oder als Nachrichtenhintergrund zu **Das Bild als** *content* Bildern mit einer bestimmten Aussage formatiert werden und die großenteils auch als Vorratsware bereitbestehen, welche auf eine möglichst breite Streuung angelegt sind und durch professionelles Marketing und häufige Wiederverwendung ihre teilweise beträchtlichen Produktionskosten wieder einspielen müssen.

In der Medienwirtschaft können sogar vermeintlich belanglose Fotomotive, die sich von Anbieter zu Anbieter bis zur Verwechslung ähneln, **Marketing** durch Vertrieb in einem professionellen Verbund, zu einem Wertstoff **und Logistik** eigener Qualität gemacht werden. Möglich wird dies durch Preisunterschiede, Serviceangebote, große Kundenstämme, berechenbare Lieferzeiten, klar strukturierte Datenbanken und Webauftritte, die Lieferung in gängigen Dateiformaten und Übertragungsprotokollen. Der Verfallswert von fotografischen Bildern verlangt nach einer ausgeklügelten Logistik, die zwischen Aktualität und Haltbarkeit vermittelt.

8.3 Professionalität und Autorschaft

Profi und Amateur Mit dem Siegeszug der Fotografie nach 1850 wurde das Fotografieren, später das spontane Knipsen, zu einer Beschäftigung von Personen, die auch ohne künstlerische oder technische Ausbildung Bilder herstellten. Dies veränderte die Produktion und Konsumption von Bildern als Kommunikationsmedium und schuf – in einer frühen Form des Bürgerjournalismus und in Verbindung mit der massenhafte Bebilderung von Tageszeitungen – die Voraussetzungen für die Entstehung des Pressebildes, ein Vorgang, der sich derzeit durch die Möglichkeiten von Handyfotografie und Blogging in Teilen wiederholt.

Sobald integrierte Handykameras neben lichtstarken Teleobjektiven auf eine prominente Person gerichtet sind, werden die Dargestellten geradezu zum Experimentierfeld neuester Fototechnologien, was so weit gehen kann – wie beim hier gezeigten Empfang des niederländischen Thronfolgers und seiner aus Argentinien stammenden

Abbildung 24: Kronprinz Willem-Alexander der Niederlande und Maxima Zorreguita in Amsterdam (2001)

Braut in den Provinzen des Landes (→ ABBILDUNG 24) – dass ebenso viele Zuschauer die Kamera betätigen wie dem Besucher die Hand entgegenstrecken. Private wie professionelle Fotografen und Filmer liefern sich einen Wettlauf um die besten Aufnahmemomente.

Wettlauf um Bilder

Nicht zufällig kommt vor diesem Hintergrund das Konzept des Privatreporters, aus dem ursprünglich nach 1900 auch der Presseillustrator hervorging, wieder in die Diskussion. In deutlicher Abgrenzung zur alltäglichen Bildproduktion, bei der jeder sein Bild beisteuern kann, werden jedoch professionelle und handwerkliche Definitionen von Fotografie, Grafik oder Film weiterhin vehement aufrechterhalten. In dem Moment, in dem jeder Einzelne mit Kamera oder Handy fotografieren kann, gewinnen logistische Vorteile an Wert. Nicht erst seit dem Aufkommen ‚technischer' Bildmedien wie der Fotografie ist Professionalität daher nur zu einem Teil als Kompetenz auf dem Gebiet der Gestaltung, Material- oder Motivwahl anzusehen, sondern schließt auch die Befolgung von Vertriebs- und Marketingregeln oder die selbstbewusste Einforderung von Urheberrechten selbstverständlich mit ein. Für einen professionellen Fotografen oder Filmproduzenten, der seine Aufnahmen für Zwecke der Werbung oder Bildberichterstattung auf eigenes Risiko herstellt und gegen Gebühr weitergibt, ist die Kenntnis solcher betriebswirtschaft-

Anspruch der Professionalität

Kenntnis der Marktregeln

121

lichen Regeln keine Nebensache, sondern seit den Anfängen der Fotografie wesentlich für das eigene Handeln und Selbstverständnis.

Schon einer der Erfinder der Fotografie, William Henry Fox Talbot (1800–77, → KAPITEL 9) hat auf sein Verfahren Lizenzgebühren beansprucht; er berief sich auf das Prinzip des Patents, mit dem innovative Technologien durch zeitweilige Exklusivverwertung gefördert werden sollten. Mit dem auflagenstarken Druckwesen und der Reproduktion von Kunstwerken kam die Frage auf, ob neben der Technologie auch **Schutz des geistigen** der damit erzeugte Inhalt, als intellektuelles Eigentum, Schutz vor dem **Eigentums** Gesetz genieße. Dazu war zu definieren, welcher Teil einer grafischen Technik schützenswert ist, worin also die Leistung eines Künstlers im Unterschied zum Drucker oder Verleger genau besteht. Dies führte Mitte des 19. Jahrhunderts zur Entstehung des Kunst- und Urheberrechts, das sich in Deutschland besonders auf den kreativen Charakter des Werkes berief, im Unterschied etwa zum US-Copyright (vgl. Rose 1993; Siegrist/Löhr 2007). Besonders für die Fotografie wurde die Frage virulent, ob ein automatisch erzeugtes Bild überhaupt einen Autor habe (vgl. Dommann 2006). Seit der Vertrieb von Fotografien um 1900 zum globalen Geschäft geworden war, bestand ein Problem der Anbieter zudem darin nachzuverfolgen, wann und wo ihre Aufnahmen verwendet wurden. Gegenstand des Handels war nun nicht mehr allein ein Bild im Sinne eines bestimmten Motivs, sondern auch dessen

Abbildung 25: Bill Clinton im Präsidentschaftswahlkampf (1992)

konkrete Nutzung. Vor allem seit den späten 1970er-Jahren hat diese Rechteverwaltung mit dem Digital Rights Management selber industrielle Formen angenommen.

Um die hieraus resultierenden Probleme zwischen öffentlichem Informationsauftrag, eigenem Verwertungsinteresse und politischer Bildregie zu beleuchten, können zwei Fotografien des ehemaligen US-Präsidenten Bill Clinton angeführt werden, zum einen aus dem Präsidentschaftswahlkampf 1992/93, zum anderen bei einer Werbetournee aus dem Jahre 2006. In beiden Aufnahmen ist auf der rechten Seite des Bildes ein anzugtragender, lächelnder Mann im Profil gezeigt, der seine Hände einer Masse begeisterter Menschen entgegenstreckt und aus einer nur leicht erhöhten Perspektive zu sehen ist. Im ersten Beispiel (→ ABBILDUNG 25) ist aus der abgesenkten Perspektive hinter einem Sicherheitsbeamten ein Wahlplakat zu erkennen.

Politische Berichterstattung

Das Motiv, als „Bad in der Menge" ein eigenes Stichwort der politischen Bildsprache und Bildberichterstattung und ein fester Bestandteil des öffentlichen Zeremoniells (→ KAPITEL 7.4), wird hier in besonders intelligenter Weise inszeniert. Es zeigt den Politiker, der sich volksnah und frei von Berührungsangst gibt, als auch den Regenten, der durch seine persönliche Erscheinung, die jedoch das Produkt einer ausgeklügelten medialen Inszenierung ist, aus einer größeren Menge hervorzustechen scheint.

„Bad in der Menge"

Abbildung 26: Ex-Präsident Bill Clinton als Wahlhelfer für seine Frau, Detail aus *Der Spiegel* vom 6. 11. 2006

Dieselbe Szene scheint sich im Abstand von über zehn Jahren noch einmal zu wiederholen, nachdem Bill Clinton längst aus dem Amt geschieden ist (→ ABBILDUNG 26), und es kann hinzugefügt werden, dass auch in den dazwischenliegenden Jahren, von unterschiedlichen Agenturen und Fotografen und in unterschiedlichen Ländern, sehr ähnliche Aufnahmen Clintons hergestellt wurden. Zahlreiche Buchdeckel zu Leben und Karriere des Präsidenten ziert ein vergleichbares Motiv.

Bildstrategie

Die Frage wäre daher, wer diese politisch gewünschte Bildstrategie im globalen Umfang steuern kann. Für den Umstand, dass sich die Fotografien über Jahre hinweg formal ähneln, dass sie also mit einiger Penetranz ein vorgeregeltes Erscheinungsbild einüben, muss es neben der Prominenz des Dargestellten noch weitere Gründe geben: Die porträtierte Figur unterwirft sich selbst dem Protokoll durch Gesten und Kleidung (→ KAPITEL 7.1), die Fotografen können ihr Objekt aufgrund reger Konkurrenz oft nur noch mit Teleobjektiven und Leitern aus einem bestimmten Winkel aufnehmen.

Störungen, Pannen

Politische Ereignisse schreiben sich außerdem nicht nur in das allgemeine Bildgedächtnis ein, weil sie besonders häufig wiederholt würden, sondern auch, weil sie als Störungen auffallen. Sittenwidrigkeiten von Prominenten finden genaueste Beachtung, ebenso Zwischenfälle wie jener, der in → ABBILDUNG 22 in der Hintergrundaufnahme zu erkennen ist: Hier ist eine Rangelei zu sehen, die sich am Rande eines Besuches des deutschen Bundeskanzlers Helmut Kohl in Halle 1991 ergab, als dieser nach einem Eierwurf auf einen Störer losstürmte. Im Unterschied zu früheren Bildgebungen können bei der spontanen Aufnahme oder Live-Übertragung einzelne Fehltritte nicht mehr gelöscht werden, und es bedarf im Falle besonders relevanter Momente daher einer minutiösen Planung, um Pannen und Gefahren auszuschließen.

Protokollarische Beschränkungen

Vor allem ein rigoroses Protokoll weist daher den Pressevertretern und dem Publikum feste Standorte zu. Dies stellt keine Aufhebung der Pressefreiheit, jedoch einen erheblichen Eingriff in die Wahrnehmung einer Person durch die Anwendung und Anerkennung von Spielregeln dar, die sich erheblich auswirken, bis hin zum sogenannten *embedded journalism,* d. h. der gezielten Begleitung politischer und militärischer Interventionen durch Journalisten. Zugang zu engeren Sicherheitsbereichen erhalten in der Regel nur die akkreditierten

Nachrichtendienste

Berichterstatter der großen Nachrichtenagenturen, die als quasi-öffentliche Institutionen das Recht auf Information wahrnehmen, auch wenn sie selbst im Auftrag von privatwirtschaftlichen Verlagen und Sendern (wie die *Associated Press* seit 1846) oder durchweg kommerziell (wie ihr einstiger Konkurrent *United Press International*) tätig und quasi-monopolistische Lieferanten für Pressebilder sind.

Die ersten Nachrichtenagenturen, die eher nebenbei auch Bilder kabelten, legten eigens Wert auf die Anonymität ihrer Zulieferer, um die amtliche und gemeinnützige Eigenart ihres Verbundes zu unterstreichen. Wirkte die Agentur damit als Vermittler zwischen Fotograf und Verlag, führte die Verselbstständigung des Prinzips auch zu vertraglichen Knebelungen und zur Entmündigung von Autoren, teilweise zur Aushebelung bildjournalistischer Prinzipien (vgl. Caujolle 2002). In der gegenwärtigen Praxis erscheint der Lieferant kaum noch im Bildnachweis neben dem Namen der Agentur, auch wenn er als individueller Urheber seine Rechte über das Material behält und Tantiemen bezieht.

Die Agentur etabliert sich so als Produzent zweiter Ordnung neben dem individuellen Autor. Kollisionen von Interessen und Konzepten wie Kreativität, Autorschaft oder Authentizität sind dadurch vorprogrammiert. Auch wird für den Leser einer Zeitung unklarer, wer für die Produktion von Nachrichten verantwortlich ist, deren Auswahl durch Seh- und Kaufverhalten ja auch von ihm unbemerkt mitgetragen wird. Bilder scheinen sich im globalen Beobachtungssystem der Presse inzwischen „autopoietisch" (Luhmann 1984, S. 148; vgl. Grittmann 2007), d. h. von selbst, zu erzeugen.

Die gewerblich tätigen Fotografen bleiben durch handwerkliche und technische Kenntnisse, durch Erfahrungen und Akkreditierungen gegenüber dem Knipser im Vorsprung. Gerade hierdurch können aber z. B. Regieanweisungen des Protokolls noch weiter verstärkt werden, etwa wenn das Bedürfnis der Fotografen nach Nähe zu ‚ihren' Prominenten dafür sorgt, dass der reglementierte Standpunkt als scheinbar exklusiver Zugang wahrgenommen wird, mit dem der professionelle Fotograf sich seinem Objekt näher wähnt. Die Geschichte der Bildreportage ist aus demselben Grund seit ihren Anfängen die Geschichte des Schutzes der Privatsphäre, des Rechtes am eigenen Bild, das gegen übermäßige fotografische Ausspähung geschützt werden muss (vgl. Ortland 2007, S. 272f.; Steinhauer 2007, S. 233f.).

Die vom Markt ermittelten Bewertungen und Justierungen scheinen sich weitgehend von einer Bildlichkeit emanzipiert zu haben, die wegen ihres Kunst- oder Originalitätswertes vergütet wurde. Die Existenz bestimmter Dienstleister belegt aber auch, dass das Bild im gewerblichen Zusammenhang sein eigenes System mit bestimmten Teilnehmern, Begrifflichkeiten und Tarifen entwickelt hat, mit spezifischen Vorstellungen eines Bildmaterials, das sich in analogen und

Autor und Agentur

Produzenten zweiter Ordnung

Der Fotograf zwischen Protokoll und Profession

Gewerbliche Definition des Bildes

digitalen Medien in Form von Lizenzen verwerten lässt. Wenn Bild und Text in den Sichtbarkeitsformen der Printmedien, Werbetafeln und Displays in einer visuellen Kommunikation aufgehen, so behalten sie ihre unterschiedlichen Kanäle und Formate. Es wird zugleich zum juristischen Streitfall, was genau an der Vertriebsform „Bild" in den Bereich des Eigentums (oder des Plagiats, des Zitats) fällt.

Bilder stehen symbolisch für den gemeinsamen Betrieb aus Herstellern, Vertriebswegen, Softwareanbietern, Rechtsabteilungen, Großverlagen und Fernsehsendern, die einen Rohstoff und seine Veredelungsstufen definieren und die gegenüber Initiativen zur Gemeinfreiheit von Bildern (z. B. als *creative commons*) ihre Position verteidigen. In solchen Verwertungsvorgängen mögen überholte Bildkonzepte weiterleben, sie ziehen aber eine Geschäftigkeit nach sich, die umgekehrt auch mitbestimmt, was unter Bildern aktuell verstanden wird.

8.4 Ausschuss, Bearbeitung, Zensur

Selbst technische Bedingungen, Regieanweisungen und professionelle Praktiken zusammengenommen sind aber nicht stark genug, um bei einer milliardenfachen Tagesproduktion von Fotografie und Film für wiederkehrende Bildstile zu sorgen. Angesichts der redaktionellen Engpässe auf dem modernen Bildmarkt kann hier, trotz eines faktischen Überangebots an Fotografien, von einer virtuellen Knappheit gesprochen werden, die alle täglichen Abläufe und Auswahlprozesse kennzeichnet, sobald ein bestimmtes Material zu bestimmten Konditionen vorliegen muss. Durch die Bündelung von Agenturen und Archivbeständen, Nachrichtenredaktionen oder Medienkonzernen und deren Bewerbung (*pushing*) treten weitere Konvergenzen auf, die sich filternd auswirken, bis hin zu dem Extremfall, dass mehrere Redaktionen unabhängig voneinander aus Tausenden von Angeboten dasselbe Motiv auswählen. Aus der Perspektive des Bildredakteurs, der für vorbereitete Textbeiträge nach formal und handwerklich geeigneten Motiven suchen soll, stellt sich der Begriff der „Bilderflut" (→ KAPITEL 3.3) daher völlig anders dar.

Wie auch im Falle der politischen Propaganda (→ KAPITEL 7) oder des Kunstsektors (→ KAPITEL 6) ist es eine Frage der systematischen Bestimmung, ob eine Nachricht als Information oder Wirtschaftsgut behandelt wird. Bilder werden auf Wunsch hergestellt und auf kollektiv verankerte geschmackliche Ideale hin optimiert. Selbst ein Tatortfoto,

Konvergenzen statt Bilderflut

Kollektive Sehweisen

das als objektiver Beweis in einem Gerichtsverfahren dienen soll, wird aussehen wie ein Tatortfoto und dabei professionalisierte Bildstile freilegen. Wo die entsprechenden Aufnahmen nicht vorhanden sind, werden sie per Montage und Nachbearbeitung bereitgestellt.

In einem solchen Falle könnte von Manipulation, Zensur oder Fälschung gesprochen werden. Jedoch bleibt fraglich, ob es Bildlichkeit jenseits von Manipulation (im Sinne von Handhabung) überhaupt geben kann. So ist überspitzend behauptet worden, dass „alle Bilder lügen" (Titel von Schreitmüller 2005): Jede Aufnahme setzt eine Technik der Wiedergabe voraus, die den Gegenstand in konstruierter Form zeigt.

Bearbeitung und Manipulation

Zudem bleiben selbst in einer größeren Perspektive, welche die Ränder eines Ereignisses einbezieht, weite Teile einer Szene ausgeblendet, solange Bildlichkeit Fokussierung verlangt. Die tägliche Praxis der Bildpresse erzwingt Entscheidungen und damit Ausblendungen. Jedes Bild, nicht anders als jeder schriftliche Bericht eines Ereignisses, ist standortgebunden, ausschnitthaft und lenkt Aufmerksamkeiten. Die Blicksteuerung wird erst ersichtlich, sobald Aufnahmen nebeneinander liegen (→ KAPITEL 10.1). Randbereiche der Berichterstattung, vermeintlich uninteressante oder fehlgegangene Aufnahmen, Bilder ohne Symbolkraft, Zeitwert oder Witz werden vom Fotografen, von der Agentur oder der Bildredaktion ausgeschossen, vom Leser überblättert.

Bildentscheidungen

Vor diesem Hintergrund wäre der Einsatz von manipulativen Mitteln zu sehen, die heute zur Alltagspraxis eines jeden gehören, der über eine digitale Fotokamera oder einen Scanner und einen PC mit Bildbearbeitungssoftware verfügt. Auch durch die digitalen Medien sind Manipulationen nicht grundlegend möglich geworden, sondern lediglich wahrscheinlicher, weil sie in digitaler Form leichter und günstiger sind und Fotografie sich durch Mausoperationen mit Bildeditorprogrammen wieder einem Zeichnungsverfahren annähert. Fototheoretiker wie Geoffrey Batchen (vgl. Batchen 1994) oder William J. Mitchell (vgl. Mitchell 1992b, nicht identisch mit W. J. T. Mitchell) gehen daher vom Eintritt in ein „post-fotografisches Zeitalter" aus, dessen Nachträglichkeit aus dem Umstand folgt, dass es auf dem Verweischarakter der Fotografie beruht (→ KAPITEL 9.4), diesen aber infrage stellt.

Post-fotografisches Zeitalter

In der digitalen und logistischen Neubestimmung von echt und falsch, Original und Kopie sind Unterscheidungen nicht völlig aufgehoben. Ein Pressebild gewinnt durch Reproduktion in einem Massenmedium eine weitere Sinndimension, wenn es schon aufgrund der

Einbettung in einen Bericht als Beitrag zu diesem begriffen wird. Sobald aus einer gelieferten Aufnahme durch redaktionelle Überarbeitung einzelne Teile herausgeschnitten werden, ohne dass dies vermerkt wird, oder sobald eine nachträgliche Beschriftung die Behauptung aufstellt, ein retuschiertes Bild werde unretuschiert abgedruckt, ergibt sich Missbrauch. Von politisch motivierter Fälschung kann gesprochen werden, wenn Bilder sich einem Publikum erkennbar als Dokumentation eines Ereignisses zu verstehen geben, ihre Ausschnitte und Motive jedoch so wählen, dass sie den Eindrücken des anwesenden Publikums zuwiderlaufen, ohne dass es hierfür eine künstlerische oder satirische Motivation gäbe.

Dies erklärt, warum sich Fotografenverbände (wie der deutsche FreeLense) weiterhin für die Unterscheidung von geliefertem und überarbeitetem Material einsetzen und die technischen Entwicklungen (etwa bei Daten- und Übertragungsformaten) im Detail beobachten müssen. Die Definition einer korrekten oder zulässigen Abbildung muss sich den Gegebenheiten der Technik oder des Marktes kontinuierlich anpassen. Sie ist auch Anlass, sich der dezenten Verschiebungen von Bildstrategien anzunehmen, welche die Bild- wie Politikmärkte durchlaufen (so Werckmeister 2005 am Beispiel des *Post-September-Eleven*).

Trotz der Macht einzelner Medienkonzerne oder der politischen Missbrauchsmöglichkeiten beim Einsatz von Bildmedien sind Produzenten ebenso wie Konsumenten Teil einer Maschinerie und ihrer Kontingenzen und verfügen mit Bildern über ein Austauschmedium. Der Steuerung von Bildern durch Monopolisten oder durch politische Interventionen stehen alltägliche Praktiken der Archivierung, Bearbeitung und Beschriftung gegenüber, die sich bild- und erinnerungspolitisch weit umfassender auswirken. Ein Großteil von Bildinformationen wie z. B. Schlag- und Suchwörter, Bildtitel oder Autoren-, Zeit- und Ortsangaben, die in speziellen Formaten mitgeliefert werden und entscheidende Bedeutung bei Auswahl und Übermittlung einer Nachricht haben, werden nicht mitgedruckt und im weiteren Verlauf der Nutzung unsichtbar (vgl. Scott 1999, S. 99ff.), genauso wie ihre nachträgliche Veränderung.

Eine zeitgemäße Analyse von Bildformen sollte daher technische, logistische und rechtliche Konditionen von Fotografie und anderen Medien einschließen. Es könnte z. B. gefragt werden, wieso Fotoreporter in der Regenbogenpresse unter Berufung auf das Urheberrecht immense Honorare verlangen, während sie sich unter Berufung auf die Pressefreiheit weigern, ihre Opfer schriftlich um Vorabgenehmigung für Aufnahme und Abdruck zu fragen. Bildkünstler, denen an der ästhetischen oder politischen Bildung ihres Publikums gelegen ist, haben einen Anspruch auf Vergütung, so wie sich eine Zeitung, die diese nicht zahlen will, auf ihren gesetzlich geschützten Informationsauftrag beruft, mit dem sie zugleich Geld verdient. Es wäre zu fragen, warum Qualitätszeitungen, die sich auf die Mehrstimmigkeit ihrer Quellen berufen, ihre gedruckten Beiträge fast immer nur mit der Aufnahme eines Anbieters versehen und immer seltener dessen Namen oder die vollständigen Bildinformationen geben. Es wäre zu fragen, wieso Modehersteller ganze Städte tapezieren, aber unter Berufung auf Urheberrechte den Abdruck eines einzelnen Plakats untersagen können. Die Berücksichtigung solcher und anderer bildwirtschaftlicher Gewohnheiten und Konflikte böte Gelegenheiten, die tiefere historisch-gesellschaftliche Einbettung des Bildermarktes zu verstehen.

Fragen und Anregungen

- Der französische Kurator und Fotograf Christian Caujolle hat vor einigen Jahren aufgefordert, zur Frage zurückzukehren, warum Bilder in der Presse überhaupt Verwendung finden. Was würden Sie ihm antworten?

- Wann und wodurch wurde Fotografie zu einem maßgeblichen Mittel der Berichterstattung?

- Welche Bedeutung kommt der ökonomischen Ordnung der Medienwelt und ihren Vertriebsstrukturen für die Gestaltung und Wahrnehmung politischer Inhalte zu?

- Inwiefern ist gerade im Zeitalter omnipräsenter Massenmedien der Begriff der Bilderflut unzutreffend?

Lektüreempfehlungen

- **Bodo von Dewitz (Hg.): Kiosk. Eine Geschichte der Fotoreportage, 1839–1973.** Begleitband zur Ausstellung Museum Ludwig / Agfa-Foto-Historama Köln (29.6.–16.9.2001) und Altonaer Museum in Hamburg (24.4.–11.8.2002), Göttingen 2001. *Reich bebilderter Ausstellungsband als Querschnitt durch ein Jahrhundert Presse-*

*geschichte, mit seltenen Beispielen aus der Sammlung des Foto-
reporters Robert Lebeck.*

- **James Lull: Media, Communication, Culture. A Global Approach,**
Neuausgabe, New York 2000. *Eine der zahlreichen Einführungen
in die visuelle Kommunikation der Informations- und Unterhal-
tungswelt, die jedoch die Ebene der globalisierten Zeichenproduk-
tion konsequent in den Blick nimmt.*

- **Claus Pias u. a. (Hg.): Kursbuch Medienkultur. Die maßgeblichen
Theorien von Brecht bis Baudrillard,** 2. Auflage, Stuttgart 2000.
*Dieser Band versammelt Auszüge aus klassischen Texten, thema-
tisch gebündelt und eingeleitet, die aus der medientheoretischen
Perspektive auch Fragen des Marktes, seines Publikums, seiner
Medien und Motive verhandeln.*

- **X für U – Bilder, die lügen.** Begleitbuch zur Ausstellung im Haus
der Geschichte der Bundesrepublik Deutschland u. a., Ausst.-Kat.,
3. Auflage, Bonn 2003. *Klar aufgebauter, mehrfach aktualisierter
Begleitband zu einer Wanderausstellung des Bonner Hauses der
Geschichte, der die verschiedenen Techniken der Fälschung und
Manipulation im Pressebereich systematisch vorstellt.*

9 Ikonen, Urbilder, Vorbilder

Benit Mussolini (signature)

Abbildung 27: Benito Mussolini, Fotopostkarte aus dem Atelier Caminada, Mailand (1919)

Griest hestu bailiges antlit vnsers bebalters · Jn dē
da schinet die gestalt des götlichen glanczes. Gedzu
ket in ain schöne wisses diechtlin Vñ gegebē veronice
czü ainem zaichen der liebe. Griest hestu geczierd der
welte ain spiegel der bailigen · Den da begerend czü
schowen die hymelschē gaiste. Künige vns von alle
finde· Vnd sieg vns zü der selige gesellschafft. Griest
hestu vnser glori in disem hertten hintliessenden vnd
schwachem leben · Fier vns czü dem vatterland o du
selige figure. Zü sehend das wöneuglich antlit cristi
vnsers herren · Bis vns ain sichere hilff ain siesse erkie
long trost vnd ain schirme. Das vns nit schade müg
die beschwerong vnser fünde Sonder das wir niess
send die ewige rüo amen
　So fil sind gegeben tag applas vnd karen disem
gebet das ich sy hie nit künd wol begriffen

Abbildung 28: Konrad Dinckmut: Ablassbrief
(um 1482), Holzschnitt

131

Der italienische „Duce" Benito Mussolini hat sich in zahllosen Bildformen verewigen lassen, darunter auch im populären Medium der Bildpostkarte. In einem Beispiel um 1919 ließ er sein Gesicht von einem Fotografen als dämonische Erscheinung inszenieren, die er per Unterschrift als ein Abbild seiner Person und der Idee, für die diese steht, beglaubigte. Als Werbekarte für seine damals neue, rechtsgerichtete Bewegung ließ er dieses Bild in großer Zahl und in mehreren Varianten zirkulieren.

Das Postkartenmotiv lässt sich auf das christliche Bildmotiv der Ikone beziehen, die im Spätmittelalter ebenfalls in einem reproduzierbaren Medium Verbreitung fand, nämlich im Ablassbrief. Der hier gezeigte Brief wurde von Konrad Dinckmut gestaltet und entstand vermutlich 1482. Der kleinformatige, schlichte Holzschnitt setzt sich zusammen aus einem Bildfeld (hier mit dem Bild Christi mit Dornenkrone) und einer längeren schriftlichen Erklärung mit dem Versprechen, dass derjenige, der diesen Druck erwirbt, mit einem Ablass der Sünden rechnen könne. Die Erläuterung mahnt aber, dass die Wirkung nicht von der Geldleistung ausgehe, sondern von der innigen Betrachtung des Bildes.

Die Darstellungen verbindet ihre frontale Wiedergabe des Gesichts, welches den Betrachter in seinen Bann ziehen soll; in beiden Fällen wird durch einen schriftlichen Zusatz unterstrichen, dass das Bildnis echt sei. Die Idee des „echten Bildes" lag schon der spätantiken Ikone zugrunde, welche durch göttliches Einwirken entstand und damit als unveränderliche Form geprägt wurde. Sie kehrt hier wieder in der stereotypen Form des Holzschnittes oder der fotografischen Aufnahme und ihrer Reproduktion. Am Beispiel Mussolinis, der gleichermaßen auf das historische Kultbild der Ikone, auf die Prinzipien der Massenmedien wie auf neue psychologische Forschungen und Techniken gesetzt haben könnte, wird außerdem deutlich, dass bildliche Darstellungen selbst dann an Traditionen und Konventionen anschließen, wenn sie eigenständige Muster und Ideen zu prägen beabsichtigen.

9.1 Erscheinung und Echtheit

9.2 Abzeichnung, Ablichtung

9.3 Image, Prägung, Typ

9.4 Das Bild als Zeitdokument

9.1 Erscheinung und Echtheit

Im Jahr 1919, kurz nach dem Ende des Ersten Weltkrieges, hat der spätere italienische Diktator Benito Mussolini (1883–1945) in Mailand das Foto-Atelier Caminada am Corso Vittorio Emanuele 13 aufgesucht und mit einer Reihe von Porträtaufnahmen beauftragt. Von den Aufnahmen wurden einige für den Bilddruck im Postkartenformat ausgewählt, andere wurden für unterschiedliche Werbezwecke weiterverarbeitet. Mit seiner Fotokampagne, der später etliche folgen sollten, trat Mussolini gezielt an die Öffentlichkeit.

Auftritt

Nachdem er in pathetischen Zeitungsartikeln und als überzeugter Pazifist der sozialistischen Partei über viele Jahre Gehör verschafft, dann aber für einen Kriegseintritt Italiens geworben hatte, war er aus der Partei gedrängt worden. Der durchaus bürgerliche Mussolini vollzog daraufhin eine politische Wende, die er vor allem mit dem Ziel verfolgte, im Auftrage seiner Unterstützer Einfluss auf den italienischen Staat zu gewinnen. 1919 verband er die Fasci d'Azione Rivoluzionaria (Bünde der Revolutionären Aktion) mit anderen politischen Gruppen zu den Fasci di combattimento (Kampfbünde), die dem Faschismus seinen Namen gaben und in denen er bürgerliche wie proletarische Kräfte organisierte.

Politische Wende

In jener Phase veränderte sich auch seine äußere Erscheinung: Die Miene wurde härter, er nahm sich den schmalen Oberlippenbart ab, Mitte der 1920er-Jahre ließ er sich auch den Kopf kahlscheren und bot damit jene markante stiernackige Erscheinung, in der er heute zumeist in Erinnerung geblieben ist: als kaltschnäuziger Machtmensch, dessen Botschaft die Aktion als solche war. Die Fotografie in → ABBILDUNG 27 zeigt die Enface-Aufnahme eines Mannes mittleren Alters in Nahsicht. Das Gesicht erscheint als Punkt inmitten einer dunklen Umgebung, beleuchtet durch ein vordergründiges Licht. Der Lichtkontrast und der Blick von unten verleihen dem Porträtierten einen ernsten und zugleich unbestimmten Ausdruck, der Skrupellosigkeit als auch Unterwerfung bedeuten könnte. Das Gesicht spielt mit einer Mischung aus Direktheit und Undurchdringlickeit, welche an die Coolness moderner Popstars erinnert, die mit ihrer Hässlichkeit kokettieren. Die starren, dunklen Augen vermitteln Entschlossenheit, Selbstverausgabung und Melancholie zugleich, Details wie das leuchtende Ohrläppchen wirken vor dem dunklen Grund fast diabolisch.

Ein neues Image

Auch Adolf Hitler hatte sich in einem seiner frühen Wahlplakate dem Betrachter in dieser Weise radikal von vorn zugewandt, als

,mondhaftes' Gesicht vor dunklem Grund ohne Hals und Kragen, das durch wenige Merkmale und durch den Namen des Porträtierten in Großbuchstaben zu einem Wahlplakat ausgebaut wurde, welches den Betrachter weniger umwerben als vielmehr anblicken und einschwören sollte.

Dieses Helldunkel-Schema des Porträts ergab sich nicht allein aus spontanen Entscheidungen, aus den Regeln des Formats oder den technischen Zwängen der fotografischen Belichtung und Fokussierung, sondern ist auch das Resultat historisch eingeübter Bildmuster, **Ikone** insbesondere derjenigen der Ikone. Bezeichnet wird damit die in der byzantinischen Welt tradierte Gattung des Kultbildes mit Christuskopf oder Muttergottes, die spätestens im 7. Jahrhundert als Bildform etabliert war und sich über die orthodoxe Kirche hinaus verbreitet hat.

Kernidee ist, dass das darin gezeigte Bildnis göttlichen Ursprungs ist und von Bild zu Bild ununterbrochen tradiert wird. Durch seine unveränderliche Form wird bezeugt, dass es auf einen Offenbarungs- **Acheiropoieton** akt zurückverweist, in dem jenes erste Bild entstand, das als Acheiro- oder Autopoieton (griechisch für: das ohne Hand / das von selbst geschaffene) dem Wort der Heiligen Schrift gleichkommt. Das Antlitz des Erlösers oder der Madonna konnte deshalb als heilig verehrt werden, da die Anbetung diesem Ursprung und nicht dem Idol gilt (vgl. Wolf 2002; → KAPITEL 2.2).

Für die Entstehung des ersten Bildes gibt es gleichwohl verschiedene Ursprünge. Neben dem *vera ikon* (lateinisch / griechisch für: das **Vera ikon** wahre Bild), welches Jesus nach dem Bericht der *Legenda aurea*, einem Volksbuch des 13. Jahrhunderts, auf seinem Leidensweg hinterlassen haben soll, indem es sich im Schweißtuch einer barmherzigen Frau namens Veronika abdrückte, ist auch das sogenannte Grabtuch Christi, das in Turin aufbewahrt wird, ein solches Urbild. Die Bildstiftung konnte ebenso über unmittelbare Anschauung erfolgen (wie im Falle des legendären Königs Abgar von Edessa, dem ein Bildnis Jesu überbracht wurde, daher „Abgarbild") oder durch Eingebung, so beim heiligen Lukas, Schutzpatron der Künstler, der die vor ihm erschienene Madonna gemalt haben soll.

Das Motiv der Ikone wurde durch ein Printmedium der Frühen Neuzeit popularisiert, durch den Ablassbrief (→ ABBILDUNG 28). So wird mit dem hier gezeigten Brief, der vom Drucker Konrad Dinckmut aus Ulm um 1482 ausgegeben wurde, das Versprechen aus- **Ablassbild** gedrückt, dass derjenige, der den Bilddruck erwerbe, mit einem Ablass seiner Sünden rechnen könne. Der beigefügte Text besagt, dass die Wirkung nicht von der Geldleistung ausgehe, sondern von der innigen Betrachtung des Bildes: Dessen Echtheit oder Authentizität ist damit stets Glaubenssache (vgl. Belting 2005). Die reformatorische Kritik sah darin hingegen einen schnöden Handel, wie er heute noch im Begriff des ‚Bußgeldes‘ weiterlebt.

9.2 Abzeichnung, Ablichtung

Das Bildnis Mussolinis präsentiert sowohl eine Person als auch eine Technik, nämlich das Medium Fotografie. Deren Einsatz ist insofern eine Geste, als eine politische Persönlichkeit einem Fotografen die maschinelle Reproduktion ihres Konterfeis zubilligt und zugleich suggeriert, mithilfe eines Apparates eine höhere Erscheinung zu fixieren. Die fotografische Technik beruht auf unterschiedlichen apparativen wie intellektuellen Grundlagen, und seit ihrer Entstehung sind in ihr Mythen der Echtheit und Naturtreue fest eingeschrieben. In dieser Hinsicht schloss die Fotografie insgesamt an das Konzept des echten Bildes an, wie das Bildnis des Politikers dazu auch die formale Nähe gesucht zu haben scheint.

Für Mussolini gilt noch, was die Fotografie ein halbes Jahrhundert zuvor verheißen hatte: Sie „erscheint als Kontaktreliquie des Realen, die das Original durch mediale Metamorphose zu ersetzen vermag, und gibt damit das Thema der Wirkungsmacht der technischen Bilder bis heute vor." (Hesse 2006, S. 54). Mit dem Begriff der „Kontaktreliquie" wird zum Ausdruck gebracht, dass Form und Material in einer physischen Beziehung stehen, die es gestattet, auch das reproduzierte Bildnis als Abdruck des Echten zu verstehen und zu verehren. Dies gilt besonders für die Reliquien und Bildnisse von Heiligen, kann aber auch auf andere Motive und Medien übertragen werden, in denen sich etwas scheinbar von selbst und unverfälscht einschreibt: Totenmaske, Siegelwachs und Stempel sollen eine Form unverändert weitergeben, wie der Philosoph Nikolaus Cusanus (1401–64) konstatierte, und damit Erinnerungsbilder festhalten.

Auch die *Silhouette* (Umrisszeichnung), der *Physionotrace* (eine Technik zur mechanischen Übertragung von Profilen in Kupferstiche) und der Scherenschnitt, die um 1800 eine große Mode erfuhren, standen für eine scheinbar direkte Fixierung eines Wesens oder Antlitzes, das dauerhaft in Erinnerung behalten werden soll. Obwohl Silhouette und Schattenwurf keine Binnenzeichnung zeigen, wurden sie zu einem Symbol von Bildlichkeit schlechthin (→ KAPITEL 10.3). Als sol-

Das Foto als Kontaktreliquie

Umriss und Idealform

135

che wurden sie auch in Karl Friedrich Schinkels Gemälde *Die Erfindung der Malerei* von 1830 gefeiert (→ ABBILDUNG 29), in dem die Menschen des Altertums einander als Schatten abzeichnen – ein Gemälde, das zwar auf Plinius' Bericht in der *Naturalis historia* (Plinius 1997, S. 20–23) Bezug nimmt, wonach die Malerei mit dem Abmalen der Schatten begonnen habe, das aber nicht zufällig auch in die Zeit der intensiven fotografischen und drucktechnischen Experimente fällt. Trotz Verwendung von Hand und Zeichenstift wird hier ein Bild reproduziert, das durch Schattenwurf bereits vorgezeichnet ist.

Ursprung der Malerei

Die Protagonisten der neuen Bildtechniken, wie die Franzosen Nicéphore Niépce, Louis Daguerre und Hippolyte Bayard oder die Briten Tom Wedgwood, John Herschel und William Henry Fox Talbot, operierten mit verschiedenen Chemikalien (zumeist Silbersalze) und Apparaturen, um lichtempfindliche Träger wie Metall, Glas und Papier zur Selbstaufzeichnung von Helligkeitswerten zu nutzen, teilweise mithilfe von Kameras, teilweise durch direkten Abklatsch von Ge-

Experimente zur Lichtzeichnung

Abbildung 29: Karl Friedrich Schinkel: *Die Erfindung der Malerei* (Ausschnitt) (1830), Öl auf Leinwand

genständen. Diese frühe Fotografie war weder einfach noch preiswert zu haben, noch war sie schnell (→ KAPITEL 13.2). Typisch für die meisten fotografischen Verfahren, die nach 1839 allgemein bekanntgemacht wurden, blieb, dass es sich um latente (verborgene) Bilder handelt, die erst nach Behandlung einer Emulsion mit Chemikalien, durch Umdruck oder nach Überblendung mehrerer Schichten ein Muster erkennen lassen; sie waren in der Regel auch nicht reproduzierbar. Ein erstes erfolgreiches Kopierverfahren, Talbots *Kalotypie* aus der Mitte der 1840er-Jahre, verwendete transparentes Papier als Negativ und gestattete wie die anderen neuen Verfahren auch Retuschen, Übermalungen und ähnliche Eingriffe. Auch traten durch die neue Form der Aufzeichnung technisch bedingte Phänomene wie extreme Tiefenschärfen, Körnungen, Weichzeichnungen und Fehlbelichtungen auf, die eigentlich mehr über die neuen Aufzeichnungsmethoden als den Gegenstand verrieten. Dementsprechend gab es eine große Zahl an Varianten, Bildqualitäten und Fehlschlägen (vgl. Batchen 1997, S. 50). **Das latente Bild**

Die Rhetorik, mit der sowohl die mit Linse und Kamera erzeugten Fotografien als auch alle Formen des kamerafreien Kontaktbildes begrüßt wurden (Naturselbstdruck, Fotogramm, Strahlenbild), hat dessen ungeachtet die immaterielle Lichtzeichnung mit der Idee des ikonischen Urbilds verbunden, das für Echtheit bürgt; sie hat den Triumph der Technik über die Natur gefeiert – wie Talbot in seiner berühmten Abhandlung von 1839 (bei Newhall 1980, S. 25) –, obwohl sich Natur im neuen Medium gerade selbst zeigen sollte. Die frühe Theorie der Fotografie zeigt sich überdies dem naturphilosophischen Denken des 18. Jahrhunderts verhaftet, wonach die Welt in ihrem stetigen Wandel dauerhaften Urbildern folge, die es zu fassen gelte (vgl. Nickel 1998, S. 20f.; → KAPITEL 4.2). Künstler waren daher keineswegs irritiert durch die neuen Erfindungen, da deren Ästhetik und Wahrheitsanspruch in weiten Teilen durch Kunst selbst eingeübt war (vgl. Stelzer 1966; Scharf 1968) und die ersten Fotografen zudem durchweg zeichnerisch ausgebildet waren. **Verbindung intellektueller Traditionen**

Obwohl die Fotografie, für deren Erfindung es kein exaktes Datum gibt, demnach eher der Kulminationspunkt einer längeren intellektuellen Entwicklung gewesen ist, hat sich die Idee der unverfälschten Abbildung als einer ihrer dauerhaften Gründungsmythen etabliert. Die Naturwahrheit des Bildes als Erfassung des Typischen, Wesentlichen oder Idealen, wie sie vorher gerade dem geistigen Blick der Malerei und Zeichnung überlassen war, wurde im 19. Jahrhundert zunehmend mit dem fotografisch erzeugten Bild, im Sinne automatischer, interventionsfreier Selbstreproduktion, gleichgesetzt. Die- **Gründungsmythen der Fotografie**

ses Wechselspiel von Naturstudium und Medientransformation führte im Laufe des 19. Jahrhunderts zu einem Begriff von ‚Objektivtät‘,
der zur Richtschnur wissenschaftlicher Arbeit und maßgeblich für
den Einsatz von Darstellungs- und Reproduktionstechniken wurde.
Im Wandel der Medien haben sich dabei unmerklich Begriffe der
Echtheit, Wahrheit, Naturtreue oder Objektivität verschoben (vgl.
Daston/Galison 2007). Auch der Begriff der Evidenz (seit dem Mittelalter gebräuchlich für Anschaulichkeit, Ersichtlichkeit oder juristische Beweiskraft) erfuhr durch den Aufstieg der Fotografie erhebliche
Deutungsverschiebungen (vgl. Ginzburg 2002; Tucker 2005).

Naturwahrheit und Objektivität

Zum Ende des 19. Jahrhunderts und unter dem Eindruck der neu-
en Röntgen'schen Strahlenbilder wurden Fotografien als Beweis dafür erachtet, unsichtbare Ausstrahlungen des Menschen – dessen
„Aura" – oder Geistererscheinungen auf der Bildplatte festhalten zu
können (vgl. Guibert 1993; Ausst.-Kat. Mönchengladbach 1997).
Während Fotografie und bildende Kunst stets in einem produktiven
und dialektischen Wechselverhältnis gestanden haben, wurde womöglich erst durch diese Mode die geltende Idee, dass ein Bild als
Artefakt dort auftrete, wo der Künstler kreativ in die Form der Natur eingreife, infrage gestellt.

Geisterbilder

Auch vor dem Hintergrund derartiger Phänomene und Debatten
erhielt die Fotografie bei ihren Kritikern im frühen 20. Jahrhundert
den negativen Beiklang, nichts als Geister und Schatten hervor-
zubringen (vgl. Kracauer 1927, S. 30). Besonders mit dem Begriff des
Massenmediums Fotografie verbinden sich seither hartnäckige Vorstellungen über die Abwertung der Bildkunst durch Abklatsch, Wiederholung und vermassungsbedingte Trivilialisierung, denen die Originalität schöpferischer Individualleistung oder die Wahrhaftigkeit
des Einzelstücks gegenüberstehen. Von Georges Didi-Huberman ist
auch bemängelt worden, dass die kunsthistorische Sicht auf Bilder
ein eingeengtes Verständnis des ‚Ikonischen‘ erkennen lasse, welches
in dieser fotografischen Tradition stehe und die Körperlichkeit des
Abdrucks und somit eine wichtige Dimension der Bildgeschichte un-
terschlage (vgl. Didi-Hubermann 1999).

Gleichförmigkeit des Massenmediums

9.3 Image, Prägung, Typ

Es ist jedoch kein Zufall, wenn der Begriff der „Ikone" in der All-
tagssprache transformiert worden ist zu einem Ausdruck für bestimmte formale Muster mit Wiedererkennungswert, die eine phy-

sisch entfernte Welt zugleich bildlich näherrücken – etwa in der „Pop-Ikone" oder in bestimmten „Images" massenmedial präsenter Personen. Als ‚ikonisch' bezeichnet werden auch historische Motive, staatliche oder sogar abstrakte Symbole, die sich durch Einspeisung in die Bildzirkulation zu dauerhaften Mustern verdichten können, für welche Pörksen die alternative Bezeichnung „Visiotyp" vorgeschlagen hat (Pörksen 1997, S. 24).

„Images"

Viele dieser Motive sind das Ergebnis von Prägemechanismen, die auf älteren Traditionen beruhen und vor allem im Bereich hoheitlicher Bild- und Bildnisformen mit Währungscharakter anzutreffen sind. So findet in der Welt der Münzen, Banknoten und Briefmarken häufig eine enge Kopplung von Bildnis, Gedenk- und Währungsfunktion statt, da das Zahlungsmittel auf breiter sozialer Anerkennung beruht, die durch die Wiedergabe von Gesichtern visuell besiegelt wird.

Münzprägung

Insofern solche Zahlungsmittel an tradierte Muster anschlossen, um Geltung zu erlangen, wurden sie zugleich zu Projektionsflächen für weitergehende historische und politische Ansprüche. Sie trugen z. B. das Konterfei von Regenten in die entlegenen Bereiche eines Landes und festigten dadurch Bild und Territorium gleichermaßen, bedienten sich dabei aber auch eingeübter Bildformen, um den Währungscharakter ihrer Bilder zu unterstreichen. Auf diese Weise entwickelte sich im Bereich der Münzen und Medaillen (vergleichbar dem Enface-Bildnis der Ikone) vor allem das Profilbildnis weiter, das seit der Antike geläufig war und nach dem Vorbild des römischen Kaiserbildnisses für einen imperialen Anspruch stand (→ KAPITEL 7.1).

Profilbildnis als klassische Form

Die deutschen Kaiser des Mittelalters haben sich dagegen thronend und in frontaler Ansicht auf ihren Münzen und Siegeln präsentiert, in Anlehnung an das christliche Motiv des Weltenrichters, um ihr Gottesgnadentum zu belegen (→ ABBILDUNG 30 zeigt Kaiser Friedrich II. in dieser Haltung, nach seiner Krönung zum König von Jerusalem im Jahre 1229). Die Person wird dabei zum Stoff eines übergeordneten Bildregimes mit wechselnden Beschriftungen und Attributen. Durch die Herstellung solcher Siegelbilder formt sich ein Typus (griechisch/lateinisch für: Gestalt, Rolle, → KAPI-

Abbildung 30: Kaiserbulle Friedrichs II., Cividale (April 1232), Goldbulle

Typus

TEL 4.2) oder Charakter (griechisch für: Einritzung), der nicht das individuelle, sondern das ‚wesentliche‘ oder unveränderliche Erscheinungsbild meint, das über die Person hinausweist. Im Gegenzug hält der Regent sein Bildmonopol aufrecht, das durch Reproduktion noch befestigt wird.

Ein solcher kollektiver Mechanismus der Bildprägung ist auch im Begriff des Stereotyps (griechisch für: festes, plastisches Bild) enthalten, welches vom Wortsinn her die Prägeform und damit die Unveränderlichkeit einer Abbildung – gleich ob in gemalter, gestempelter oder fotografierter Form – meinte. Deren Reproduktion im Holzschnitt, als Siegel oder in der Fotografie erhöht nur den Währungscharakter, da sie für die Zirkulation und Weitergabe von Mustern und so für die Verankerung im Bildgedächtnis sorgt. Hierauf beruht auch der Begriff einer ethnologischen oder komparatistischen Stereo- **Imagologie** typforschung (oder Imagologie), die nach den kollektiven Werturteilen fragt, die sich in Bildern und Texten abzeichnen, etwa nationalen oder ethnischen Charakteren, und die sich in der gesellschaftlichen Erinnerung sedimentieren.

Wenn Mussolini eine Bildnisfotografie als Postkarte austeilt und auf deren weißen Sockelbereich seinen Namenszug setzt, so schließt er an Traditionen der Geldprägung und andere hoheitliche Bildfor- **Beglaubigung** men an (→ KAPITEL 7.3). Wie in der Umschrift von Medaillen wird durch namentliche Kennzeichnung das Bildnis noch zusätzlich beglaubigt. Daneben macht sich bereits der Name des Fotografen vernehmlich, der als Urheber direkt unterhalb der Fotografie erscheint (und in anderen Varianten seine Signatur direkt in die Aufnahme geschrieben hat). Die konkurrierenden Unterschriften lassen zugleich erkennen, wie sich der Begriff der ‚Echtheit‘ oder ‚Eigenhändigkeit‘ der Bildherstellung im Medienwechsel immer weiter verschiebt.

9.4 Das Bild als Zeitdokument

Es wird in der Literatur weiterhin als Wesensmerkmal der fotografischen Technik angesehen, dass diese einen verweisenden Charakter besitze und im Unterschied zur Malerei, die ihren Gegenstand frei erfinden könne, immer auf diesen bezogen bleibe, indem sie dessen Lichtwirkung lediglich aufzeichne. Hierfür wurde der Begriff der In- **Indexikalität** dexikalität geprägt (vgl. Krauss 1998, S. 79), der meist unter Hinweis auf das logische Schema *Ikon – Index – Symbol* verwendet wird, mit dem der amerikanische Philosoph Charles S. Peirce

(1839–1914) unterschiedliche Verweisformen beschrieben hat (Peirce 1998, S. 307). Diese bezogen sich nicht auf Bilder, konnten aber darauf angewendet werden: Ein Schuh lässt sich abzeichnen (Ikon), er hinterlässt eine Spur (Index), er kann als Ding für seinen Träger stehen (Symbol). In enger Verbindung mit der Indexikalität der Fotografie als Spurbild steht die Frage, ob sie durch digitale Medien grundlegend verändert wurde (vgl. Amelunxen u. a. 1996, → KAPI-TEL 8.4).

In Entsprechung dazu hat sich in der Semiotik der Begriff der Ikonizität für nicht-indexikalische Verweise (z. B. in der Gebärdensprache, in der Wortsprache) durchgesetzt, d. h. für nicht-fotografische Darstellungen. Wie die Beispiele der autopoietischen Ikone und der Münze darlegen sollten, sind zeigende, stellvertretende oder kontaktmäßige Beziehungen jedoch in so vielfältiger Weise überblendet, dass die Übertragung eines semiotischen Schemas auf eine einzelne Technologie wie die Fotografie fehlgehen kann.

Ikonizität

Zum Beispiel lässt sich auch mit einem präzisierten Begriff des Verweises allein noch nicht zeigen, ob Medium, Form und Erscheinungscharakter des Mussolini-Fotos auf besondere zeitliche, intellektuelle oder technische Bedingungen reagierten. Der spätere Diktator bedient sich der Fotografie, obwohl ihm Scharen von Porträtmalern zur Verfügung gestanden hätten, um das Konterfei auf eine bestimmte Idealvorstellung hin zu sublimieren. Nur in dem Wissen um seinen Aufstieg konnte das Foto Mussolinis von Verehrern als Aufzeichnung einer messianischen Erscheinung gelesen, als eine Reliquie be- und gehandelt werden. Zum Zeitpunkt der Aufnahme ist Mussolinis späteres Image, das ihn meist als kahlköpfig untersetzten Militär und als Volksredner in übertriebener theatralischer Pose zeigt, uneinheitlich, die Wandlung vom Aktionisten zum Alleinherrscher noch nicht abgeschlossen – damit ist er auch ein Abbild der widersprüchlichen Strömungen, die unter dem Oberbegriff Faschismus zusammengefasst sind.

Mussolini als historisches Bild

Die konkrete Art der Präsentation des Gesichtes konnte außerdem nur darauf spekulieren, dass sie tatsächlich überzeugt. Ähnliche Postkarten der Zeit zeigen Mussolini in derselben Position und Beleuchtung, jedoch durch Überarbeitungen vollkommen umgestaltet. Die hier gezeigte Aufnahme wurde sogar durch Montage in einen Napoleon Bonaparte verwandelt (vgl. Sturani 1995, Nr. 122). Dem entsprechen zahlreiche andere Aufnahmen, die Mussolini mal als Privatperson, Ratspräsident, Sportler oder Musiker verkleidet zeigen. Einen roten Faden bildet lediglich das Bedürfnis nach Präsenz in den

Nachbearbeitungen

Massenmedien, die zum Austragungsort geltungsbedürftiger Politiker wurden. Mussolini bediente sich eines großen Film- und Fotokonzerns (dem Istituto Luce), der mithilfe von Kino, Lichtshows und anderen Effekten, teilweise aus der Experimentalkunst übernommen, die Bevölkerung zu stimulieren und steuern versuchte. Im Gegenzug **Telekratie** unterwarf sich auch Mussolini den Regeln der Telekratie (griechisch für: Herrschaft der Fernmedien), um als Bildnis möglichst omnipräsent zu sein (→ KAPITEL 8.3).

Sein Porträt von 1919 setzt außerdem, nicht anders als die meisten individuellen Bildnisse, auf die Neigung des Betrachters, sich von einem Gesicht angeschaut zu fühlen. Womöglich knüpft es bereits an **Das Gesicht als** die seinerzeit neue Gestaltpsychologie, an Erkenntnisse der Verhal- **Urbild und Schema** tensforschung oder der Massensoziologie an. Es sucht hier zudem in fast gewaltsamer Weise die Blickbeziehung zum Betrachter, um aus der neuen Fülle von täglichen Bildern weiterhin hervorzustechen – jene spontane Berührung, die Roland Barthes in seinem Essay *Die helle Kammer* 1980 als das *punctum* bezeichnet hat, durch das selbst ein verblichenes Alltagsfoto den Betrachter anspreche (im Unterschied zum *studium* als der historischen Betrachtung von Motiven und Gegenständen, vgl. Barthes 1980). Mussolini baut gezielt auf jene „starke Organisation" des Gesichts (vgl. Löffler/Scholz 2004; Krause 1995), die heute von einem Großteil der modernen Zeitschriftentitel und Produktanzeigen für die appellative Ansprache an den Betrachter genutzt wird (→ KAPITEL 10.2).

Dass der künftige *Duce* – wie auch Hitler durchweg – mit sinistrer Miene in eine Kamera schaute, entsprach einem historisch eingeübten Bild des entschlossenen und verlässlichen Herrschers, das nach dem Zweiten Weltkrieg schrittweise verdrängt wurde. Der heutige Blick auf Mussolinis Fotografie kann jedoch nicht umhin, in der Ausdrucksqualität des Bildes schon das Ende des *Duce* vorweggenommen zu sehen. Außerdem lässt sich bei einer Darstellung, die immerhin einige hundert oder tausend Zuschauer gefunden hat, selbst bei gründlicher Recherche nicht bestimmen, welche Wirkung früher von ihr ausgegangen ist. Ob die verschiedenen Betrachter aus dem Gesicht einen Charakter oder ein Programm herausgelesen ha- **Rezeptionen** ben, einen Retter oder Dämon darin erkannten, blieb der individuellen Rezeption überlassen. Fest steht, dass eine große Zahl von Wählern dem Bild offenkundig eine Botschaft entnommen hat, die sie faszinierte, ohne das kommende Grauen darin wahrzunehmen.

Fragen und Anregungen

- Welche verschiedenen Verbindungen gehen das Porträt Mussolinis und die verwendete Technik ein?

- Definieren Sie den Begriff „Acheiropoieton". Gibt es jenseits einer religiösen Offenbarung noch andere Acheiropoieta? Inwiefern wäre eine solche Bezeichnung gerechtfertigt oder auch nicht gerechtfertigt?

- Dient die Wiedergabe des Herrscherbildes in Münzen oder auf Standbildern lediglich einer Erkennungsfunktion, oder hat sie rein symbolischen Charakter?

- Recherchieren Sie, woher der Begriff „Objektivität" stammt. Was bedeutet er heute und woher hat er seine (zumeist positive) Konnotation erhalten?

Lektüreempfehlungen

- **Geoffrey Batchen: Burning with Desire. The Conception of Photography,** Cambridge, Mass. 1997 (Paperback 1999). *Eine anspruchsvolle und dichte englischsprachige Darstellung, die aber die Darstellungen zur Erfindung und Theorie der Fotografie in vielen Punkten korrigiert*

- **Hans Belting: Das echte Bild. Bildfragen als Glaubensfragen,** München 2005. *Belting, dessen Buch „Bild und Kult" gleichfalls grundlegend ist, schlägt mit dieser Sammlung von Studien einen Bogen von den Anfängen der Bildverehrung zu den neuesten, durch Medien, Kunst und Wissenschaft provozierten Fragen der Bildkritik.*

- **Andreas Beyer: Das Porträt in der Malerei,** München 2002. *Am Anfang war das Porträt – an dieser Gattung lassen sich die historischen Wandlungen von Form, Technik und Funktion besonders deutlich erkennen. Eine so materialreiche wie gut lesbare Einführung, die ein gewaltiges Forschungsfeld bewältigt.*

- **David Ganz / Thomas Lentes: Ästhetik des Unsichtbaren. Bildtheorie und Bildgebrauch in der Vormoderne (KultBild Bd. 1),** Berlin 2004. *Die Herausgeber haben Tagungsbeiträge versammelt, die* *das Verhältnis von Bildverbot und Ästhetik, die Rolle von materiellen Bildern für die Definition des Undarstellbaren und die oft vernachlässigten weltlichen Themen mittelalterlicher Kunst in Fallstudien nachzeichnen.*

- **Peter Geimer (Hg.): Ordnungen der Sichtbarkeit. Fotografie in Wissenschaft, Kunst und Technologie,** Frankfurt a. M. 2002. *Ein zum Standardwerk aufgestiegener, praktischer Sammelband, der ältere und neue Studien zu den Kontexten wissenschaftlicher Fotografie in deutscher Übersetzung vereint.*

10 Bild neben Bild: Vergleich, Kombination, Übersicht

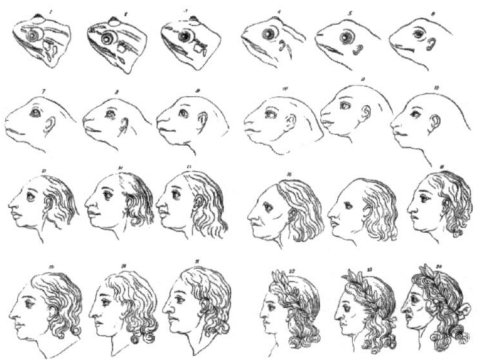

Abbildung 31: Morphologische Transformation vom Froschkopf zum Idealtypus des Gottes Apoll, aus: Johann Caspar Lavater, *Essays on Physiognomy* (1855)

Abbildung 32: John Berger u. a.: *Sehen. Das Bild der Welt in der Bilderwelt* (1974), Cover der deutschen Taschenbuchausgabe

Auch speziellere anthropologische Theorien wie die Ausdruckslehre des 17. Jahrhunderts oder die Physiognomik des 18. Jahrhunderts erfuhren durch den illustrierten Buchdruck größere Popularität. Sie stellten ihren Gegenstand in anschaulichen Bildserien und Übersichtstafeln vor, die auf formalen Vergleichen beruhten und dabei Merkmale oder Tendenzen erkennen lassen sollten. Besonderen Erfolg hatte ein Bildschema, dass die Frage zu lösen schien, wie eine Entwicklung des Tiers zum Menschen vorzustellen wäre, wie in dem hier gezeigten Beispiel aus einem physiognomischen Lehrbuch. Vergleichende und reihende Praktiken erwiesen sich als durchschlagendes Mittel der Argumentation.

Der Umschlag der deutschen Ausgabe eines Buches von John Berger von 1974 zeigt eine nackte, leicht verschleierte Frauenfigur auf einem Gemälde des 16. Jahrhunderts neben einer Werbeanzeige des 20. Jahrhunderts. Durch die simultane Präsentation wird eine zeitübergreifende Kontinuität suggeriert, welche das mythologische Bild einer Venus durch die moderne, oft voyeuristisch geprägte Perspektive zu lesen nahelegt und welche umgekehrt das neue Motiv in eine alte Bildtradition einordnet. Um zu betonen, dass die beiden Bilder weit auseinanderliegenden Epochen und Kontexten entstammen, werden sie zusätzlich datiert und betitelt.

Bilder können in unterschiedlichster Form zu Paaren, Arrangements, Reihen oder Übersichten montiert und durch bildliche wie textliche Mittel weiter verknüpft werden. Hierdurch werden formale Aspekte betont, Aufmerksamkeiten gesteuert und komplexere Aussagen im Sinne von kausalen Zusammenhängen oder Entwicklungen ausgebreitet. Als argumentative oder rhetorische Techniken sind Bildkombinationen oder -gegenüberstellungen gleichermaßen grundlegend für die Vermittlung, Werbung oder auch Propaganda wie für den formalen Vergleich oder die medizinische Diagnose. Sie stellen neue Zusammenhänge und Assoziationen her und können auch neue Fehlermöglichkeiten produzieren. Umso mehr bedarf das vergleichende Sehen, auf der Grundlage von Bildern und darüber hinaus, einer bewussten Anwendung und Übung.

10.1 Die Unmittelbarkeit des Vergleichs

Es ist eine tägliche Erfahrung, dass Bilder, die nebeneinander gestellt werden, eine neue Wirkung entfalten. Manchmal scheint es dabei, dass Bilder sich in einem unmittelbaren Vergleich von selbst in den Vordergrund drängen, bestimmte Lesarten oder Anordnungen geradezu erzwingen. Selbst in einer noch so kleinen Objektmenge stechen Merkmale hervor, die eine Auswahl oder assoziative Lesart begünstigen oder an eine verbalsprachliche Syntax erinnern. Dabei greifen tiefer liegende kognitive Mechanismen, die zunächst unabhängig von bildlicher Repräsentation sind (wie bei der alltäglichen Unterscheidung eines Gesichtes von einem anderen oder bei der Unterscheidung von Geruchs- oder Geschmackswahrnehmungen). Dass nebeneinander stehende Bilder oder Muster Beziehungen eingehen oder auf verschiedenen Ebenen eine höhere Einheit ergeben, ist aber eines der typischen Phänomene visueller Kultur und zugleich die Bedingung für die Deutung jedes einzelnen Bildes, dessen Binnengliederung sich ja ebenfalls aus diversen Untereinheiten aufbaut.

Intuitive Bindungen

Die Herstellung solcher Bindungen sagt nicht nur etwas über die betreffenden Bilder aus, sondern legt vor allem die Einstellung ihrer Betrachter frei. Individuelle Aufmerksamkeit, Erinnerungsvermögen, Schulung, physische Ermüdungserscheinungen entscheiden darüber mit, ob Muster und Motive in einem Buch, auf einem Leuchttisch oder in einer Kartei tatsächlich als Elemente einer Menge mit vergleichbaren Eigenschaften erkannt werden. Motivtraditionen, Moden, Aufführungsorte, Wiederholungseffekte, professionelle Routinen kommen hinzu. Wenn Fotografen oder Bildredakteure aus einer Reihe von Bildern ein Titelbild auszusuchen haben, bewerten sie Aufnahmen bereits vor dem Hintergrund, dass eine Auswahl unvermeidlich ist. Der prüfende Blick bringt dabei durchaus auch spontane Kriterien des Vergleichs und der Beschreibung hervor, die erst in einem zweiten Schritt rationalisiert werden können. Allenfalls in einem automatisierten Vergleich am Rechner, z. B. bei einer Suche nach bestimmten Mustern in einer Bilddatenbank oder nach der Abweichung zweier Datensätze, sind bestimmte Kriterien schon vorab definiert, die über den Erfolg einer Suche oder über den diagnostischen Wert eines Ergebnisses entscheiden sollen.

Eingeübte und gesteuerte Sehweisen

Selbst das scheinbar zufällige Nebeneinander zweier Abbildungen in einem Archiv oder Album ist nie ohne Geschichte und Voraussetzung. Was aus einem Meer von Abbildungen von selbst hervorzutreten scheint, ist nicht nur von Ausdrucksqualitäten und Bewertungen

Bildkultur und Vergleich

abhängig, sondern auch von der Überlieferungssituation, von technischen und medialen Bedingungen wie Verfügbarkeit und Betrachtbarkeit (→ KAPITEL 3.2).

Zusammenwirkung von Bildern

Sämtliche Arten von mehrteiligen Bildern, Bildserien sowie Ausstellungen und Sammlungen beruhen aber im weitesten Sinne auf den assoziativen Wechselwirkungen von Bildereignissen und Sehweisen und werden so zum Abbild kognitiver Vorgänge: die spielerische oder esoterische Kombinatorik des Barock, das Zufallsexperiment des Surrealismus, die künstlerische Collage mit ihren verfremdenden Schnitten oder die Montagetechnik des Films. Diese fügen Einzelbilder zu Ensembles, Reihen oder Abfolgen zusammen und fokussieren dadurch auf bestimmte formale Aspekte, die eine Beziehung herstellen. Sie visualisieren Entwicklungen, Trends, Zeitraffer, Oppositionen oder Gemeinsamkeiten, Vorher-Nachher-Effekte und kausale Zusammenhänge. Hieraus entstehen in einem weiteren Sinne komplexere Argumente und Erwartungen. Dieser Komplex, in dem ein Bildobjekt gezeigt wird, kann selber argumentative Bedeutung gewinnen und die Wirkung des Objektes in erheblichem Maße verändern.

Als Beispiel kann hier das Titelbild der deutschen Ausgabe von John Bergers Buch *Sehen. Das Bild der Welt in der Bilderwelt* von 1974 dienen (→ ABBILDUNG 32), das ein leicht verschleiertes weibliches Ganzfigurenbildnis aus der Werkstatt Lucas Cranachs aus dem 16. Jahrhundert neben eine Werbeanzeige des 20. Jahrhunderts stellt, bei der eine Frau beim Duschen zu sehen ist. Bildüberschriften sollen die Darstellungen zeitlich einordnen. Dass beide Figuren nackt sind, verbindet sie – dass sie ihre Nacktheit unterschiedlich zeigen, trennt

Historische Reihen

sie. Indem die moderne Bilderwelt des Anzeigen-Fotos in eine Reihe mit namhaften Werken der Kunstgeschichte gestellt wird, kann eine solche Gegenüberstellung einen historisch ununterbrochenen Bedarf nach voyeuristischen Darstellungen suggerieren. Aus der Paarbildung ergäbe sich durchaus die Deutung, dass Bilder zu allen Zeiten Ausdruck des Voyeurismus waren. Eigentlich zielt das Buch jedoch gerade auf die unstillbare Konsumsucht der modernen Welt ab.

10.2 Text gegen Bild?

Ob unterschiedliche Betrachter tatsächlich aus einer Bildpaarung Analogien, Differenzen, Typen oder Stilreihen herausfiltern und ob ein bestimmter Vergleich bei einem unbekannten Rezipienten auch die gewünschte Wirkung erzielt, bleibt unwägbar. Wie die zahllosen

Beispiele der politischen Reklame zeigen, bei denen sich Herrscher in eine Reihe mit politischen Ahnen stellen, um sich darüber zu legitimieren (etwa die Kopfreihe *Marx / Engels / Lenin / Stalin*), wirkt mancher genealogische Vergleich eher bemüht und entlarvend. Eben darum bleibt es selten bei der reinen Zusammenfügung von Abbildungen, sondern kommen andere Elemente wie Beschriftungen, Pfeile oder Farben zum Einsatz.

Zusätzliche Mittel

Selbst die Kombination von Bild und Text geht noch von der Syntheseleistung aus, wonach einzelne Objekte in montierter Form ein neues Bild, hier genauer: ein neues Bildkonzept ergeben. Hierunter fällt auch die nach 1900 aufgekommene Idee, Industrieprodukte wie Kosmetika, Kleidung, oder Nahrungsmittel mit Porträts von Personen, insbesondere mit positiv gestimmten weiblichen Gesichtern, zu verknüpfen, und zwar unabhängig von jeglichem inhaltlichen Zusammenhang. Dies beruhte auf der Idee des Blickfangs oder *eye catchers,* bei dem ein auffälliges Bild (Schild, Signal) auf ein weniger auffälliges Produkt verweist, mit welchem es möglichst eng verkoppelt wird.

Blickfang

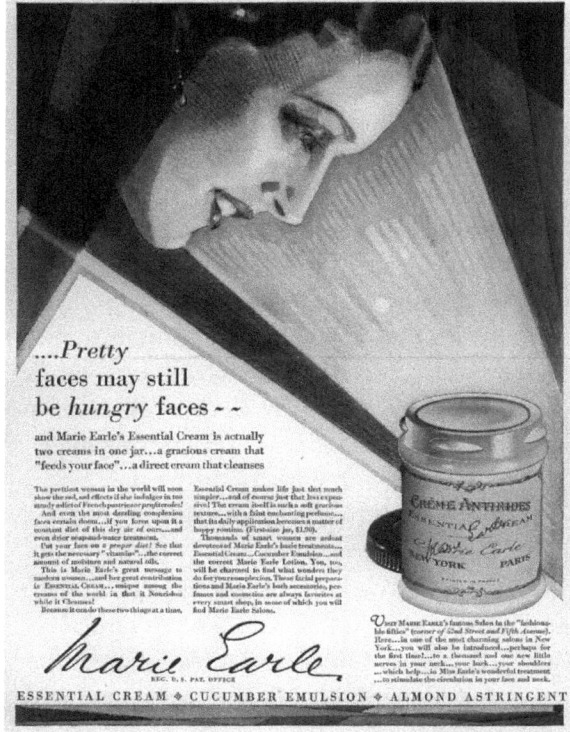

Abbildung 33: Werbeanzeige für *Marie Earle Kosmetikcreme*, USA (späte 1920er-Jahre)

Verbindung von grafischen Mitteln

→ ABBILDUNG 33 zeigt die Anzeige aus einer US-amerikanischen Unterhaltungszeitschrift der späten 1920er-Jahre. Darin ist der Kopf einer modisch geschminkten Frau im Profil mit einfachen und lavierten Strichen, nach Art einer Modezeichnung, dargestellt, der von unten rechts von einem abstrakten Lichtmuster beleuchtet wird. Lichtbänder gehen von einer Kosmetikdose aus, die aus der Anzeige plastisch hervorzutreten scheint, und ziehen den Blick der Frau auf sich. Zeichnerische Mittel kommen zum Einsatz, um den Gegenstand in plakativer Form darzubieten.

Texte werden knapper

Während das Motiv des Mangels ein Leitthema der Schönheitsindustrie geblieben ist, hat sich das Verhältnis von Bild und Text in den letzten hundert Jahren kontinuierlich verschoben zugunsten kurzer und möglichst selbsterklärender Botschaften. In der Anzeige dient die Grafik noch der Überredung, einen Text zu studieren, der an die Vernunft der Kundin appelliert. Die Benennung konkreter Vorzüge konnte später mehr und mehr entfallen, die Kombination von Bildern mit Objekten oder Texten wurde so alltäglich, dass kaum noch bemerkt wurde, dass sie überhaupt erst einmal erfunden werden musste.

In der heutigen Anzeigenwerbung oder Produktgestaltung ist die Bindung von Bild und Text untrennbar. Die freie Assoziation von unabhängigen Bildmotiven und Produktgestalt wurde aber aufgrund ihres Erfolges zum Standardmittel der visuellen Kommunikation, die eine wachsende Produktferne der Werbung mit sich gebracht hat. Immer seltener spielen die zu verkaufenden Gegenstände in Anzeigen die Hauptrolle. Wie die Werbepsychologie zu belegen versucht, die

Kundenbindung

um 1900 in den USA aufkam (vgl. Scott 1908), ist Werbung in der Aufmerksamkeitsökonomie nicht mehr Überzeugungsarbeit, sondern Kontaktintensivierung und emotionale Vereinnahmung. Im vorliegenden Fall sollte der (damals noch neue) Einsatz des Frauengesichtes den Betrachter vor allem assoziativ einbinden und außerdem eine Botschaft verständlich machen, die sprachlich reichlich sperrig wirkt: „Schöne Gesichter können durchaus hungrige Gesichter sein, weil sie der Nährstoffe bedürfen." Nach demselben Prinzip soll der Konsument seit den 1940er-Jahren *corporate identities* von Großkonzernen verinnerlichen und Loyalität zu einem Produkt entwickeln, indem er dessen ‚Produktwelt' selbsttätig weiterdenkt.

Jeder Gang durch einen Supermarkt legt diese Regeln der industriell verfassten Konsumgesellschaft frei, wo Verpackungen nach den Regeln der Industriegestaltung so aufgebaut sein sollen, dass sie die Kauflust wecken, aber auch schnell auszumachen sind, sodass sich aus Materialien, farblichen, typografischen und anderen Mitteln ein komplexeres Produktbild ergibt. Eine zentrale Bedeutung hat dabei die Ausstattung der Packung selbst mit Abbildungen gewonnen, seit die Hersteller im Kampf um Marktplätze und Regalmeter über Packungsinhalte informieren wollen oder ihre Markenprodukte zu größeren Serien ausbauen, deren Inhalte bildlich angezeigt werden sollen (etwa die Geschmacksrichtungen von Joghurts in Form von Früchten).

(Randnotiz: Bilder auf der Packung)

Dabei ist bezeichnend für die symbolische Funktion von Bildern, dass solche Zusätze selbst dann instruktive oder informative Funktion haben können, wenn der Gegenstand, den sie zeigen, hochgradig abstrahiert ist oder in nichtgrafischer Form kaum noch angetroffen wird. So kommt seltenen Tierarten wie dem Hasen oder Adler die zweifelhafte Ehre zu, das Logo oder die Grundform für Lebensmittel zu sein (Schokoladenhasen), obwohl sie dem Großstädter nur noch durch den Zoo oder das Fernsehen bekannt sind. Kunden können darin gleichwohl ein kulturell fest verankertes Symbolwesen erkennen.

(Randnotiz: Verselbstständigung von Zeichen)

Diese enge Bindung von Motiv und Bezeichnung wurde durchaus als problematisch angesehen. Die sozialpsychologische Forschung Frankfurter Prägung der 1920er-Jahre und die aus ihr folgende Kritische Theorie, ebenso die anglo-amerikanischen *cultural studies,* denen der angeführte Band Bergers noch zuzurechnen ist, konstatierten, dass die sogenannte Kultur- oder Bewusstseinsindustrie durch Musterbilder gekennzeichnet ist, welche die entfremdete und geistig verarmte Lebens- und Denkweise des modernen Menschen spiegeln, der bei einem Bildmotiv nur noch den Kaufanreiz verspürt (vgl. Horkheimer / Adorno 1944; Enzensberger 1964, → KAPITEL 3.3). In der vereinfachenden und unreflektierten Kopplung von Begriffen und Bildern äußert sich eine konsumorientierte Form der Blindheit gegenüber der Umwelt und ihrer Vielfalt.

(Randnotiz: Bilder der Entfremdung)

Diese enge Beziehung von Bildern und ihren Texten ist im Bereich der Werbung und des Journalismus (→ KAPITEL 8.4), wo Bildinformationen fast immer von Texten begleitet oder durch sie verändert werden, jedoch so elementar, dass es in den 1920er-Jahren auch Vorstöße zur ‚Befreiung‘ des Bildes von textlichen Barrieren gab, um die Eigenständigkeit des Visuellen zu stärken, etwa in der Bildreportage und in der Plakatgestaltung. Illustrierte Zeitschriften experimentierten mit Bildpaarungen (z. B. Alfred Flechtheims Monatsblatt *Querschnitt* unter dem Titel *Juxtaposition*). Auch die kritische Beschäftigung der 1970er-Jahre mit visuellen Massenphänomenen führte

(Randnotiz: ‚Befreiung‘ des Bildes vom Text)

dazu, den gewerblichen Bildgebrauch zu revidieren. Es wurde anerkannt, dass ein Bild auch in der Massenreproduktion Träger kulturellen Wissens bleibt (ob als Mittel der Werbung oder Instruktion) und eine eigenmächtige symbolische Bedeutung besitzt, die durch Betextung weder ersetzt noch überschrieben wird (vgl. Ehmer 1971).

Verbraucherschutz im Bild

Inzwischen muss selbst die Konsumgüterindustrie Sorge dafür tragen, dass die Werbefunktion ihrer Verpackungsbilder nicht in Konflikt gerät mit ihrer Orientierungsfunktion. Dies äußert sich z. B. darin, dass Packungsabbildungen aus gesetzlichen Gründen mit Hinweisen wie „Serviervorschlag" versehen werden, um sicherzustellen, dass eine Verpackung nicht verspricht, einen Löffel mitzuliefern, wenn dieser auf einem Joghurtbecher gezeigt wird. Kaum rechtliche Probleme bereitet dagegen die Farbe, die – wie bei *food photography* üblich – vom Produkt deutlich abweicht und dies in den meisten Fällen auch darf. Angesichts der zunehmenden Komplexität internationaler Märkte und ihrer Wettbewerbsbestimmungen entsteht ein so fiktiver wie widersprüchlicher globaler Betrachter-Konsument, dessen Bildkompetenz nur noch im Sinne der Marktforschung definiert wird.

10.3 Visuelle Dialoge: Sammlung, Tafel, Lehrbuch

Bildliches Wissen

Das Sammlungswesen der Frühen Neuzeit kennt verschiedene Räume, in denen sich ästhetisches Wissen unterschiedlichster Bestimmung zusammengefunden hat. Die höfischen Kunst- und Wunderkammern mit ihren *artificialia* und *naturalia* (lateinisch für: künstliche und natürliche Dinge) dienten bis ins 18. Jahrhundert einer systematischen Fragestellung: Hier waren in Schubladen und Vitrinen die ‚merkwürdigen' Objekte der Welt zusammengetragen, deren Ordnung mit jedem neuen Eintrag neu zu bestimmen war. Die Aufnahme in die Sammlung bedeutete eine Hervorhebung des einzelnen Objekts (vgl. Pomian 1988); hierdurch wurde das Sammelwesen zugleich zum Vorläufer des modernen Labors (vgl. Bredekamp 1993). Dem prüfenden Forscherblick sollten sich in einer vergleichenden Betrachtung die geheimen Spielregeln der Natur zu verstehen geben (→ KAPITEL 4.2). Auch in Lehranstalten neueren Typs, etwa der Humboldt'schen Berliner Universität von 1810, behielten die Kunst- und Natursammlungen ihre Bedeutung für die Schulung des vergleichenden Blicks und die Produktion von Wissen.

Das heutige Museum (griechisch für: Ort der Musen), das zunächst nur einen gedruckten Katalog von Objekten bezeichnete, beruht auf solchen Kabinetten, auf fürstlichen und kirchlichen Kunstschätzen sowie auf den Kollektionen humanistischer Gelehrter, die sich aus Neugier (lateinisch *curiositas*) dem Studium antiker Monumente, den Naturalien oder dem Münzwesen verschrieben hatten. Im 17. und 18. Jahrhundert erarbeiteten sogenannte Dilettanten, Laien oder Amateure meist adliger Herkunft auf dem Gebiet der Botanik, Mineralogie oder Chemie auf diese Weise ein beträchtliches Wissenskapital, zumal sie sich in königlichen Gesellschaften (etwa der englischen Royal Society von 1660), in Akademien oder bürgerlichen Vereinen („Fruchtbringenden" oder „Naturforschenden" Gesellschaften) organisierten und ihre Sammlungen zum Zwecke der Kommunikation unter Gleichgesinnten abzeichnen und drucken ließen.

Größte Bedeutung kam folglich den Fortschritten im Buchdruck und in der grafischen Illustration zu, um die mannigfaltigen Phänomene der Natur zu konservieren und weiterzugeben. Für die taxonomische Bestimmung von Tier- und Pflanzenarten, Mineralien oder Fossilien und für anthropologische Zwecke wurden reproduzierbare Abbildungen, meist auf gesonderten Tafeln, im internationalen Austausch als Medium unerlässlich. Der Zuwachs an Wissen erforderte Ordnung durch Listen, Tableaus und gliedernde Begriffstafeln (Synopsen). Der Bildvergleich, zu instruktiven oder argumentative Zwecken, in anatomischen oder zoologischen Schautafeln, in Enzyklopädien und Almanachen, wurde dadurch weiter forciert und eingeübt, ebenso die Typisierung von Formen durch die Umsetzung für den Druck und den Kupferstich.

Umgekehrt erwuchs der wissenschaftlichen Illustration seit dem Buchdruck mehr Gewicht bei der Definition und Verbreitung wissenschaftlicher Theorien. Die Physiognomik, seit dem 16. Jahrhundert auch in illustrierten Lehrwerken behandelt, hätte sich ohne das Druckwesen nicht durchgesetzt. So war den *Physiognomischen Fragmenten zur Beförderung der Menschenkenntnis und Menschenliebe* (1775–78) von Johann Caspar Lavater (1741–1801) gewaltiger Erfolg als Bildband und in separaten Kupferstichen beschieden, mit mehr als einhundert folgenden Handausgaben (→ ASB KOŠENINA). 1804 erschien in englischer Übersetzung eine posthume Bearbeitung mit über 400 Darstellungen.

Lavater versprach, dem Wesen menschlicher Charaktere durch die Sammlung von Kopfstudien (zunächst nur Porträts, Silhouetten) auf die Spur zu kommen und fügte seinem Werk auch Muster der Proportions- und Ausdruckslehre aus der Zeit seit Dürer hinzu. Die Grafik der *Transformation vom Froschkopf zum Idealtypus des Gottes Apoll* (→ ABBILDUNG 31) war in den ersten Ausgaben und Übersetzungen noch nicht enthalten, beruhte aber auf Vorstudien Lavaters. Sie sollte den abstrakten Umstand anschaulich machen, dass sich bei bleibender Grundform aus dem Affen der Mensch ‚entwickeln' kann, und so die antike Ideenlehre als unsichtbare Einheit von Naturerscheinungen weiterdenken. Die Reihe führte ein Schema des Arztes Peter Camper (1722–89, → KAPITEL 4.2) weiter, das dieser 1791 in Bildform vorgestellt hatte. Lavaters Schautafeln ebenso wie die schädelkundlichen Studien Franz Joseph Galls (1758–1828), in denen ebenfalls Evolutions- und Rassenkonzepte ausgebreitet waren, zeichneten sich durch ein diskriminierendes Denken aus und konnten später wegen ihrer Demonstrations- und Diffamierungskraft auch ohne Weiteres ideologisch vereinnahmt werden (vgl. Wechsler 1993).

Im Zeitalter der Fotografie hat die Sammlung und Reihenbildung von massenhaft hergestellten Aufnahmen den kriminalistischen Blick weiter befördert, das Bild geriet in den Sog industrialisierter Informationsbeschaffung. Bekannte Beispiele sind die Experimente zur Ermittlung von Gesichtstypen durch Foto-Überblendungen (vgl. Galton 1878) und die anthropometrischen Studien von Alphonse Bertillon 1879/80, ebenso die Fotografien von Psychiatriepatienten (*Iconographie photographique de la Salpêtrière*, Paris 1876–80), welche psychische Auffälligkeiten typenhaft repräsentieren sollten. Die Vielfalt der Erscheinung wurde stets zugunsten der klassifizierenden Sicht reduziert, den Gesetzen der Apparatur und der Kartei unterworfen, die sich an Mehrheiten orientiert (vgl. Regener 1999). Das vermeintlich interventionsfreie Bild war dabei in vielen Fällen präpariert, retuschiert, nachgestellt (→ KAPITEL 9.3), das Gesicht *en face* wurde zum Leitbild der Merkmalerkennung, deren Individualität vor allem Identifizierung meint (→ KAPITEL 6).

Im 19. Jahrhundert entstanden, begünstigt durch höhere Druckzahlen und Drucktechniken wie die Lithografie, größere illustrierte Werke mit einheitlich aufgebauten Abbildungsteilen, die als *Icones* (etwa *Icones plantarum*, Bilder der Pflanzenwelt) und seit Mitte des 19. Jahrhunderts als *Bild-Atlas* oder *Bilder-Conversations-Lexika* (wie von Brockhaus 1838) bezeichnet und angeboten wurden. Als Tafelwerk meinte der „Atlas" keine Kartensammlungen, wie im heutigen Wortgebrauch, sondern ein Medium mit großer Abbildungszahl. Die entsprechenden Werke lieferten Abbildungen zur Erdkunde, zur Baugeschichte, Botanik und anderen Fachgebieten, indem sie ihren Gegenstand mithilfe von Kupferstichen und Farbtafeln in eine

Individuum und Apparat

Atlanten, Karten, Übersichten

Übersichtsform brachten (als Bildatlas der Medizin, Bildatlas der Baukunst usw.), darin an Comenius' Vorbild des *orbis pictus* (→ KAPI-TEL 1), die Bildungsideen der Aufklärung oder den Anschauungsunterricht nach Johann Heinrich Pestalozzi (1746–1827) anschließend. Durch die Lithografie befördert wurde auch die Herstellung von farbigen Schulwandtafeln, darunter seinerzeit weltweit gefragt die Arbeiten des Zoologen Rudolf Leuckart (1822–98).

Um Atlanten in einem der Kartierung vergleichbaren Sinne handelte es sich aber insofern bei vielen Druckwerken, als mit der Anordnung von Bildern zu Reihen und Musterkatalogen auch Lücken und Fehlstellen nahegelegt und damit Ziele weiterer Forschung, im Sinne von *missing links*, bezeichnet wurden. Der Fokussierung dienten sie durch Systematisierung und Anreicherung des Materials, das in seiner Kompaktheit auf der Tafel eine eigene Logik entfaltet. Im Laufe von gut 200 Jahren wurden wissenschaftliche und Lehrinhalte in diesem Prozess zunehmend konventionalisiert. Anordnungen, Skalen und Legenden, Kombinationen von Auf- und Seitenansichten, Grafiken und Fotografien dienten der erhöhten Anschaulichkeit und ermöglichten die heute durchgängige wissenschaftliche Argumentation mit Statistiken und fotografischen Belegen (→ KAPITEL 11.4).

Missing links

Die verbilligte und vereinfachte Reproduktion von Kupferstichen und Abgüssen, später von Fotografien und Bilddrucken begünstigte seit dem 18. Jahrhundert ebenso die Zusammenführung und Veröffentlichung entlegener Objekte der Archäologie, Kunst oder Ethnologie in virtuellen Ausstellungen, indem Reproduktionen von Artefakten in Fototheken, Katalogen, Korpuswerken und Ausstellungen vereint und vergleichend vorgeführt werden konnten. Die Gestaltung von illustrierten Atlanten und Lehrtafeln gestattete auch, die fixierte Anordnung von Bildern aufzubrechen zugunsten von offenen Arrangements, wie in Aby Warburgs Atlasprojekt *Mnemosyne* zum Nachleben der antiken Bildwelt (vgl. Warburg 2008). Der Kunsthistoriker André Malraux sah mit der Reproduktion ein „imaginäres Museum" entstehen (so sein berühmter Aufsatz von 1947), ein Museum ohne Wände, in dem die Wandlungen und Kontinuitäten der Kunst sichtbar werden (→ KAPITEL 4.3).

Das Museum ohne Wände

10.4 Techniken der Aufmerksamkeit

Jedes visuelle Ordnungsschema, ob als Lehrtafel oder Polizeikartei, übt Sehweisen ein. Eine größere Auswahl bleibt eine Auswahl. Schon

Übersicht
als Machtform

die bloße Übersicht über eine bestimmte Menge nebeneinander-gestellter Daten, als Schautafel, Datenbank oder Aktensammlung, ist ein mächtiges Dispositiv: Die darin vorgenommene Ausrichtung, Auswahl und Anordnung visueller Informationen auf ein konsisten-tes Gesamtbild hin geht einher mit selektiven Wahrnehmungen und interpretationsverstärkenden Kanalisierungen. Der Vergleich wird durch seine visuelle Präsentation auch dort wirksam, wo er nicht the-matisiert ist, zumal wenn er jenseits von Objekten abstrakte Sachver-halte (z. B. Evolution) vor Augen führt.

(Schein)Kausalitäten

Die suggestive Kraft der visuellen Argumentation geriet dadurch für den wissenschaftlichen Gebrauch in Misskredit, weil sie in der Werbung oder in der Politik zu propagandistischen oder diffamieren-den Zwecken eingesetzt worden ist, auch kam es immer wieder zu Fälschungsversuchen und Nachbearbeitungen, wenn Bilder als Beleg-material dem postulierten „Gesetz der Serie" (Kammerer 1919) nicht genügten oder nicht vollständig vorlagen.

Andererseits kann selbst ein Vergleich, der in die Substanz seiner Objekte eingreift, Wissen stiften und Evidenzen erzeugen, die auf an-derem Wege nicht zu gewinnen sind. Die Experimente des Medizi-ners Camillo Golgi (1844–1926) haben beispielsweise mithilfe von

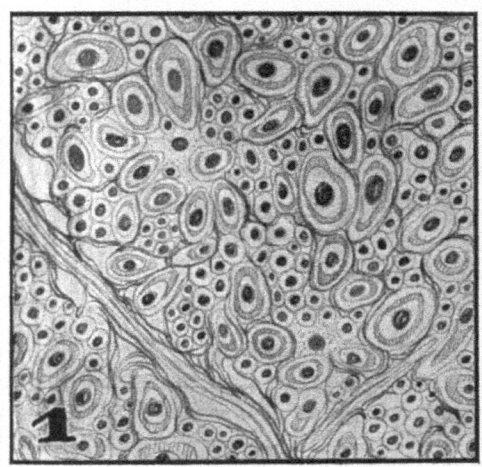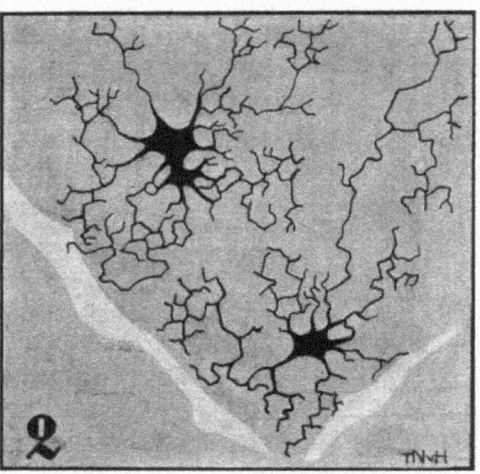

Abb. 26. Die Golgimethode. Ein und dasselbe Nervenpräparat. 1 nach der allgemeinen Färbemethode, 2 nach dem Golgischen Chromsilber-Imprägnations-Verfahren behandelt. Die in Präparat 1 kaum auffindbaren Zellen treten in Präparat 2 isoliert in voller Schönheit zutage.

Abbildung 34: Darstellung der Färbemethode von Neuronen nach Golgi in Fritz Kahns medi-zinischem Lehrbuch *Das Leben des Menschen* (1929)

Silbersalzen Nervenzellen durch Einfärbung sichtbar gemacht und sie damit quasi den Techniken der Fotografie unterworfen. In einem Lehrbuch der 1920er-Jahre hat der Mediziner Fritz Kahn daraus eine Vorher-Nachher-Opposition gemacht (→ ABBILDUNG 34), die einerseits zeigen sollte, dass ein Nervengewebe unterschiedliche Ansichten bieten kann, andererseits auf dem Kontrast von Bildern gleicher Größe und Position gründete, die der Betrachter als verwandt lesen sollte. Beides sind gefilterte Ansichten eines Dinges, das als Bild immer einen Hersteller, eine Technik und damit eine kulturelle Einbettung hat. Vorher-Nachher

Als Technik der Wissenserzeugung und Aufmerksamkeitssteuerung ist der aktive und argumentative Bildvergleich – wie er auch in dem hier vorliegende Band in unterschiedlichen Formen praktiziert wird – unersetzlich geblieben. Durch die Fokussierung auf Bilder und deren Anordnung wird eine spezifische Rhetorik genutzt, welche vorhandene Wahrnehmungsmechanismen oder eingeübte Leserichtungen nutzt und dadurch kausale oder syntaktische Beziehungen herstellt, Merkmale in Form von Kontrasten oder Ähnlichkeiten herausarbeitet. Techniken wie die Projektion von Diapositiven im Unterricht oder die Bebilderung von Büchern haben kontrastive und polarisierende Gegenüberstellungen zum zentralen Instrument eines formanalytischen Sehens in Kunstgeschichte und Archäologie gemacht. Mittel der Wissenserzeugung

Der Vergleich ist nicht unbedingt bildlich, und nicht immer geht es beim Vergleich um formale Ähnlichkeit. Die Komparatistik versucht zu bestimmen, inwieweit sich z. B. bildliche und nicht-bildliche (etwa literarische oder musikalische) Werke sinnvoll auf Strukturähnlichkeiten hin untersuchen lassen. Unsichtbare Merkmale wie eine genetische Information können sichtbare Befunde (etwa morphologische Merkmalsbestimmungen) ergänzen oder korrigieren (→ KAPITEL 11.2). Nichtbildliche Vergleiche

Techniken der Kombinatorik und des Spiels der Bilder, als Teil der bildlichen Imaginations- und Gedächtniskultur (lateinisch *ars imaginatica, ars memorativa*) der Frühen Neuzeit, wurden im Verlauf der Reformation und vor allem mit der fortschreitenden Numerisierung und Mathematisierung im 19. Jahrhundert weiter zurückgedrängt. Der Bildvergleich, bei dem funktional ausdrücklich verschiedene Entitäten aufgrund äußerlicher Ähnlichkeit miteinander zu assoziativen und auch kreativen Zwecken in Beziehung gesetzt werden, bleibt jedoch eine grundlegende Kulturtechnik, denn auch der Vergleich z. B. von vermeintlich reinen Zahl- oder Messwerten beruht an irgend- Vergleich als Kulturtechnik einer Stelle wieder auf morphologischen Kriterien. Auch wenn assoziative oder spielerische Vergleiche und Kombinationen Gefahr laufen, „Verlegenheitslösungen" der Wissenschaft zu sein oder unter dem Verdacht der Vorwissenschaftlichkeit stehen, waren sie ein Mittel zur Gewinnung des neuzeitlichen Wissens und sind ebenso eine Grundlage seiner stetigen Aktualisierung (vgl. Bredekamp 1993; Stafford 1999).

Visuelle Befunde

Einer Auffassung, wonach visuell begründete Schlüsse wissenschaftlich unbrauchbar oder rein suggestiv sind, kann entgegengehalten werden, dass der Vergleich als Technik der Erkennung sich bis in gegenwärtige Disziplinen und Technologien fortsetzt und seine zentrale Bedeutung als Mittel der Diagnose oder Datenerhebung behält (z. B. in der Medizin, in der Biologie, in der Auswertung von Satellitenaufnahmen). Auch hier gilt, dass Theorien und Interpretationen von visuellen Erfahrungen ihren Ausgang nehmen, selbst wenn sie selber nicht visuell sein mögen oder sich vom Visuellen entfernen.

Übung im Vergleichen

Es gehört zur Dialektik jeglicher Wahrnehmung – nicht nur der visuellen –, dass diese nur ausschnitthaft sein kann und auf weiteren Voraussetzungen beruht. Die Gegenüberstellung von Objekten führt zu einer gesteigerten Aufmerksamkeit für Unterschiede und Differenzen, für mediale Eigenarten wie für eigene Wahrnehmungen. Sofern diese Gegenüberstellung durch Bildmedien und Darstellungstechniken möglich wird, dient das vergleichende Sehen zugleich deren kritischer Analyse. Daher sollte es mit umso mehr Nachdruck geübt werden, um die darin vollzogenen Operationen bewusst und als Kulturtechnik nutzbar zu machen.

Fragen und Anregungen

- Wodurch kommt es zu „Gesetzen der Serie" und wodurch erhalten diese ihre Plausibilität?

- Nennen Sie Beispiele für Disziplinen, in denen der visuelle Vergleich wichtig oder unerlässlich ist.

- Gibt es Beispiele, in denen der visuelle Vergleich fehl am Platz ist? Unter welchen Bedingungen wäre dies der Fall?

- Wodurch hat das Aufkommen der Fotografie das vergleichende Sehen grundlegend verändert?

Lektüreempfehlungen

- Horst Bredekamp: Antikensehnsucht und Maschinenglauben. Die Geschichte der Kunstkammer und die Zukunft der Kunstgeschichte, 3. Auflage, Berlin 2007. *Bredekamps Essay ist eine programmatische Studie, die über kultursoziologische oder wissenschaftshistorische Ansätze hinaus die kreative Bedeutung jenes ‚physischen Wissens‘ erkannt hat, das in den Wunderkammern und Naturalienkabinetten manifestiert ist und hier als Grundlage des modernen Denkens und Wissens begriffen wird.*

- Andreas Grote (Hg.): Macrocosmos in Microcosmo. Die Welt in der Stube. Zur Geschichte des Sammelns 1450 bis 1800, Opladen 1994. *Der Band von 1994 vereint museologische und wissenschaftshistorische Beiträge, die jeweils als Einführungen in das Sammelwesen, die historische Museumskultur oder die Entstehung der frühneuzeitlichen Wissenschaften dienen können.*

- W. J. T. Mitchell: Bildtheorie, Frankfurt a. M. 2008. *Mitchell liest Bilder wie Wörter, um deren sprachähnliche Eigenmacht gegenüber dem Textlichen zu analysieren. Text-Bild-Bezüge, das Problem des Vergleichs und andere Aspekte werden in diesem Aufsatzband behandelt, der trotz seines Titels nicht so sehr eine Bildtheorie vorlegen als deren Möglichkeit und Grenzen diskutieren will.*

- Marion G. Müller: Grundlagen der visuellen Kommunikation. Theorieansätze und Methoden, Konstanz 2003. *Im Unterschied zu anwendungsbezogenen Gestaltungslehren entwickelt der Band die analytischen Verfahren und den Beitrag verschiedener Disziplinen zum Studium visueller Kultur einschließlich von Werbung und Presse.*

- Jens Thiele: Das Bilderbuch. Ästhetik – Theorie – Analyse – Didaktik – Rezeption, 2. Auflage Bremen / Oldenburg 2003. *Der Band gibt eine Übersicht in Geschichte und Themen der Bildpädagogik.*

11 Das Bild als Wissensmodell: Diagramme

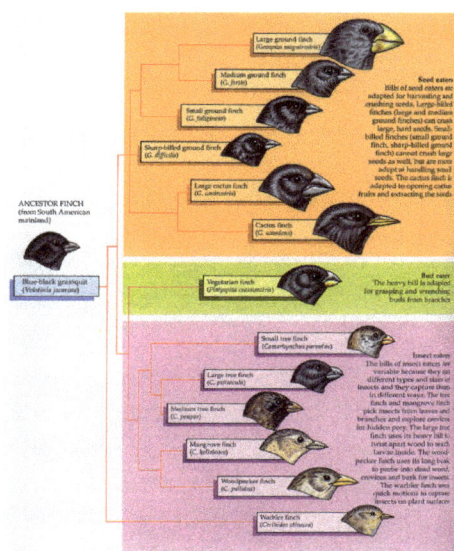

Abbildung 35: Adaptive Radiation:
Anpassung der Schnabelform der soge-
nannten Darwinfinken auf Galapagos
(Ausschnitt) (1995)

Abbildung 36: Frühes Berliner U-Bahn-Netz mit unterlegtem Stadtplan (1914)

Die bildliche Erläuterung für die Abstammung oder Ausdifferenzierung von Tierarten in biologischen Lehrbüchern ist trotz einer Vielzahl von Varianten in seiner Grundstruktur recht einheitlich geblieben: Aus einer bestimmten Tierart, oft in einer zentralen Position am Bildrand durch eine vereinfachende Typendarstellung vertreten, entwickeln sich Unterarten, die über ein bestimmtes Merkmal voneinander geschieden werden können. Linien oder Pfeile verbinden die einzelnen Elemente miteinander, Benennungen geben zusätzliche Unterscheidungshilfen. Zeitskalen am Rand oder Ausrichtungen der Grafik von links nach rechts, oben oder unten können Leserichtungen vorgeben.

Durch Linien sind auch die Punkte im historischen Berliner U-Bahn-Plan miteinander verbunden, wobei hier jedem Punkt ein Bahnhof zugeordnet ist, der durch die blass unterlegte Zeichnung eines Stadtplans weiterhin auf eine geografische Umwelt bezogen bleibt.

Obwohl in ihrer Darlegung von Sachverhalten, in ihrer Funktion und auch in ihrer Bezugnahme zur Umwelt unterschiedlich angelegt, verbindet beide Grafiken, dass in ihnen mit einfachen Mitteln eine Beziehung dargestellt wird, die in abstrakter und schematischer Weise den Austausch oder die Weitergabe von Merkmalen oder Informationen (z. B. Erbeigenschaften) oder den Verkehr (von Menschen, Fahrzeugen, Gütern) zum Gegenstand hat. Die Bildaussage kalkuliert in beiden Fällen mit bestimmten Sehgewohnheiten und enthält eine ganze Reihe von unausdrücklichen Nebenbedeutungen. Grafische Vereinfachungen werden mit textlichen Zusätzen oder Zahlenangaben ausbalanciert oder ergänzt. In der Ausarbeitung solcher Visualisierungsformen, im „Diagramm", bildet sich eine eigene Formensprache und Bildgattung aus, die nicht nur den Wandel eines Wissensstandes oder die Veränderung von Bildstilen wiederspiegelt, sondern auch das visuelle Denken und Arbeiten innerhalb vorhandener Formen erkennen lässt, welches auch jenseits der Aufgabenstellung der betreffenden Fachdisziplinen von bildgeschichtlichem Interesse ist.

11.1 Anschauungsbilder: Stammbaum und Kladogramm

Zumeist aus dem Schulunterricht wird vielen Lesern eine Darstellung geläufig sein, in der die Abstammung oder Ausdifferenzierung von Tierarten bildlich erläutert wird. Dabei wird der Stellvertreter einer bestimmten Art durch Linien oder Pfeile mit weiteren Symbolen verknüpft, die repräsentativ für davon abhängige Arten und deren Merkmale stehen. Beliebt wurde dieses Schema in Lehrbüchern vor allem als Mittel, um eine revolutionäre Beobachtung des Biologen Charles Darwin (1809–82) zu erläutern. Dieser hatte auf mehreren der Galapagos-Inseln Vögel eingefangen, die er zunächst für verschiedene Arten hielt und bei seiner Rückkehr durchweg als Finken bestimmte, welche lediglich eine auffällige Feindifferenzierung entwickelt hatten. Seine Beobachtung führte schließlich zur Überwindung der strikten Trennung von Arten und Varietäten und zu einem fließenden Evolutionsmodell, das er dann in der *Entstehung der Arten* 1859 publizierte.

Abstammungs-schemata

Die Geschichte dieser Theorie beruht auf einer Reihe von Legenden, die sich erst im Nachhinein aus der epochalen Bedeutung von Darwins Werk ergaben, was sich schon darin zeigt, dass die ihm zu Ehren Darwinfinken genannten Tiere streng genommen keine Finken, sondern Ammern sind (vgl. Steinheimer/Sudhaus 2006, S. 409; Voss 2007), aber vor allem im Erfolg des Abstammungsschemas, das in dieser grafischen Form nicht von Darwin herrührte. Unabhängig von der biologiegeschichtlichen Bedeutung wird am Beispiel des diagrammatischen Musters deutlich, in welchem spannungsvollen Verhältnis sich Prozesse, Gesetzmäßigkeiten oder Sachverhalte und deren bildliche Darstellung zueinander befinden.

Erfolg eines Modellbildes

Das hier gezeigte Schema visualisiert ein Modell der Spezialisierung durch räumliche Ausstreuung, die an der Anpassung von Schnabelformen an Umgebung und Ernährung oder anderen Merkmalen ablesbar ist (adaptive Radiation). Das Schema beruht auf der Grundüberlegung, dass sich aus einem Punkt durch Verzweigung neue Punkte ergeben. Seine Struktur wird daher auch als dendritisch (baumartig) bezeichnet. Dahinter verbirgt sich das Bildmodell des Stammbaums, der seit Jahrhunderten dazu dient, Verwandtschaftsverhältnisse anzuzeigen, und der daher auch verwendet wurde, um Auskunft über Vererbungsregeln oder Prinzipien der biologischen Artentstehung zu geben (vgl. Berns 2000a). Alternative Modellbilder wie dasjenige der Wurzel sind in diesem übertragenen Sinne ebenfalls ‚baumartig‘.

Dendritische Struktur

Alternative Formen

Ob die dendritische Struktur geeignet ist, die Entstehung oder Entwicklung von Arten angemessen zu beschreiben, ist aus biologischer Sicht umstritten. Mit wachsenden naturwissenschaftlichen Detailerkenntnissen wurden Darstellungen erwogen, die einen stärkeren tabellarischen oder Formelcharakter hatten und auf eine grafische Darstellung verzichteten. Auf einer modellhaften Betrachtungsebene wurde der Baum auch durch Strukturen ersetzt, in denen andere Beziehungen und Kreuzungspunkte denkbar sind, so durch die Koralle, die Darwin selbst vorgeschlagen hat (vgl. Bredekamp 2006), aber auch durch Wurzelwerke (Rhizome), Netze und andere. Nicht immer waren diese Versuche erfolgreich, wie die langjährigen Überlegungen Darwins belegen, der letztlich doch beim Bild des Stammbaums blieb.

Vorgaben der bildlichen Darstellung

Wer jedoch einen genealogischen Stammbaum wählt, um damit ein Modell (der Vererbung, der Artentstehung) zu visualisieren, akzeptiert dessen motivische, grafische und metaphorische Traditionen und Konnotationen und muss seine fachlichen Aussagen entsprechend anpassen. So werden bestimmte Regelmäßigkeiten und Ordnungstendenzen der Darstellung fortgeschrieben: Der im Finken-Diagramm gezeigte linke Tierkopf ist isoliert, die mutmaßlichen Abkömmlinge rechts haben Köpfe unterschiedlicher Größe, wobei unklar bleibt, ob dies aus Platzgründen so dargestellt wird oder aus symbolischen Gründen, z. B. weil die damit bezeichneten Populationen eine bestimmte Größe haben. Dennoch ergeben sich aus solchen Elementen bereits Interpretationsmöglichkeiten. In der konkreten Gestaltung als Baum würden derartige Lesarten und Übertragungen noch weiter verstärkt (so ist ein Stammbaum meistens aufsteigend, eine Ahnentafel dagegen absteigend).

Das Bild als Aussage

Evolution ist das zentrale Thema der Darstellung, selbst wenn sie dies nicht ausdrücklich mitteilt. Mit dem Schema soll vorstellbar gemacht werden, wie sich Tierarten auseinander differenzieren und aufeinander folgen. Dabei gibt es offensichtlich Richtungen, Beziehungen und Verwandtschaften. Unklar bleibt aber, was genau durch die dargestellten Exempla oder ihre Namen repräsentiert wird. Denn die Finkenarten meinen nicht individuelle Tiere, sondern Gruppen von Tieren. Diese haben jedoch in sich wiederum Variationen, und es ist offen gelassen, ab wann die Variation eine Tierart (Biospezies) beschreibt. Anders als zum Beispiel bei einem sogenannten ,Tortendiagramm‘, das Anteile oder Wahlergebnisse in Prozentzahlen repräsentiert, kann hier außerdem nicht gesagt werden, ob die Aufzählung vollständig ist.

Im Finken-Diagramm geht es ausdrücklich um eine vereinfachende Darstellung und nicht um die Herstellung präziser Naturwiedergaben. Trotz verfügbarer fotografischer Abbildungen kommen weiterhin Grafiken zum Einsatz, die das mutmaßlich Typische, Wesentliche oder Charakteristische einer Art betonen sollen (→ KAPITEL 5.3, 9). Die Typendarstellung beruht wiederum auf einer Tradition, wonach das Einzelne als Anzeiger eines größeren Zusammenhangs, als Exemplar gedeutet wird und als Abbildung oder ausgestopftes Exponat der Repräsentant einer größeren Gruppe von Lebewesen oder Dingen ist. In der Sichtweise Carl von Linnés (→ KAPITEL 4.2, 6.2) hat jedes Naturobjekt seinen Platz in einem gegliederten System der Natur. Zugewiesen wird ihm dieser Platz über äußerliche Merkmale. Obwohl Linnés rein äußerlicher Blick von seinen Nachfolgern widerlegt wurde, welche stattdessen auch innere Bauformen und funktionale Aspekte einbezogen, ist die typisierende Sehweise weiterhin in jede Darstellung eingeschrieben, die so etwas wie eine ‚Tierart‘ repräsentieren soll.

Selbst in einfachsten Darstellungen verbirgt sich damit schon eine Fülle von Annahmen und Voraussetzungen. In der biologischen Morphologie hat dies dazu geführt, dass je nach Bestimmungsmethode in einzelnen Lehrbüchern auf Bilder sogar vollständig verzichtet wird. Alternativ ist in der Evolutionsbiologie eine andere Darstellungsform der Differenzierung von Tierarten vorgeschlagen worden, die als Kladogramm bezeichnet wird (griechisch *klados*, Ast). Es zeigt ebenfalls Verzweigungen, wird aber anders arrangiert. Hier gehen nur von einem einzigen Hauptstrang Nebenlinien ab. Diese kommen somit alle aus demselben Stamm und stehen für Arten oder *taxa* (lateinisch für: Einträge, Zähler).

In der beispielhaften → ABBILDUNG 37 wird das Prinzip durch unterschiedlich gefärbte Felder unterstützt, in denen von innen nach außen Abzweigungen zu Mengen zusammengefasst sind. Rote Balken markieren auf dem Hauptpfad gemeinsame Merkmale. Damit wird eine Beziehung zwischen der Hauskatze (im Kreisinneren) über Leopard und Wolf zum Pferd, andeutungsweise auch darüber hinaus bis zur Schildkröte hergestellt, die vom Betrachter schrittweise erschlossen werden kann. Weil die Abzweigung (eine sogenannte Artabspaltung) vom Stamm aus nur in eine Richtung geht, stehen an jedem Abzweig zwei Richtungen zur Auswahl. Dabei zählt lediglich die Richtung der Linie, nicht die Linie selber; sie ist eine virtuelle Verbindung mit Verweisfunktion auf einen gemeinsamen genetischen Nenner.

Der Unterschied zum eingeübten Stammbaumbild scheint zunächst gering, ist aber erheblich: Das Schema zeigt nicht mehr an, welche

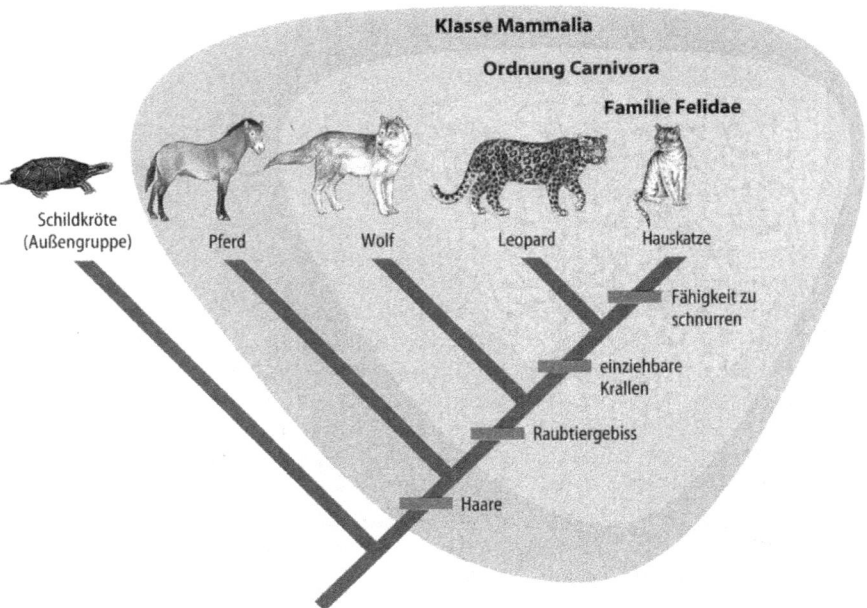

Abbildung 37: Kladistische Darstellung der Säugetierverwandtschaft (2002), Pearson Education

Art sich aus welcher anderen gebildet hat und damit – etwa im Sinne einer besseren Anpassung an die Umwelt – höher entwickelt wäre. Das Schema sieht weder eine Rangordnung noch Epochen oder Zeitfolgen vor. Probleme der Entstehung von Arten (z. B. durch Isolationsmechanismen und Fortpflanzungsbarrieren) bleiben offen. Durch Merkmale werden lediglich Abspaltungen definiert, nicht die Arten als solche. Dennoch wird es möglich auszudrücken, dass Hauskatze und Leopard (und diese beide wiederum mit dem Wolf) gemeinsame genetische Wurzeln haben.

Neutralisierung der Aussage

Trotz seiner geringen Veränderung gegenüber dem Stammbaum erteilt das Kladogramm damit eine deutliche bildstrategische Absage gegenüber Modellen, in denen der Stamm für eine klar definierbare Chronologie steht oder etwa der Mensch als Sieger der Evolution die anthropozentrische ‚Krone‘ der Schöpfung ausmacht und insofern in der Bildaussage als ‚erfolgreicher‘ dargestellt wird als andere Lebensformen, z. B. die zahl- und variantenreicheren Insekten (vgl. Gould 1990).

Die Modellierung von Wissen zwischen Begriff und Bild löst damit das Ziel einer zweidimensionalen visuellen Umsetzung des Problems der Artenentstehung ein, bei der die Verwandtschaft zwischen

Arten in anderer Form gruppiert werden kann, und zwar in Beziehung zu einem gemeinsamen Stamm und nicht nur in Abhängigkeit zu übergeordneten oder vorangegangenen Arten. Die Bildaussage wird so weit wie möglich neutralisiert. Ein natürlicher Prozess wie die Entstehung biologischer Vielfalt erweist sich als grafisches Problem, und zwar nicht nur, weil es in Bildform ausgedrückt wird, sondern auch, weil es über die Dekonstruktion von eingeübten Modellen erfolgt, als ein gemeinsames Arbeiten an Formen.

11.2 Bilder des Transports: Netzpläne

Es ist immer eine Kunst gewesen, Diagramme anzulegen und sie den sich ändernden Gegebenheiten und Sehkonventionen anzupassen. Ein Blick auf die Geschichte der Grafiken, die zur Anzeige von U-Bahn-Netzen angelegt worden sind, lässt die Probleme bei der Visualisierung dieser verwickelten, größtenteils unterirdischen Netze schnell erkennen. Insbesondere bei Metropolen mit zahlreichen sich kreuzenden Linien entstanden dabei schwer durchschaubare Knäuel von Linien und Punkten, bei denen es nicht immer möglich war, überhaupt noch Stations- oder Liniennamen einzusetzen. Es bedurfte daher schon aus grafischen Gründen einer ‚Entflechtung‘ des Bildlichen. Doch vereinfachte Schemata entstanden nicht allein aus (typo)grafischen Zwängen oder aus dem Bedarf nach einer größerer Klarheit, sondern auch, weil die zugrundeliegenden Verkehrswege zentral geregelt werden sollten. Dafür wurden, wie in Kraftwerken, Radiostationen und anderen technischen Einrichtungen, entsprechende Schalttafeln (Displays) mit Lämpchen und Verbindungslinien verdrahtet, die im Falle der U-Bahn über Weichenstellungen Auskunft gaben und später auch zur unmittelbaren Bedienung eingesetzt wurden. Die entsprechenden Tafeln boten wahrscheinlich später die Vorlage für Darstellungen, bei denen die Wegenetze großer Städte wie London, Berlin oder New York nachgezeichnet wurden.

Der frühe Plan der Berliner Hochbahngesellschaft (→ ABBILDUNG 36) stammt aus dem Jahr 1914. Anders als das evolutionsbiologische Diagramm und im Unterschied zu moderneren Gestaltungen ist der Plan noch eng auf die in ihm vorgeführte Umwelt bezogen: Hier gibt es eine Straßenkarte, die dem Streckennetz als blasse Unterzeichnung hinterlegt ist, um anzugeben, wo sich die U-Bahn-Haltepunkte im Stadtraum befinden. Die eingezeichneten Bahnstrecken folgen der realen Krümmung der Trassen. Busverbindungen sind ebenfalls ein-

Visualisierung von Wegenetzen

Netzplan und Stadtplan

getragen. Der Umgang mit dem Stadtplan ist so habitualisiert, dass sich das Wegenetz der U-Bahn nicht ohne Weiteres von ihm lösen lässt. Nur schrittweise kam es danach zu Vereinfachungen, um einer erhofften Lesbarkeit willen, angefangen mit der Fortlassung des unterlegten Straßennetzes.

Auch die berühmten Londoner Pläne folgten bis 1919 der Stadttopografie. Einen radikalen Einschnitt stellt die Schematisierung der *Tube Map* durch Harry Beck dar, bei der 1933 alle Linienrichtungen vereinheitlicht und Linien durch Farben unterschieden wurden. Sämtliche Strecken verlaufen seither in einem Winkel von 45 oder 90 Grad zueinander. Solche Muster sind heute für U-Bahnpläne weltweit üblich. Dass es sich dabei um Konventionen handelt, wird ersichtlich, wenn der Abgleich mit der kartografierten oder real erfahrbaren Welt gesucht wird – bis heute wird z. B. eine lange Verbindungslinie zwischen zwei Stationen, die sich aus rein gestalterischen Gründen ergibt, von vielen Passagieren als längere Fahrzeit gelesen. Die grafische Modellbildung löst damit Formen von der Wirklichkeit ab, um sie für eine spezifische Aussage zuzurichten oder um sie als reduzierendes *less is more* für eine bestimmte Lesart effizienter zu gestalten (vgl. Arnheim 1969, S. 154–155).

Grafische Modellbildung

Auch die Evolutionstheorie erwies sich nach Darwin als kartografisches Problem, da die enge Beziehung von Arten und ihren Räumen im Stammbaum nur mühsam darstellbar war. Im Unterschied zum Diagramm der Tierarten-Differenzierung bezeichnet in einem Wegenetz jeder Punkt eindeutig einen Bahnhof. Auch die beiden Diagrammen zugrunde liegende Kombination aus Linien, Knoten und Legenden unterscheidet sich deutlich: Während das biologische Diagramm eine dendritische Struktur vorstellt, wirft ein Verkehrsnetz in Anlehnung an den Stadtgrundriss meist Fäden in Strahlenform aus, mit aus dem Zentrum führenden Linien, die dann mit Querverbindungen versehen werden. Deshalb wurden entsprechende U-Bahnpläne im Berliner Fachjargon als „Netzplan" oder „Netzspinne" bezeichnet.

Stammbaum und Netzspinne

Von einem Diagramm lässt sich beim U-Bahn-Schema insofern sprechen, als es zur Abstraktion dient und auch auf andere Verkehrsprozesse übertragen werden kann, in denen reale Orte oder Waren mit bestimmten Vorgängen verknüpft sind. Parallel zu den Versuchen der Visualisierung von U-Bahn-Netzen ergaben sich auch andere Varianten zur Darstellung von Steuerungsmechanismen und Transportwegen. Wahrscheinlich haben die U-Bahn-Karten jedoch mit ihren konkreten Gestaltungsvorschlägen und aufgrund ihrer wachsenden Popularität auch stilistisch in andere Bereiche hineingewirkt, in denen Abläufe oder Prozesse der Steuerung oder des Transports (z. B. in Logistik und Informatik) in schematischer Form lesbar gemacht werden.

11.3 Das Diagramm als Bildgattung

Das Diagramm hat im Verlauf der Kulturgeschichte eine eigene Formensprache ausgebildet, die auch für das Verständnis von Bildlichkeit insgesamt aufschlussreich ist (vgl. Gorman 2000). Diagramme sind Geometrisierungen oder Typisierungen von Formen, deren Vereinfachung durch die Kombination mit Texten und Zahlen aufgefangen wird.

Politische Diagramme etwa, welche den Aufbau von Organisationen oder Verfassungsformen erklären sollen, zeigen mit Pfeilen Abhängigkeiten, Stimmrechte oder Hoheiten an. Statistische Umfrageergebnisse werden zur besseren Vergleichbarkeit als Balken oder Figuren dargestellt und ergänzend mit Prozentzahlen beziffert und mit Legenden erläutert. Austauschprozesse und Abläufe in der Elektronik, in ökologischen oder kybernetischen Modellen, in der Biochemie, Militärtaktik, Neurologie, in der Unternehmensorganisation und in etlichen anderen Bereichen, in denen systemische oder strategische Übersichten Anwendung finden, werden inzwischen so selbstverständlich als Diagramme mit einer festen Semantik von Linien, Pfeilen, Skalen oder Markierungen beschrieben, dass sie kaum noch anders gedacht werden können: Ein Linienrahmen um einen Begriff in einer Folienpräsentation bezeichnet einen Bereich, der von einem anderen Kasten unterschieden wird.

Wenn derlei Mittel im parteipolitischen, ökonomischen oder naturwissenschaftlich-technischen Kontext zum Einsatz kommen (wie auch im Beispiel des biologischen Entwicklungsmodells), wird diskutiert, inwieweit die bildliche Umsetzung von Informationen Aussagen unzulässig vereinfacht (→ KAPITEL 10.3). Jede bildliche Darstellung verändert den Wert einer Information, sie kann ihn mehren aber auch mindern. Sie hat zudem eine modellierende und konzipierende Funktion.

In gewisser Hinsicht antworten Diagramme mit ihrer visuell-rhetorischen Form auch auf Metaphern, die als bildhafte sprachliche Ausdrücke abstrakte Themen veranschaulichen und damit verdeutlichen. Wissenschaften erweisen sich seit jeher als metaphernreich (vgl. Maasen / Weingart 2000; Brown 2003). Wenn Aristoteles

Kombination von Bild, Text, Zahl

Anwendungsbereiche

Veränderung des Informationswertes

Metapher und visuelle Analogie

(384–322 v. Chr.) die Lebewesen und ihre Entwicklung in einer Rangordnung sieht, so argumentiert er, dass die Natur ihre Übergänge (etwa von der unbelebten zur belebten Natur oder von einer Gattung zur anderen) stets nur „schrittweise" mache (Aristoteles 1957, S. 324). Hieraus entstand das Sprachbild einer ‚Stufenleiter der Natur‘ (lateinisch *scala naturae),* die z. B. vom französischen Naturhistoriker Georges-Louis Leclerc de Buffon umgedeutet wurde (in seiner *Histoire naturelle générale et particulière,* 1749–88), um zu erklären, wie die Lebewesen der Erde sich schubweise über Jahrtausende (aufgrund von Katastrophen, Klimaeinflüssen) weiter entwickelt haben könnten. Buffon hielt Linnés Arteneinteilung für willkürlich, musste schließlich aber die Existenz von Fortpflanzungsbarrieren, als Abstufungen zwischen Tierarten, anerkennen.

Bilder als Wissensträger

Die bildsprachliche Metaphorik in den Naturwissenschaften ist ein Indiz ihrer langen kulturgeschichtlichen Bindung. Aus Gründen angeblich mangelnder Objektivität wird sie zwar oft bestritten, bleibt aber eine unumgängliche Grundlage wissenschaftlicher Erkenntnis, die es entsprechend zu reflektieren gilt. Die bildhistorische Forschung muss sich daher darauf richten, sowohl das kritische wie kreative Potenzial dieser Analogien und Ikonologien zu rekonstruieren, um nicht nur den Wissensgehalt historischer Motivwelten begreifbar zu machen, sondern vor allem die Strukturen des eigenen Wissens zu überprüfen.

Dia-gramm: durch Zeichnung sichtbar

Der Begriff des Diagramms (griechisch für: durch Zeichnung, d. h. mithilfe einer Zeichnung gegeben) deckt eine ganze Reihe unterschiedlicher grafischer Lösungen ab (z. B. Pfeil-, Balken- oder Tortendiagramm, Organigramm) und wird auch alternativ zu „Grafik", „Bildsymbol", „Illustration" und anderen Begriffen verwendet, die ihrerseits einen Bildcharakter zum Ausdruck bringen. Verwandte Formen sind das „Piktogramm" (als Orientierungsmittel), die „Bildstatistik" (zur Anzeige von Mengen) oder die „Bildreihe" (mit Pfeilen zur Anzeige von Kausalitäten, Fortsetzungen, Domino-Effekten usw., → KAPITEL 10).

Bildfunktionen

Der Begriff des Diagramms gestattet eine feine Differenzierung von Bildfunktionen: So hat etwa der Netzplan trotz seiner Emanzipation von geografischen Gegebenheiten weiterhin eine kartierende Funktion, die bestimmte Abstände und Relationen spiegelt. Damit unterscheidet er sich z. B. von jenen kosmologischen Modellen der Antike und des Mittelalters, in deren Weltordnung Gott den Nabel der Erde bildet, aus dem die Weltströme quellen: Lange Zeit als unbedarfte Form der Weltkarte interpretiert, sollten diese Modelle nicht den räumlichen Aufbau der Natur wiedergeben, sondern das Verhältnis von Mensch, Erdkreis und Gott in eine schlüssige Ordnung (griechisch *kosmos*) bringen. Als Modell zeigen sie demnach eine Ordnung und haben selbst eine solche – und insofern diagrammatischen Charakter. Diagramme reagieren nicht nur auf Sehgewohnheiten, sondern entwickeln diese selbsttätig weiter.

Es zeigt sich, dass weit mehr Darstellungen dem Prinzip diagrammatischer Darstellungen unterliegen als nur diejenigen, welche allgemein dafür gehalten werden. Eine Vielzahl von Darstellungsmöglichkeiten ergibt sich aus grafischen Elementen oder aus der Kombination von Bild, Text und Zahl (Tabelle, Bauplan, Formel, Muster usw.). Sie alle unter dem Begriff „Diagramm" zu fassen, sollte nicht dazu führen, die mit ihnen möglichen Differenzierungen auszuschlagen: Ein Diagramm ist keine Bildfolge (Serie, Comic) und keine Montage (Film, Dada), auch wenn es darin vergleichbare grafische Elemente oder Strukturen gibt. Ebenso ist es auch möglich, dass diagrammatische Strukturen in hybriden Bildformen zu finden sind (etwa in der Lehrtafel oder im Bildatlas), die gewöhnlich nicht als Diagramm aufgefasst werden, in denen aber ähnliche schematische oder grafische Beziehungen zwischen Objekten hergestellt werden.

Begriffsverwendung

Vergleichbar der möglichst neutralen Bildform wissenschaftlicher Darstellungen, empfiehlt sich daher auch für die Wahl der Begriffe, eher von „Diagrammen" (oder alternativ von „Grafiken" oder „Modellen") zu sprechen, wenn bestimmte Bezeichnungen wie „Stammbaum" oder „Karte" aufgrund ihrer Konnotationen zu voraussetzungsreich sind.

11.4 Arbeit am Modellbild

Ein Diagramm steht immer für eine Abstraktionsleistung, die in gedachter Form oder durch konkrete Symbole auf Objekte oder Sachverhalte Bezug nimmt. Die Schwierigkeit der Diagrammbildung besteht nicht nur in der Frage, wie sich komplexe Themen wie z. B. Entwicklung oder Umwelt in ein Schema bringen lassen, sondern auch darin, ob für die Darstellung eines Sachverhaltes überhaupt ein Diagramm zu wählen ist. Jede Information textlicher wie bildlicher Art ist eine In-Form-Bringung und Verkürzung von Wirklichkeit, die mit jeweils anderen Leistungen oder Zusätzen ausgeglichen werden kann. Weder das bildliche Schema noch die umfassende Textstudie

Diagramm als Abstraktionsleistung

allein machen die Adaption von Tierformen oder den Aufbau von Transportwegen erklärlich oder ersichtlich.

Jedoch erhält derselbe Sachverhalt, als Satz, als Formel oder als Grafik dargeboten, unterschiedliche Überzeugungskraft oder Ersichtlichkeit. Hierbei gibt es formale Regeln, die sich in Traditionen und Grammatiken bewegen – im bildlichen Bereich etwa, dass ein Pfeil für Richtungen, Kräfte oder Wirkungen steht. **Probleme der Visualisierung** durch Formeln oder Grafen sind damit immer auch Probleme des Bildgebrauchs und der Kommunikation selbst, da ihre einzelnen Elemente bestimmte Lesarten fördern und eine Symbolkraft besitzen, die über das rein Darstellende weit hinausgehen können (im Extremfall so weit, dass die äußere Umwelt an ein bildliches Schema angepasst wird).

Im Prozess der individuellen Herausarbeitung von informativen Grafiken – oftmals ausgehend von einer groben Skizze – wird deutlich, dass im Diagramm eine enge Verbindung von Gedanke und grafischer Darstellung besteht, das ein „Anschauliches Denken" (so der Titel von Arnheim 1969) eigener Art freisetzt, welches rein begrifflich nicht zu gewinnen ist. Es ist keineswegs Illustration, sondern bezeichnet die wissensgenerierende Dimension wissenschaftlicher Darstellung. Zu dieser Darstellung gehört, dass sie im Laufe der Zeit bei bleibender Grundstruktur epochale Veränderungen erfahren kann. Stil- und Aussageveränderungen (z. B. in den U-Bahn-Plänen, die sich von geografischen Gegebenheiten lösen) sind ebenso beobachtbar wie Tendenzen einer eigenständigen diagrammatischen Stilistik.

Visualisierungsformen müssen auf Grundlage des neuesten Erkenntnisstandes weitergedacht werden, dienen aber auch selber dieser Weiterentwicklung. Das Nachdenken über passende oder instruktive grafische Formen wird damit zu einer Forschungsfrage und meint nicht nur deren didaktische Vermittlung. Sie ist vor allem fachspezifisch. Daher liegen Handbücher zu ihrer Gestaltung im technischen (vgl. Ballstaedt 1999) oder naturwissenschaftlichen Kontext vor (vgl. Wood 1978; Hodges 2003), die sich z. B. von Anweisungen für kartografische Zwecke deutlich unterscheiden (vgl. Bertin 1967; MacEachren 1995).

Bei ihrer Ausarbeitung wird versucht, ein etabliertes Bild zu korrigieren oder sogar zu ersetzen: So musste die evolutionsbiologische Frage nach der Art, ob als Stufe, Zweig oder Blatt gedacht, mit der Erkenntnis in Einklang gebracht werden, dass Populationen sich dynamisch und umweltbedingt entwickeln, dass es Anachronismen und Katastrophen gibt, die den natürlichen Lauf verändern usw. Heraus-

Probleme der Visualisierung

Anschauliches Denken

Bildfragen als Forschungsfragen

Vom Baum zur dendritischen Struktur

gefordert wird die Visualisierungsform auch durch Beobachtungen wie diejenige, dass morphologisch definierte Artverwandtschaften und genetische Verwandtschaften divergieren können und die entsprechenden Stammbäume unterschiedlich ausfallen müssten (vgl. Maddison 1997, S. 524). Der Stammbaum wurde daher zum Kladogramm weitergeführt, was ein visuelles Argument für neue Evolutionstheorien liefern konnte (z. B. bei Hennig 1982). Motive wie der ‚Baum' entfalten andererseits, unabhängig von der biologischen Funktion, eine eigene Logik, die Vorstellungen des Organischen oder Selbstreproduzierenden mit sich bringen, welchen sich selbst diejenigen nicht entziehen können, die das Baum-Motiv eigentlich für ungeeignet halten. Auch die inzwischen gängige Alternative, das Kladogramm, ist unter dieser Bedingung eine dendritische, also verästelte Struktur geblieben.

Ein allgemeineres, aber erhebliches Problem ist die Gefahr, dass solche Modellbilder aufgrund von eingeübten Sehmustern und lokalen oder kulturellen Voreinstellungen falsch interpretiert werden könnten, gerade wenn sie einen Gegenstand scheinbar eindeutig und allgemein verständlich, unter Verzicht auf eine Fülle von Informationen, wiedergeben sollen. Sie transportieren unbewusst Bilder des vermeintlich Allgemeinverständlichen. Mit den unterschiedlichen Versuchen, geeignete Darstellungsformen zu finden – z. B. im Bereich international verständlicher Bedienungsanleitungen (vgl. Schwender 1999), wird zugleich deutlich, dass sämtliche wissenschaftlichen Disziplinen, die über Ordnungen und Entwicklungen spekulieren (darunter die biologische Systematik) stets innerhalb bildlicher Möglichkeiten und Grenzen operieren.

Fehldeutungen

Diese gilt es im Gegenzug in stilistischer wie kontextueller Hinsicht zu studieren. Ein scheinbar rein biologisches Thema wie die Evolution ist daher ebenso gut auch ein Problem der Bild- oder der Kunstgeschichte (vgl. Bonhoff 1993; Bogen / Thürlemann 2003). Mit einer visuellen Darstellung werden Relationen vorgestellt, die in dieser Form nicht existieren, die aber zum Operationsfeld von Deutungen, Vermutungen und Anschauungen werden können.

Das Diagramm als kunstgeschichtliches Thema

Dem Bild als Argument und Informationsträger muss daher eine entsprechende Kompetenz gegenübergestellt werden, die sich nicht von selbst ergibt. Die Diskussionen um die Rolle des Bildes als Wissens- oder Informationsträger brachten auch dem Diagramm wieder verstärkte Aufmerksamkeit. Mit der Bedeutung antiker und mittelalterlicher Weltmodelle, der Bedeutung visuellen Denkens für die Arbeit mit neuen Medien, mit der Gestaltung von internationalen Zei-

Sehkompetenz

chensystemen oder der normierenden Funktion von Infografiken und bildstatistischen Mitteln befasst sich die kunst- und wissenschaftsgeschichtliche, pädagogische oder soziologische Forschung derzeit umso intensiver, je weniger sie in ihnen bloße Nach- oder Abbilder der Wirklichkeit erkennt.

Fragen und Anregungen

- Welche Eigenschaften sprechen für oder gegen ein vereinfachtes Schema bei U-Bahn-Plänen?

- Wieso könnte sich das vereinfachte Schema trotz möglicher Probleme international durchgesetzt haben?

- Was bedeutet eine Linie in einem Kladogramm, was unterscheidet sie von einer Linie in einem Stammbaum?

- Was unterscheidet ein Diagramm von einer Grafik?

- Gibt es Elemente in Diagrammen (Linien, Pfeile, Farbwerte, Proportionen usw.), die voraussetzungslos, d. h. ohne besonderes Vorwissen verstanden werden können?

- Gibt es Bereiche, in denen sich diagrammatische Darstellungen grundsätzlich ausschließen?

Lektüreempfehlungen

- Horst Bredekamp / Birgit Schneider / Vera Dünkel (Hg.): Das Technische Bild. Kompendium für eine Stilgeschichte wissenschaftlicher Bilder, Berlin 2008. *Das Handbuch enthält Fallstudien des gleichnamigen Forschungsprojektes zu Bereichen wie Diagrammatik, technische Zeichnung oder Bildgebung, ergänzt um Einführungstexte, Begriffserläuterungen und kommentierte Bildbeispiele.*

- Sebastian Gießmann: Netze und Netzwerke. Archäologie einer Kulturtechnik, 1740–1840, Bielefeld 2006. *Die Studie verfolgt neben der Begriffsgeschichte des Netzes auch verschiedene Nachrichten- und Transportmedien, für die der Begriff des Netzwerkes in Anspruch genommen wurde (wie z. B. die optische Telegrafie) und setzt sich kritisch mit der Konjunktur auseinander, die er in den letzten Jahren erfahren hat.*

- David Gugerli / Barbara Orland (Hg.): **Ganz normale Bilder. Historische Beiträge zur visuellen Herstellung von Selbstverständlichkeit,** Zürich 2002. *Interdisziplinärer Sammelband zur Frage der Normalisierung von Informationen durch Grafiken, mit lehrreichen Fallstudien aus Bereichen wie Wirtschaft, Politik, Statistik.*

- Martin Kemp: **Bilderwissen. Die Anschaulichkeit naturwissenschaftlicher Phänomene,** Köln 2003. *Kemp, bekannt durch seine Studien zur naturwissenschaftlichen Leistung Leonardo da Vincis, liefert hier ein Überblickswerk zum historischen Zusammenwirken von Kunst, Technik und Wissenschaft und zur wissensgenerierenden Bedeutung von Zeichnung und modellhaftem Denken.*

- Edward R. Tufte: **Envisioning Information,** 10. Auflage, Cheshire, Conn. 2005. *Klassische und materialreiche Studie zur Frage der Aufbereitung von Daten in grafischer Form.*

12 Standpunkte der Betrachtung

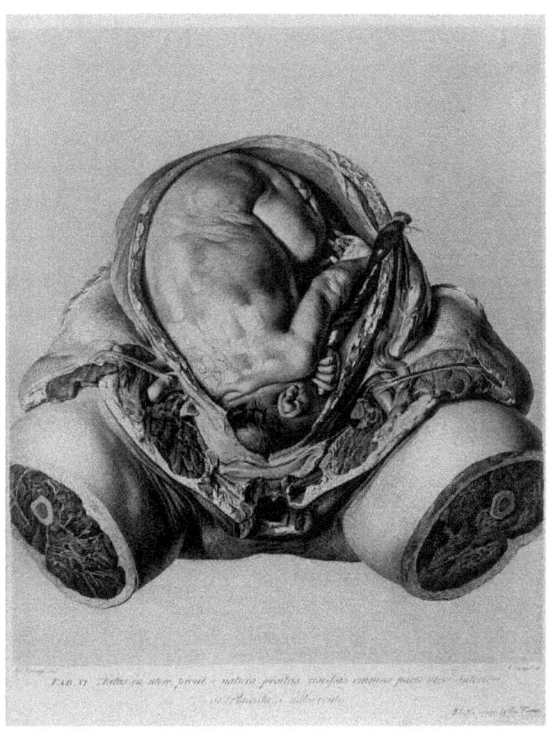

Abbildung 38: Robert Strange: *Foetus in utero* (1774), Kupferstich nach Jan van Riemsdyk

Abbildung 39: Blick durch die Kolonnade zum Ruinenberg im Park von Schloss Sanssouci, Potsdam (2003)

*Ein Kupferstich aus einem berühmten medizinischen Lehrbuch des
18. Jahrhunderts zeigt den Unterleib einer Schwangeren, der mit einem Schnitt geöffnet wurde und den Blick freigibt auf einen Fötus im
fortgeschrittenen Entwicklungsstadium. Der Leib der Mutter ist sitzend oder liegend dargestellt, der Bauch schräg von unten zu sehen.
Der Körpereinschnitt legt in drastischer Schärfe Haut- und Gewebeschichten frei. Besonders auffällig ist dabei die Darstellung der Beine als Stümpfe, die durchaus unabhängig von perspektivischen Regeln und anatomischen Notwendigkeiten gewählt wird.*

*Im Ausblick vom Potsdamer Schloss Sanssouci nach Norden finden
sich in Sichtweite auf einem Hügel einige eigentümliche Architekturelemente, die von weitem an antike Ruinen erinnern. Die Ansicht
spielt mit den Grenzen der Fernwahrnehmung und der festgelegten
Betrachterposition, indem sie ein bildhaftes Ensemble vorführt, welches sich zwar in freier Natur befindet, aber auch an idyllische Land-
schaftsmalereien erinnert. Sie ruft dabei Bilder vergangener Größe
auf und ist eine Reminiszenz an Reisen in die klassische Welt des Sü-
dens, die in einem preußischen Arkadien weiterleben soll.*

Eine der alltäglichen Leistungen der Bildkunst ist die Überführung
räumlicher Ansichten und plastischer Objekte in eine zweidimensionale Form. In der Szenografie und Reliefkunst der Antike fand die
Raumdarstellung in dem Versuch, Ansichten perspektivisch zu verkürzen, einen ersten Höhepunkt. Seit der italienischen Renaissance
wurde die zentralperspektivische Konstruktion von Fluchtlinien zur
verbindlichen Regel. Besonders durch die Fotografie schien diese
Darstellungsweise als quasi natürliche und damit korrekte Sehweise
bestätigt. Demgegenüber bleibt das flächige Bild gleich welcher
Machart eine Kunstform mit eigenen Gesetzen. Darstellungen aus
dem Bereich der Anatomie und Naturwissenschaft, der Architektur-
oder Maschinenzeichnung, optische Experimente und Täuschungen,
aber auch neue Spielwelten zeigen, dass es eine Vielfalt alternativer
Möglichkeiten der Visualisierung und Erfahrung von Raum in zweidimensionalen Bildern gibt.

12.1 Der medizinische Blick

Der königliche englische Leibarzt William Hunter (1718–83) hat durch den Zeichner Jan van Riemsdyk (1750–88) eine große Zahl von anatomischen Körperdarstellungen anfertigen lassen, die er als Kupferstichtafeln seinem berühmten Buch *The Anatomy of the Human Gravid Uterus* aus dem Jahre 1774 beigefügt hat. → ABBILDUNG 38 zeigt eine Seite mit dem Unterleib einer Schwangeren, der durch einen Schnitt geöffnet wurde und den Blick freigibt auf einen Fötus im fast vollendeten Entwicklungsstadium.

Die Perspektive weist dem Betrachter einen Standpunkt zu, durch den er den Körper gleichsam mit dem Auge seziert. Vom Leib der Mutter, der sitzend oder liegend aufgebahrt ist, sind nur der Bauch und die Beinansätze zu sehen, in einer Vorderansicht schräg von unten in den Bauchraum hinein. Offenbar sieht der Betrachter einen Leichnam vor sich. Der Körpereinschnitt legt in absichtsvoller Drastik und Detailgenauigkeit, für die Hunters Buch berühmt wurde, die Haut- und Gewebeschichten frei, hinter denen sich das Ungeborene verbirgt. Die Scham bleibt unverstellt, im Unterschied zu einigen anderen Tafeln. Besonders auffällig ist die Wiedergabe der Beine als Stümpfe, die mit dem Hackbeil eines Schlachters abgetrennt zu sein scheinen.

Drastik der Darstellung

Das Blatt zeigt auch die Meisterschaft des Kupferstechers Robert Strange (1721–82) an, der in wogenden, unmerklich an- und abschwellenden und mikrometerdünnen Schraffuren die Zeichnung Riemsdyks umgesetzt und mit allen Mitteln der schwarzen Druckkunst den Körper modelliert hat. Um zwischen Umrisslinien, Binnenstrukturen und Schattierungen zu unterscheiden, bot er in der Gravur der Linien ein Höchstmaß an Verfeinerung auf.

Übertragung für den Druck

Die Illustration ist Teil eines Lehrwerkes zum Bau des menschlichen Körpers und beruht auf einer langen Tradition. Seit Jahrhunderten zeigen anatomische Darstellungen in gezeichneter, gedruckter oder modellierter Form Körperteile, Skelette oder Präparate. Immer wieder geschieht dies in einer als makaber oder überzogen empfundenen Weise, etwa wenn ein Skelett seine abgezogene Haut lässig auf dem Arm trägt, wie in Andreas Vesalius' *De humani corporis fabrica* (1539–42). Darin äußern sich jedoch kunsthistorische Anspielungen (bei Vesalius etwa an die Häutung des heiligen Bartholomäus), die sich auch der Beteiligung namhafter Zeichner, Maler und Bildhauer verdanken, welche die Darstellungen ausgeführt und durch den großen Erfolg ihrer Illustrationen den medizinischen Blick ihrerseits mit-

Anatomische Bildtradition

179

geformt haben. Auch Hunters Bild eines Unterleibs ohne Beine und Arme ist eine Reminiszenz an Vorbilder wie den antiken Torso des Belvedere, und spielt mit dem Kontrast von blutiger Körperlichkeit, klassischer Skulptur und lupenreiner Präzision.

Anatomische Testbilder

Die Geschichte der anatomischen Darstellung ist außerdem reich an Motiven des auf- und abgeschnittenen Körpers oder an dreidimensionalen Modellen aus Holz oder Wachs, die sich zerlegen lassen. Hunters Werk ist darin grundsätzlich nicht ungewöhnlich, jedoch von besonderer Drastik. Das Bild inszeniert nicht nur Naturnähe, sondern auch einen Gewaltakt, denn theoretisch hätten Arzt und Zeichner den sezierten Unterleib auch zeigen können, indem sie die Beine in Linien auslaufen, am Bildrand verschwinden lassen oder als unscharfe Fortsätze zeigen. Abgeschnittene Beine zu zeigen, welche die Knochen freilegen, war für die Darstellung der Schwangerschaft nicht erforderlich, konnte sich jedoch aus dem Wunsch erklären, den Betrachter vor rohe Tatsachen zu stellen, deren Blutigkeit nicht jeder gewachsen ist. Da in vielen europäischen Gesellschaften bei der Niederkunft nur Hebammen zugelassen waren (für deren Ausbildung Hunter als Leiter zuständig war), gewährten die Abbildungen außerdem dem männlichen Publikum der Medizinstudenten eine visuelle Form des Studiums des weiblichen Körpers und der Schwangerschaft.

Mitempfinden und Mitleidlosigkeit

Die Brutalität der anatomischen Darstellung der Neuzeit ist eine Geste therapeutischer Art, wonach die Wissenschaftlichkeit der Naturforschung die Mitleidlosigkeit des sezierenden Blicks verlangt: Der Mensch ist als Maschine zu sehen, dessen Funktionen sehen und Disfunktionen nur heilen kann, wer sie ohne Zögern zergliedert. Die Kunst stellt ihm absichtsvoll ihr Vermögen an die Seite, mit Motiven des Schocks und der Schmerzempfindung den Betrachter zu prüfen (wie es in Rembrandts bekannter *Anatomie des Dr. Tulp* von 1632 selber zum Bildthema wird), oder zeigt den geöffneten Leib wie eine Maschinenzeichnung im unbarmherzigen Schnittbild. Damit füllt das Bild jene ästhetische Stelle aus, die durch das Fehlen von Schmerzerfahrung, Lärm, Gerüchen und anderen Sinnesreizen freibleibt.

Die Anatomie bleibt eine Projektionsfläche, in der sich Neugier, zeitgenössische Körpervorstellungen und hochspezialisierte Bildtechniken zu einem eigentümlichen medizinischen Blick verdichten. Zumeist wird dabei der männliche Körper als Maßform und anatomischer Typus, der Frauenkörper als Operationsfeld wahrgenommen (vgl. Hentschel 2001, S. 21).

12.2 Schnitt und Perspektive

Der Begriff Anatomie leitet sich vom griechischen Wort für Schnitt, *tomé*, ab. Anatomie eröffnet den Blick in den Körper, indem sie ihn aufschneidet, sei es mit chirurgischen oder bildlichen Mitteln. Insofern Bilder einen Raum in zweidimensionale Form überführen, handelt es sich ebenfalls um einen Schnitt, jedoch in visueller Form. Hierin berühren sich Kunst und Wissenschaft auf das engste.

Auch Hunters Darstellungen waren in diesem eingeübten Sinne anatomisch, dass sie den chirurgischen Schnitt über eine Perspektivdarstellung visualisierten. Die Darstellung wählt wie die meisten Vorgänger eine frontale Ansicht auf den Unterleib, während in einigen anderen Darstellungen bei Hunter auch die Seitenansicht vorkommt. Daraus folgt, dass es zu einer bildlichen Darstellung durchaus Alternativen gibt, die nicht gewählt wurden und darum eine bewusste Wahl der Mittel erkennen lassen. Es wird eine Perspektive gewählt, die den Betrachter aufgrund des festen Standpunktes zum Bestandteil des Bildes macht.

Wahl des Standpunktes

In der griechischen Philosophie der Antike wurde unter dem Begriff der Optik die Frage erörtert, ob der Blick des Menschen die Welt abtaste, indem er Sehstrahlen aussende, oder ob das Auge eine aufnehmende Funktion habe. Einige Klärung verschafften hier um das Jahr 1000 die arabischen Schriften, die Abu Ali al-Hasan Ibn Al-Haitham (latinisiert Alhazen) zugeordnet werden, welcher das Prinzip der Vergrößerung durch Glaskrümmung erkannte und das Auge als optisches Instrument begriff. Mit der darin enthaltenen Idee der Lichtbrechung konnte auch der Bildträger zu einem Medium der Lichtaufzeichnung werden. Der Architekt und Theoretiker Leon Battista Alberti (1404–72) verglich in der Folge die Bildtafel mit einem Schleier oder einer „Durchsicht" (lateinisch *per-spicere*), welche diese Strahlen gleichsam auffange (→ KAPITEL 1.2), analog zur Rückwand der Lochkamera, wo eine Gaze den Lichtstrahl empfangen, streuen und betrachtbar machen kann.

Bilder als Projektionen

Aus der Bildgestaltung der Frühen Neuzeit, vor allem der Malerei der Florentiner Frührenaissance, mit welcher der Aufstieg der zentralperspektivischen Darstellung verbunden wird, lässt sich rekonstruieren, wie einzelne Maler und Architekten systematisch nach den Regeln der Verkürzung suchten, aus denen sich zweidimensionale Ansichten von Gebäuden oder Straßenzügen konstruieren lassen müssten. Dabei entstand die Idee, die vor dem Betrachter fliehenden Seitenlängen an einen gemeinsamen Fluchtpunkt zu ziehen und dadurch den Anschein von Räumlichkeit zu erwecken, weil der Betrachter des Bildes einen ähnlichen Effekt auch in der Natur zu erleben meint, wo ihm durch seinen individuellen Standpunkt Fluchtlinien in einer bestimmten Konstellation gekrümmt oder verkürzt erscheinen müssen.

Als die entsprechenden Prinzipien durch Andrea Masaccio (1401–28) und seinen Umkreis auch auf vornehme Themen der Kirchenkunst angewandt wurden (*Trinitätsfresko* von 1427 in Santa Maria Novella), entwickelte sich aus der zentralperspektivischen Darstellung ein neuer Darstellungsstil, der später zu einem regelrechten ästhetischen Dogma zweidimensionaler Konstruktion von Raumbildern werden sollte und der nachhaltig die Idee untermauerte, wonach das Bild vor allem Abbild eines Dinges sei, weil es einen gegebenen Raumeindruck, eine Scheinansicht, imitiere oder simuliere. In der Folge kamen Perspektivtechniken aller Art unter Einsatz von Gittern, Visieren oder Glasscheiben zum Einsatz. Einflussreich waren Albrecht Dürers Vorschläge zum Bau von Apparaturen in seiner *Unterweisung der Messung* von 1525, mit denen die detailgenaue Übertragung ins Bild möglich werden sollte.

In den Traktaten des 17. Jahrhunderts und mit der Entstehung von Akademien und Ingenieursschulen wurde die perspektivische Verkürzung von Körpern, Räumen und Schattenwürfen zum Lehrinhalt und so die Vorstellung weiter untermauert, dass nur durch eine in sich widerspruchsfreie Perspektivkonstruktion die Wiedergabe räumlicher Objekte korrekt möglich sei. Die Modelle gingen von einem zentralen Standpunkt und einer ungekrümmten Projektionsfläche aus, was keineswegs der individuellen Sicht des Menschen entspricht. Diese Vorstellung blieb aber insofern einflussreich, als die Mehrzahl der Menschen bis heute keine Linearperspektive konstruieren kann, dennoch diese Art der Raumabbildung im Vergleich z. B. zu einer isometrischen Darstellung (mit gleichen Kantenlängen) als ‚naturnäher‘ ansieht.

Mit der Weiterentwicklung der Mikroskopie im 18. und 19. Jahrhundert, die mit verbesserter Linsentechnik allmählich vom Studium makroskopischer Objekte (wie Fliegen, Würmern und Pflanzenfasern) zu hauchdünnen Mikrotomen (Feinschnitten) überging, welche durchleuchtet werden konnten, wurde das Betrachtungsobjekt selber durch einen Schnitt so zugerichtet, dass es den Regeln der Betrachtung unterworfen war. Durch Projektion auf lichtempfindliche Träger ließen sich die Schnitte auch fotografieren, wie die Aufnahmen von William Henry Fox Talbot um 1840 oder der (nachgestochene) Atlas von Albert Donné und Leon Foucault von 1845 dokumentieren (vgl.

Zentralperspektive als Stil

Aufstieg der Zentralperspektive

Mikroskopie als Schnittbild

Brons / Bredekamp 2005). In der Mikroskopie wurde so eine alternative zweidimensionale Form der Sichtbarkeit üblich, bis im 20. Jahrhundert die elektronischen und computergestützten Verfahren zu virtuellen Raumansichten auf der Grundlage von Messdaten zurückkehrten (vgl. Hennig 2004).

Mit der erfolgreichen Anwendung der nach Wilhelm Conrad Röntgen (1845–1923) benannten Strahlen nach 1895 wurde es außerdem möglich, den Körper auf nicht-chirurgischem Wege zu durchdringen. Bei deren fotografischer Aufzeichnung ergab sich aus der Wirkung der Strahlungsquelle auf den Bildträger eine neue Art der ‚Perspektive‘, da z. B. eine Schädelaufnahme im Profil (medizinisch: in der sagittalen Ebene) nur deshalb eine Umrisslinie hinterlässt, weil die Strahlen die seitlichen Kopfpartien durchdringen, während sie in der Längsachse absorbiert werden. Da außerdem vor allem das feste Knochengewebe sichtbar wurde, schien die Radiografie einen Blick in den Kern des Menschen selbst zu bieten. Die eigentümliche Verbindung von flächiger Projektion und dezenten Tiefenabstufungen verlangten nach einer besonderen Übung in der Interpretation von Röntgenbildern, um diese auch medizinisch zu nutzen; als ein fotografisches Verfahren wurde das Röntgenbild jedoch als Mittel der Repräsentation nicht nur schnell akzeptiert, sondern zu einer regelrechten Mode innerhalb wie außerhalb der Medizin (vgl. Dommann 2003; Dünkel 2008).

Röntgenbilder zwischen 2-D und 3-D

Als Schnittbild neuen Typs bot die Röntgentechnik die Grundlage für die Entwicklung tomografischer Verfahren zur Aufzeichnung elektromagnetischer Wellen (CT, MRT, PET). Weil sich anhand der Röntgenfotografie nicht immer bestimmen lässt, ob ein hellerer Bereich auf die Eigenschaften oder nur auf die Dicke eines Gewebes zurückzuführen ist, wurde auf elektronischem Wege und mithilfe neuer Scanner versucht, Mehrfachaufnahmen des Körpers zusammenzuführen und am Bildschirm auszugeben (Computer-Tomografie). Die CT war damit ein erweitertes Röntgenverfahren. Die Magnetresonanztomografie kann auf den ersten Blick ebenfalls als Röntgenbild erscheinen, verwendet jedoch keine Röntgenstrahlung. Mittels starker Magnetfelder richtet sie Atomkerne aus und regt diese an, wobei sich bei der Reorientierung der Kerne unterschiedliche Gewebearten in großer Abstufung differenzieren und im Bild als Kontraste wiedergeben lassen (→ ABBILDUNG 12).

Errechnung und Rendering von 3-D-Bildern

Die erzeugten Bilder sind ein Produkt von Summationen und Umrechnungen für die Bildschirmdarstellung aus mehreren Koordinaten, darin etwa dem CAD vergleichbar (→ KAPITEL 14.1); sie erhalten ihren spezifischen Sinn jedoch erst durch das erfahrene Auge und den am Objekt geschulten Mediziner, verlangen Übung in der Steuerung des Geräts, in der Dosierung von Strahlungen und beruhen auf zusätzlichen Informationen. Erst aus deren Kombination ergibt sich ein Gesamtbild von der Gestalt und Lage eines Fötus oder eines Körperorgans. Die Betrachterperspektiven unterscheiden sich auch im Hinblick auf die Anwendung der Ergebnisse. Anders als z. B. bei minimal-invasiven Verfahren wie der Endoskopie, deren Schnitt über Lichtkanäle oder Kamerasonden angeleitet ist, kann der Chirurg bei einer Tomografie in einer Weise in den Körper vordringen, wie es allein die Aufnahme gestattet. Für ihn hat die anatomische Darstellung, in welcher MRT- oder CT-Werte mit dem Körperbau korreliert sind, daher eine andere Bedeutung als z. B. für den Neurologen, der funktionelle Bildgebungsverfahren einsetzt (fMRT, fCT), um damit Stoffwechsel-Aktivitäten darzustellen.

Operationen am Bild

Besonders die von der Medizin verwendete anatomische Repräsentation gestattet es, an kulturell eingeübte Sehweisen anzuschließen (vgl. Cartwright 1997), z. B. in der Pränataldiagnostik, bei der Ultraschallechos zu plastischen Ansichten umgerechnet werden. In Verbindung mit Geräuschwiedergaben entstehen hieraus sogenannte 4-D-Visualisierungen des Embryos, die seit einigen Jahren im industriellen Maßstab als Baby-TV von Arztpraxen angeboten werden. Das Rendering der Körperform ist dabei von eigentümlicher Kälte und Unschärfe, der Bedarf nach derartigen Darstellungen speist sich aber offenkundig aus dem Bedürfnis, die technisch möglich gewordene Darstellung festzuhalten und anderen vorzuführen. Aus diesem Grunde werden auch diagnostisch unbrauchbare Ultraschallaufnahmen in der Presse zirkuliert, um die Öffentlichkeit in symbolischer Form über die Schwangerschaft prominenter Personen zu informieren (vgl. Orland 2005). Joseph Dumit hat für spezialisierte radiologische Verfahren wie das PET gezeigt, wie stark hier Sehkonditionen in das Bildergebnis hineinwirken (vgl. Dumit 2004).

Baby-TV

Die heute eingesetzten Verfahren haben unterschiedliche Leistungen und Grenzen und kommen bei unterschiedlichen Problemen zum Einsatz. Eine MRT hat bei marktüblichen Systemen eine Auflösung von ca. einem Millimeter (was im neurologischen Bereich eine erhebliche Unschärfe darstellt), und ihre Reaktionszeit ist geringer als z. B. beim klassischen EEG. Eines der zentralen medizinischen Forschungsziele besteht weiterhin darin, aus der Kombination solcher Technologien neue Zugänge zum Körper und seinen Funktionen zu entwickeln.

Kombination von Technologien

12.3 Raum und Illusion

Aus der Vielfalt von Mitteln, mit denen Bildräumlichkeit erzeugt wird, lassen sich historisch vor allem deren Unterschiede erkennen, selbst wenn die betreffenden Bilder von Zeitgenossen jeweils als korrekt oder natürlich gelobt wurden (vgl. Snyder 1998). Dass ein Pfeil nach oben auf einem Verkehrsschild als „in Fahrtrichtung geradeaus" gedeutet wird, ist nur möglich aufgrund der Erfahrungen mit perspektivischen Darstellungen, während derselbe Pfeil an einem Fahrstuhl als „nach oben" interpretiert würde.

Kulturell geprägte Raumdarstellungen

Erwin Panofsky (1892–1968) hat in diesem Sinne die Perspektive als eine kulturabhängige, „symbolische Form" bezeichnet (Panofsky 1927, vgl. Neher 2005) und daran weitergehende transzendentalphilosophische Überlegungen angeknüpft. Darauf aufbauend hat Ernst H. Gombrich (1909–2001) in *Art and Illusion* gezeigt, dass zweidimensionale Bilder gerade dann als überzeugend angesehen wurden, wenn ihnen die Illusion gelang, natürlich zu sein, je mehr sie sich von der dreidimensionalen Raumerfahrung unterscheiden (vgl. Gombrich 1960). Der Begriff der Illusion verliert dadurch seinen pejorativen Sinn, dass nur als korrekt gelten kann, was Räumlichkeit vortäuscht.

Perspektive als symbolische Form

Die Geschichte der Bildkunst ist daher auch voller Beispiele von Perspektiven, die einem bestimmten Regelwerk widersprechen und von symbolischen und narrativen Vorstellungen zeugen, welche bewusst auf den Eigenwert des malerischen Raumes setzen (vgl. White 1957) oder die, wie die Blickführungen des Films, mit räumlichen und zeitlichen Wechseln operieren (→ KAPITEL 13). Weit erfolgreicher als die Zentralperspektive waren Licht- und Luftperspektiven, die in der Folge entwickelt wurden, bis hin zum Rendering von Landschaften am Computer. Innerhalb eines Bildrahmens gelten Gesetze, die nicht auf die Umwelt allein bezogen sind, sondern die sich aus dem Umgang mit Bildmedien ergeben und Bildstile erzeugen.

Räume des Bildes

Aus der kulturhistorischen Konditionierung heraus kann es der Betrachter als korrekt empfinden, dass in der Zentralperspektive Linien, die vom Betrachter wegführen, in einem einzigen Fluchtpunkt münden, querverlaufende Linien (vgl. die Linien in → ABBILDUNG 40) dagegen parallel zum unteren Bildrand und somit unverkürzt bleiben. Stünde der Betrachter vor einer Wand und drehte den Kopf nach links, müsste sich auch diese in der Ferne verjüngen. In einer entsprechenden Konstruktion mit mehreren Fluchtpunkten, wie sie schon Leonardo da Vinci gefolgt hat, ergäbe sich daraus zwangs-

Grenzen der Perspektive

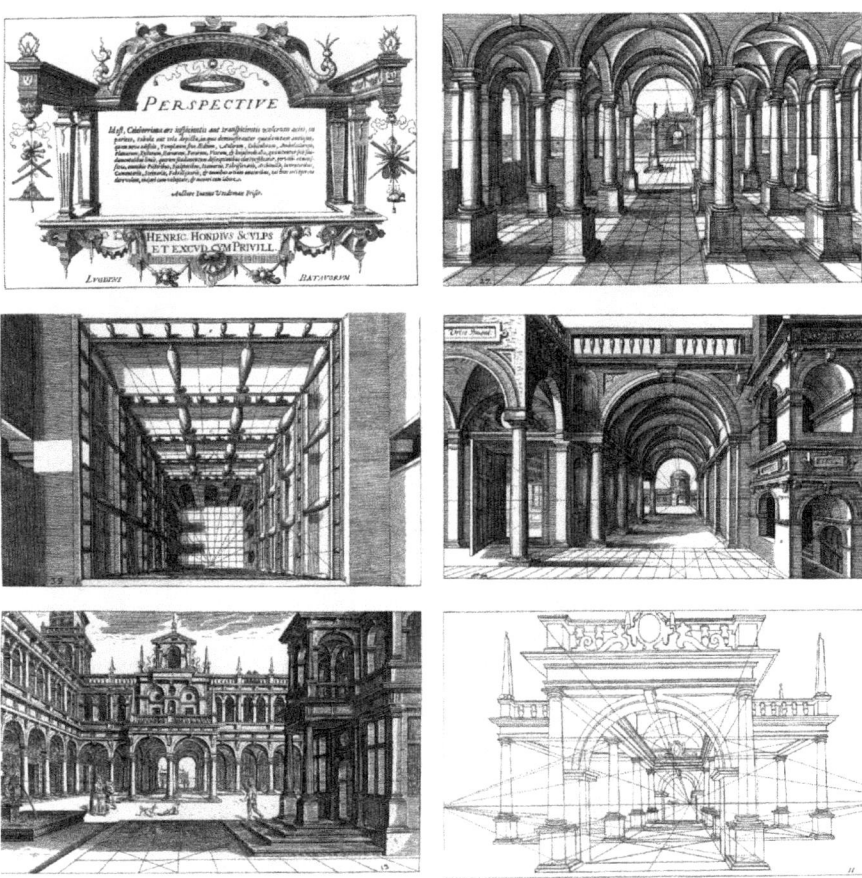

Abbildung 40: Hendrick Hondius: Muster für Architekturperspektiven (1604–05), Kupferstiche, Lemgo, Weserrenaissance Museum Schloß Brake

läufig eine gewölbte Darstellung nach Art eines ,Froschauges', die in der Mal- und Zeichenkunstkunst jedoch ebenso die Ausnahme geblieben ist wie die fotografische Aufnahme mit extremen Weitwinkeln. Ein Sonderfall ist die Verwendung von Objektiven zur Vermeidung von stürzenden Linien bei Gebäuden oder für Messbilder, mit denen Baumaße ablesbar gemacht werden.

Wie die Geschichte der Bauzeichnung und des Architekturmodells zeigt, gibt es außerdem fließende Übergänge von flächigen Bildmedien zu räumlichen Bildern, Modellen und Plastiken. → ABBILDUNG 41 reproduziert eine kombinierte Plan- und Gewölbezeichnung des römischen Malers Andrea Pozzo, der als Freskenmaler im späten **Illusionismus** 17. Jahrhundert zu großem Ruhm gekommen ist. Im vorliegenden

Abbildung 41: Andrea Pozzo: Kuppel in Ansicht von unten / Grundriss mit Perspektive (1693)

Fall hat er den kreisförmigen Grundriss einer Kuppel in eine Gebäu-
deuntersicht überführt, die das mögliche Aussehen des Bauwerkes als
Tiefenblick mit Licht- und Schattenwürfen und starken Verkürzun-
gen vorstellt. Die elegante Verbindung von Schnittbild und Raumeindruck verdeutlicht zugleich, dass dieser Raumeindruck auch im aus-
geführten Gebäude durch eine Malerei hervorgerufen werden
könnte, die sich mit einem bestimmten Standpunkt im Kirchenraum
deckt – eine Technik, in der Pozzo die größte Meisterschaft besaß.

Die Wahrnehmung der Umwelt findet zumeist in binokularer
Form statt, unter Einbeziehung haptischer und akustischer Reize.
Augen und Kopf bewegen sich, das Gehirn rechnet Reize auch nichtvisueller Art in ein Gesamtbild ein; ein Betrachter wird nicht alle
Dinge gleichmäßig scharf und gleichermaßen aufmerksam wahrnehmen (→ KAPITEL 4.3). Testbilder, deren Betrachter über Größenverhältnisse, Abstände oder Blickrichtungen getäuscht werden, lassen erkennen, wie sehr grundlegende psychologische Faktoren an der
Konstruktion räumlicher Verhältnisse beteiligt sind. Hier äußert sich

Sehen in Räumen

eine evolutionär geprägte Präferenz für das Sehen in Volumina und Distanzen.

Dieselben kognitiven Faktoren sind aber auch grundlegend für eine Akzeptanz räumlicher Darstellungen in zweidimensionaler Form. Ein statisches Bild trägt den veränderten Bedingungen der Fläche im Prozess der Aufzeichnung und seines Lernens Rechnung, so wie auch die konkrete Bauweise eines Fotoapparates ihren bildhistorischen Vorlauf hat. Selbst in der plastischen Kunst zeigt sich, dass die **Von allen Seiten** Skulptur nicht immer „von allen Seiten schön" sein muss (so der Ti- **schön** tel einer Ausstellung in Berlin 1995), sondern zuweilen auf einen bestimmten Standpunkt hin bezogen ist, um ihre Wirkung zu entfalten. Zeichnungen von Bildhauern der Zeit um 1800 (wie Antonio Canova oder Johan Tobias Sergel) lassen erkennen, dass ihre Figuren eine bevorzugte Ansicht haben, auf die bestimmte Gesten, Mienen oder Schattenwürfe hin konzipiert sind und die sich mit Bildauffassungen der Malerei decken.

Fernansicht Auch die Fernansicht auf Gebäude, Gebäudeteile oder Landschaften, welche die Welt szenografisch auffasst, wird als Spiel mit Bildformen und Sehgewohnheiten weitergetrieben. Als Beispiel ist in → ABBILDUNG 39 eine Anlage reproduziert, die sich in Potsdam nördlich von Schloss Sanssouci befindet und bei der einige Architekturelemente (eine Kolonnade aus drei Säulen und Architravbalken, ein umgestürzter Säulenschaft, weiter dahinter ein kantiger Turm und eine ringförmige Wand) von weitem an antike Ruinen erinnern. Das Ensemble, das ab 1748 im Auftrage König Friedrichs II. gestaltet wurde, lässt vom Schloss her betrachtet ein Tableau entstehen, welches im Spiel von Tageslicht und Wetter einen lebendigen Eindruck hinterlässt; den Zuschauer lädt es noch heute zum Nachsinnen über untergegangene Welten oder den Reiz südländischer Reiseziele ein.

Durch die Malerei des Barock waren Friedrich und seine Zeitgenossen an antikisierende Landschaften gewöhnt, wie sie die italienischen oder italienreisenden Künstler in großer Zahl auch für den **Entdeckung der** nordeuropäischen Markt produzierten. Diese Entdeckung der „Natur **Landschaft** als Landschaft" (so der Titel der Studie von Fechner 1986) hatte in der Malerei schon in der Renaissance mit der Emanzipierung des Bildhintergrundes von religiösen Hauptszenen und historischen Motiven eingesetzt. Im 16. Jahrhundert hatte sich die Landschaft als Bildmotiv weiter autonomisiert, befördert durch die Italienreisen von Künstlern aus ganze Europa, denen im 18. Jahrhundert auch gebildete Bürgerschichten folgten. Mit der veränderten Wahrnehmung von Landschaft als Kunstgattung und der vermehrten Reisetätigkeit wurde schließlich auch die Gestaltung der Umwelt selbst als höhere Kunstform gewürdigt, etwa in Christian Hirschfelds *Theorie der Gartenkunst* von 1779.

Die Anlage von Sanssouci entstand zu einer Zeit, in der die Landschaft in der Gattungshierarchie der Bildkunst allmählich zu einem eigenen Thema aufstieg. In der ästhetischen Theorie wurde anerkannt, dass es „malerische" (englisch *picturesque*) Erscheinungen der Landschaft gebe, die sich von selbst einzurahmen und als Ansicht (französisch *vue*, englisch *view*) und Motiv der Kunst zu empfehlen scheinen, womöglich unterstützt durch Experimente mit der Lochkamera und anderen optischen Geräten. Die Landschaft wurde unterdessen auch zur Projektionsfläche für gesellschaftliche Modelle der Natürlichkeit und Freiheit oder für Reflexionen über die Größe oder den Niedergang der eigenen Kultur. Das Spiel mit der auf einen fernen Punkt bezogenen Perspektive diente in diesem Diskurs der Transformation eines Raumes ins zweidimensionale Bild und stellte selber einen geistreichen Diskurs über die Frage dar, was Bilder sind und welchen Anteil der Betrachter an ihnen hat.

Das Pittoreske: Von der Natur zum Bild

12.4 Immersion, Simulation

Mit der weiteren physiologischen und psychologischen Erforschung des räumlichen Sehens nach 1800 (→ KAPITEL 4) nahmen auch die experimentellen und spielerischen Darstellungen ihren Aufschwung, die das konstruierende Sehen von Räumen systematisch zu nutzen versuchten. Dies äußerte sich z. B. im Erfolg der Stereoskopie, die in den 1830er-Jahren durch Experimente des Arztes Charles Wheatstone zur Binokularität vorangetrieben und dann in der zweiten Hälfte des 19. Jahrhunderts systematisch verwertet wurde. Die Firmen Keystone und Underwood wurden Weltmarktführer für stereoskopische Fotokarten von bekannten wie exotischen Reisezielen, Ereignissen und Unterhaltungsmotiven. Da fast jeder Haushalt der USA am Ende des Jahrhunderts über ein zur räumlichen Betrachtung erforderliches brillenartiges Gerät verfügte, kann hier von dem ersten Bildmassenmedium gesprochen werden.

Stereoskopie

In der militärischen Aufklärung oder in der Weltraumfahrt sind stereoskopische Aufnahmen unverändert in Gebrauch, um dem Betrachter eine dritte Dimension zu bieten, welche der Differenzierung von Oberflächen und Volumina dient. In der Morphometrie werden aus zweidimensionalen Sichten, z. B. elektronenmikroskopischen Ge-

Sichtung und Vermessung

webeaufnahmen, dreidimensionale Strukturen stereografisch errechnet. Gebräuchlicher sind jedoch Techniken zur virtuellen Abformung von Räumen und Objekten mithilfe von Laservermessungen (wie in der Archäologie und Geodäsie oder im Bodyscanning) oder mehrseitigen Fotoaufnahmen, die durch Animation oder Interpolation zu räumlichen und bewegten Bilder umgerechnet werden.

Diorama, Panorama Noch vor dem Stereoskop wurde das Panorama, das zunächst nur ein Rund- oder Rollenbild meint, zu einem Mittel der plastischen Wiedergabe eingesetzt und ausgehend von England durch Robert Barker, Thomas Girtin u. a. zu einem international erfolgreichen Unterhaltungsmedium mit eigenen Theaterbauten gemacht, die auf bestimmte Tribünen durchschritten werden konnten oder sogar über bewegliche Sitze nach Art einer Geisterbahn verfügten (vgl. Buddemeier 1970; Oettermann 1980). Die optische Umhüllung des Betrachters wurde noch intensiviert durch den Einsatz von Durchlichtbildern (Dioramen) aus der Hand des geschäftstüchtigen Malers und Fotopioniers Louis Jacques Mandé Daguerre (1787–1851), dessen Technik europaweit Furore machte. Eine Fortsetzung sollte die Technik später noch einmal im 3-D-Kino erhalten, die sich aus Kostengründen nicht durchgesetzt hat; sie ist im Breitformatfilm und den gekrümmten Leinwänden von Großkinos aber noch präsent.

Immersion In Verbindung mit Bewegung und Lichteffekten (→ KAPITEL 13) ergaben sich für den Betrachter Möglichkeiten einer weitgehenden Immersion (lateinisch für: Eintauchen) in den Bildraum. Die linearperspektivisch konstruierten Darstellungen des Panoramas blieben jedoch trotz gewisser Bewegungsmöglichkeiten an feste Standpunkte gebunden. Im 20. Jahrhundert wurde daher erneut versucht, die Position des Betrachters in das Bild einzubeziehen, mit intensiven Forschungen im industriellen wie im künstlerischen Bereich. Durch die Verbindung von körpernahen Steuergeräten und Displays (Datenhelm, Datenhandschuh), leistungsstarken Grafikprozessoren und neuen mathematischen Modellen wurde es möglich, dynamische und fließende Bewegungen zu simulieren, die in Echtzeit auf Bewegungen von Hand oder Auge ansprechen. Durch gezielte Einschränkungen der individuellen Sicht durch Helme oder dunkle Räume und das Ersetzen äußerer Reize z. B. durch künstliche Vibrationen und Geräusche ergab sich daraus, in Verbindung mit der eigenen Imaginationsleistung, ein neuer Eindruck autonomer Realität. Im Computerspiel oder in Lehreinrichtungen (Cockpit-Simulator) wird dieser Eindruck durch die Anpassung an Reaktionsgeschwindigkeiten und andere Lerneffekte weiter verinnerlicht.

Aus der digitalen Synthetisierung von Raumwirkungen und unter dem Eindruck der beginnenden gesellschaftlichen Computerisierung folgte die Diskussion, inwieweit die menschliche Wirklichkeit vollständig simulierbar und referenzlos werde, was Jean Baudrillard als Aufgehen des Menschen im Scheinhaften, als Entgrenzung von Wirklichkeit und Einbildung beschrieben hat (Baudrillard 1981; vgl. Weibel 1984). Dies schien bestätigt durch die Kreation künstlicher Akteure, der sogenannten *Avatare,* oder die Verschmelzung von Körper und Technik in Maschinenmenschen, den *Cyborgs* (dazu kritisch Haraway 1991). Durch den Begriff der ‚Virtualität‘ wurde einerseits die Verwischung von Realitätsebenen ausgedrückt, andererseits aber auch, dass visuelle Simulationen (im Unterschied zu rechnerischen Simulation, → KAPITEL 14) durch die Nutzung von Bildschirmmedien und die Einbeziehung des Betrachters weiterhin zahlreiche historische Anbindungen erkennen lassen (vgl. Hemken 2000; Grau 2002).

Virtualität und Simulation

Die neue Visualität, auch in Verbindung mit einem erweiterten Körperschema durch Steuerknüppel oder Displays, führt weder zur Abschaffung der individuellen Erfahrung im Bildraum noch zum kompletten Verlust von Referenzialität. Programmgrenzen und Programmabstürze sind Teil der technischen Wirklichkeit. Lev Manovitch hat die Perfektion der Nachahmung daher in *The Language of New Media* als einen der „Mythen“ der Digitalität und Interaktivität bezeichnet (Manovitch 2001, S. 52ff.).

Mythen der Neuen Medien

Fragen und Anregungen

- Wieso bleiben Bilder stets an Regeln der Perspektivität gebunden?

- Welche Bedeutung hat räumliches Vorstellungsvermögen für die Interpretation von anatomischen Bildern oder bildgebenden Verfahren?

- Was könnten die Gründe sein, dass das Stereoskop, das erfolgreichste Bildmedium des 19. Jahrhunderts, im 20. Jahrhundert wieder verschwand?

- Woher rührt es, dass nur Fotografien, die mit einer bestimmten Brennweite aufgenommen sind, als natürlich empfunden werden?

- Definieren Sie den Begriff der „Landschaft“.

Lektüreempfehlungen

- Kenneth Bryson Roberts / J. D. W. Tomlinson: The Fabric of the Body. European Traditions of Anatomical Illustration, Oxford u. a. 1992. *Bis zum Erscheinen dieses beeindruckenden Bandes zur medizinischen Illustration war Marielene Putschers „Geschichte der medizinischen Abbildung" (München 1972) alleiniger Standard; trotz seines Titels enthält der Band auch einige außereuropäische Beispiele.*

- Klaus Peter Dencker (Hg.): Weltbilder – Bildwelten. Computergestützte Visionen. Interface 2, Hamburg 1995. *Der Band versammelt wichtige Autoren und Pioniere der Theorie virtueller Welten und dient damit trotz seines Alters als nützlicher Einstieg in das inzwischen unüberschaubare Gebiet.*

- Samuel Y. Edgerton: Die Entdeckung der Perspektive, München 2002. *Im Amerikanischen erstmals erschienen 1975 („The Renaissance rediscovery of linear perspective"), liegt dieser Band seit kurzem in deutscher Übersetzung vor und bietet eine gute Einführung in die Geschichte der Zentralperspektive seit der Renaissance.*

- Martin Kemp: The science of art. Optical themes in western art from Brunelleschi to Seurat, New Haven u. a. 1990. *Reich bebilderte, thematisch gegliederte Darstellung raumbezogener künstlerischer Techniken der Perspektive, Geometrie, Stereoskopie u. a., die auch in die Geschichte der Ästhetik und Erkenntnistheorie eingeordnet werden.*

- Konrad Lischka: Spielplatz Computer. Kultur, Geschichte und Ästhetik des Computerspiels, Hannover 2002. *Eine Darstellung des Computerspiels, die über dessen ästhetisch-stilistischen Eigenarten und spezifische Neuerungen hinaus auch industrielle, technische und soziale Faktoren behandelt.*

13 Bewegte Bilder

Abbildung 42: Filmstill aus Michael Hanekes Film *Caché* (2005)

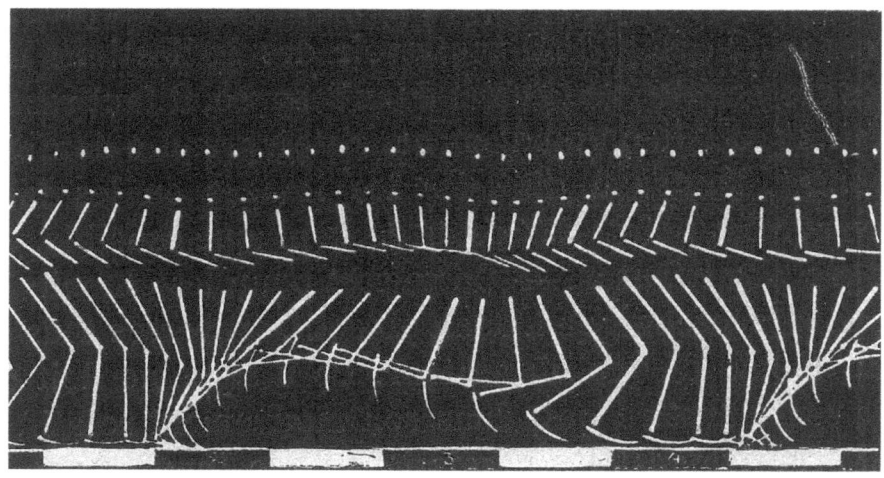

Abbildung 43: Etienne-Jules Marey: Chronofotografie einer menschlichen Schrittbewegung
(1882)

Das Filmstill zeigt den Blick einer Kamera aus einer engen Seitengasse auf eine Straßenkreuzung und die dahinterliegende dichte Bebauung. Die Einstellung wird erst in dem Moment als bewegtes Bild erkennbar, als Personen und Automobile die Straße passieren. Aus dem Hintergrund sind Stimmen zu hören. Nach einer kurzen Wartezeit wird die Sicht plötzlich beeinträchtigt, das Bild beginnt zu flimmern und Bewegungsdetails erscheinen in schnellem Rückwärtslauf – der Betrachter beginnt, die Ansicht als laufende Videoaufzeichung einer Überwachungskamera zu deuten. Damit beginnt der Spielfilm „Caché" (Versteckt) von Michael Haneke.

Das Festhalten von Bewegung wird auch als eine besondere Stärke der Fotografie angesehen, die mit fortschreitender Technik kürzere Belichtungszeiten zuließ und dadurch unter anderem für wissenschaftliche Studien interessant wurde. In der Chronofotografie von Etienne-Jules Marey sind mit einer speziellen Kamera 40 Phasen einer Schrittbewegung als Sequenz aufgezeichnet, die ein Mann, der in ein schwarzes Kostüm mit weißen Seitenlinien gekleidet wurde, auf einer Fotoplatte hinterlassen hat. Als fließendes Bewegungsmuster ist diese Aufzeichnung aber mehr als nur Messergebnis oder Sichtbarmachung; sie steht zugleich für eine technische Ästhetik.

Die Medien des Films und der Videoaufzeichnung werden gemeinhin als bewegte Bilder verstanden, und zwar selbst dann, wenn sie keine Bewegung vorführen oder vollziehen. Darüber hinaus ist Bewegung, als Handlung, Spannungsbogen oder erzählerischer Moment, als Performanz, als Sequenz von Augenblicksaufnahmen oder Reportagestrecken, auch in den unbewegten Bildformen der Fotografie, der Malerei und Grafik präsent und nicht notgedrungen an technische, insbesondere „neue" Medien gebunden. Bewegung beschreibt mehr als nur die Flüchtigkeit oder Zeitlichkeit eines bestimmten Mediums und erfährt auch jenseits von Chronofotografie, Kinematografie oder elektronischen Medien historisch wandelbare Darstellungsweisen.

13.1 Das Bewegtbild

Die Eingangssequenz zu Michael Hanekes Film *Caché* spielt mit dem Überraschungsmoment, dass ein scheinbar statisches Bild einer Straßenschlucht unvermittelt von Passanten und Autos durchkreuzt wird, ehe sich plötzlich die Ansicht selbst in flimmernde Bewegung versetzt. Hintergrundstimmen lassen vermuten, dass es sich bei den Aufnahmen um eine Videoaufzeichnung handelt, die gerade von zwei Unbekannten analysiert und dazu vor- und zurückgespult wird.

Unerwartete Bewegungen

Die simple Szene spielt auf das Thema der filmischen Überwachung an. Sie kann aber auch verdeutlichen, auf welchen unterschiedlichen Ebenen ‚Bewegung‘ im Filmischen enthalten ist und Elemente der Gestaltung, der Aufzeichnung und Wiedergabe beschreibt. Im Zeitalter digitaler Medien und nach einer über 100-jährigen Geschichte des Kinos muss die Diskussion von Bildlichkeit zwar das bewegte Bild selbstverständlich einschließen, allerdings ist aufgrund der sich wandelnden Medien, Erscheinungsformen und Sehweisen immer wieder neu zu bestimmen, wo die Grenze verläuft, die den Unterschied von bewegten und nichtbewegten Bildern markiert.

Film und Video werden im Unterschied zur Malerei oder Fotografie als Bewegtbilder angesehen. Dies liegt zunächst schon begründet in der Frühgeschichte des Films, der seit seiner Anfangszeit als technische Animation des stehenden Bildes begriffen wurde, was sich auch in der Bezeichnung der ersten Apparaturen wie Kinetoskop oder Cinématographe niederschlug, die auf *kinesis* (griechisch für: Bewegung) verweisen, während alternative Gerätebezeichnungen wie Biograph oder Vitascope sich nicht durchgesetzt haben.

Kinesis, Kinetik

Wie in der Fotografie gibt es kein Datum der Erfindung des Films, außer in industriegeschichtlicher Hinsicht mit der Patentierung des Cinématographe durch die Brüder Auguste und Louis Lumière 1895. Die Lumières waren in Frankreich jedoch nicht allein, und auch in England, Deutschland und den USA wurde längst mit Bewegtbildmedien experimentiert, allerdings mit unterschiedlichem Erfolg. Das Optische Theater Émile Reynauds von 1892 beruhte wie die Dioramen des Malers und Fotopioniers Louis Daguerre (→ KAPITEL 12.4) auf gemalten Bildern, das etwa zeitgleich entwickelten Kinetoskop von Thomas Edison und William K. Dickson war ein schrankförmiges Gerät, das als Guckkasten nur von Einzelbetrachtern genutzt werden konnte und daher sowohl die Neugier der Betrachter als auch die Zensur auf den Plan rief.

Erfindungen des Films

Kino vor dem Kino

Bewegtbilder im Sinne der *kinesis* gab es zudem schon vor dem Kino als Bildkultur: im Schattenspiel und als optische Apparatur. Die *laterna magica*, als Projektionsapparat mit auswechselbaren Lichtbildern, erfreute sich im 17. und 18. Jahrhundert großer Beliebtheit, um die Scheinhaftigkeit von Bildern zu demonstrieren; wegen ihrer bildrhetorischen Kraft wurde sie im 19. Jahrhundert jedoch auch als evidenzstiftendes Mittel in der Wissenschaft, Bildung, im Militär und vor Gericht entdeckt (vgl. Ruchatz 2003). Sie lässt sich ebenso wie die inszenierten Bewegungen, die Maschinen und Lichteffekte des Theaters zu Vorläufern des bewegten Filmbildes zählen (vgl. Berns 2000b; Gronemeyer 2004).

Einen ersten Höhepunkt erlebte das Lichtspiel mit den Dioramen nach 1830, die in aller Welt Nachahmer fanden und zur Errichtung eigener Spielstätten für die Massenbesucher führten. Hier kamen

Lichtwirkung auf das Publikum

halbtransparente und übergroße Wände zum Einsatz, die unter wechselndem Lichteinfall neue Ansichten und Gegenstände zeigten und auch Schattenspielereien oder die Überblendung mit realen Gegenständen zuließen. Schaulust wurde dadurch zu einem breiten gesellschaftlichen Thema (→ KAPITEL 3.4). Auch wurden Sehweisen und -techniken des Kinos in anderen Kunstformen wie Malerei und Literatur vorbereitet (vgl. Segeberg 1996).

Edisons Kinetoskop-Salons hatten ihren Zenit in den USA und Europa um 1895 bereits überschritten, Edison entwickelte nun ebenfalls projektionsfähige Verfahren, in Frankreich löste das Großatelier von Georges Méliès die Brüder Lumière als Produzenten ab. Wanderkinos kamen auf, um steigende Produktionskosten einzuspielen, und es wurde mit Ton und Farbe experimentiert. Um 1900, innerhalb

Kino als Bildkultur

von fünf Jahren, hatte das Kino als Kulturform damit einen ersten Strukturwandel erlebt. Vergleichbare Wandel ergaben sich in den 1920er-Jahren durch den immensen Erfolg der Lichtspielhäuser und den Aufstieg der großen Studios. Im Kino samt Projektionstechnik, Aufführungsorten und Distributionskanälen etablierte sich ein industrielles, arbeitsteiliges Unternehmen, an dem Regisseure und Kameraleute, Editoren und Musiker beteiligt waren. Kino wurde so zur ‚Technologie' im breiteren Sinne, indem es zu einer eigenen Bildkultur mit neuen Stilmitteln und Erlebnisräumen fand. Formen und Stoffe der filmischen Erzählung wurden in Verbindung mit dem zahlenden Publikum zu einem Phänomen mit Massenwirkung; die Vorführdauer erreichte die Länge von Theaterstücken.

Die Strahlkraft des Filmbildes im dunklen Kinosaal wurde verstärkt durch Einbindung pathetischer Musik und Rede, die dem sichtbaren Bild eine zusätzliche psychologische Tiefenwirkung zu geben vermochte. Das Kollektiverlebnis des temporär aufgeführten Leinwandbildes, das die Theaterbesucher miteinander verband, konnte schließlich die Aura des Originals ersetzen, wie Walter Benjamin behauptete (vgl. Benjamin 1936). Die landläufige Idee des bewegten Bildes und seiner Motive ist seit der Hochphase der Kinos in der ersten Hälfte des 20. Jahrhunderts mit der Idee verknüpft, dass Kinematografie eine grundsätzlich neue mediale Erfahrung beschreibe.

Aura

Da der populäre abendfüllende Spielfilm jedoch eine definierbare zeitliche Länge hat und aus Elementen besteht, die meist zu erzählbaren Handlungen zusammenlaufen, wurde der ‚Inhalt' eines Kinofilms zu einer Narration, die ihre Akteure und *settings* hat, sich in der Filmkritik nacherzählen lässt und eine große Nähe zu Strukturen des Dramas und des Romans aufweist, noch dazu im Tonfilm mit gesprochenen Dialogen. Dadurch entstanden einerseits spezifische Probleme der Kinoproduktion (z. B. die Frage der Landessprache und ihrer Übersetzung), andererseits etablierte sich das Kino als Kunstform neben Theater oder Roman.

Narration des Films

Verbunden mit dem Faszinationsraum des Kinos ist die Diskussion um die größere Kraft der Illusion oder Emotionalisierung durch bewegliche und leuchtende Medien, durch die der Begriff der Illusion neue Bedeutung annahm, insbesondere im Zusammenhang mit der politisch-propagandistischen Indienstnahme des Films in der ersten Hälfte des 20. Jahrhunderts (vgl. Koch 2006). Im psychotechnischen Filmexperiment dieser Zeit wurde daher auch die Betrachterwirkung von Filmbildern, unabhängig von konkreten politischen Botschaften und Motiven, erforscht, indem Probanden Lichtreizen ausgesetzt wurden – ein Forschungsfeld, das viele Brührungspunkte zur Avantgardekunst besaß (vgl. Vöhringer 2007; Reichert 2007).

Medienemotionen

13.2 Bilder des Bewegten

Der Erfolg des Cinématographen beruhte neben seiner Lichtwirkung im dunklen Saal, von dem sich der Film im Laufe der letzten Jahrzehnte wieder emanzipiert hat, auf unterschiedlichen technischen Erfindungen. So war das sogenannte Malteserkreuz erforderlich, um die Filmspule nicht nur zu transportieren, sondern vor allem für Sekundenbruchteile zu verzögern, um die kurzzeitige Durchleuchtung zu ermöglichen. Die Erfindung dieses Bauteils im Projektor hatte ihre

Optische Spielzeuge

Vorgeschichte in den kinetischen Apparaten und optischen Spielzeugen, die nach 1800 allgemein beliebt wurden, etwa in Form des Thaumatrop (einem zweiseitig bemalten Kippbild), des Zoetrop (einer rotierenden Trommel mit Einzelbildern), des Phenakistikop (rotierende Scheibe). Alle diese Spielzeuge, deren Fülle zum Teil nur durch Initiativen von Sammlern und Filmkünstlern wie Werner Nekes rekonstruiert werden konnte (vgl. Dewitz / Nekes 2002), zerlegten Bewegung nach Art des Daumenkinos in Einzelmomente und setzten sie durch Bewegung der Apparatur wieder zusammen. Zur selben Zeit stieg das Interesse an der Physiologie und Wahrnehmungsgeschwindigkeit des Auges und konnten sich fotografische Aufnahmeverfahren etablieren.

Schon William Henry Fox Talbot (→ KAPITEL 9.2) hatte mit Blitzlicht operiert, um die Aufnahmedauer seiner Aufnahmen zu verkürzen. Selbstbewusst verkündete er in der Beschreibung seines Verfahrens der *Photogenic Drawings* 1839, dass es ihm möglich geworden sei, den Schatten selbst, der sprichwörtlich sei für alles Flüchtige, in einem dauerhaften Bild festzuhalten. Das bis dahin vergebliche Festhalten des Schattens war ein Topos der fotografischen Literatur (→ KAPITEL 9.2), der bis in die Ästhetik der Antike zurückverwies, etwa auf das kaiserzeitliche römische Gedicht vom geliebten Earinus, dessen Bildnis von einem Spiegel eingefangen worden sei (vgl. Eder 1932, S. 6). Mit der Idee des Festhaltens war eine Auffassung von Bewegung ausgedrückt, die ihrerseits in der Tradition der bildenden Künste stand.

Festhalten des Flüchtigen

Die fototechnische Aufnahme und Wiedergabe von Bewegung wurde im Laufe des 19. Jahrhunderts schrittweise verbessert durch höhere Empfindlichkeit der Träger, praktischere Trockenverfahren (seit Mitte der 1850er-Jahre, sehr erfolgreich die Gelatine-Beschichtung von Richard Learch Maddox 1871), durch schnellere Blendentechniken, lichtstärkere Linsen, elektrische Blitzerzeugung und tragbare Kameras (Kodak, Goertz ab 1888, Leitz ab 1924). Hierdurch wurde es möglich, kurze Momente auch ohne größeren Aufwand einzufangen.

Aufnahmetechnik

Mit weitergehenden Techniken der Momentaufzeichnung experimentierte Eadweard Muybridge (1830–1904), dem es erstmals 1877 gelang, durch Einsatz von Kameras in Reihenschaltung die Galoppbewegung des Pferdes aufzuzeichnen; später wurde er berühmt durch Einzelaufnahmen von nackten Personen beim Gehen oder Treppensteigen, die durch Einsatz elektrischer Lichtquellen möglich wurden und die er als *electro-photographic investigations* veröffentlichte.

Bewegungsfotografie

Auf derselben stroboskopischen Technik mit Lichtblitzen beruhten auch die späteren Aufnahmen von Wassertropfen durch Harold E. Edgerton (1903–90), in ihrer Ästhetik vergleichbar den Experimenten mit Projektilen durch Ernst Mach (1838–1916), in denen Luftwellen und Flugbewegungen in bislang ungeahnter Weise sichtbar gemacht wurden. Auch Muybridges Zeitgenosse, der französische Arzt Etienne-Jules Marey (gleichfalls 1830–1904), wurde bekannt für Phasenbilder eines Mannes in schwarzem Anzug mit Markierungen, die durch kurze Belichtung ein regelmäßiges Muster auf der Fotoplatte hinterließen und symbolisch für eine neue technische Form standen (→ ABBILDUNG 43).

Marey hatte 1883 für seine Studien zudem ein ‚fotografisches Gewehr‘ konstruiert, bei dem ein Verschluss über dem lichtempfindlichen Träger rotierte und so kürzeste Aufnahmefolgen zuließ, und ein neuartiger Schlitzverschluss gestattete es Ottmar Anschütz (1846–1907) im Jahre 1888, unter freiem Himmel Aufnahmen von Vogelbewegungen und anderen Motiven zu machen.

,Fotografische Gewehre‘

Ein Merkmal der meisten Bewegungsstudien blieb jedoch die Fixierung der Kamera, die durch Elemente wie Fußbodenraster oder Uhrzeitanzeigen unterstützt wurde und damit den wissenschaftlichen, interventionsfreien und überprüfbaren Charakter der Aufnahmen belegen sollte. Dies unterstützte noch die hohen Erwartungen an Bewegungsstudien. Kurz nach 1900 wurden sie als Mittel der ergonomischen Forschung bei Frank B. Gilbreth (1868–1924), aber auch zur Optimierung der Arbeitsleistung entdeckt, so in der von Frederick W. Taylor (1856–1915) begründeten Arbeitswissenschaft. Meist wurde mit Lichtspuren in Langzeitbelichtungen gearbeitet, um Bewegungen in ihrem Fluss darstellen zu können (darin den heutigen Linienmustern vergleichbar, die sich aus der Aufzeichnung von Bewegungen des Augapfels ergeben (*eyetracking*)). Die Aufnahmen reproduzierten in dieser Anordnung jedoch vor allem ein bestimmtes Verständnis von Bewegung und Kontinuität in stillgestellter Form und waren trotz ihrer neuartigen Muster nicht unbedingt zweckmäßig.

Laborbilder der Bewegung

13.3 Standbild, Moment, Augenblick

Im Archivwesen oder auf dem Bildmarkt haben Bewegtbilder spezifische Aufzeichnungs- und Speicherformate und stellen besondere konservatorische Ansprüche. Der bisher größere Teil der TV-Geschichte ist z. B. aufgrund fehlender Aufzeichnung unwiederbringlich

Speicherprobleme des Bewegtbildes

verloren, ebenso wie zahllose Kinofilme, bei denen aufgrund hoher Kopienzahlen davon ausgegangen wurde, dass stets weitere Kopien existieren, und daher Bestände vernachlässigt oder sogar gezielt vernichtet wurden, wie es im Filmverleih üblich war. Der in → ABBILDUNG 5 gezeigte Ausschnitt aus dem beschädigten Asta-Nielsen-Werk *Afgrunden* kann als Echo dieser Phase des Filmbildes gesehen werden, das für diverse Verfalls- und Zersetzungsprozesse anfällig ist.

Allerdings ist in diesem Zusammenhang auch eine allmähliche terminologische Gleichordnung jüngerer Medien zu beobachten. Wenngleich Bewegtbilder anders bezeichnet und gespeichert werden als z. B. Fotografien, wird von diesen nicht minder als „Bildern" gesprochen (so im englischen Begriff *motion pictures*). Die begriffliche Klammer wurde begünstigt durch die gemeinsame Mediengeschichte **Standbilder** und den Umstand, dass Film oder Video Bilder im Sinne eines stehenden Motivs liefern können, sofern sie sich ablichten oder anhalten lassen, im Unterschied zu einem abgespielten Musikstück, bei dem durch Druck auf die Pausentaste der Ton verstummt.

Durch die vergleichsweise leichtere Reproduktion stehender Bilder wird ermöglicht, dass diese sich in anderer Weise verbreiten und überliefern. Zeitungsredaktionen verwenden unterschiedslos Fotografien, Screenshots von Fernsehern oder Computermonitoren und Videostandbilder parallel nebeneinander. Daher kommt es, dass einzelne, zum Teil willkürlich ausgewählte Aufnahmen als Schlüsselbilder **Schlüsselbilder der** der Presse fungieren, obwohl sie einer Sequenz entstammen und so **Reportage** gesehen einen Ausschnitt aus einer Serie von Varianten zeigen. Das Auftauchen des führenden Attentäters vom 11. September 2001, Mohammed Atta, wurde zu einem Motiv der Printmedien, weil die Videosequenz von Überwachungskameras im Fernsehen übertragen, abfotografiert und dann von verschiedenen Seiten ins Internet gestellt wurde (vgl. Geimer 2006, S. 36).

Die Nachnutzung einzelner Motive in Presse und Buch dehnt die Zeit des Einzelmoments und erweitert damit den Blick für einzelne Momente einer Situation. Hieraus könnte gefolgert werden, dass **Präferenz für** dem Standbild eine höhere ikonische Qualität, Intensität oder Nach- **statische Bilder?** wirkung innewohnt, so wie es filmische Szenen gibt, die sich als Kulminationspunkt einer Bewegung geradezu für die Reproduktion als *still* anbieten. Wo elektronische Displays (Bildschirmhintergründe) rein technisch gesehen animiert werden könnten, unterbleibt dies meist, wenn die Bewegung als irritierend empfunden wird. Pressefotografie wie Kunstfotografie setzen sich von der Filmaufzeichnung gelegentlich ab mit dem Hinweis, durch ihre Auswahl, ja Inszenierung des entscheidenden Momentes Wirklichkeit zu sublimieren, so wie der stehenden Darstellung aufgrund ihrer ununterbrochenen Betrachtbarkeit ein höherer Erinnerungswert zugesprochen wird. Dieser Wert ist jedoch gerade aus der Wechselwirkung von stehenden Bildern und Bewegung, von festen Bildformen und deren imaginativer Vervollständigung zu verstehen.

Dass der Wunsch nach Bewegung im Bild nicht immer besteht, hat mehrere Gründe. Fließende Übergänge zwischen dem bewegten und stehenden Bild zeigen die Skizzen, die als *storyboards* zur Konzeption von Filmen entstehen, oder Fotografien, die zur Auswahl von Drehorten dienen. Filmstudios lassen Fotos berühmter Akteure in werbetauglichen Situationen anfertigen, weil diese sich besser für die Plakatproduktion eignen. Wie der Buntdruck im Buch (→ KAPITEL 5.2) ist Bewegung zudem ein unverändert quantitativer und kostenrelevanter Faktor für Produktion, Übertragung und Archivierung von Bilddaten. Trotz Webcam oder Handy-Videotelefonie mit großen Übertragungsraten werden Bildbotschaften nicht unbedingt beweglich verschickt.

Vor dem Hintergrund einer globalen alltäglichen Praxis des Fotoalbums oder der Zeitungsillustration scheint es immer noch keine Selbstverständlichkeit, sich ausschließlich mit bewegten Bildern zu umgeben. Nur eine Minderheit dürfte ihr privates Videoarchiv konsequent aus einminütigen Sequenzen bestehen lassen, die als einzeln abgelegte, herausoperierte „filmische Gesten" (so ein Sammlungsprojekt des Filmkünstlers Harun Farocki) eher einen analytisch-wissenschaftlichen Wert haben. Zu den wenigen dafür ausgewiesenen Orten, an denen sich eine Darstellung ununterbrochen bewegen darf, gehören Fernsehgeräte oder Werbetafeln im Stadtraum.

Filmische Gesten

Die vergleichende Diskussion um die Bewegung im Bild rührt aus der Entstehungsphase einer modernen Ästhetik – die zugleich die Zeit intensiver fotografischer Experimente war. Georg Wilhelm Friedrich Hegel (1770–1831) hat in seiner Vorlesung zur Ästhetik unter der Überschrift *Die Handlung* (Hegel 1832, Bd. 1, S. 266) beschrieben, wie statische Figuren eines Gemäldes oder Skulpturen durch Gesten, Haltungen oder Mimik eine Bewegung zum Ausdruck bringen, die im Unterschied zum Theater nicht performativ, als zeitlicher Ablauf vor dem Auge des Zuschauers aufgeführt werde. Eine Bildhandlung lebe durch Narration und Ausdruckskraft und nicht dadurch, dass sie in einer bestimmten Bewegung festgehalten werde. Andernfalls müsste diese Bewegung – wie in einem Gemälde von Hegels Zeitgenossen Ludwig Schnorr von Carolsfeld (→ ABBILDUNG 44) zu beobachten – schwebend oder eingefroren wirken.

Handlung als Darstellungsproblem

Abbildung 44: Ludwig Schnorr von Carolsfeld: *Der Sturz vom Felsen* (1833), Öl auf Holz

Fruchtbarer
Augenblick

Hegel schloss mit seiner Darstellung an klassische Theorien der Historienmalerei und insbesondere an die Idee des „fruchtbaren Augenblicks" an, die Gotthold Ephraim Lessing (1729–81) in seiner Schrift *Laokoon oder Über die Grenzen der Malerei und Poesie* (1766) am Beispiel einer antiken Skulpturengruppe entwickelt hatte. In dieser kämpfen der Priester Laokoon und seine Söhne vergeblich gegen Schlangen, die ihnen als göttliche Strafe für ihre Warnung der Trojaner vor den Griechen geschickt wurden.

Lessing beschreibt den Zenit einer Handlung, den ein Künstler in seiner Darstellung aufspüren und fassen müsse, um in der mechanisch unbewegten Figur den Eindruck von Lebendigkeit aufrechtzuerhalten, welcher sich aus dem Weiterdenken des Geschehens ergebe. In der künstlerischen Darstellung wird durch die Komposition eine innere Bildspannung erzeugt. Damit entwickelt auch Lessing seine *Innere Bildspannung* Ästhetik noch in vergleichender Weise aus dem Verhältnis der Gattungen (Bildkunst, Text, Theater) weiter, wendet Begriffe der Dramentheorie auf das Bild an und umgekehrt.

Leben, Bewegung und Entwicklung, auch im Sinne weltgeschichtlicher Umbrüche oder sozialer Mobilität, wurden zu dieser Zeit als Themen entdeckt und später durch neue Erfahrungen von Bewegung wie z. B. durch die Eisenbahnreise noch verstärkt (vgl. Schivelbusch 2007). Auch begann die physiologisch-psychologische Forschung zu begreifen, dass selbst im vermeintlich stehenden Bild stets ein unwiederholbarer Moment enthalten ist, insofern jede Betrachtung durch ein bewegliches Auge, unter inneren wie äußeren Einflüssen erfolgt und individuell unterschiedliche Längen und Intensitäten hat (→ KAPITEL 4.2; vgl. Gombrich 1984). Der Begriff der „Animation" hat seit *Animation* dieser Phase zahlreiche Wandlungen durchlaufen, als er zunächst die Bewegung im Zuschauer und die Beseelung von Gegenständen und Bildern und später auch die konkrete Bildbewegung (wie im Zeichentrick) bezeichnete.

13.4 Analysen

Bewegung ist so gesehen nicht zu reduzieren auf die mechanische Seite des Bildträgers, bildinterne Bewegung nicht identisch mit dem zeitlichen Ablauf des Films und der Positionsbewegung durch Montage von Kamera-Einstellungen. Indem sich aus der Montage erst durch den Betrachter eine Bewegung ergibt, ist auch der Film ein imaginativer Akt (vgl. Wulff 1993; Currie 1997), so wie im neuen Medium *Film als imaginativer Akt* Film auch ästhetische Erfahrungen der vorkinematografischen Zeit weitergeführt werden.

Der französische Philosoph und Filmtheoretiker Gilles Deleuze (1925–95) sah gerade in dieser Besonderheit des „zeitlichen" und des „bewegten" Bildes das Paradox, das den Film ausmacht. Deleuze war fasziniert von der Frage, was filmische Bewegung überhaupt sei, *Theorie der* wenn doch zahlreiche Filmszenen nur unbewegte Motive zeigten, *Bewegung* wenn das „große Bild" einer Totale oder eines Erstarrungsmomentes als eingefrorene Form wahrgenommen werde oder wenn die Illusion von Bewegung nur dadurch möglich sei, dass Einzelaufnahmen in Folge von Sekundenbruchteilen und nur mithilfe von Licht und Motorantrieb eine Bewegung suggerierten, die letztlich aus der Trägheit des Auges entstehe (vgl. Deleuze 1998).

Um diese Frage zu klären, ging er zurück auf den Begründer der sogenannten Lebensphilosophie Henri Bergson (1859–1941), der eine Definition von „Bewegung" versucht hatte, welche auf die radikale Umdefinition von Raum und Zeit, Maße und Relationen in den modernen Naturwissenschaften reagierte. Eine Person, die sich durch einen Raum begibt, weiß allein durch den Abgleich mit vorangegangenen

Kino als Denkform

Eindrücken, dass sie sich bewegt; Bergson sah daher im kinematografischen Bild eine geeignete Metapher, um die Seinserfahrung als Ablauf gespeicherter Bewegung zu beschreiben. Sein Interpret Deleuze fand darüber hinaus, in der Tradition der Kunstwissenschaft, Grundformen menschlicher Erkenntnis im Kino aufgezeichnet: Kino in seinen Strukturen zu studieren, heiße, das menschliche Denken selbst zu studieren.

In seiner zweibändigen Analyse hat sich Deleuze dabei an den prominenten Autorenwerken der Filmkunst wie z. B. den Montagen Sergej Eisensteins orientiert. Es wäre gleichwohl auch möglich gewesen, den umgekehrten Weg zu beschreiten, da sich in der Führung

Ethnografie des Amateurfilms

oder Montage von privaten Filmen aus Urlaub und Freizeit bestimmte Muster und Techniken der Filmsprache (Schnitte, Längen, Geschwindigkeit der Kameraführung) mit persönlichen Erwartungen und den Bedingungen semi-professioneller Film- und Videotechnik verbinden, die einen ethnografischen Einblick in ein Medium und seine Handhabung gestatten.

Damit ließe sich umgekehrt zeigen, welche kulturtechnische Leistung die Einsicht darstellte, ab wann eine Bildfolge oder Bildmontage als gelungen gelten kann und wieso ein Film auch als Unterhaltungsmedium eine Kunst ist. Nach Ansicht des Filmeditors Walter S. Murch gibt es den Schnitt als Blinzeln oder Zwinkern schon in der Natur, wobei das Auge sich vor großen Schwenkbewegungen schließt

Elegante Schnitte

(Murch 2004). Daher ist es möglich, auch eine gelungene Montage als natürlich zu empfinden, wenn sie durch Schnitte unterbrochen wird. Der unmerkliche Schnitt ist demnach zwingend, weil sonst die Filmbetrachtung gestört würde. Andererseits sind radikale zeitliche und örtliche Sprünge in der Kameraposition oder Montage für das Auge verträglich, weil sie im Kino über längere Zeit eingeübt wurden oder mit anderen Erfahrungen (Fortbewegungsmittel, Traum, Sinnestäuschungen, auch Literatur) kongruieren.

Die Analyse bewegter Bilder in Film, Fernsehen und Video zielt darauf ab, die Phänomene und Stilmittel des Films auch jenseits erzählbarer Handlungen darzulegen, begreift quantitative Faktoren wie Schnittzahlen und Einstellungslängen, Auftritte von Akteuren oder mengenmäßige Text-Bild-Verhältnisse als filmische Stile jenseits des rein Bildhaften. Unumgehbar wird dies bei dem Versuch, die schnellen Schnitte vieler Musikvideos und TV-Sendungen zu differenzieren. Die Analyse muss berücksichtigen, dass das elektronische Bild der Videoaufzeichnung im Hinblick auf Technik, Nutzung, Speicherung und Ästhetik nicht mit dem physischen Medium Film gleichzusetzen ist, oder dass in der Übertragung des Live-Fernsehens nicht geschnitten, sondern geschaltet wird. Die „bewegten Bilder" des Kinos sind nicht gleichzusetzen mit der sozialen und kommunikativen Bedeutung der Massenausstrahlung von Fernseh-Livebildern, die kollektive Wirkung des Kinos wird durch den privaten DVD-Konsum verändert, dafür ersetzt durch die Ansteuerung und Wiederholung von Einzelszenen oder leichtere Herstellung von Standbildern. Im Austausch dieser Medien und in ihrer gesellschaftlichen Diffusion haben sich zugleich gemeinsame Formatierungsregeln, Maße und Motive eingespielt.

Quantitative Stilanalysen

Das Filmbild als *still* zu repräsentieren, wie es auch in → ABBILDUNG 42 geschieht, könnte sich vor dem Hintergrund der Deleuze'schen Frage als unzulässig erweisen (vgl. Koebner 2006). Aus der längeren Erfahrung mit dem Film und seiner Geschichte betrachtet, zeigt sich aber immerhin, dass auch das bewegte Bild seine fruchtbaren gestischen Momente und Einstellungen sucht und die darin gezeigte Handlung nicht weniger imaginativ vervollständigt wird als im stillstehenden Foto in der Zeitung. Umgekehrt setzt jede Wahrnehmung in Stadträumen Bewegung voraus, erzeugen Lichtspiele und Theater auf ihre Weise bewegte Bilder (→ KAPITEL 12). Bewegung ist damit kein exklusives Problem des Films und kein Unterscheidungsmerkmal zu Medien wie Video, Fernsehen oder digitalen Bildgebungsverfahren. Die Bilder sind vielmehr so still oder so bewegt wie die Betrachter, von denen sie angeschaut werden.

Bewegung ist nicht filmspezifisch

Fragen und Anregungen

- Welche Faktoren führen dazu, dass Bewegtbilder – trotz ihrer Alltäglichkeit und Verfügbarkeit – nicht immer zum Einsatz kommen?

- Was unterscheidet Video- und Filmtechnik? Ist diese Unterscheidung in digitalen Medien noch aufrechtzuerhalten?

- Im Comic wird mit diversen Stilmitteln, etwa durch Linien, Bewegung dargestellt. Was könnte dafür verantwortlich sein, dass diese Striche als Ausdruck von Bewegung verstanden werden können?

- Was unterscheidet die Bildhandlung des Gemäldes von der (durchaus bildhaften) Handlung auf der Theaterbühne?

Lektüreempfehlungen

- Georg Füsslin: **Optisches Spielzeug oder: Wie die Bilder laufen lernten**, Stuttgart 1993. *Füsslin hat sämtliche Apparate und Spielzeuge erläutert und auch ihre Hersteller und Vertriebswege recherchiert. Das Buch ist ein wichtiges und lesbares Handbuch zur Vorgeschichte bewegter Bilder.*

- Thomas Hensel / Klaus Krüger / Tanja Michalski (Hg.): **Das bewegte Bild. Film und Kunst,** München 2006. *Der Tagungsband versucht die Bedeutung der kunsthistorischen Herangehensweise für die Analyse des Filmbildes neu zu beleuchten.*

- Knut Hicketier: **Film- und Fernsehanalyse,** 4. Auflage, Stuttgart 2007. *Aus dem Vergleich von Film und Fernsehen und vor dem Hintergrund ihrer gesellschaftlichen Wahrnehmung entwickelt der Autor Kriterien zur Beschreibung von Bewegtbildern und informiert über die Grundbegriffe und methodischen Ansätze ihres Studiums.*

- Daniela Kloock (Hg.): **Zukunft Kino. The end of the reel world,** Marburg 2008. *Angesichts der technischen Bedingungen des Kinos, deren (digitale) Veränderungen sich grundlegend auf die Mediennutzung auswirken, beschreiben Filmpraktiker und -kritiker hier den aktuellen Strukturwandel des Kinos.*

14 Ein- und Ausblicke

Datei Bearbeiten Ansicht Werkzeuge Optionen Lesezeichen Fenster Hilfe

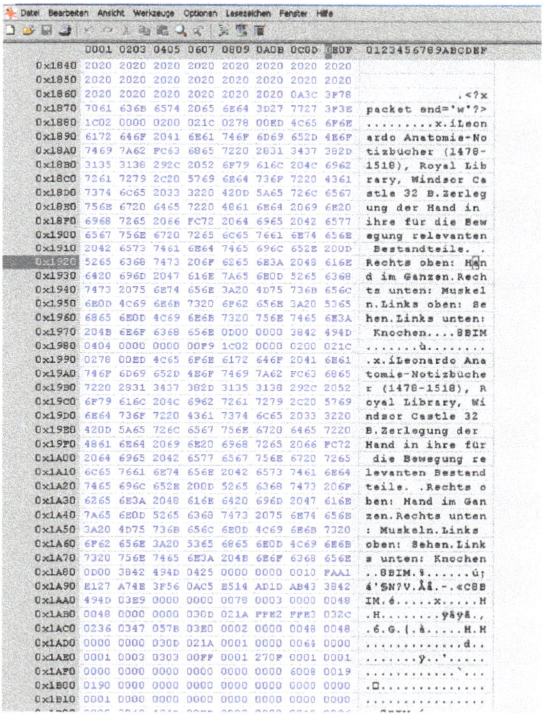

Abbildung 45: Ausgabe einer Bilddatei in Hexadezimalzahlen, erstellt auf Grundlage einer anatomischen Zeichnung Leonardo da Vincis (2006)

High Resolution Stereo Camera (HRSC) auf Mars Express
Mars: Bildabdeckung nach drei Jahren
10.01.2004 - 10.01.2007 (Orbits 10 - 3800)
10-20 m/Pixel
20-30 m/Pixel
30-50 m/Pixel
50-100 m/Pixel
Auflösung der Bilddaten am Marsboden (Nadirkanal)

Abbildung 46: Bildabdeckung des Planeten Mars nach 3 Jahren (2007), ausgeführt mithilfe der HRS-Kamera an Bord des Mars Express

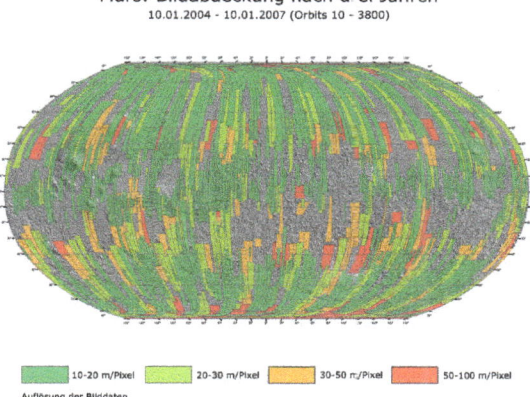

Grundlage für den Quellcode eines Scans, der von einem Computer-editor in Hexadezimalwerten ausgegeben wurde, war eine Zeichnung von Leonardo da Vinci aus einem anatomischen Notizbuch, das sich im Besitz der Royal Library im englischen Windsor Castle befindet. Ein Bearbeiter hat weitere Informationen am Computer eingetragen, die in der Datei gespeichert wurden. Der Anfang der Zahlenkolonnen besteht daher nun aus versteckten Angaben zu Herkunft und Urheberrecht der Abbildung, die am Rand in Text übersetzt sind. Diese Angaben sind aber normalerweise weder in der Bildschirmansicht noch im Ausdruck sichtbar.

In einer kartografischen Projektion wird die flächendeckende fotografische Erfassung des Planeten Mars, die sich schrittweise aus dem Umlauf einer ESA-Sonde ergibt, vorgestellt. Ein begleitender Text führt aus, dass der Mars durch Einsatz verschiedener fotografischer, spektroskopischer und Lasertechnologien inzwischen gründlicher vermessen ist als der Planet Erde.

Je nach Anwendungsgebiet sind die visuellen Darstellungsmittel, die in Naturwissenschaft und Technik zum Einsatz kommen, höchst ausdifferenziert und werden je nach Gebiet auch anders bezeichnet, etwa als Sichtbarmachung und Visualisierung. Besonders im wissenschaftlich-technischen Kontext der Aufzeichnung, Messung und Beobachtung erweisen sich Bildgebungen als Produkte digitaler Technologien, die in unterschiedlichen Ausgabeformaten vorliegen können und je nach Zweck oder Zielgruppe weiter verdichtet und aufbereitet werden. Bilder werden daher oft als rein illustrative Mittel der Kommunikation zwischen Wissenschaft und Öffentlichkeit angesehen, die eine Fülle von Daten unterschlagen. Eine der größten Herausforderungen der wissenschaftlichen Beschäftigung mit der modernen Bildkultur besteht derzeit darin, diese überhaupt in ihrer vollen technologischen Komplexität zu beschreiben.

14.1 Sichtbarmachung
14.2 Redundanz, Komplementarität
14.3 Neue Ikonologien
14.4 Das Recht der Bilder

14.1 Sichtbarmachung

Ein Großteil der heutigen Bilddaten läuft nicht in den Massenmedien   **Bilddaten**
oder im privaten Mediengebrauch auf, sondern in Naturwissenschaft
und Technik sowie im militärischen und sicherheitspolitischen Sektor.
Die Kosten für die Bildproduktion in der Raumfahrt und Astrophysik, maßgeblich auch in der militärischen Aufklärung und in der polizeilichen Überwachung (vgl. Levin u. a. 2002; Hempel / Metelmann
2005) dürften, wenn sie unter dem Aspekt der Bildproduktion betrachtet würden, die Mengen und Produktionskosten der Massenmedien, der Film- oder Spieleindustrie weit überschreiten.

Angesichts der komplexen Wissensgebiete wie Meteorologie,
Medizin oder Sozialwissenschaften, in denen Bildmedien vielfältig
zum Einsatz kommen und auch angesichts des fließenden Übergangs
von numerischen und visuellen Daten ist die Menge, die bei einer
Analyse moderner Bildformen zu untersuchen wäre, selbst mit Datenbanken nicht mehr überschaubar. Paul Virilio hat daher in den
Visierungen durch Kameras und Spionagesatelliten, die vor allem
dem militärisch motivierten Bedürfnis nach Information geschuldet
sind, eine größere „Sehmaschine" am Werk gesehen, die nur noch   **„Sehmaschine"**
um ihrer selbst willen Sichtbarkeit produziert und dabei ihre Be-
schleunigung kontinuierlich erhöht (Virilio 1989). Die heutigen industrialisierten Gesellschaften sind für eine bestimmte Form der
Sichtbarmachung offensichtlich mehr Geld auszugeben gewillt als jede vorangegangene. Die technologische Aufrüstung von Armeen, die
Ausstattung mit Sichtgeräten für Soldaten und Sprengkörper gleichermaßen, machen Bildtechnologien zu einer neuen Bewaffnung.
Daneben sind Millionen von Menschen bereit, für Spielkonsolen
oder Fernsehangebote auf Großbildschirmen Monatsgehälter zu in-
vestieren.

Die Teilchenbeschleuniger internationaler Großprojekte und For-   **Bildreaktoren**
schungsverbünde der Hochenergiephysik verstehen sich vor diesem
Hintergrund zurecht als die „größtmöglichen Mikroskope" (Eigen-
werbung des CERN in Genf), da sie Anordnungen zur Ausrichtung
und Aufzeichnung von Spuren sind, welche dann auch visuell umgesetzt werden können. Obwohl das primäre Ziel solcher Anlagen keine Bild-Produktion mehr ist, äußert sich noch in der Aufzeichnung
von Experimentalanordnungen dieser Größenordnung ein Bedürfnis,
in den Aufbau der Natur oder des Weltalls mit anderen als rechnerischen Mitteln hineinzusehen, um dabei trotz aller Vorausberechnung
auf unerwartete Wirklichkeiten zu stoßen.

Datenbilder

Wie die Rede von der „ultimativen Mikroskopie" verdeutlicht, wird hier ausdrücklich an Bildtraditionen angeschlossen, bei denen es um Sichtbarkeit in einem optisch-technischen und kulturell geläufigen Sinne geht. Die sogenannten ‚Bilder' der Wissenschaft bestehen allerdings in der Regel aus numerischen Informationen und sind das Ergebnis von Messungen und Rechenprozessen (vgl. Galison 2002; Kittler 2004), die in langen Beobachtungsphasen und unter Anwendung automatisierter Verfahren erfolgen. Radioteleskopische Daten der Weltraumforschung oder tomografische Verfahren der Medizin (→ KAPITEL 12) ergeben erst dadurch anschauliche Ansichten, dass sie zusammengesetzt, interpoliert und aufbereitet und den technischen Bedingungen des Bildschirms soweit angepasst werden, dass sie darüber hinaus auch den Anforderungen z. B. eines Arztes oder den Sehgewohnheiten eines Zeitungs- oder Fernsehpublikums entgegenkommen.

Sichtbarmachung

Hinter dem aufbereiteten einheitlichen Erscheinungsbild können sich unterschiedliche Techniken der Sichtbarmachung verbergen. Sie kann sowohl durch Übersetzung von Zahlwerten aus Messungen erfolgen, wie sie in bildgebenden Verfahren stattfindet. In empirischer Hinsicht unterscheidet sie sich dadurch von Visualisierungen, bei denen Objekte von Grund auf generiert oder Prozesse simuliert werden, z. B. bei der Prognose von Klimaentwicklungen oder Wetterverhältnissen, die einen zukünftigen Zustand modellieren und bis zu dessen Eintreten ungesichert bleiben.

Virtuelle Objekte

Unabhängig vom Verfahren entstehen jedoch durch Umrechnung von Zahlwerten in Bildanzeigen neue Artefakte eigener Realität, deren Gegenstand nur rechnerisch gegeben ist, die aber dennoch zum Gegenstand von Diskursen werden können und nicht nur Fakten schaffen, sondern womöglich auch der einzige Zugang zum Gegenstand bleiben. Es formieren sich virtuelle Objekte entrückter oder mikroskopischer Welten, die nur in aufbereiteter Form zu studieren sind. Ihre Virtualität erhält diese verfahrensgebundene Sichtbarkeit dadurch, dass auch Operationen auf der Grundlage von Visualisierungen erfolgen (vgl. Hacking 1983). Entgegen den häufigen Bekundungen von Labormitarbeitern, wonach die eigene Arbeit unmittelbar „an Daten" stattfinde und Visualisierungen z. B. in Form von Diagrammen, gefärbten Ansichten oder räumlichen Modellen lediglich zur öffentlichen oder didaktischen Vermittlung dienen, erfordert die Komplexität der Verfahren an irgendeiner Stelle zumeist den Einsatz von Software und grafischen Techniken, die den Daten eine analysierbare Struktur gibt. Die Forschung muss sich auf dem Weg zum

Forschungsobjekt durch diese diversen Bildgebungsstufen wieder „hindurcharbeiten" (Knorr-Cetina / Amann 1990, S. 272), um zu sehen, an welcher Stelle welche Daten wie gefiltert wurden.

Die zugrunde liegenden Messdaten sind aber ebenfalls nicht das Ding als solches, sondern nur die Echos von etwas, das innerhalb bestimmter Versuchsanordnungen im Labor ein Auslöser für instrumentelle Aufzeichnungen war. Die Soziologen Bruno Latour und Steve Woolgar haben daher in ihrem einflussreichen Aufsatz von 1979 unter dem Titel *Laboratory Life* konstatiert, dass die Visualisierungen innerhalb der vermeintlich exakten Wissenschaften Einschreibungen (*inscriptions*) von Messungen sind, insofern diese sich soweit zu Objekten des Diskurses verdichten, dass sie das Objekt selbst repräsentieren und sich dadurch auch allmählich in der Wissenschaft fortschreiben (vgl. Latour / Woolgar 1979). Der Begriff der Repräsentation wurde in diesem Zusammenhang auf die Probe gestellt und teilweise umgedeutet, da er einerseits die Idee der Wiedergabe von etwas enthält, andererseits durchaus einschließt, dass Daten am Bildschirm ‚aufgeführt' werden (vgl. Rheinberger 1997; → KAPITEL 7.4).

Einschreibungen, Aufführungen

14.2 Redundanz, Komplementarität

Die Formen der Sichtbarmachung, die in den diversen Disziplinen zur Anwendung kommen, sind zudem funktional höchst verschieden, und sie genießen nicht durchweg die gleiche gesellschaftliche Aufmerksamkeit. Ein Blick in das wissenschaftliche Publikationswesen offenbart eine Vielfalt an unspektakulären Grafen, Skizzen und Formeln, an umfangreich kommentierten Detailaufnahmen und Bildserien. Die einzelnen Illustrationsformen, Anzeige- und Dokumentationstechniken sind dabei je nach Fachgebiet und Nutzung genauestens ausdifferenziert. Die Archivierung von Bildern als Messdaten, Ausdruck, Fotoplatte, Modell, Notiz, Bildreihe oder Scan wird je nach Disziplin verschieden gehandhabt und formatiert. Innerhalb der unter Naturwissenschaften subsumierten Disziplinen und Publikationsorte unterscheiden sich Bedeutung und Ausprägung visueller Medien, und zwar nicht weniger als die Fächer selber.

Fachunterschiede

Besondere Beachtung findet die Astrophysik, sobald sie aus ihren Datenlieferungen plastische und farblich ansprechende Ansichten generieren kann. Das Deutsche Zentrum für Luft- und Raumfahrt (DLR) hat z. B. für die ESA-Marsmission *Mars Express* die HRSC entwickelt, die offiziell eine „optische" (da mit lichtbrechenden Lin-

Bilder vom Mars

sen ausgestattete) Kamera ist und die aus ca. 250 km Höhe die Marsoberfläche systematisch ablichtet. Die Kamera kann Punktaufnahmen bis zu zwei Meter Auflösung pro Pixel anfertigen. Die schematisierte → ABBILDUNG 46 kartiert die abgetasteten Zonen des Mars je nach Auflösung.

Komposite Bilder

Die Daten der HRSC sind Scans von Lichtintensitäten, wie sie prinzipiell jede Digitalkamera erzeugen würde, ihre elektronischen Sensoren filtern aber in mehreren Aufnahmeschritten gezielt nur jene Spektralbereiche heraus, die es ermöglichen, unterschiedliche Auflösungen zu erhalten, Oberflächenbeschaffenheiten zu prüfen und Höhenunterschiede zu registrieren (vgl. Uhlig 2006). Es handelt sich also auch hier um aus mehreren Spektralbereichen zusammengesetzte Aufnahmen, deren Daten auf der Erde über mehrere Stationen geleitet und dekomprimiert, kalibriert und entzerrt werden. Das Ergebnis sind keine Einzelaufnahmen von Marsoberflächen, sondern Sets von Bildvarianten, die in mehreren Qualitäten vorliegen. Ganz explizit vermerkt das DLR zu jeder Veröffentlichung, dass es sich um Datenkompositionen handele, deren endgültige Farbgebung nachträglich erfolge (→ KAPITEL 5.4). Damit sind derlei visuelle Produkte durchweg „komposite Bilder" (Hinterwaldner/Buschhaus 2006), da sie nicht nur errechnet sind, sondern aus unterschiedlichen Messtypen wie Laser-, Radio- und Spektrografie bestehen und durch Differenz- und Kontrastverfahren, durch optische Analysen und automatische Bilderkennungen ausgewertet werden. Heutige Sternwarten, die mit konventionellen optischen Spiegelteleskopen arbeiten, verwenden Geräte, die sich zwecks Lichtausbeute aus Hunderten von Einzelspiegeln zusammensetzen und deren Aufnahmen auf rechnerischem Wege von erdatmosphärischen Störungen bereinigt werden.

Komplementarität

Eine belastbare Repräsentation eines Gegenstandes ergibt sich damit erst aus der Summe von Verfahren, die nebeneinander zum Einsatz kommen. Dies können unterschiedliche Messungen sein, aber auch fotografische Aufnahmen, die durch Beschreibungen und Zahlwerte ergänzt oder erläutert werden. Schon wenn ein Geologe des 18. Jahrhunderts durch das Land ging, umfassten seine Aufzeichnungen Angaben über Orte, Höhen und Entfernungen, Schächte und Flöze, Fossilien und Mineralien und bestanden aus losen Blättern, Tagebucheinträgen, Marginalien und Briefen, die in alle Welt gingen. Die Hybridität und Beweglichkeit ist ein Merkmal dieser auf Wissensvermehrung ausgerichteten Aufzeichnungen, die später aus Druck- oder Archivgründen oftmals vereinfacht, separiert oder unterschlagen werden.

In vergleichbarer Weise bleiben auch heute, trotz prinzipieller elektronischer Speichermöglichkeiten, viele dynamische und interaktive Bildformen, die der Operation am Bildschirm dienen, unaufgezeichnet oder verschwinden nach temporärem Gebrauch. Es wird ein Überschuss erzeugt, der von vornherein auf Redundanz und Komplementarität angelegt ist, um laufende Forschungen, Rechnungen und Diskussionen zu dokumentieren oder für spätere unerwartete Fragestellungen offen zu bleiben. Aufzeichnende und bildgebende Verfahren verwenden beispielsweise Bitraten, die über der Verarbeitungsgrenze herkömmlicher Prozessoren und analoger Aufzeichnungsgeräte liegen können, weil von einer zukünftigen Leistungssteigerung dieser Basistechnologie auszugehen ist.

Diese Mechanismen der Vorauswahl sind für ein konkretes Bildergebnis ebenfalls konstitutiv. An ihnen wird ersichtlich, welch großer Bereich von Gegenständen einer kulturellen wie wissenschaftlichen Aufmerksamkeit entgeht, wenn diese nicht als kulturell relevante und geprägte Bilder verstanden werden. Aus alltäglichen Bildpraktiken wie der Fotoreportage, Werbung oder Kunst ist bekannt, dass es zu jedem prominent gewordenen Motiv unzählige Bilder gibt, die verworfen oder unterdrückt werden, weil sie bestimmte Anforderungen oder Bildformate unterlaufen, prognostizierten Verwertungszielen nicht genügen oder ihre Herstellung misslingt. Den wenigen bekannten Zeichen der wissenschaftlichen Bildkommunikation steht ebenfalls ein gewaltiger Vorrat ungezeigter Formen gegenüber, ohne die sie nicht existieren würden und der ihnen zusätzliche Bedeutung verleiht.

Aus den oben genannten Gründen könnte es müßig sein, von Bildern „der" Wissenschaft oder „in der" Wissenschaft zu sprechen. Eine bildkritische Analyse kann aus der Betrachtung naturwissenschaftlicher und technischer Visualisierungen dennoch Rückschlüsse ziehen auf Eigenschaften visuellen Denkens. Denn mit der Vielfalt an Optionen und Medien zur Gewinnung, Visualisierung und Vermittlung von Information, in der größeren Breite von Phänomenen und angesichts unüberschaubarer Betrachtergruppen machen sich auch im sogenannten wissenschaftlichen Bereich bestimmte Stile, Konventionen und Präferenzen bemerkbar (→ KAPITEL 3.4, 10.2). In jeder Bildpraxis äußert sich unausgesprochen ein visuelles Denken von Experimentatoren, Programmierern und Ingenieuren, das über die Komplementarität und Komplexität von Bildgebungen nachsinnt (vgl. Krohn 2006). Wenn Abbildungen von naturwissenschaftlichen Lehrbüchern mit jeder Auflage ausgetauscht werden, so sind sie als Lehrmittel

Redundanz

Bilder als Kulturerbe

Stile der Visualisierung

überholt, als Forschungsquelle werden sie unerlässlich. Die analytische Schwierigkeit besteht vor allem darin, die quellenmäßigen Grenzen dieses flüchtigen Gegenstandes zu bestimmen.

Spezifizierung

Ein bildwissenschaftliches Studium, wie es derzeit erwogen wird, müsste sich dieser gegenwärtigen Bildkultur und ihrer gewollten Informationsfülle soweit verschreiben, dass sie die verwendeten Techniken ebenso wie deren kulturellen Mehrwert bestimmt. Dazu muss die Analyse mehr leisten als nur die in den Medien zirkulierten, häufig abgedruckten Aufnahmen historisch einzuordnen und zu unterfüttern. Sie muss analysieren, in welchen unsichtbaren Formaten visuelle Daten vorliegen, welche Komplexität Bilder und Bildgebrauchsformen erreichen. In einer logistischen und epistemologischen Analyse wäre zu fragen, wie sich Wissenschaften darin organisieren, globalisieren und spezialisieren. Mit Bildmedien zu arbeiten, ist kein Spezifikum einer einzelnen Wissenschaft. Umso schwieriger ist die Frage nach ihren Spezifika.

14.3 Neue Ikonologien

Da sich Visualisierungen in einem figuralen, mimetischen Sinne nur ergeben, indem Codes und Algorithmen von einem Grafikprozessor aufbereitet werden, sind ‚Bilddateien‘ lediglich Digitalisate in einem bestimmten Dateiformat. Sie können ebenso gut über nicht-visuelle Kanäle und Wandler ausgegeben werden und in andere mediale Aggregatzustände übergehen. Die Existenz digitaler Bilder ist daher, terminologisch strenggenommen, bestritten worden (vgl. Pias 2003; Schröter / Böhnke 2004).

Gibt es digitale Bilder?

Auch lässt sich im Digitalen die Schirm- oder Druckerausgabe unterscheiden von jener Code-Ebene, auf der zum Beispiel Scans bearbeitet, durchsucht oder anderweitig prozessiert werden können, ohne dass dies die visuelle Oberfläche verändert. Die Leonardo-Zeichnung (→ ABBILDUNG 45), die als Zahlenfolge ausgegeben wurde, zeigt, wie sich Daten – schlimmstenfalls auch böswillige Anhänge – in eine Datei einschreiben. Solche Einschreibtechniken, bei denen Zusatzinformationen (Metadaten) dem Bild beigefügt werden, sind wiederum Teil des Dateiformats, etwa im verbreiteten EXIF-Format für Digitalkameras, und eine Grundlage für eine strukturierte Datenübertragung und -archivierung.

Code-Ebene der Bilder

Als Kompositum ist das am Computer generierte Bild jedoch nicht virtueller als analoge Darstellungen. Seh- und Vorführgewohnheiten,

Jedes Bild ist komposit

Aufmerksamkeiten, Bildgattungen und -motive bleiben im Digitalen erhalten und begegnen sich dort mit den veränderten Möglichkeiten der Produktion und Bearbeitung. Deutlich wird dies an den vorgefertigten Filtern von Bildbearbeitungsprogrammen, die dem zumeist privaten Nutzer im digitalen Raum die Imitation traditioneller Bildstile aus Malerei, Druckwesen oder Fotografie erleichtern (vgl. Heidenreich/Ernst 1999).

Doch auch im wissenschaftlich-technischen Sektor haben neue Bildgebungsverfahren und universelle Digitaltechnologie nicht dazu geführt, dass tradierte Begriffe wie Kamera, Fotografie oder Bild abgelegt wurden. Bezeichnungen wie Visualisierung oder Sichtbarmachung transportieren weiterhin bestimmte Vorstellungen von bildlicher Präsenz. Sobald in einer beliebigen Datenmenge aufgrund von rechnerischen Aufbereitungen, von Programmierweisen, Konventionen oder individuellen Wahrnehmungen gewisse Muster oder Regelmäßigkeiten festgestellt werden, fügen sich kulturtechnische Traditionen der Wahrnehmung und Beschreibung ein (→ KAPITEL 2, 4). **Bildliche Präsenz**

Auch am Rechner ergeben modellhafte Darstellungen ein gemeinsames Bild für Konzepte wie „Staat", „Zelle" oder „Gen", als Konglomerate aus mehreren Sinnschichten, in denen architektonische oder mechanische Vorstellungen mit historischen Vorbildern und metaphorischen Übertragungen unterschiedlicher Epochen überblendet sind (→ KAPITEL 11.2). Selbst eine unrichtige Darstellung kann sich in diesen Transformationsvorgängen als Baustein einer kompakten Idee festsetzen, z. B. die Vorstellung der DNA als Doppelhelix, die zu einem prominenten Gegenstand der Bildkritik geworden ist und von Stephen J. Gould als „kanonische Ikone" der Wissenschaft bezeichnet wurde (Gould 2002, S. 326; vgl. Pörksen 1997, S. 122–135). **Sinnschichten**

Bilder sind auch hier als Teil einer größeren Sprache, als Zusammenhang von Grafiken, Texten, Praktiken und kollektiven Echos zu verstehen: „Jedes Bild muss ein Glied einer Bilderkette sein, denn stünde es nicht in einer Tradition, wäre es nicht zu entziffern" (Flusser 2000, S. 17). In diesem Sinne wurde eine neue Ikonologie vorgeschlagen (Mitchell 1986; vgl. Bolvig/Lindley 2003), die auf die neue Ordnung der elektronischen Medien reagiert und zugleich ältere kulturtheoretische Ansätze aufnimmt. **Neue Ikonologie**

Denn als vermeintlicher ‚Ersatz' für eine Sache sind Bilder auch in der technisierten Welt eine verwirklichte Vorstellung, ein Mittel der Reflexion und Distanznahme, der Überzeugung oder Abwehr, ein kollektiv wirksames Symbol. Dies war bereits Leitgedanke der Symbolphilosophie Ernst Cassirers (vgl. Cassirer 1923) und Susanne K. **Symbole**

Langers (vgl. Langer 1942) oder auch der Ethnologie von Clifford Geertz (vgl. Geertz 1973). Bruno Latours ethnologische Studie von 1979, die auf Praktiken der Wissenschaften abzielte, lässt sich ebenfalls dieser symbolphilosophisch-ikonologischen Richtung zuordnen, da sie belegt, wie sich bestimmte Muster und Objekte innerhalb eines Diskurses im Austausch mit der Geschichte wechselseitig stabilisieren. Insofern die ‚Inskriptionen' der Laborwelt einerseits nur das indirekte und gefilterte Abbild von ansonsten unbekannten Zuständen ist, sie andererseits aber in der wissenschaftlichen Kommunikation Echtheitsstatus annehmen können, wiederholt sich in ihnen der Mechanismus des authentischen, autopoietischen Bildes (→ KAPITEL 9.1).

Clash of technologies

Auch in ihren avanciertesten Formen bleiben die *technosciences* Teil menschlicher Erkenntnisproduktion mit den dazugehörigen Symbolisierungsformen von Bild, Schrift und Zahl und ihren kulturtechnischen Möglichkeiten und Grenzen. Daher wäre anstelle eines *iconoclash*, der den Glaubensstreit um Bilder in die Gegenwart weiterdenkt (vgl. Latour / Weibel 2002; → KAPITEL 2.2), im Kontext der heutigen Technik und Naturwissenschaften vielleicht besser von einem *clash of technologies* zu sprechen, der mächtige Bilder produziert.

14.4 Das Recht der Bilder

Globale Bilder

Der Prozess der Industrialisierung lässt sich seit mindestens einem Jahrhundert nicht mehr in den Grenzen lokaler Gesellschaften fassen. Durch die systematische Produktion ästhetischer Güter werden nicht nur Traditionen des Sehens und Abbildens weitergetragen oder weiterentwickelt, also beispielsweise europäische oder amerikanische Sehmuster in andere Teile der Welt exportiert, sondern dem jeweiligen Produkt auch die unsichtbaren Praktiken der Verwaltung und Bewertung beigefügt. *Google Earth* und Fotoblogs, *embedded journalism* und konfektionierte Nachrichteninhalte, Bollywood-Kino und Statussymbole bewirken die weitere Kurzschließung globaler Bildangebote.

Ikonosphären

In den Motivwelten und Bildfolgen der Massenmedien breiten sich seit mehreren Jahrzehnten Zeichen- und Gedächtniswelten aus, die eine eigene Realität oder Ikonosphäre (Bilderwelt) formen (vgl. Burgin 1981; Metz 1992). Charaktere in Werbung oder *Telenovelas* übernehmen eine autonome Stellvertreterfunktion mit katalogartigen Identifikationsfiguren, welche die Durchschnittsgesellschaft repräsentieren und im symbolischen Auftrag ihrer Betrachter miteinander kommunizieren. Bilder unterschiedlichsten Formats entfalten eine visioaktive Funktion, indem sie durch die Berührung mit anonymen Zuschauergruppen und in der unvorhersehbaren persönlichen wie kollektiven Betrachtung und Erinnerung unerwartete Wirkungen haben (→ KAPITEL 8.4). Diese Prozesse erreichen eine derartig große Komplexität, dass die darin ablaufenden Vorgänge des Erinnerns und Vergessens nur noch systemisch, in Begriffen der Ökonomie oder Ökologie, zu beschreiben sind. Produktion und Konsumption von Bildern fließen ineinander und begünstigen den Eindruck, dass Bilder ein selbstreproduzierendes Eigenleben führen, welches biologisch (vgl. Mitchell 2005b), ja sogar epidemisch (vgl. Sperber 1985; Ross 1994) erklärt worden ist.

Weder die Kontingenz der Bildproduktion noch der angeblich flüchtige Erinnerungswert von Medien wie Fotografie oder Video führt jedoch zu einer nachlassenden Herstellung neuer Bilder. Eine vorbehaltlose Kritik der visuellen Kultur, die diese als Konsumrealität und Realitätskonsum begreift, kommt dadurch wieder zu einem Bildbegriff, der aus der Pluralität der Perspektiven und Medien die erinnernden, abwehrenden, voyeuristischen wie aggressiven Funktionen herausliest und dadurch auch anthropologische Aspekte des Bildgebrauchs und der Bildwirkung neu entdeckt, die Hans Belting als Hauptgebiet der zukünftigen bildwissenschaftlichen Forschung definiert hat (vgl. Belting 2006).

Bildgebrauch und Anthropologie

Besonders im Bildnis, welches ein künstliches Gegenüber des Betrachters zeigt, finden dabei Sehen und Gesehenwerden, Zeigen und Besitzen in leibhaftiger Gestalt zusammen. Die europäische Kunsttheorie sieht in der Erinnerungsfunktion an Personen eine wesentliche Leistung von Bildern, indem sie Abwesendes real vertreten, zu dinglichen Erinnerungen werden. Bild, Körper und Bildmedium treten in ein untrennbares Wechselverhältnis. Der große Anteil an Bildnissen in Archiven, Fotoalben und Gemäldegalerien belegt dies bis heute, ebenso die juristische Verteidigung des Rechts am eigenen Bild, der Angriff auf Bilder als Stellvertretern (durch symbolische Verbrennung von Fotografien), die Unkenntlichmachung von Bildern durch schwarze Balken.

Das Bild als Bildnis

Mit einem anderen Akzent wird jener leibliche und intellektuelle Akt der Aufzeichnung und Bildkonstruktion untersucht, in dem sich Denken visuell äußert und formiert (Bredekamp 2004). Der ‚Bildakt' stiftet Bilder, die ein Spiegel ihrer Herstellung bleiben und zugleich dauerhafte Wirkung zeitigen. Dieser Ansatz führt die ikonologische Tradition Warburg'scher Prägung fort, in der die Bildherstellung als

Bildakt

Akt der Vergegenwärtigung und als Zugang zu psychophysischen Prozessen begriffen worden ist. Als Gegenüber des Betrachters leben Bilder für denjenigen, der sie leben lässt. In ihnen sind sämtliche Erwartungen, Sehweisen und Potenziale in vollem Umfang präsent, und zwar auch unabhängig von ihrer mimetischen Funktion und Qualität.

Facettierung

Sichtbarkeit wird in der facettierten Medienwelt und im Kampf um Bandbreiten und Übertragungsraten, in Technologien wie Videotelefonie und MMS oder der Umgestaltung von Stadträumen und Wohnwelten durch großformatige Displays und Visuals auf globaler Ebene neu formatiert. Aus der digitalen Fotografie ergeben sich durch die Möglichkeiten der Sofortbetrachtung, -filterung und -löschung und durch die digitale Speicherung spürbare Nutzungsveränderungen. Wenn von gegenwärtigen Fotoherstellern die fortgeschrittenste digitale Bilderkennung dazu verwendet wird, um Gesichter zu fokussieren – seit kurzem auch mit automatischer Lächel-Erkennung – wird damit der uralten Porträtfunktion entgegengekommen.

Technik und Kultur

Die dabei erzeugte fotografische Aufnahme erfüllt nicht mehr notwendig eine Bildungsaufgabe, sie schließt diese aber auch nicht aus, insofern jede Bildherstellung neu gelernt werden muss und dies im Austausch mit vorhandenen Technologien geschieht, auch in trivialer Form. Auf einer technisch fortgeschrittenen Ebene ergibt sich ein quantitativ wie qualitativ verändertes Zusammenspiel von Sinneswahrnehmung und Motorik, Technik und Kultur. Das Foto im privaten Portemonnaie, in dem sich Papier und Zeichen begegnen und das scheinbar einem magischen Glauben folgt, wonach der Betrachter sich dem Abgebildeten näher wähnt, ist ebenso ein Zeugnis der industriellen Bildproduktion, wie es Ausdruck der individuellen Bildgebung bleibt.

Visuelle Kompetenz

Der beträchtlichen individuellen Kompetenz im Umgang mit digitalen Medien steht allerdings heute keine vergleichbare „visuelle Kompetenz" (Huber 2002) auf dem Gebiet der Fototechnik, Visualisitik, Optik oder Druckvorstufe, in Motivkunde oder Pressegeschichte gegenüber. Diese bleibt nach wie vor Spezialisten überlassen. Schlechte Reproduktionen, untaugliche Präsentationsmittel, oberflächliche Bildbetrachtungen, unvollständige oder unrichtige Beschriftungen sind Produkte preiswerter Technologien und Gewöhnungseffekte und Anzeichen eines nachlassenden Interesses am Kommunikations- und Werkzeugcharakter von Bildern. Wie schon in der Zeit der Druckkunst oder der jungen Fotografie sind diese Medien, auch als Kulturtechniken des Sehens, immer wieder neu zu lernen.

218

Zur Geschichtlichkeit bildlichen Ausdrucks und bildlicher Kommunikation gehört es, dass Aufmerksamkeiten sich in den verschiedenen Zivilisationen strukturell wie intellektuell wandeln. Das Bild begleitet sämtliche Entwicklungen der Gesellschaft und spiegelt damit auch die in ihr bestehenden Widersprüche. Demgegenüber gibt es weiterhin, tief in die Sinnschichten der ästhetischen Diskussion eingeschrieben, einen Grundverdacht gegen das intellektuelle Vermögen von Bildern als Stellvertreter einer Sache oder als Ersatz für die verbale Sprache.

Verdacht gegen Bilder

Gegen die verallgemeinernde und entmaterialisierende Rede vom Bild, welche die Vielfalt der Bildgeschichte und Bildtechniken und ihrer individuellen Betrachtungen und Nutzungen systematisch eingrenzen will zugunsten einiger abstrakter Lehrsätze, empfehlen sich begriffliche Unterscheidungen von Soft- und Hardware, Pinsel und Leinwand, Foto und Kamera, Bearbeitungsstufen und Ausgabekanälen, welche die Unterschiedlichkeit der visuellen Äußerung deutlich machen können – denn es wird keine Theorie des Bildes geben, solange diese nur den Zweck verfolgt, die nähere Betrachtung zu erübrigen.

Für die Vielfalt

Fragen und Anregungen

- Wissenschaftler unterschiedlichster Fachgebiete, etwa Biologen oder Archäologen, verwenden bei Feldstudien gleichzeitig Fotografien, Materialproben und andere Techniken der Aufzeichnung, um Beobachtungen zu dokumentieren. Inwieweit wird die Bedeutung von Bildmedien als Arbeitsmittel dadurch relativiert oder befördert?

- Was meint der Begriff der Sichtbarmachung im Unterschied zum Begriff der bildgebenden Verfahren?

- Inwiefern beruht die gegenwärtige Dominanz bildbasierter Technologien vor allem auf den Möglichkeiten elektronischer Produktion und Reproduktion?

- Videobotschaften oder Pressebilder entfalten ungeahnte globale Wirkungen. Wäre es deshalb gerechtfertigt, von ‚epidemischen‘ Prozessen zu sprechen?

- Inwiefern wiederholen sich in neuen Bildtechnologien ältere Bildpraktiken, und worin unterscheiden sie sich von jenen?

Lektüreempfehlungen

- Wolfgang Ernst / Ute Holl / Stefan Heidenreich (Hg.): Suchbilder. Visuelle Kultur zwischen Algorithmen und Archiven, Berlin 2003. *Der medienwissenschaftliche Sammelband geht den Automatisierungen der Bilderkennung in unterschiedlichen Bereichen (z. B. der Satellitenüberwachung) nach.*

- Bettina Heintz / Jörg Huber: Mit dem Auge denken. Strategien der Sichtbarmachung in wissenschaftlichen und virtuellen Welten, Wien / New York 2001. *Unverändert aktuelle Aufsatzsammlung, die wissenschaftshistorische und kulturwissenschaftliche Beiträge zur technischen Konstruktion von Bildern in den Naturwissenschaften zusammenbringt.*

- Inge Hinterwaldner / Markus Buschhaus: The picture's image. Wissenschaftliche Visualisierungen als Komposit, Paderborn / München 2006. *Auf den Summations- und Kompositcharakter von Bildern und Bildgebungen ausgerichteter Band, der eine Übersicht über jüngere Forschungen zur Bedeutung der Bildmedien in Naturwissenschaften, Medizin und Technik bietet.*

- Bruno Latour / Peter Weibel (Hg.): Iconoclash. Gibt es eine Welt jenseits des Bilderkrieges?, Berlin 2002. *Der voluminöse Begleitband zur gleichnamigen Karlsruher Ausstellung geht in interdisziplinären Aufsätzen der Frage nach, wie sehr in der wissenschaftlichen Bilddiskussion ein älterer Bilderstreit wiederholt wird.*

- Christa Maar / Hubert Burda (Hg.): Iconic Turn. Die neue Macht der Bilder, Köln 2004. *Veröffentlichung einer Vortragsreihe zur gesellschaftlichen Bedeutung von Bildern aus interdisziplinärer Perspektive. Inzwischen in einem thematisch etwas offeneren Nachfolgeband „Iconic Worlds" fortgesetzt.*

15 Serviceteil

15.1 Weitere Monografien und Sammelbände

Bildtheorie / Bildwissenschaft

- The Anthropological Turn. Gender Studies als Kunstgeschichte. Kritische Berichte 4, 2001. *Die Themenausgabe der Zeitschrift „Kritische Berichte" legt dar, dass eine anthropologisch motivierte Bildwissenschaft stets einen Geschlechteraspekt einschließt.*

- Jean Baudrillard: Das Ereignis, Weimar 2007. *Baudrillards Reaktion auf den 11. September 2001 war höchst umstritten. Sie wurde jüngst als kurzes Bändchen in Übersetzung vorgelegt, die andeuten mag, in welcher Weise der französische Bilddiskurs (vgl. Régis Debray, Jacques Rancière) Medium und Gesellschaft verknüpft.*

- Gottfried Boehm (Hg.): Homo pictor, München / Leipzig 2001. *Im Anschluss an seine Anthologie historischer Texte zur Bildfrage von 1994 hat Boehm hier kunst- und bildgeschichtliche Aufsätze zur Funktion der Bildgebung in Kunst und Wissenschaft versammelt.*

- Gernot Böhme: Theorie des Bildes, 2. Auflage, München 2004. *Böhme zeigt in seinem Essay, dass die Behandlung des Bildthemas selber bereits seit der Antike Argumente für weitergehende philosophische Fragen geliefert hat und ohne diese nicht zu verstehen ist.*

- Thomas Knieper / Marion G. Müller (Hg.): Kommunikation visuell. Das Bild als Forschungsgegenstand – Grundlagen und Perspektiven, Köln 2001. *Der Sammelband bietet einen Querschnitt zur Bedeutung von Bildern in Soziologie, Politik, Werbung u. a. sowie eine Einleitung in zukünftige bildwissenschaftliche Forschungsfelder.*

- Martin Schulz: Ordnungen der Bilder. Eine Einführung in die Bildwissenschaft, München 2005. *Liefert eine Zwischenbilanz der anhaltenden Diskussionen um Ziele und Felder einer ‚Bildwissenschaft'.*

- Lambert Wiesing: Artifizielle Präsenz. Studien zur Philosophie des Bildes, Frankfurt a. M. 2005. *Lambert hat mit „Sichtbarkeit des Bildes" die Geschichte der formalen Ästhetik behandelt, die einen neuen Bildbegriff begründete. Mit diesem Aufsatzband unter-*

streicht er seine These, dass Bilder selbst Gegenstand sind und diesen nicht nur wiedergeben.

Bild und Wissenschaft

- Andreas Beyer / Markus Lohoff (Hg.): Bild und Erkenntnis. Formen und Funktionen des Bildes in Wissenschaft und Technik, München / Berlin 2005. *Der Sammelband zum Jubiläum der RWTH Aachen gibt einen beeindruckenden Querschnitt über fast alle Einsatzfelder wissenschaftlich-technischer Visualisierung.*

- Olaf Breidbach: Bilder des Wissens. Zur Kulturgeschichte der wissenschaftlichen Wahrnehmung, München 2005. *Der Autor versteht ‚Bildwissenschaften' als Disziplinen, die Bilder nicht nur thematisieren, sondern mit und in ihnen arbeiten, und gibt hier einen historischen Abriss.*

- Lorraine Daston / Peter Galison: Objektivität, Frankfurt a. M. 2007. *Daston und Galison forschen dem Wandel eines Begriffs nach, der sich im Austausch mit wissenschaftlichen und künstlerischen Praktiken entwickelt hat und der zugleich die Bedeutung des Bildthemas für die Wissenschaftsgeschichte belegt. Sehr anspruchsvoll, aber trotzdem gut lesbar und materialreich.*

- Felice Frankel: Envisioning Science. The Design and Craft of the Scientific Image, Cambridge / London 2002. *Frankel behandelt das wissenschaftliche Bild aus der Sicht einer Fotografin und Grafikerin, die den modernen Experimentatoren die sinnstiftende Qualität von Visualisierungen vermittelt.*

- Wolfgang Lefèvre / Jürgen Renn / Urs Schoepflin (Hg.): The Power of Images in Early Modern Sciences, Basel u. a. 2003. *Ein wissenschaftsgeschichtlicher Band, der nachweist, wie die Entstehung der modernen Wissenschaft und Technik auf Bildern beruhte.*

Zeichen, Sprache, Kommunikation

- Wolfgang Harms (Hg.): Text und Bild, Bild und Text. DFG-Symposium 1988, Stuttgart 1990. *Die Studien zur visuellen Argumentation, zur Sprachlichkeit des Bildes oder zur Tradition der Emblematik sind bis heute einschlägig.*

- Manfred Faßler: Bildlichkeit. Navigationen durch das Repertoire der Sichtbarkeit, Wien 2002. *Faßler sondiert in diesem kompakten Band die Vielzahl von medien- und funktionsspezifischen Zeichen, die die visuelle Kommunikation der heutigen Welt gliedern.*

- Barbara Naumann (Hg.): Vom Doppelleben der Bilder. Bildmedien und ihre Texte, München 1993. *Der Sammelband behandelt historiografische Fragen der Beschriftung, Text-Bild-Bezüge im Comic, Musikvideo und anderen Medien und dient somit als guter Einstieg in Fragen, die sich inzwischen weit ausdifferenziert haben.*

- Felix Thürlemann: Vom Bild zum Raum. Beiträge zu einer semiotischen Kunstwissenschaft, Köln 1990. *Die Einleitung erläutert die Entwicklung der Bildsemiotik und ihre Bedeutung für die Betrachtung von Kunstwerken, gefolgt von historischen Fallstudien.*

Kulturen des Sehens, Techniken des Bildes

- Ausst.-Kat. Sehsucht. Über die Veränderung der visuellen Wahrnehmung. Kunst- und Ausstellungshalle der Bundesrepublik Deutschland, Göttingen 1995. *Materialreicher Ausstellungskatalog, und als Buch zugleich Kompendium ausgezeichneter Aufsätze zur Geschichte von Kernmedien visueller Information und Unterhaltung (Panorama, Guckkasten, Spionagefotografie u. a.).*

- David C. Lindberg: Auge und Licht im Mittelalter. Die Entwicklung der Optik von Alkindi bis Kepler, Frankfurt a. M. 1987. *Im Englischen weiterhin aufgelegt, ist die deutsche Ausgabe jedoch vergriffen. Lindberg hat die jahrhundertelange Theoriebildung zum Sehen aufgearbeitet, um Bedeutung und Probleme der Optik seit Kepler verständlich zu machen.*

- Susanne Regener: Fotografische Erfassung. Zur Geschichte medialer Konstruktionen des Kriminellen, München 1999. *Die wohl umfangreichste Studie zum Verhältnis von Fotografie und kriminalistischer Fahndung, die auch mit der Geschichte des Porträts, der visuellen Kommunikation, mit anthropologischen und psychologischen Forschungen u. a. vielfältig verwoben ist.*

- Christine Strothotte / Thomas Strothotte: Seeing Between the Pixels. Pictures in Interactive Systems, Heidelberg / Berlin 1997. *Eine kombinierte theoretische und praxisorientierte Darstellung der Computervisualistik.*

- **Wolfgang Ullrich: Die Geschichte der Unschärfe**, 2. Auflage, Berlin 2003. *Ullrich liefert hier eine alternative Geschichte der Fotografie von den Anfängen bis heute, deren Zweck keineswegs in der exakten Reproduktion von Wirklichkeit bestanden hat.*

15.2 Periodika und Zeitschriften

Printmedien

- **Ästhetik und Kommunikation**, hg. vom Ästhetik & Kommunikation e. V., Berlin 1970ff. *Ein Beispiel für die konsequente Auseinandersetzung mit den Formen der Massenmedien, Alltagskultur und des urbanen Raums.*

- **Bildwelten des Wissens. Kunsthistorisches Jahrbuch für Bildkritik**, hg. von Horst Bredekamp / Matthias Bruhn / Gabriele Werner, Berlin 2003ff. *Die Bildwelten des Wissens versammeln in halbjährlichen Themenbänden die avancierten Beiträge aller Disziplinen zur Bedeutung visueller Form für die Wissenschaft.*

- **Critical Inquiry**, hg. von W. J. T. Mitchell, Chicago 1974ff. *Wie die „Kritischen Berichte" ist auch der Critical Inquiry ein gesellschaftspolitisches und kulturtheoretisches Projekt, das in den letzten Jahren verstärkt Aufgaben und Unterschiede von ‚Visueller Kultur' und ‚Bildwissenschaft' diskutiert hat.*

- **Isis**, hg. von Bernard Lightman, Chicago 1912ff. *1817–48 erschien schon einmal eine Zeitschrift dieses Namens „für Naturgeschichte, vergleichende Anatomie und Physiologie" bei Brockhaus – die US-Variante widmet sich, in dieser Tradition und inzwischen seit fast einhundert Jahren, der Beziehung von Wissenschaft, Technik und Kultur. Sie ist das Organ der History of Science Society.*

- **Fotogeschichte**, hg. von Anton Holzer, Marburg 1981ff. *Die „Beiträge zur Geschichte und Ästhetik der Fotografie" setzen sich zum Ziel, Fotografie als eigenständige künstlerische und technische Bildform zu analysieren und dokumentieren. Die Publikation erscheint in vierteljährlichen Themenbänden.*

- **tische Berichte**, hg. von Ulrike Gehring u. a., Marburg 1972ff. *Organ des Ulmer Vereins für Kunst- und Kulturwissenschaft* *behandeln die Kritischen Berichte die gesellschaftspolitische Bedeutung von Kunst- und Bildproduktion. Der Titel geht zurück auf die gleichnamige kunstwissenschaftlich ausgerichtete Zeitschrift, die ab 1927 erschien.*

- **Leonardo,** hg. von Roger Malina (Executive Editor), Cambridge, MA 1968ff. *In Paris gegründet vom Künstler Frank Malina, wird Leonardo heute am Massachusetts Institute of Technology (MIT) als Zeitschrift der International Society for the Arts, Sciences, and Technology weitergeführt, mit einem Schwerpunkt in der Beziehung von Technik und zeitgenössischer Kunstpraxis.*

- **Media, Culture & Society,** hg. von Raymond Boyle u. a., London u. a. 1979ff. *Die Zeitschrift für Medien und Kommunikationstheorie hat einen Schwerpunkt in den historischen und sozioökonomischen Technologiefolgen und behandelt regelmäßig visuelle Medien.*

- **Perception,** hg. von Richard Gregory, London 1972ff. *Interdisziplinäre Zeitschrift zur Geschichte, Theorie und Praxis der Wahrnehmung im psychologischen und technischen Bereich.*

- **Visual Resources (VR),** hg. von Helene E. Roberts (†) / Christine Sundt im Auftrag der Visual Resources Association, Philadelphia, PA 1980ff. *Die VR sind ein internationales Journal für Dokumentation mit einem Schwerpunkt auf Bildarchiven.*

- **Rundbrief Fotografie,** hg. von Wolfgang Hesse / Klaus Pollmeier, Neue Folge Esslingen 1994ff., Stuttgart 2005ff. *Der in Dresden betreute Rundbrief, seit 1994 als gedruckte Quartals-Zeitschrift, informiert über die Geschichte, Konservierung und Archivierung von Fotografie und gibt Hinweise auf neueste Online-Ressourcen u. a. (www.rundbrief-fotografie.de).*

Elektronische Zeitschriften:

- **Image – Journal for Interdisciplinary Image Science,** hg. von Klaus Sachs-Hombach, Web-Adresse: www.image-online.info. *Elektronische Zeitschrift für Bildwissenschaft.*

- **Images re-vues,** hg. von Giuseppe Di Liberti u. a., Web-Adresse: www.imagesrevues.org. *Eine gemeinsame französischsprachige online-Publikation der Pariser Forschungseinrichtungen für Kunsttheorie und Anthropologie.*

15.3 Internetressourcen in Auswahl

Weitere Angaben (insbesondere zu Sonderforschungsbereichen sowie zu weiteren Text- und Bilddatenbanken) finden Sie unter *www.akademie-studienbuch.de*.

Deutschsprachige Zentren mit bildwissenschaftlichem Schwerpunkt

- **Bauhaus-Universität Weimar, Fakultät Medien, Schwerpunkt Medienkultur,** Web-Adresse: www.uni-weimar.de/cms/medien/medienkulturemkefms.html. *Die Universität hat einflussreiche Studiengänge zu Medienkultur, Design, Film und Computer installiert und ist eine der zentralen Anlaufstellen im Bereich Neue Medien.*

- **Das Technische Bild,** Web-Adresse: www2.hu-berlin.de/kulturtechnik/dtb.php. *Im Jahr 2000 gegründetes Projekt am Hermann von Helmholtz-Zentrum für Kulturtechnik (Humboldt-Universität zu Berlin), das die Visualisierungsformen von Naturwissenschaft, Medizin und Technik aus einer kunst- und bildgeschichtlichen Perspektive untersucht und dokumentiert.*

- **Institut für Bildtheorie der Universität Rostock,** Web-Adresse: www.ifi.uni-rostock.de. *Das 2007 gegründete Institut gehört zur theologischen Fakultät der Universität Rostock und soll Veranstaltungen zum Verhältnis von Bildtheorie und –praxis ausrichten.*

- **Institute for Studies in Visual Culture e. V.,** Web-Adresse: www.isvc.org. *Das von den Kulturtheoretikern Tom Holert und Mark Terkessidis aus privater Initiative gegründete Forum steht hier als Beispiel für eine alternative Form von Bildwissenschaft und widmet sich der Kritik moderner Bildproduktion und -konsumption.*

- **Kunsthochschule für Medien Köln,** Web-Adresse: www.khm.de. *Kunsthochschulen können sich grundsätzlich als bildwissenschaftliche Einrichtungen verstehen; diese staatliche Kunsthochschule ist ein Beispiel für die fließenden Übergänge zu akustischen und anderen Ausdrucksformen, die sich aus dem Medium selber und insbesondere aus seiner elektronischen Form ergeben.*

- **Schweizer Nationaler Forschungsschwerpunkt eikones,** Web-Adresse: www.eikones.ch. *Der Forschungsschwerpunkt eikones ist ein Verbund bildwissenschaftlicher Projekte der Schweizer Hochschu-*

len mit Sitz an der Universität Basel und richtet seit 2006 internationale Veranstaltungen und Vortragsreihen aus.

- **Stiftung imai – Inter Media Art Institute,** Web-Adresse: www.imaionline.org. *2006 gegründete gemeinnützige Stiftung mit Sitz im Düsseldorfer Ehrenhof, die sich dem Vertrieb und der Konservierung von Medienkunst (v.a. Bewegtmedien) widmet und hierzu auch Informationsveranstaltungen sowie eine Datenbank für Medienkunst anbietet.*

- **Virtuelles Institut für Bildwissenschaft Magdeburg,** Web-Adresse: www.bildwissenschaft.org. *Vom Bildtheoretiker Klaus Sachs-Hombach an der Magdeburger Universität gegründetes virtuelles Diskussionforum mit Rezensionen, Linksammlungen und Tagungsberichten und der Zeitschrift Image (→ Periodika). Die Seite ist derzeit im Umzug begriffen.*

- **Zentrum für Bildwissenschaft der Donau Universität Krems,** Web-Adresse: www.donau-uni.ac.at/zbw. *Das Kremser Zentrum bietet einen gebührenpflichtigen berufsbegleitenden Studiengang Bildwissenschaften an und veranstaltet auch öffentliche international besetzte Tagungen. Die Webseite bietet außerdem Veranstaltungskalender zum Thema u. a. weiterführende Informationen.*

- **Zentrum für Literatur- und Kulturforschung Berlin,** Web-Adresse: www.zfl.gwz-berlin.de. *Das Zentrum für Literatur- und Kulturforschung versteht sich als interdisziplinäre Einrichtung mit Schwerpunkten in Kulturwissenschaft und Wissenschaftsgeschichte, mit zahlreichen öffentlichen Veranstaltungen und Publikationen zu bildwissenschaftlichen Fragestellungen.*

- **ZKM Karlsruhe – Zentrum für Kunst und Medientechnologie,** Web-Adresse: www.zkm.de. *Das ZKM wurde 1997 vom Kunsthistoriker Heinrich Klotz gegründet, um den Brückenschlag von Kunst, Design und neuen Medien zu befördern und hat seither durch Sammlungstätigkeiten, Ausstellungen, Tagungen und assoziierte Studiengänge seine Vorreiterstellung in der deutschsprachigen Kultur- und Wissenschaftslandschaft behauptet.*

Netzwerke, Portale, Blogs

- **Virtueller Katalog Kunstgeschichte,** Web-Adresse: www.artlibraries.net. *Ein Verbund des Zentralinstituts für Kunst-*

geschichte München mit anderen Fachbibliotheken, für die Meta-
suche über alle wichtigen deutschen sowie zahlreiche internationale
kunsthistorische Bibliotheken (insbesondere Aufsätze).

- **Deutsches Filminstitut, Frankfurt a. M.,** Web-Adresse:
www.deutsches-filminstitut.de. *Die Webseite verweist auf Filme
und deren Hersteller, auf Archive, sowie auf andere nationale Institute und bietet sich als Einstieg an.*

- **Deutsche Gesellschaft für Publizistik und Kommunikation,** Web-
Adresse: www.dgpuk.de. *Webseite für Ankündigungen und Mit-
gliederarbeit der. In diesem Zusammenhang besonders relevant die
Fachgruppe Visuelle Kommunikation* (vgl. auch die **Deutsche Gesellschaft für Ästhetik,** Web-Adresse: www.dgae.de).

- **Humanities and Social Sciences Online,** Web-Adresse:
www.h-net.org. *Das größte internationale Netzwerk zu allen geis-
tes- und sozialwissenschaftlichen Feldern, mit Sitz an der Michigan
State University, bietet Mailinglisten samt Datenbank aller Listen-
beiträge. Hier besonders zu erwähnen die bildwissenschaftlichen
Rezensionen der Liste www.arthist.net.*

- **International Association of Word and Image Studies,** Web-Adresse:
www.iawis.org, *Webseite der mitgliederstarken Association of
Word and Image Studies, die sich mit Text-Bild-Beziehungen befasst und eine gleichnamige Zeitschrift herausgibt.*

- **Iconic Turn – Das neue Bild der Welt,** Web-Adresse:
www.iconic-turn.de. *Eine Webseite der Hubert Burda Stiftung
München, die von einer erfolgreichen Vorlesungsreihe ausgegangen
ist und nun als Informationsportal mit Rezensionen, Interviews
u. a. weitergeführt wird.*

- **Infosthetics,** Web-Adresse: www.infosthetics.com. *Infosthetics (Ab-
kürzung für information aesthetics) ist ein Beispiel für private Dis-
kussionsforen, hier zum Thema Datenvisualisierung, und bringt
regelmäßig neue wissenschaftliche Bildbeispiele.*

Bildressourcen

- **Bildarchiv der Kunst und Architektur in Deutschland,** Web-Adresse: www.bildindex.de. *Vom Dokumentationszentrum Foto Marburg unterhaltene Datenbank.*

- **Bildarchiv Preußischer Kulturbesitz (BPK)**, Web-Adresse: www.bpkgate.picturemaxx.com. *Gewerbliche Bildstelle der Staatlichen Museen zu Berlin, die z. T. Bildmaterial aus anderen Sammlungen vertreibt und mit diesen kooperiert, z. B. mit den staatlichen französischen Museen; die Bildrecherche ist kostenfrei.*

- **Deutsches Filminstitut,** Web-Adresse: www.deutsches-filminstitut.de. *Das Deutsche Filminstitut ist ein Beispiel für zahlreiche nationale Einrichtungen zur Bewahrung des filmischen Kulturerbes und bietet auf der Webseite u. a. Hinweise auf Archive und Sammlungen.*

- **Deutsches Historisches Museum Berlin,** Web-Adresse: www.dhm.de. *Unter „Sammlungen" „Objektdatenbank" bietet das Deutsche Historische Museum Berlin eine Bilddatenbank mit zahlreichen Objektansichten.*

- **dfd Deutscher Fotodienst,** Web-Adresse: www.dfd-images.de. *Datenbank für Pressebilder, die aber auch die Recherche über Bilder aus der Film- und Fernsehgeschichte (defd movies) gestattet. Die kostenfreie Recherche gibt wasserzeichengeschützte Aufnahmen aus.*

- **ECHO (European Cultural Heritage Online),** Web-Adresse: www.echo2.mpiwg-berlin.mpg.de/home. *Das Projekt verbindet Webangebote mit wissenschaftsgeschichtlichen Quellen, darunter zahllose illustrierte Werke in gescannter Form.*

- **Fotostoria,** Web-Adresse: www.fotostoria.de. *Eine von der Kunsthistorikerin Ruth Goebel betreute Seite mit Neuigkeiten zur Fotogeschichte und einer umfangreichen Liste von Online-Bildarchiven, nach Ländern sortiert.*

- **George Eastman House – International Museum of Photography and Film,** Web-Adresse: www.eastmanhouse.org. *Das Eastman House residiert im historischen Wohnsitz des Gründers der Eastman Kodak Company und beherbergt heute eine der wichtigsten fotohistorischen Sammlungen, die auch online recherchierbar sind.*

- **Joconde,** Web-Adresse: www.culture.gouv.fr/documentation/joconde/fr/pres.htm. *Online-Katalog der französischen Museen.*

- **Kabinette des Wissens. Die Sammlungen der Humboldt-Universität zu Berlin,** Web-Adresse: www.sammlungen.hu-berlin.de. *Die Humboldt-Universität verfügt über eine der größten wissenschaftlichen Objektsammlungen weltweit, die in diesem Pilotprojekt schrittwei-*

se erschlossen werden; über die Projektseite werden außerdem auch alle Universitätssammlungen weltweit zentral indiziert.

- **Library of Congress Washington DC,** Web-Adresse: www.loc.gov/ rr/print/catalog.html.: *Online-Katalog der zur Sammlung der Drucke und Fotografien.*

- **New York Public Library Digital Gallery,** Web-Adresse: www.digitalgallery.nypl.org/nypldigital *Die NYPL verfügt über wertvolle Bestände der US-amerikanischen Fotogeschichte und insbesondere der großen Zeit der Reportage, die sie so weit wie möglich und in sehr guter Qualität digitalisiert und online stellt.*

- **Pictura Paedagogica Online,** Web-Adresse: www.bbf.dipf.de/ VirtuellesBildarchiv. *Suchmaschine des Sammlungsprojektes „Pictura Paedagogica Online" zur Geschichte der Lehrillustration (abgeschlossen, aber weiter unterstützt).*

- **Prometheus – Das verteilte digitale Bildarchiv für Forschung und Lehre,** Web-Adresse: www.prometheus-bildarchiv.de. *Vereinsmäßiger Verbund von Diatheken, für Privatpersonen gegen Gebühr, für Mitglieder teilnehmender Institutionen in der Regel kostenfrei. Es lassen sich auch die beteiligten Datenbanken aufrufen.*

- **Sächsische Landes- und Universitätsbibliothek, Portal der Virtuellen Fachbibliothek Kunst,** Web-Adresse: www.vifaart.slubdresden.de. *Gestattet auch die Recherche über Bilder (z. B. der Deutschen Fotothek Dresden).*

- **Science & Society Picture Library, London,** Web-Adresse: www.scienceandsociety.co.uk. *Angebot mit über einer Million Bildern aus dem Bestand des Science Museum, des National Museum of Photography, Film & Television und des National Railway Museum.*

- **Smithsonian Institution,** Web-Adresse: www.photography.si.edu/ Search.aspx. *Suchmaschine über die Bildarchive und Sammlungen der Smithsonian Institution (US-amerikanischer Museumsverbund).*

- **Visual Arts Data Service, The University College for the Creative Arts,** Web-Adresse: www.vads.ahds.ac.uk/about. *Eine englische Bild- und Textdatenbank für das gesamte Feld der visuellen Kultur.*

- **Warburg Electronic Library Hamburg,** Web-Adresse: www.welib.de. *Personalisierbare Forschungsdatenbank zur Politischen Ikonographie (Anmeldung erforderlich).*

16 Anhang

→ ASB

Akademie Studienbücher, auf die der vorliegende Band verweist

ASB D'APRILE / SIEBERS Iwan-Michelangelo D'Aprile / Winfried Siebers: Das 18. Jahrhundert. Zeitalter der Aufklärung, Berlin 2008.

ASB KELLER Andreas Keller: Frühe Neuzeit. Das rhetorische Zeitalter, Berlin 2008.

ASB KOŠENINA Alexander Košenina: Literarische Anthropologie. Die Neuentdeckung des Menschen, Berlin 2008.

Informationen zu weiteren Bänden finden Sie unter www.akademie-studienbuch.de

16.1 Zitierte Literatur

Alberti 1435 Leon Battista Alberti: Die Malkunst (de pictura), in: ders, Das Standbild – Die Malkunst – Grundlagen der Malerei (lateinisch / deutsch), hg. von Oskar Bätschmann / Christoph Schäublin, Darmstadt 2000, S. 194–314.

Alberti 1440 Leon Battista Alberti: Das Standbild (de statua), in: ders, Das Standbild – Die Malkunst – Grundlagen der Malerei (lateinisch / deutsch), hg. von Oskar Bätschmann / Christoph Schäublin, Darmstadt 2000, S. 142–181.

Altmann 2006 Jan Altmann: Färbung, Farbgestaltung und früher Farbendruck am Ende der Naturgeschichte, in: Bildwelten des Wissens 4,1: Farbstrategien, Berlin 2006, S. 69–77.

Amelunxen u. a. 1996 Hubertus von Amelunxen / Stefan Iglhaut / Florian Rötzer: Fotografie nach der Fotografie. Die Umgrenzung des Fotografischen durch das Digitale, Dresden 1996.

Anders 1956 Günther Anders: Die Antiquiertheit des Menschen. Über die Seele im Zeitalter der zweiten industriellen Revolution, München 1956.

Aristoteles 1957 Aristoteles: Tierkunde (Die Lehrschriften 8,1), hg. von Paul Gohlke, Paderborn 1957.

Aristoteles 1991 Aristoteles: Metaphysik, 3. Auflage, Hamburg 1991.

Arnheim 1943 Rudolf Arnheim: Gestalt and Art, in: Journal of Aesthetics and Art Criticism 2, 1943, S. 71–5.

Arnheim 1954 Rudolf Arnheim: Kunst und Sehen [engl. 1954], 3. Auflage, Berlin 2000.

Arnheim 1969 Rudolf Arnheim: Anschauliches Denken. Zur Einheit von Bild und Begriff [englisch: Visual Thinking, 1969], 8. Auflage, Köln 2001.

Ausst.-Kat. Berlin 1983 Skulptur und Macht. Figurative Plastik im Deutschland der 30er und 40er Jahre, Berlin 1983.

Ausst.-Kat. Karlsruhe 2005 Exit – Ausstieg aus dem Bild, Zentrum für Kunst- und Medientechnologie Karlsruhe, Karlsruhe 2005.

Ausst.-Kat. Köln 1981 Farbe im Photo. Die Geschichte der Farbphotographie von 1861 bis 1981, Josef-Haubrich-Kunsthalle Köln, Köln 1981.

Ausst.-Kat. Mönchengladbach 1997 Im Reich der Phantome. Fotografie des Unsichtbaren, Städtisches Museum Abteiberg u. a., Ostfildern-Ruit 1997.

Baigrie 1996 Brian S. Baigrie (Hg.): Picturing Knowledge. Historical and Philosophical Problems Concerning the Use of Art in Science, Toronto 1996.

Ballstaedt 1999 Steffen-Peter Ballstaedt: Die bildliche Darstellung von Handlungen in technischen Dokumenten, in: Clemens Schwender (Hg.), Zur Geschichte der Gebrauchsanleitung, Frankfurt a. M. 1999, S. 67–94.

Bann 2001 Stephen Bann: Parallel Lines. Printmakers, Painters and Photographers in Nineteenth-Century France, New Haven 2001.

Barthes 1980 Roland Barthes: Die helle Kammer. Bemerkung zur Fotografie [1980], Frankfurt a. M. 2007.

Batchen 1994 Geoffrey Batchen: Phantasm. Digital Imaging and the Death of Photography, in: Aperture 136, 1994, S. 47–50.

Batchen 1997 Geoffrey Batchen: Burning with Desire. The Conception of Photography, Massachusetts 1997.

Baudrillard 1981 Jean Baudrillard: Simulacres et Simulation, Paris 1981.

Baxandall 1999 Michael Baxandall: Die Wirklichkeit der Bilder. Malerei und Erfahrung im Italien des 15. Jahrhunderts, Berlin 1999.

Becker 1982 Howard S. Becker: Art Worlds, Berkeley 1982.

Belting 2004 Hans Belting: Bild und Kult. Eine Geschichte des Bildes vor dem Zeitalter der Kunst, 6. Auflage, München 2004.

Belting 2005 Hans Belting: Das echte Bild. Bildfragen als Glaubensfragen, München 2005.

Belting 2006 Hans Belting: Bild-Anthropologie. Entwürfe für eine Bildwissenschaft, 3. Auflage, München 2006.

Belting / Blume 1989 Hans Belting / Dieter Blume (Hg.): Malerei und Stadtkultur in der Dantezeit. Die Argumentation der Bilder, München 1989.

Benjamin 1931 Walter Benjamin: Kleine Geschichte der Photographie (1931), in: Walter Benjamin, Gesammelte Schriften, hg. von Rolf Tiedemann und Hermann Schweppenhäuser, Bd. 2,1, Frankfurt a. M. 1991, S. 368–385.

Benjamin 1936 Walter Benjamin: Das Kunstwerk im Zeitalter seiner technischen Reproduzierbarkeit [1936, deutsch 1955], Frankfurt a. M. 2007.

Bense 1969 Max Bense: Ästhetik und Informationstheorie, Reinbek 1969.

Berger 1990 Ursel Berger: Georg Kolbe – Leben und Werk, mit dem Katalog der Kolbe-Plastiken im Georg-Kolbe-Museum, Berlin 1990.

Bermingham 2000 Ann Bermingham: Learning to draw. Studies in the cultural history of a polite and useful art, New Haven u. a. 2000.

Berns 2000a Jörg Jochen Berns: Baumsprache und Sprachbaum. Baumikonographie als topologischer Komplex zwischen 13. und 17. Jahrhundert, in: Kilian Heck / Bernhard Jahn (Hg.), Genealogie als Denkform in Mittelalter und Früher Neuzeit, Tübingen 2000, S. 155–176 und S. 230–246.

Berns 2000b Jörg Jochen Berns: Film vor dem Film. Bewegende und bewegliche Bilder als Mittel der Imaginationssteuerung in Mittelalter und Früher Neuzeit, Marburg 2000.

Bertin 1967 Jacques Bertin: Graphische Semiologie. Diagramme, Netze, Karten, 2. Auflage, Berlin / New York 1983.

von Beyme 1998   Klaus von Beyme: Die Kunst der Macht und die Gegenmacht der Kunst, Frankfurt a. M. 1998.

Beyrodt 1991   Wolfgang Beyrodt: Kunstgeschichte als Universitätsfach, in: Peter Ganz u. a. (Hg.), Kunst und Kunsttheorie 1400–1900, Wiesbaden 1991, S. 313–332.

Birren 1945   Faber Birren: Selling with Color, New York 1945.

Blumenberg 1999   Hans Blumenberg: Die Lesbarkeit der Welt, 4. Auflage, Frankfurt a. M. 1999.

Blümle / von der Heiden 2005   Claudia Blümle / Anne von der Heiden (Hg.): Blickzähmung und Augentäuschung. Zu Jacques Lacans Bildtheorie, Berlin / Zürich 2005.

Boehm 1994   Gottfried Boehm: Die Wiederkehr der Bilder, in: Gottfried Boehm (Hg): Was ist ein Bild?, München 1994, S. 11–38.

Bogen / Thürlemann 2003   Steffen Bogen / Felix Thürlemann: Jenseits der Opposition von Text und Bild. Überlegungen zu einer Theorie des Diagrammatischen, in: Alexander Patschovsky (Hg.), Die Bildwelt der Diagramme Joachims von Fiore. Zur Medialität religiös-politischer Programme im Mittelalter, Stuttgart 2003, S. 1–22.

Bolvig / Lindley 2003   Axel Bolvig / Philipp Lindley (Hg.): History and Images. Towards a New Iconology, Turnhout 2003.

Bonhoff 1993   Ulrike Maria Bonhoff: Das Diagramm. Kunsthistorische Betrachtung über seine vielfältige Anwendung von der Antike bis zur Neuzeit, Münster 1993.

Bourdieu 1985   Pierre Bourdieu: Eine illegitime Kunst. Die sozialen Gebrauchsweisen der Photographie [1985], Nachdruck Hamburg 2006.

Brakensiek / Rockel 1993   Stefan Brakensiek / Irina Rockel: Alltag, Klatsch und Weltgeschehen. Neuruppiner Bilderbogen. Ein Massenmedium des 19. Jahrhunderts, Bielefeld 1993.

Brassat 2005   Wolfgang Brassat (Hg.): Bild-Rhetorik, Tübingen 2005.

Bredekamp 1975   Horst Bredekamp: Kunst als Medium sozialer Konflikte, Frankfurt a. M. 1975.

Bredekamp 1993   Horst Bredekamp: Antikensehnsucht und Maschinenglauben. Die Geschichte der Kunstkammer und die Zukunft der Kunstgeschichte, Berlin 1993.

Bredekamp 1997   Horst Bredekamp: Das Bild als Leitbild. Gedanken zur Überwindung des Anikonismus [1997], in: Horst Bredekamp / Jörg Probst (Hg.), Bilder bewegen. Von der Kunstkammer zum Endspiel, Berlin 2007, S. 136–156.

Bredekamp 2004   Horst Bredekamp: Bildakte als Zeugnis und Urteil, in: Monika Flacke (Hg.), Mythen der Nation, Bd. 1, Mainz, S. 29–66.

Bredekamp 2005   Horst Bredekamp: Drehmomente – Merkmale und Ansprüche des iconic turn, in: Christa Maar / Hubert Burda (Hg.), Iconic Turn. Die neue Macht der Bilder, 3. Auflage, Köln 2005, S. 15–26.

Bredekamp 2006   Horst Bredekamp: Darwins Korallen. Die frühen Evolutionsdiagramme und die Tradition der Naturgeschichte, 2. Auflage, Berlin 2006.

Bredekamp 2007   Horst Bredekamp: Galilei der Künstler. Der Mond, die Sonne, die Hand, Berlin 2007.

Breidbach 2005   Olaf Breidbach: Bilder des Wissens. Zur Kulturgeschichte der wissenschaftlichen Wahrnehmung, Paderborn / München 2005.

Bringéus 1982   Nils Arvid Bringéus: Volkstümliche Bilderkunde, München 1982.

Brink 1998   Cornelia Brink: Ikonen der Vernichtung. Öffentlicher Gebrauch von Fotografien aus nationalsozialistischen Konzentrationslagern nach 1945, Berlin 1998.

Brinkmann 2007   Vinzenz Brinkmann (Hg.): Bunte Götter. Die Farbigkeit antiker Skulptur. Zugleich Ausst.-Kat. Museum für Kunst und Gewerbe Hamburg, 4. Auflage, Hamburg 2007.

Brockhaus 1837   Bilder-Conversations-Lexikon für das deutsche Volk. 4 Bde., Leipzig 1837–41.

Brons / Bredekamp 2005   Franziska Brons / Horst Bredekamp: Fotografie als Medium der Wissenschaft. Kunstgeschichte, Biologie und das Elend der Illustration, in: Christa Maar / Hubert Burda (Hg.), Iconic Turn. Die neue Macht der Bilder, 3. Auflage, Köln 2005, S. 365–381.

Brown 2003   Theodore L. Brown: Making Truth. Metaphor in Science, Urbana 2003.

Brückner 1966   Wolfgang Brückner: Bildnis und Brauch. Studien zur Bildfunktion der Effigies, Berlin 1966.

Brückner 1973   Wolfgang Brückner (Hg.): Die Bilderfabrik, Ausst.-Kat. Historisches Museum, Frankfurt a. M. 1973.

Bruhn 2007   Matthias Bruhn: Tarife für das Sichtbare. Eine kurze Geschichte der Bildagenturen, in: Fotogeschichte 27, 2007, S. 12–25.

Buddemeier 1970   Heinz Buddemeier: Panorama, Diorama, Photographie. Entstehung und Wirkung neuer Medien im 19. Jahrhundert, München 1970.

Buffon 1749–1788   Georges-Louis Leclerc de Buffon: Histoire naturelle générale et particulière, Paris 1749–1788.

Burgin 1981   Victor Burgin: Photography, Phantasy, Function, in: Victor Burgin (Hg.), Thinking Photography, London 1981, S. 177–216.

Burke 2003   Peter Burke: Augenzeugenschaft. Bilder als historische Quellen, Berlin 2003.

Buswell 1935   Guy T. Buswell: How People Look at Pictures. A Study of the Psychology and Perception in Art, Oxford 1935.

Cartwright 1997   Lisa Cartwright: Screening the body. Tracing Medicine's Visual Culture, 2. Auflage, Minneapolis u. a. 1997.

Cassirer 1923   Ernst Cassirer: Philosophie der symbolischen Formen, 3. Bde., Berlin 1923–29.

Caujolle 2002   Christian Caujolle: Die Welt im Rechteck. Eine kleine Geschichte des Fotojournalismus, in: Le Monde diplomatique, deutsche Beilage der *tageszeitung* Nr. 6852 vom 13.9.2002.

Chaplin 1995   Elizabeth Chaplin: Sociology and Visual Representation, London 1995.

Claus 1970   Jürgen Claus: Expansion der Kunst. Beiträge zu Theorie und Praxis öffentlicher Kunst, Reinbek 1970.

Crary 1996   Jonathan Crary: Techniken des Betrachters. Sehen und Moderne im 19. Jahrhundert, Dresden 1996.

Crone 1999   Robert A. Crone: A History of Color. The Evolution of Theories of Light and Color, Dordrecht u. a. 1999.

Crow 2000   Thomas Crow: Painters and Public Life in Eighteenth-Century Paris, 6. Auflage, New Haven u. a. 2000.

Currie 1997   Gregory Currie: Image and Mind. Film, Philosophy and Cognitive Science, Cambridge 1997.

Danto 1996   Arthur C. Danto: Kunst nach dem Ende der Kunst, München 1996.

Daston / Galison 2007   Lorraine Daston / Peter Galison: Objektivität, Frankfurt a. M. 2007.

Deleuze 1998   Gilles Deleuze: Kino I. Das Bewegungs-Bild, 2. Auflage, Frankfurt a. M. 1998.

Dewitz / Nekes 2002   Bodo von Dewitz / Werner Nekes (Hg.): Ich sehe was, was du nicht siehst! Die Sammlung Werner Nekes, Göttingen 2002.

Didi-Huberman 1999   Georges Didi-Huberman: Ähnlichkeit und Berührung. Archäologie, Anachronismus und Modernität des Abdrucks, Köln 1999.

Didi-Huberman 2007   Georges Didi-Huberman: Bilder trotz allem, München 2007.

Dilly 1979   Heinrich Dilly: Kunstgeschichte als Institution, Berlin 1979.

Doelker 2002   Christian Doelker: Ein Bild ist mehr als ein Bild. Visuelle Kompetenz in der Multimedia-Gesellschaft, 3. Auflage, Stuttgart 2002.

Dommann 2003   Monika Dommann: Durchsicht, Einsicht, Vorsicht. Eine Geschichte der Röntgenstrahlen, 1896–1963, Zürich 2003.

Dommann 2004   Monika Dommann: Vom Bild zum Wissen. Eine Bestandsaufnahme wissenschaftshistorischer Bildforschung, in: Gesnerus 61, 2004, S. 77–89.

Dommann 2006   Monika Dommann: Der Apparat und das Individuum. Die Verrechtlichung technischer Bilder (1860–1920), in: Martina Heßler (Hg.), Konstruierte Sichtbarkeiten. Wissenschafts- und Technikbilder seit der Frühen Neuzeit, München 2006, S. 347–367.

Dumit 2004   Joseph Dumit: Picturing Personhood. Brain Scans and Biomedical Identity, Princeton, N.J. 2004.

Dünkel 2008   Vera Dünkel: Das Auge der Radiographie. Zur Wahrnehmung einer neuen Art von Bildern, in: Matthias Bruhn / Kai-Uwe Hemken (Hg.), Modernisierung des Sehens. Sehweisen zwischen Künsten und Medien, Bielefeld 2008, S. 207–220.

Durkheim 1895   Émile Durkheim: Les règles de la méthode sociologique, Paris 1895 (deutsch: Die Regeln der soziologischen Methode, Neuwied, 5. Auflage, Berlin 2002).

Eco 1987   Umberto Eco: Semiotik. Entwurf einer Theorie der Zeichen, München 1987.

Eder 1932   Josef Maria Eder: Geschichte der Fotografie, 4. Auflage, Halle 1932.

Edwards 1997   Elizabeth Edwards (Hg.): Anthropology and Photography 1860–1920, New Haven / London 1997.

Ehmer 1971   Hermann K. Ehmer (Hg.): Visuelle Kommunikation. Beiträge zur Kritik der Bewußtseinsindustrie, Köln 1971.

Ehrenfels 1890   Christian von Ehrenfels: Über „Gestaltqualitäten". Sonderdruck der Vierteljahrsschrift für wissenschaftliche Philosophie 14, Leipzig 1890, Bl. 12–43.

Ehrenspeck / Schäffer 2003   Yvonne Ehrenspeck / Burkhart Schäffer (Hg.): Film- und Fotoanalyse in der Erziehungswissenschaft. Ein Handbuch, Opladen 2003.

Elkins 1999   James Elkins: The Domain of Images, Ithaca / New York 1999.

Elkins 2007   James Elkins: Visual Practices across the University, Paderborn / München 2007.

Elkins 2008   James Elkins (Hg.): Visual Literacy, New York / London 2008.

Enzensberger 1964   Hans Magnus Enzensberger: Bewußtseins-Industrie, Frankfurt a. M. 1964.

Fechner 1986   Renate Fechner: Natur als Landschaft. Zur Entstehung der ästhetischen Landschaft, Frankfurt a. M. 1986.

Feyerabend 1980   Paul Feyerabend: Wissenschaft als Kunst, Frankfurt a. M. 1980.

Fiedler 1887   Conrad Fiedler: Der Ursprung der künstlerischen Thätigkeit, Leipzig 1887.

Flusser 2000   Vilém Flusser: Ins Universum der technischen Bilder, 6. Auflage, Göttingen 2000.

Focillon 1939   Henri Focillon: Das Leben der Formen [1939], deutsche Ausgaben Bern und München 1954.

Foster 1988   Hal Foster (Hg.): Vision and Visuality, New York 1988.

Foucault 1971   Michel Foucault: Die Ordnung der Dinge. Eine Archäologie der Humanwissenschaften [1971], Neuausgabe Frankfurt a. M. 2003.

Franck 1998   Georg Franck: Ökonomie der Aufmerksamkeit. Ein Entwurf, München 1998.

Freedberg 1989   David Freedberg: The Power of Images. Studies in the History and the Theory of Response, Chicago/London 1989.

Freedberg 1994   David Freedberg: The Failure of Colour, in: John Onians (Hg.), Sight and Insight. Essays in Honor of E. H. Gombrich, London 1994, S. 245–262.

Freud 1899   Sigmund Freud: Traumdeutung [1899/1900], 12. Auflage, Frankfurt a. M. 2005.

Freund 1968   Gisèle Freund: Photographie und bürgerliche Gesellschaft. Eine kunstsoziologische Studie, München 1968.

Frosh 2003   Paul Frosh: The Image Factory. Consumer Culture, Photography and the Visual Content Industry, Oxford 2003.

Fry 1920   Roger Fry: Vision and Design [London u. a. 1920], Nachdruck Oxford 1981.

Gage 2007   John Gage: Kulturgeschichte der Farbe. Die Sprache der Farben, Leipzig 2007.

Galison 2002   Peter Galison: Images scatter into data, data gather into images, in: Bruno Latour/Peter Weibel (Hg.), Iconoclash, Karlsruhe/Cambridge 2002, S. 300–323.

Galton 1878   Francis Galton: Composite Portraits, in: Nature 18, 1878, S. 97–100.

Geertz 1973   Clifford Geertz: Dichte Beschreibung. Beiträge zum Verstehen kultureller Systeme [englisch 1973], Frankfurt a. M. 2002.

Geimer 2006   Peter Geimer: Gegensichtbarkeiten, in: Bildwelten des Wissens 4,2: Bilder ohne Betrachter, Berlin 2006, S. 33–42.

Ginzburg 2002   Carlo Ginzburg: Spurensicherung, Berlin 2002.

Goethe 1807   Johann Wolfgang von Goethe: „Die Absicht eingeleitet" [Fragment von 1807], in: Goethes Werke, hg. von Emil Staiger, Frankfurt a. M. 1966, Bd. 6, S. 358–362.

Goldthwaite 1993   Richard A. Goldthwaite: Wealth and the Demand for Art in Italy, 1300–1600, Baltimore u. a. 1993.

Gombrich 1960   Ernst H. Gombrich: Kunst und Illusion. Zur Psychologie der bildlichen Darstellung [englisch 1960], 2. Auflage der Neuausgabe Berlin 2004.

Gombrich 1984   Ernst H. Gombrich: Der fruchtbare Moment. Vom Zeitelement in der bildenden Kunst, in: Ernst H. Gombrich: Bild und Auge. Neue Studien zur Psychologie der bildlichen Darstellung, Stuttgart 1984, S. 40–62.

Goodman 1984   Nelson Goodman: Weisen der Welterzeugung, Frankfurt a. M. 1984.

Gorman 2000   Andreas Gorman: Imaginationen des Unsichtbaren. Zur Gattungsgeschichte des wissenschaftlichen Diagramms, in: Hans Holländer (Hg.), Erkenntnis, Erfindung, Konstruktion. Studien zur Bildgeschichte von Naturwissenschaft und Technik vom 16. bis zum 19. Jahrhundert, Berlin 2000, S. 51–71.

Gould 1990   Stephen Jay Gould: Wonderful Life, New York 1990 (deutsch: Zufall Mensch. Das Wunder des Lebens als Spiel der Natur, München 1991).

Gould 2002   Stephen Jay Gould: Ein Dinosaurier im Heuhaufen, 2. Auflage, Frankfurt a. M. 2002.

Graselli/Philipps 2003   Margaret Morgan Graselli/Ivan E. Philipps (Hg.): Colorful Impressions. The Printmaking Revolution in Eighteenth-Century France, Washington 2003.

Grasskamp 1981   Walter Grasskamp: Museumsgründer und Museumsstürmer. Zur Sozialgeschichte des Kunstmuseums, München 1981.

Grau 2002   Oliver Grau: Virtuelle Kunst in Geschichte und Gegenwart, 2. Auflage, Berlin 2002.

Grimm 1854 Artikel „Bild", in: Jacob Grimm / Wilhelm Grimm, Deutsches Wörterbuch, 16 Bde., Leipzig 1854–1960, Bd. 2, Spalten 8–16.

Grittmann 2007 Elke Grittmann: Das politische Bild. Fotojournalismus und Pressefotografie in Theorie und Empirie, Köln 2007.

Groebner 2004 Valentin Groebner: Der Schein der Person. Steckbrief, Ausweis und Kontrolle im Europa des Mittelalters, München 2004.

Gronemeyer 2004 Nicole Gronemeyer: Optische Magie. Zur Geschichte der visuellen Medien in der Frühen Neuzeit, Bielefeld 2004.

Groys 1994 Boris Groys: Der ein-gebildete Kontext, in: Peter Weibel (Hg.), Kontext Kunst. Kunst der 90er Jahre, Köln 1994, S. 257–281.

Guibert 1993 Hervé Guibert: Phantom-Bild. Über Photographie, Leipzig 1993.

Habermas 1962 Jürgen Habermas: Strukturwandel der Öffentlichkeit. Untersuchungen zu einer Kategorie der bürgerlichen Gesellschaft, Neuwied 1962 [1. Auflage der Neuauflage von 1990, Frankfurt a. M. 2006].

Hacking 1983 Ian Hacking: Representing and Intervening, Cambridge 1983 (deutsch: Einführung in die Philosophie der Naturwissenschaften, Stuttgart 1996).

Haraway 1991 Donna Haraway: Simians, Cyborgs, and Women. The Reinvention of Nature, New York 1991 (deutsch:. Die Neuerfindung der Natur, Frankfurt a. M. u. a. 1995).

Hartwig 1986 Helmut Hartwig: Die Grausamkeit der Bilder – Horror und Faszination in alten und neuen Medien, Weinheim / Berlin 1986.

Haskell 1995 Francis Haskell: Die Geschichte und ihre Bilder. Die Kunst und die Deutung der Vergangenheit, München 1995.

Hauser 1951 Arnold Hauser: The social history of art, London 1951 (deutsch: Sozialgeschichte der Kunst, München 1990).

Hegel 1832 Georg Wilhelm Friedrich Hegel: Vorlesungen über die Ästhetik, Werke, Bd. 13–15, auf der Grundlage der Werke von 1832–1845 von Eva Moldenhauer und Karl Markus Michel neu edierte Ausgabe, 3 Bde., Frankfurt a. M. 1986.

Heidenreich / Ernst 1999 Stefan Heidenreich / Wolfgang Ernst: Digitale Bildarchivierung. Der Wölfflin-Kalkül, in: Sigrid Schade / Georg Christoph Tholen (Hg.), Konfigurationen. Zwischen Kunst und Medien, München 1999, S. 306–320.

Hemken 2000 Kai-Uwe Hemken (Hg.): Bilder in Bewegung. Traditionen digitaler Ästhetik, Köln 2000.

Hempel / Metelmann 2005 Leon Hempel / Jörg Metelmann (Hg.): Bild – Raum – Kontrolle. Videoüberwachung als Zeichen gesellschaftlichen Wandels, Frankfurt a. M. 2005.

Hennig 1982 Willi Hennig: Phylogenetische Systematik, Berlin 1982.

Hennig 2004 Jochen Hennig: Vom Experiment zur Utopie. Bilder in der Nanotechnologie, in: Bildwelten des Wissens 2,2: Instrumente des Sehens, Berlin 2004, S. 9–18.

Hennig 2006 Jürgen Hennig: Farbeinsatz in der medizinischen Visualisierung, in: Bildwelten des Wissens 4,1: Farbstrategien, Berlin 2006, S. 9–16.

Hentschel 2001 Linda Hentschel: Pornotopische Techniken des Betrachtens. Raumwahrnehmung und Geschlechterordnung in visuellen Apparaten der Moderne, Marburg 2001.

Hentschel 2002 Klaus Hentschel: Mapping the Spectrum. Techniques of Visual representation in Research and Teaching, Oxford u. a. 2002.

Hesse 2006   Wolfgang Hesse: Metamorphosen. Fotografie und Reproduktion, in: Wahr-Zeichen. Fotografie und Wissenschaft. Begleitbuch zur Ausstellung Technische Sammlungen Dresden, Universitätssammlungen und Altana Galerie, Dresden, Dresden 2006, S. 54–61.

Hinterwaldner / Buschhaus 2006   Inge Hinterwaldner / Markus Buschhaus: The picture's image. Wissenschaftliche Visualisierungen als Komposit, Paderborn / München 2006.

Hodges 2003   Elaine R. S. Hodges: The Guild Handbook of Scientific Illustration, Hoboken 2003.

Hoffman 2000   Donald D. Hoffman: Visuelle Intelligenz. Wie die Welt im Kopf entsteht, München 2000.

Hoffmann 2004   Detlef Hoffmann: Bildliche und bildlose Repräsentation, in: Bettina Bannasch / Almuth Hammer (Hg.), Verbot der Bilder – Gebot der Erinnerung. Mediale Repräsentationen der Schoah, Frankfurt a. M. / New York 2004, S. 381–396.

Hofmann 1974   Hasso Hofmann: Repräsentation. Studien zur Wort- und Begriffsgeschichte von der Antike bis zum 19. Jahrhundert [1974], Neuausgabe Berlin 1990.

Hofmann 1999   Wilhelm Hofmann (Hg.): Die Sichtbarkeit der Macht – Theoretische und empirische Untersuchungen zur Visuellen Politik, Baden-Baden 1999.

Hooke 1726   Robert Hooke: An Instrument of Use to take the Draught, or Picture of any Thing [1694], in: Philosophical Experiments and Observations, London 1726.

Horaz 2008   Quintus Horatius Flaccus: Ars poetica (lateinisch / deutsch), übers. von Eckart Schäfer, Stuttgart 2008.

Horkheimer / Adorno 1944   Max Horkheimer / Theodor W. Adorno: Dialektik der Aufklärung. Philosophische Fragmente [1944], 16. Auflage der Neuausgabe, Frankfurt a. M. 2006.

Hubel 1990   David H. Hubel: Auge und Gehirn. Neurobiologie des Sehens, 2. Auflage, Heidelberg 1990.

Huber 2002   Hans Dieter Huber / Bettina Lockemann / Michael Scheibel (Hg.): Bild Wissen Medien. Visuelle Kompetenz im Medienzeitalter, München 2002.

Inhetveen / Kötter 1996   Rüdiger Inhetveen / Rudolf Kötter (Hg.): Betrachten, Beobachten, Beschreiben. Beschreibungen in Kultur- und Naturwissenschaften, München 1996.

Jäger 2000   Jens Jäger: Photographie. Bilder der Neuzeit. Einführung in die Historische Bildforschung, Tübingen 2000.

Jaritz 1996   Gerhard Jaritz (Hg.): Pictura quasi Fictura. Die Rolle des Bildes in der Erforschung von Alltag und Sachkultur des Mittelalters und der frühen Neuzeit, Wien 1996.

Jöchner 2003   Cornelia Jöchner (Hg.): Politische Räume. Stadt und Land in der Frühneuzeit, Berlin 2003.

Kammerer 1919   Paul Kammerer: Das Gesetz der Serie. Eine Lehre von den Wiederholungen im Lebens- und im Weltgeschehen, Stuttgart u. a. 1919.

Kämpfer 1997   Frank Kämpfer: Propaganda. Politische Bilder im 20. Jahrhundert. Bildkundliche Essays, Hamburg 1997, S. 8–19.

Kaschuba 2008   Wolfgang Kaschuba: Bildkonsum, in: Matthias Bruhn / Kai-Uwe Hemken (Hg.), Modernisierung des Sehens. Sehweisen zwischen Künsten und Medien, Bielefeld 2008, S. 55–65.

Kaufmann 2000   Günter Kaufmann: Neue Bücher – alte Fehler. Zur Bildrepräsentation in Schulgeschichtsbüchern, in: Geschichte in Wissenschaft und Unterricht 51, 2000, S. 68–87.

Kemp 1974   Wolfgang Kemp: Disegno. Beiträge zur Geschichte des Begriffs zwischen 1547 und 1607, in: Marburger Jahrbuch für Kunstwissenschaft 19, 1974, S. 218–240.

Kemp 1979   Wolfgang Kemp: „... einen wahrhaft bildenden Zeichenunterricht überall einzuführen." Zeichnen und Zeichenunterricht der Laien 1500–1870. Ein Handbuch, Frankfurt a. M. 1979.

Kemp 1991   Wolfgang Kemp: Kontexte. Für eine Kunstgeschichte der Komplexität, in: Texte zur Kunst 2, 1991, S. 89–101.

Kemp 1996   Wolfgang Kemp: Die Räume der Maler. Zur Bilderzählung seit Giotto, München 1996.

Kemp 2003   Martin Kemp: Bilderwissen. Die Anschaulichkeit naturwissenschaftlicher Phänomene, Köln 2003.

Kemp 2005   Martin Kemp: Leonardo, München 2005.

Kemp/Amelunxen 2006   Wolfgang Kemp/Hubertus von Amelunxen (Hg.): Theorie der Fotografie 1839–1995, 4 Bde., München 2006.

Kerbs 2001   Diethart Kerbs: Kunsterziehungsbewegung und Kulturreform, in: Kaspar Maase/Wolfgang Kaschuba (Hg.), Schund und Schönheit. Populäre Kultur um 1900, Köln/Weimar/Wien 2001, S. 378–397.

Kittler 2004   Friedrich Kittler: Schrift und Zahl. Die Geschichte des errechneten Bildes, in: Christa Maar/Hubert Burda (Hg.), Iconic Turn. Die neue Macht der Bilder, 3. Auflage, Köln 2005, S. 186–203.

Knorr-Cetina 1999   Karin Knorr-Cetina: ‚Viskurse der Physik‘. Wie visuelle Darstellungsformen ein Wissenschaftsgebiet ordnen, in: Jörg Huber/Martin Heller (Hg.), Konstruktionen Sichtbarkeiten, Wien u. a. 1999, S. 245–263.

Knorr-Cetina/Amann 1990   Karin Knorr-Cetina/Klaus Amann: Image Dissection in Natural Scientific Inquiry, in: Science, Technology, & Human Values 15, 1990, Heft 3, S. 259–283.

Köbler 1980   Gerhard Köbler: Germanisches Wörterbuch, Gießen-Lahn 1980.

Koch 2006   Gertrud Koch/Christiane Voss (Hg.): … kraft der Illusion, München/Paderborn 2006.

Koebner 2006   Thomas Koebner/Thomas Meder: Bildtheorie und Film, München 2006.

Koschorke 2007   Albrecht Koschorke: Macht und Fiktion, in: Frank Thomas u. a. (Hg.), Des Kaisers neue Kleider, Frankfurt a. M. 2007, S. 73–84.

Kracauer 1927   Siegfried Kracauer: Die Photographie [1927], in: Siegfried Kracauer: Das Ornament der Masse, Frankfurt a. M. 1977, S. 21–39.

Krämer/Bredekamp 2003   Sybille Krämer/Horst Bredekamp (Hg.): Bild, Schrift, Zahl, Paderborn/München 2003.

Krause 1995   Rainer Krause: Gesicht – Affekte – Wahrnehmung und Interaktion, in: Gertrud Koch (Hg.), Auge und Affekt. Wahrnehmung und Interaktion, Frankfurt a. M. 1995, S. 57–72.

Krause 2005   Katharina Krause/Klaus Niehr/Eva-Maria Hanebutt-Benz (Hg.): Bilderlust und Lesefrüchte. Das illustrierte Kunstbuch von 1750 bis 1920. Begleitbuch zur Ausstellung im Gutenberg-Museum Mainz, Leipzig 2005.

Krauss 1998   Rolf H. Krauss: Walter Benjamin und der neue Blick auf die Photographie, Ostfildern 1998.

Kress 2006   Gunther Kress: Literacy in the New Media Age, Nachdruck London/New York 2006.

Krieger 2007   Verena Krieger 2007: Was ist ein Künstler? Genie – Heilsbringer – Antikünstler. Eine Ideen- und Kunstgeschichte des Schöpferischen, Köln 2007.

Krohn 2006   Wolfgang Krohn (Hg.): Ästhetik in der Wissenschaft. Interdisziplinärer Diskurs über das Gestalten und Darstellen von Wissen, Hamburg 2006.

Krüger 2001   Klaus Krüger: Das Bild als Schleier des Unsichtbaren. Ästhetische Illusion in der Kunst der frühen Neuzeit in Italien, München 2001.

Krüger 2007   Matthias Krüger: Das Relief der Farbe. Pastose Malerei in der französischen Kunstkritik 1850–1890, München/Berlin 2007.

Krynitz 1775   Artikel „Bild", in: Johann Georg Krynitz: Oekonomische Encyklopädie oder allgemeines System der Staats- Stadt- Haus- und Landwirthschaft, Bd. 5, Berlin 1775, S. 291–297.

Lacan 1996   Jacques Lacan: Die vier Grundbegriffe der Psychoanalyse. Seminar XI, Weinheim/Berlin 1996.

Langer 1942   Susanne K. Langer: Philosophie auf neuem Wege. [Philosophy in a New Key, 1942], Neuausgabe Frankfurt a. M. 1992.

Latour/Woolgar 1979   Bruno Latour/Steve Woolgar: Laboratory Life. The Social Construction of Scientific Facts, Beverly Hills 1979.

Latour/Weibel 2002   Bruno Latour/Peter Weibel (Hg.): Iconoclash, Karlsruhe/Cambridge 2002.

Lavater 1775–78   Johann Caspar Lavater: Physiognomische Fragmente zur Beförderung der Menschenkenntnis und Menschenliebe, Leipzig/Winterthur 1775–78.

Lefèvre 2003   Wolfgang Lefèvre/Jürgen Renn/Urs Schoepflin (Hg): The Power of Images in Early Modern Sciences, Basel u. a. 2003.

Levin 2002   Thomas Y. Levin/Ursula Frohne/Peter Weibel (Hg.): Ctrl_Space. Rhetorics of Surveillance from Bentham to Big Brother, Zentrum für Kunst- und Medientechnologie Karlsruhe, Cambridge, Mass. 2002.

Lingner 1994   Michael Lingner: Die Krise der ‚Ausstellung' im System der Kunst, in: Betriebssystem Kunst. Kunstforum 125, Januar/Februar 1994, S. 182–187.

Lippincott 1995   Louise Lippincott: Expanding on Portraiture. The Market, the Public and the Hierarchy of Genres in Eighteenth-Century Britain, in: Ann Bermingham/John Brewer (Hg.), The Consumption of Culture, 1600–1800. Image, Object, Text, London/New York 1995, S. 75–88.

Locher 2008   Hubert Locher: Reproduktionen: Erfindung und Entmachtung des Originals im Medienzeitalter, in: Matthias Bruhn/Kai-Uwe Hemken (Hg.), Modernisierung des Sehens. Sehweisen zwischen Künsten und Medien, Bielefeld 2008, S. 38–53.

Löffler/Scholz 2004   Petra Löffler/Leander Scholz (Hg.): Das Gesicht ist eine starke Organisation, Köln 2004.

Luhmann 1984   Niklas Luhmann: Soziale Systeme. Grundriss einer allgemeinen Theorie [1984], Nachdruck Frankfurt a. M. 2006.

Luhmann 2002   Niklas Luhmann: Die Kunst der Gesellschaft, Neuausgabe Darmstadt 2002.

Lynch 1985   Michael Lynch: Art and Artifact in Laboratory Science, London u. a. 1985.

Maase 2001   Kaspar Maase: Schund und Schönheit. Ordnungen des Vergnügens um 1900, in: Kaspar Maase/Wolfgang Kaschuba (Hg.), Schund und Schönheit. Populäre Kultur um 1900, Köln/Weimar/Wien 2001, S. 9–28.

Maasen/Weingart 2000   Sabine Maasen/Peter Weingart: Metaphors and the Dynamics of Knowledge, London/New York 2000.

MacEachren 1995   Alan M. MacEachren: How Maps Work. Representation, Visualisation, and Design, New York 1995.

Maddison 1997   Wayne P. Maddison: Gene Trees in Species Trees, in: Systematic Biology 46, 1997, S. 523–536.

Malraux 1947   André Malraux: Das imaginäre Museum [1947], Neuausgabe Frankfurt a. M. 1987.

Manovitch 2001   Lev Manovitch: The Language of New Media, Cambridge, Mass. 2001.

Marin 1977   Louis Marin: Détruire la peinture, Paris 1977 (deutsch: Die Malerei zerstören, Berlin 2003).

Mayor 1971   Alpheus Hyatt Mayor: Prints & People. A Social History of Printed Pictures, New York 1971.

Mersmann / Schulz 2006   Birgit Mersmann / Martin Schulz (Hg.): Kulturen des Bildes, Paderborn / München 2006.

Messaris 1994   Paul Messaris: Visual „literacy". Image, Mind, and Reality, Boulder u. a. 1994.

Metz 1992   Christian Metz: The Imaginary Signifier, in: Gerald Mast / Marshall Cohen / Leo Braudy (Hg.), Film Theory and Criticism, Bd. 4, New York 1992, S. 782–802.

Meyer 1998   Thomas Meyer: Politik als Theater. Die neue Macht der Darstellungskunst, Berlin 1998.

Meyer 2006   Roland Meyer: Lichtbildbelehrung. Bilder im Grenzbereich, in: Bildwelten des Wissens 4,2: Bilder ohne Betrachter, Berlin 2006, S. 64–68.

Michalski 1932   Ernst Michalski: Die Bedeutung der ästhetischen Grenze für die Methode der Kunstgeschichte [1932], Berlin 1996.

Mietzner 2005   Ulrike Mietzner: Visual History. Images of Education, Oxford u. a. 2005.

Mitchell 1986   W. J. T. Mitchell: Iconology. Image, Text, Ideology, Chicago 1986.

Mitchell 1992a   W. J. T. Mitchell: The Pictorial Turn [1992], in: Christian Kravagna (Hg.), Privileg Blick. Kritik der visuellen Kultur, Berlin 1997, S. 15–40.

Mitchell 1992b   William J. Mitchell: The Reconfigured Eye. Visual Truth in the Post-Photographic Era, Cambridge, Mass. 1992 (4. Auflage 2001).

Mitchell 2005a   W. J. T. Mitchell: There are no Visual Media, in: Journal of Visual Culture 4, 2005, S. 257–266.

Mitchell 2005b   W. J. T. Mitchell: What Do Pictures Want? The Lives and Loves of Images, Chicago 2005.

Montias 2006   Michael J. Montias: Works of Art Competing with Other Goods in Seventeenth-Century Dutch Inventories, in: Neil De Marchi (Hg.), Mapping Markets for Paintings in Europe 1450–1750, Turnhout 2006, S. 54–66.

Müller 1826   Johannes Müller: Zur vergleichenden Physiologie des Gesichtssinnes des Menschen und der Thiere, Leipzig 1826.

Murch 2004   Walter Murch: Ein Lidschlag, ein Schnitt. Die Kunst der Filmmontage, Berlin 2004.

van Musschenbroek 1741   Peter van Musschenbroek: Grundlehren der Naturwissenschaft, Leipzig 1741.

Naumann / Pankow 2004   Barbara Naumann / Edgar Pankow (Hg.): Bilder-Denken. Bildlichkeit und Argumentation, München 2004.

Neher 2005   Allister Neher: How Perspective could be a Symbolic Form?, in: The Journal of Aesthetics and Art Criticism 63, 2005, S. 359–373.

Newhall 1980   Beaumont Newhall (Hg.): Photography. Essays and Images, New York (The Museum of Modern Art) 1980.

Nickel 1998   Douglas R. Nickel: Nature's Supernaturalism. William Henry Fox Talbot and Botanical Illustration, in: Kathleen S. Howe (Hg.), Intersections. Lithography, Photography, and the Traditions of Printmaking. Albuquerque 1998, S. 15–23.

Nickelsen 2006   Kärin Nickelsen: The challenge of colour. Eighteenth-century botanists and the hand-colouring of illustrations, in: Annals of Science 63, 2006, S. 3–23.

Nordhofen 2001   Eckhard Nordhofen (Hg): Bilderverbot. Die Sichtbarkeit des Unsichtbaren, Paderborn 2001.

Oettermann 1980 Stephan Oettermann: Das Panorama. Die Geschichte eines Massenmediums, Frankfurt a. M. 1980.

Oexle 1997 Otto G. Oexle (Hg.): Der Blick auf die Bilder. Kunstgeschichte und Geschichte im Gespräch, Göttingen 1997.

Orland 2005 Barbara Orland: Artifizielle Körper – lebendige Technik. Technische Modellierungen des Körpers in historischer Perspektive, Zürich 2005.

Ortland 2007 Eberhard Ortland: Urheberrecht als Bildregime, in: Jean-Baptist Joly / Cornelia Vismann / Thomas Weitin (Hg.), Bildregime des Rechts, Stuttgart 2007, S. 268–288.

Pächt 1977 Otto Pächt: Methodisches zur kunsthistorischen Praxis. Ausgewählte Schriften, München 1977.

Panofsky 1924 Erwin Panofsky: Idea. Ein Beitrag zur Begriffsgeschichte der älteren Kunsttheorie [1924], 7. unveränderte Auflage, Berlin 1993.

Panofsky 1927 Erwin Panofsky: Perspektive als symbolische Form [1927], in: Erwin Panofsky, Perspektive als symbolische Form und andere Aufsätze, Frankfurt 2000.

Panofsky 1939 Erwin Panofsky: Ikonographie und Ikonologie. Eine Einführung in die Kunst der Renaissance [englisch 1939], in: Erwin Panofsky, Sinn und Bedeutung in der bildenden Kunst, Neuausgabe Köln 1978, S. 36–67.

Paul 2006 Gerhard Paul (Hg.): Visual History. Ein Studienbuch, Göttingen 2006.

Peirce 1998 Charles Sanders Peirce: New Elements of Mathematics (The Essential Peirce. Selected Philosophical Writings, hg. von Nathan Houser u. a., Bd. 1), Bloomington u. a. 1998.

Philostratos 1968 Philostratos: Die Bilder [Eikones], übers. und hg. von Otto Schönberger, München 1968.

Pias 2003 Claus Pias: Das digitale Bild gibt es nicht. Über das (Nicht-)Wissen der Bilder und die informatische Illusion, in: zeitenblicke. Journal für Geschichtswissenschaft 2 / 1, 2003 [elektronische Ressource].

Pieske 1989 Christa Pieske: Bilder für Jedermann. Wandbilddrucke 1840–1940. Begleitbuch zur Ausstellung im Museum für Deutsche Volkskunde SMB PK Berlin u. a., München 1989.

Pietsch 2006 Annik Pietsch: Ida Meyer und die Notation der Malerei, in: Bildwelten des Wissens 4,1: Farbstrategien, Berlin 2006, S. 40–42.

Platon 2004 Platon: Politeia (Der Staat), in: ders., Sämtliche Werke, Bd. 2, übers. von Friedrich Schleiermacher, hg. von Ursula Wolf, 30. Auflage, Reinbek bei Hamburg 2004.

Plinius 1997 Gaius Plinius Secundus: Naturalis historia (lateinisch / deutsch), Bd. 35, übers. von Roderich König, 2. Auflage, Düsseldorf 1997.

Pomian 1988 Krzysztof Pomian: Der Ursprung des Museums. Vom Sammeln, Berlin 1988.

Pörksen 1997 Uwe Pörksen: Weltmarkt der Bilder. Eine Philosophie der Visiotype, Stuttgart 1997.

Puttfarken 2000 Thomas Puttfarken: The Discovery of Pictorial Composition. Theories of Visual Order in Painting 1400–1800, New Haven u. a. 2000.

Reck 2002 Hans Ulrich Reck: Mythos Medienkunst, Köln 2002.

Regener 1999 Susanne Regener: Fotografische Erfassung. Zur Geschichte medialer Konstruktionen des Kriminellen, München 1999.

Reichert 2007 Rámon Reichert: Im Kino der Humanwissenschaften, Bielefeld 2007.

Reichle 2005 Ingeborg Reichle: Kunst aus dem Labor, Wien 2005.

Rheinberger 1997   Hans-Jörg Rheinberger / Michael Hagner / Bettina Wahrig Schmidt (Hg.): Räume des Wissens. Repräsentation, Codierung, Spur, Berlin 1997.

Riegl 1901   Alois Riegl: Spätrömische Kunstindustrie [1901], Nachdruck Berlin 2000.

Rose 1993   Mark Rose: Authors and Owners. The Invention of Copyright, Cambridge, Mass. 1993.

Ross 1994   Andrew Ross: For an Ecology of Images, in: Norman Bryson / Michael Ann Holly / Keith Moxey (Hg.), Visual Culture. Images and Interpretations, Hanover, NH 1994.

Ruchatz 2003   Jens Ruchatz: Licht und Wahrheit. Eine Mediumgeschichte der fotografischen Projektion, München 2003.

Runge 1982   Philipp Otto Runge: Die Begier nach der Möglichkeit neuer Bilder. Briefwechsel und Schriften zur bildenden Kunst, 2. Auflage, Leipzig 1982.

Sachs-Hombach 2005   Klaus Sachs-Hombach (Hg.): Bildwissenschaft. Disziplinen, Themen, Methoden, Frankfurt a. M. 2005.

Scharf 1968   Aaron Scharf: Art and Photography [London 1968], Nachdruck London 1986.

Schivelbusch 2007   Wolfgang Schivelbusch: Geschichte der Eisenbahnreise. Zur Industrialisierung von Raum und Zeit im 19. Jahrhundert, 4. Auflage, Frankfurt a. M. 2007.

Schmitt u. a. 1997   Hanno Schmitt / Jörg W. Link / Frank Tosch (Hg.): Bilder als Quellen der Erziehungsgeschichte, Bad Heilbrunn 1997.

Scholz 2004   Oliver R. Scholz: Bild, Darstellung, Zeichen. Philosophische Theorien bildhafter Darstellung, 2. Auflage, Frankfurt a. M. 2004.

Schreitmüller 2005   Andreas Schreitmüller: Alle Bilder lügen. Foto, Film, Fernsehen, Fälschung, Konstanz 2005.

Schröter / Böhnke 2004   Jens Schröter / Alexander Böhnke (Hg.): Analog-Digital. Opposition oder Kontinuum?, Bielefeld 2004.

Schuck-Wersig 1993   Petra Schuck-Wersig: Expeditionen zum Bild. Beiträge zur Analyse des kulturellen Stellenwertes von Bildern, Frankfurt a. M. u. a. 1993.

Schwender 1999   Clemens Schwender (Hg.): Zur Geschichte der Gebrauchsanleitung, Berlin / Bern / New York 1999.

Scott 1908   Walter D. Scott: The Theory and Practice of Advertising. A Simple Exposition of the Principles of Psychology in Their Relation to Successful Advertising, Boston 1908.

Scott 1999   Clive Scott: The Spoken Image. Photography and Language, London 1999.

Segeberg 1996   Harro Segeberg (Hg.): Die Mobilisierung des Sehens. Zur Vor- und Frühgeschichte des Films in Literatur und Kunst, München 1996.

Sekula 1989   Allan Sekula: Der Körper und das Archiv [1989], in: Herta Wolf (Hg.), Diskurse der Fotografie. Fotokritik am Ende des fotografischen Zeitalters, Bd. 2., Frankfurt a. M. 2003, S. 269–334.

Semper 1860   Gottfried Semper: Der Stil in den technischen und tektonischen Künsten oder Praktische Ästhetik [1860–63], Nachdruck Mittenwald 1977.

Siegrist / Löhr 2007   Hannes Siegrist / Isabella Löhr: Kulturelle Handlungsrechte in der Moderne. Die Geschichte des geistigen Eigentums und der Urheberrechte, in: Matthias Middell (Hg.), Dimensionen der Kultur- und Gesellschaftsgeschichte, Leipzig 2007, S. 223–234.

Snyder 1998   Joël Snyder: Nineteenth-century photography of sculpture and the rhetoric of substitution, in: Geraldine A. Johnson (Hg.), Sculpture and Photography. Envisioning the Third Dimension, Cambridge, Mass. u. a. 1998, S. 21–34.

Sontag 2003:   Susan Sontag: Das Leiden anderer betrachten, München 2003.

Sperber 1985   Dan Sperber: Anthropology and Psychology. Towards an Epidemology of Representations, in: Man 20, 1985, S. 73–89.

Spielmann 2005   Yvonne Spielmann: Video. Das reflexive Medium, Frankfurt a. M. 2005.

Stafford 1999   Barbara Maria Stafford: Visual Analogy. Consciousness as the Art of Connecting, Cambridge, Mass. u. a. 1999.

Steadman 2002   Philip Steadman: Vermeer's Camera. Uncovering the Truth Behind the Masterpieces, 2. Auflage, Oxford 2002.

Steinhauer 2007   Fabian Steinhauer: Das Recht am eigenen Bild in den Kollisionen der Sichtbarkeit, in: Jean-Baptiste Joly / Cornelia Vismann / Thomas Weitin (Hg.), Bildregime des Rechts, Stuttgart 2007, S. 222–246.

Steinmeier / Sudhaus 2006   Frank D. Steinmeier / Walter Sudhaus: Die Speziation der Darwinfinken und der Mythos ihrer initialen Wirkung auf Charles Darwin, in: Naturwissenschaftliche Rundschau 59 (8), 2006, S. 409–422.

Stelzer 1966   Otto Stelzer: Kunst und Photographie [1966], Neuausgabe München 1978.

Stoichita 1998   Victor I. Stoichita: Das selbstbewusste Bild. Vom Ursprung der Metamalerei, München 1998.

Sturani 1995   Enrico Sturani: Otto milioni di cartoline per il duce. Turin 1995.

Tagg 1988   John Tagg: The Burden of Representation, Basingstoke 1988.

Thompson 1917   D'Arcy W. Thompson: Über Wachstum und Form [engl. 1917], Frankurt a. M. 2006.

Tolkemitt / Wohlfeil 1991   Brigitte Tolkemitt / Rainer Wohlfeil: Historische Bildkunde. Probleme – Wege – Beispiele, Berlin 1991.

Tucker 2005   Jennifer Tucker: Nature Exposed. Photography as Eyewitness in Victorian Science, Baltimore 2005.

Uhlig 2006   Franziska Uhlig: Ready-made Farbe. Vom Mond aus betrachtet, in: Bildwelten des Wissens 4,1: Farbstrategien, Berlin 2006, S. 25–33.

Veblen 1899   Torstein Veblen: Theorie der feinen Leute [New York 1899], deutsche Neuausgabe Frankfurt a. M. 2007.

Vester 1975   Frederic Vester: Denken, Lernen, Vergessen [1975], 32. Auflage, München 2007.

Virilio 1989   Paul Virilio: Die Sehmaschine, Berlin 1989.

Vischer 1857   Friedrich Theodor Vischer: Ästhetik oder Wissenschaft des Schönen, 6 Bde., München 1846–57.

Vöhringer 2007   Margarete Vöhringer: Avantgarde und Psychotechnik. Wissenschaft, Kunst und Technik der Wahrnehmungsexperimente in der frühen Sowjetunion, Göttingen 2007.

Volk 1996   Andreas Volk (Hg.): Vom Bild zum Text. Die Photographiebetrachtung als Quelle sozialwissenschaftlicher Erkenntnis, Zürich 1996.

Voss 2007   Julia Voss: Darwins Bilder. Ansichten der Evolutionstheorie 1837–1874, Frankfurt a. M. 2007.

Voßkamp / Weingart 2005   Wilhelm Voßkamp / Brigitte Weingart (Hg.): Sichtbares und Sagbares. Text-Bild-Verhältnisse, Köln 2005.

Wagner 2004   Monika Wagner: Die ‚tabula rasa‘ als Denk-Bild. Zur Vorgeschichte bilderloser Bilder, in: Barbara Naumann / Edgar Pankow (Hg.), Bilder-Denken. Bildlichkeit und Argumentation, München 2004, S. 67–86.

Waldenfels 2001 Bernhard Waldenfels: Spiegel, Spur und Blick. Zur Genese des Bildes, in: Gottfried Boehm (Hg.), Homo Pictor. München / Leipzig 2001, S. 14–31.

Warburg 1905 Aby M. Warburg: Der Tod des Orpheus. Bilder zu dem Vortrag über Dürer und die italienische Antike, Hamburg 1905.

Warburg 2008 Aby M. Warburg: Der Bildatlas Mnemosyne, hg. von Martin Warnke, in: ders.: Gesammelte Schriften 2.1, 3. Auflage, Berlin 2008.

Warnke 1996 Martin Warnke: Hofkünstler. Zur Vorgeschichte des modernen Künstlers, 2. Auflage, Köln 1996.

Warnke 2002 Martin Warnke (Hg.): Bildersturm. Die Zerstörung des Kunstwerkes, 2. Auflage, Frankfurt a. M. 2002.

Warnke 2005 Martin Warnke: Bildwirklichkeiten, Göttingen 2005.

Wechsler 1988 Judith Wechsler (Hg.): On Aesthetics in Science, Boston / Basel 1988.

Wechsler 1993 Judith Wechsler: Lavater, Stereotype, and Prejudice, in: Ellis Shookman: The Faces of Physiognomy. Interdisciplinary Approaches to Johann Caspar Lavater, Columbia, SC 1993, S. 104–125.

Weibel 1984 Peter Weibel: On the History and Aesthetics of the Digital Image, in: Ausst.-Kat. Ars Electronica, Linz 1984.

Wenk 1997 Silke Wenk: Henry Moore. Large Two Forms, Frankfurt a. M. 1997.

Wenzel 1995 Horst Wenzel: Hören und Sehen, Schrift und Bild. Kultur und Gedächtnis im Mittelalter, München 1995.

Werckmeister 2005 Otto Karl Werckmeister: Der Medusa-Effekt. Politische Bildstrategien seit dem 11. September 2001, Berlin 2005.

White 1957 John White: The Birth and Rebirth of Pictorial Space [London 1957], Nachdruck London 1987.

Whyte 1951 Lancelot Law Whyte (Hg.): Aspects of Form [1951], 3. Auflage, Bloomington 1966.

Wind 1960 Edgar Wind: Kunst und Anarchie [1960], Neuausgabe Frankfurt a. M. 1994.

Wolf 2002 Gerhard Wolf: Schleier und Spiegel. Traditionen des Christusbildes und die Bildkonzepte der Renaissance, München 2002.

Wölfflin 1915 Heinrich Wölfflin: Kunstgeschichtliche Grundbegriffe. Das Problem der Stilentwicklung in der neueren Kunst [1915], 19. Auflage, Basel 2004.

Wood 1978 Phyllis Wood: Scientific illustration. A guide to biological, zoological, and medical rendering techniques, design, printing, and display, New York 1978.

Wulff 1993 Hans J. Wulff: Bilder und imaginative Akte, in: Zeitschrift für Ästhetik und allgemeine Kunstwissenschaft 38, 1993, S. 185–205.

Zimmermann 1971 Albert Zimmermann (Hg.): Der Begriff der repräsentatio im Mittelalter. Stellvertretung, Symbol, Zeichen, Bild, Berlin 1971.

16.2 Abbildungsverzeichnis

Abbildung 23: Der französische Präsidentschaftskandidat Nicolas Sarkozy zu Pferde (2007). The Associated Press.

Abbildung 24: Kronprinz Willem-Alexander der Niederlande und Maxima Zorreguita in Amsterdam (2001). Foto: Raymond Rutting / De Volkskrant.

Abbildung 25: Bill Clinton im Präsidentschaftswahlkampf (1992). Scott Applewhite / The Associated Press.

Abbildung 26: Ex-Präsident Bill Clinton als Wahlhelfer für seine Frau, Detail aus *Der Spiegel* vom 6.11.2006. Foto: The Associated Press.

Abbildung 27: Benito Mussolini, Fotopostkarte aus dem Atelier Caminada, Mailand (1919).

Abbildung 28: Konrad Dinckmut: Ablassbrief (um 1482), Holzschnitt. Germanisches Nationalmuseum.

Abbildung 29: Karl Friedrich Schinkel: *Die Erfindung der Malerei* (Ausschnitt) (1830), Öl auf Leinwand. Von der Heydt-Museum, Wuppertal.

Abbildung 30: Kaiserbulle Friedrichs II., Cividale (April 1232). Staatsarchiv Würzburg, Mainz, Domkapitel Urkunden, 1232 April/I.

Abbildung 31: Morphologische Transformation vom Froschkopf zum Idealtypus des Gottes Apoll, aus: Johann Caspar Lavater, Essays on Physiognomy, London 1855, Tafel 78 / 79.

Abbildung 32: John Berger u. a.: *Sehen. Das Bild der Welt in der Bilderwelt* (1974), Cover der deutschen Taschenbuchausgabe. Rowohlt Verlag GmbH, Reinbek bei Hamburg 1974.

Abbildung 33: Werbeanzeige für *Marie Earle Kosmetikcreme,* USA (späte 1920er-Jahre). Ad*Access On-Line Project-Ad#BH1524. John W. Hartman Center for Sales, Advertising & Marketing History, Duke University Rare Book, Manuscript, and Special Collections Library. http://library.duke.edu/digitalcollections/adaccess.

Abbildung 34: Darstellung der Färbemethode von Neuronen nach Golgi, aus: Fritz Kahn, Das Leben des Menschen. Medizinisches Lehrbuch, Band 4, Stuttgart 1929, Abb. 26.

Abbildung 35: Adaptive Radiation: Anpassung der Schnabelform der sogenannten Darwinfinken auf Galapagos (Ausschnitt) (1995), aus: William K. Purves u. a., Life. The Science of Biology, 4. Aufl., Sunderland, Mass. 1995, S. 450.

Abbildung 36: Frühes Berliner U-Bahn-Netz mit unterlegtem Stadtplan (1914). BVG Archiv.

Abbildung 37: Kladistische Darstellung der Säugetierverwandtschaft (2002), Fig. 25.12 aus: Neil A. Campbell / Jane B. Reece (Hg.): Biology, 6. Aufl., S. 498. 2002 Pearson Education, Inc., mit freundlicher Genehmigung.

Abbildung 38: Robert Strange: *Foetus in utero,* Kupferstich nach Jan van Riemsdyk, aus: William Hunter, The Anatomy of the Human Gravid Uterus, Birmingham 1774 (Tafel 6). Staatsbibliothek zu Berlin.

Abbildung 39: Blick durch die Kolonnade zum Ruinenberg im Park von Schloss Sanssouci, Potsdam (2003). Stiftung Preußische Schlösser und Gärten Berlin-Brandenburg / Hagen Immel.

Abbildung 40: Hendrick Hondius: Muster für Architekturperspektiven (1604–05), Kupferstiche. Lemgo, Weserrenaissance Museum Schloß Brake.

Abbildung 41: Andrea Pozzo: Kuppel in Ansicht von unten / Grundriss mit Perspektive (1693), aus: ders., Prospettiva de' pittori a architetti, Rom 1693, Bd I, Abb. 91.

Abbildung 42: Filmstill aus Michael Hanekes Film *Caché* (2005). WEGA Filmproduktionsges. m.b.H., Wien.

Abbildung 43: Etienne-Jules Marey: Chronofotografie einer menschlichen Schrittbewegung (1882).

Abbildung 44: Ludwig Schnorr von Carolsfeld: *Der Sturz vom Felsen* (1833), Öl auf Holz, aus: Ausst.-Kat. Klassizismus und Romantik in Deutschland. Gemälde und Zeichnungen aus der Sammlung Georg Schäfer, Schweinfurt, Germanisches Nationalmuseum Nürnberg [1. Juli–2. Oktober 1966], Schweinfurt 1966. Museum Georg Schäfer, Schweinfurt.

Abbildung 45: Ausgabe einer Bilddatei in Hexadezimalzahlen (2006), erstellt auf Grundlage einer anatomischen Zeichnung Leonardo da Vincis. Matthias Bruhn, Berlin.

Abbildung 46: Bildabdeckung des Planeten Mars nach 3 Jahren (2007), ausgeführt mithilfe der HRS-Kamera an Bord des Mars Express.

(Der Verlag hat sich um die Einholung der Abbildungsrechte bemüht. Da in einigen Fällen die Inhaber der Rechte nicht zu ermitteln waren, werden rechtmäßige Ansprüche nach Geltendmachung ausgeglichen.)

16.3 Sachregister

16.4 Glossar

Acheiropoieton   (griechisch für: ohne Hand erzeugt) Selbsttätig entstandenes Bild.

Adaptive Radiation   Bezeichnung der Evolutionsbiologie für die Ausbreitung einer Tierart unter Anpassung (Adaption) von Körperformen, Ernährungsweisen u. a. an die gegebenen Umweltverhältnisse.

Ädikula   (lateinisch für: Häuschen) Kleiner Bau, Vorbau oder Nische zur Aufnahme von Grabmälern oder Figuren, später auch gliederndes Bauelement von Wänden und Fassaden.

Autopoiesis   Begriff für die Selbsterhaltung eines Systems (Biologie, Soziologie) oder eine äußerlich scheinbar ursachenlose Selbsterzeugung.

Avatar   Künstliche Person, in virtuellen Welten auch als Figur zur Stellvertretung einer Person. Aus der hinduistischen Lehre entlehnt (dort für eine leibliche Erscheinung göttlicher Kräfte), wurde der Begriff Anfang der 1990er-Jahre in der Science-Fiction-Literatur popularisiert.

Cyborg   Aus den Anfängen der bemannten Raumfahrt stammender Begriff zur Bezeichnung einer menschlichen Lebensform, die sich unwirtlichen Umwelten anpasst (abkürzend für Kybernetischer Organismus), u. a. durch technische Prothesen, die den Körper funktional verändern. Zentraler Terminus einer trans- oder posthumanen Anthropologie.

Daguerrotypie   Fotografische Metallplatte, die in einem von Louis Jacques Mandé Daguerre (1787–1851) entwickelten Positiv-Verfahren mit Jod- oder Bromdämpfen behandelt, anschließend belichtet und schließlich mit Quecksilber bedampft wird, wobei sich eine detailreiche Zeichnung aus schwärzlichem Silber bildet. In der ersten Hälfte des 19. Jahrhunderts trotz Nichtkopierbarkeit und eingeschränkten Betrachtungsmöglichkeiten führendes Verfahren.

Diorama   Durchsichtbild, bei dem (besonders im Verfahren Louis Jacques Mandé Daguerres) zur Erzeugung spektakulärer Bildeffekte eine transparente Wand beidseitig bemalt und abwechselnd beleuchtet wird; im Ausstellungswesen auch Vitrinen mit räumlicher Tiefengliederung, etwa in der Naturkunde zur Nachgestaltung von Lebensräumen.

Drip painting (oder *action painting*)   Malweise, bei der Farbmengen großzügig auf eine Bildfläche getropft oder gespritzt werden und bei der sich das Ergebnis aus dem Zusammenspiel von Körpereinsatz, Bildbegrenzung und Materialverhalten ergibt (besonders forciert durch den US-amerikanischen Maler Jackson Pollock).

Effigie   (lateinisch *effigies* oder *effigia*) Bild, Nachbildung. Zumeist plastische Figur oder Sache zur Stellvertretung einer Person, z. B. im rechtlichen Kontext zur Verurteilung einer Person in Abwesenheit.

Elektroenzephalogramm (EEG)   Besonders in der Medizin und Neurologie gebräuchliches grafisches Verfahren, das die Aktivität des Gehirns an der Kopfoberfläche mittels Elektroden aufzeichnet (Hirnstrommessung) und als kurvenmäßige Darstellung der Spannungsverläufe ausgibt.

Elektrokardiogramm (EKG)   Vergleichbar dem EEG, registriert das EKG mittels Elektroden die Aktivität der Herzmuskelfasern und gibt diese in grafischer Form als Spannungskurven aus.

Idol   (griechisch *eidolon*) Seelenbild, Bild der Verstorbenen, später auch Götzenbild (Idol).

Idolatrie   Götzendienst.

Ikonizität   Zeichentheoretischer Begriff zur Beschreibung von Beziehungen – auch nichtvisueller Art – auf Grundlage gemeinsamer Eigenschaften (in der Semiotik, der nonverbalen Kommunikation u. a.).

Ikonografie   Historische Beschreibung von Bildmotiven und deren formaler Wandlung zur Bestimmung von Bildinhalten, besonders im Hinblick auf größere, z. B. mythologische Zeichensysteme (als klassische Ikonografie, christliche Ikonografie u. a.).

Ikonologie Im 16. Jahrhundert Bezeichnung für Handbücher der Emblematik, in der Kunst- und Kulturwissenschaft um 1900 für eine Bilderlehre, die neben der → ikonografischen Formtradition auch die kollektiven Funktionen von Bildern und damit ihre Einbettung in größere (zum Beispiel literarische oder religiöse) Kontexte untersucht; später auch auf die Funktion bestimmter Bauformen u. a. übertragen.

Ikonosphäre Bildwelt; in der Semiotik der Massenmedien absolutes Wachstum von Bildmengen.

Indexikalität Verweis- oder Zeigecharakter, aus dem Bereich der Sprachwissenschaft (dort verstanden als Kontextabhängigkeit) in einem semiotischen, d. h. zeichentheoretischen Sinne übertragen auf Bildformen, insbesondere die Fotografie, um deren Beziehung von Bildmedium und Bildinhalt zu diskutieren.

Informel Westliche Kunstrichtung, die nach dem Zweiten Weltkrieg im Anschluss an Expressionismus und Surrealismus auf den spontanen Ausdruck mit den Mitteln von Farbe und Zeichnung experimentiert hat.

Kalotypie Fotografisches Negativverfahren von William Henry Fox Talbot (1800–77), bei dem Papiere mit Silbernitrat und Kaliumiodid bestrichen werden und sich zu einer lichtempfindlichen Schicht verbinden. Nach der Belichtung können sie auf Papier transparent gemacht und so umkopiert werden.

Liturgie (griechisch für: Gemeinschaftlicher Dienst, vor allem Gottesdienst). Bezeichnet die Gesamtheit von Handlungen, Gegenständen oder Räumen, die für die religiöse Zeremonie erforderlich sind.

Makroskopie Studium der mit bloßem Auge erkennbaren Gegenstände und Strukturen.

Manierismus (italienisch *maniera*, Handhabung, Manier) Begriff der europäischen Stilgeschichte zur Charakterisierung der (vor allem durch das Werk Michelangelos markierten) Übergangsphase von einer auf das klassische Vorbild zielenden Hochrenaissance zu barocken Formen, gekennzeichnet durch verspieltere, teilweise verdrehte und überdehnte Proportionen und Figurationen in Bildkunst und Architektur.

Medium (lateinisch Mitte, mittel-, Übermittler) Der Begriff kann je nach Fachgebiet eine Reihe unterschiedlicher Übermittler oder Träger bezeichnen, z. B. das technische Medium zur Übermittlung von Signalen oder Informationen oder auch (z. B. in der Parapsychologie) eine lebende Person, die eine Nachricht überträgt, in der Bildkunst den Bildträger, in der Biologie oder Chemie auch die Nährlösung oder das Lösungsmittel.

Morphologie (griechisch für: Formenlehre) Ausgehend von den Überlegungen Johann Wolfgang Goethes zur äußeren Entwicklung von Individuen und Arten, die sich als Kritik eines statischen Formdenkens verstand, meint Morphologie heute die formale Beschreibung von Strukturen, Prozessen und Funktionen unterschiedlicher Komplexität.

Panorama Rundbild, das durch eine kreisförmig geschlossene Ansicht den Eindruck einer natürlichen Umgebung simuliert. Populäre europäische Bildform um 1800 mit eigenständigen Theaterbauten.

Piktogramm Bildschrift, abstrahierende grafische Darstellung mit Lehr-, Erkennungs- oder Leitfunktion

Polyptychon Bezeichnung für eine mehrteilige Tafel (Schreibtafel, Altaraufsatz), mit mehr als zwei (Diptychon) oder drei (Triptychon) Gliedern.

Predella Schmale und querformatige Sockelzone unterhalb des Hochaltars, oft mit szenischen Darstellungen.

Radiografie Bildaufzeichnung mithilfe von Röntgenstrahlen, wichtiges Hilfsmittel der Radiologie (Strahlenmedizin).

Stereoskopie Räumliches Sehen auf Grundlage von Darstellungen, die durch Überblendung zweier standpunktverschiedener Aufnahmen und mithilfe geeigneter Betrachtungsgeräte, Farbbrillen u. a. Techniken das binokulare Sehen imitieren und eine (medienbedingt eingeschränkte) Tiefenwirkung hervorrufen.

Stroboskopie (von griechisch *strobos*, Wirbel) Phänomen, bei dem eine Bewegung eingefroren wirkt, etwa weil sie in extrem kurzen Abständen durch eine Blende hindurch sichtbar oder blitzartig belichtet wird. Zunächst als optische Täuschung entdeckt, findet der Effekt gezielte Anwendung bei der Aufzeichnung von Bewegungen und ist auch grundlegend für die Projektion von Filmbildern.

Tableau vivant (französisch für: lebendes Bild) Eine von Personen kurzzeitig nachgestellt Szene, die sich zumeist an bekannte Vorbilder der Bildkunst anlehnt.

Tachismus (französisch *tache*, Fleck) Kunstrichtung des → Informel, die in der Malkunst auf den expressiven Eigenwert des pastosen oder geklecksten Malstoffes gesetzt hat.

Taxonomie (griechisch *taxon*, Rang / *nomos*, Wort, Gesetz) Ordnung von Einträgen in einer Sammlung oder einem Klassifikationsschema, insbesondere in der biologischen Artbestimmung.

Telekratie Politische Herrschaft durch Telemedien (Fernsehen, Radio), zugleich Dominanz derselben innerhalb des politischen Systems, die zur gesteigerten Bedeutung sichtbarer Erscheinungsbilder und Handlungen führt.

Tympanon (griechisch für: Trommel) Meist plastisch geschmücktes Giebelfeld oder Feld über dem Türsturz, typisches Ausstattungselement des mittelalterlichen Kirchenbaus zur bildlich-programmatischen Gestaltung der Haupteingänge.

www.ingramcontent.com/pod-product-compliance
Lightning Source LLC
Chambersburg PA
CBHW052137170526
45162CB00004B/39